SURVIVAL GUIDE TO
ORGANIC CHEMISTRY

Bridging the Gap from General Chemistry

SURVIVAL GUIDE TO
ORGANIC CHEMISTRY

Bridging the Gap from General Chemistry

Patrick E. McMahon
Bohdan B. Khomtchouk
Claes Wahlestedt

CRC Press
Taylor & Francis Group
Boca Raton London New York

CRC Press is an imprint of the
Taylor & Francis Group, an **informa** business

CRC Press
Taylor & Francis Group
6000 Broken Sound Parkway NW, Suite 300
Boca Raton, FL 33487-2742

© 2017 by Taylor & Francis Group, LLC
CRC Press is an imprint of Taylor & Francis Group, an Informa business

No claim to original U.S. Government works

Printed on acid-free paper
Version Date: 20160908

International Standard Book Number-13: 978-1-4987-7707-0 (Hardback)

Library of Congress Cataloging-in-Publication Data

Names: McMahon, Patrick E. | Khomtchouk, Bohdan B. | Wahlestedt, Claes.
Title: Survival guide to organic chemistry : bridging the gap from general chemistry / Patrick E. McMahon, Bohdan B. Khomtchouk, and Claes Wahlestedt.
Description: Boca Raton, FL : CRC Press, [2017] | Includes bibliographical references and index.
Identifiers: LCCN 2016030171| ISBN 9781498777070 (hardcover) | ISBN 9781315380186 (ebk : alk. paper)
Subjects: LCSH: Chemistry, Organic--Textbooks.
Classification: LCC QD253.2 .M34 2017 | DDC 547--dc23
LC record available at https://lccn.loc.gov/2016030171

Visit the Taylor & Francis Web site at
http://www.taylorandfrancis.com

and the CRC Press Web site at
http://www.crcpress.com

Printed and bound in the United States of America by
Edwards Brothers Malloy on sustainably sourced paper

To my wife, Rosemary, without whom nothing could have been accomplished, my 7-year-old twins, Patrick and Grace, and the faculty and students of Benedictine University who have made teaching both rewarding and enjoyable.

Patrick E. McMahon

To my son, Sviatoslav Bohdanovich Khomtchouk.

Bohdan B. Khomtchouk

To my highly talented and dedicated coauthors.

Claes Wahlestedt

Contents

SECTION I—Organic Practice Exams

SECTION II—Multiple-Choice Practice Exams

Preface

This book evolved over 23 years of teaching both general chemistry and organic chemistry at Benedictine University in Lisle, Illinois. The overall goal of the topics and descriptions in this book is to serve as a guideline for students to navigate the troubled waters from general chemistry to a thorough understanding of organic concepts. I have frequently had the opportunity to guide classes of students starting from the first semester of general chemistry through a completion of organic II. It is from this perspective that this book is written.

The focus of this book is not to review or reexamine concepts well described in the available comprehensive texts. Rather, the chapters in this book represent specifically selected topics which provide expanded and/or alternative concept explanations, new and useful organizational principles, and in almost all cases, specific procedures and step-by-step guidelines for solving problems. The descriptions in the book are designed to help the student "get the right answer."

Chapters 1, 2, 3, 6, and 10 describe how to apply and then extrapolate general chemistry skills that are necessary for working with organic molecules. Chapters 8 and 9 provide a summary and method for applying thermodynamics and kinetics to organic reactions. Other chapters covering organic skills, such as structure notation (Chapter 4) and organic nomenclature (Chapter 5) provide expanded explanations along with detailed stepwise procedures. The chapters dealing directly with organic reaction systems are designed to provide similar step-by-step processes guiding students through the sequential determination of reaction identification, mechanism, and the final correct regio- and stereoisomer product. Chapter 18, in particular, presents a complete organizational method and procedure for solving the multitude of nucleophilic reactions at carbonyl carbons.

To augment the stated goal, this book includes a large compendium of practice problems, all with complete solutions. Most chapters have internal solved examples and additional solved practice problems. In addition, a complete variety of practice exams covering all major organic topics is included; each exam has separate corresponding answer pages. All exams comprise questions and problems given during organic chemistry semester coursework at Benedictine University.

Patrick E. McMahon

Authors

Patrick E. McMahon has taught for 23 years both general chemistry and organic chemistry at Benedictine University in Lisle, Illinois. Prior to that, he was a research scientist for Amoco Chemical Company at the Naperville, Illinois, campus. His major area of research is organic reaction catalysis. He has also taught at several institutions in the Chicago area, including Elmhurst College, Dominican University, and Triton Community College. His teaching recognitions include the B. J. Babler Award for Outstanding Contribution to Undergraduate Instruction from the University of Illinois, the Deans Award for Teaching Excellence from Benedictine University, and the first recipient of the Shining Star Award from Benedictine University Student Senate for outstanding service to students. He is a member of the American Legion and served in the United States Army from 1970 to 1972.

Bohdan B. Khomtchouk is a triple-major summa cum laude (BS molecular biology and biochemistry, BS mathematics, and BS physics). He is currently a United States Department of Defense NDSEG Fellow at the University of Miami Miller School of Medicine.

Claes Wahlestedt, MD, PhD, is the Leonard M. Miller Professor at the University of Miami Miller School of Medicine and is working on a range of chemistry-related translational efforts in his roles as associate dean and center director for therapeutic innovation. The author of some 250 peer-reviewed scientific publications with 30,000+ citations, his ongoing research projects concern chemical biology, epigenetics, genomics, and drug/biomarker discovery across several therapeutic areas.

General Concepts for Covalent Bonding and Constructing Lewis Structures for Organic Molecules

1.1 GENERAL CONCEPTS

1.1.1 COVALENT BONDING

1. *A **covalent bond** is a shared pair of electrons between two atoms.* The shared electron pair can **most simply** be described as occupying a region of space formed by orbital overlap of **one** atomic orbital from **each** of the two atoms forming the bond. This overlap is called a **bonding molecular orbital** and always takes **only two** electrons.

2. Covalent bonding, when described as overlapping atomic orbitals, is an approximation of Molecular Orbital Theory. Mixing of the wave equations for atomic orbitals yields one molecular orbital for each atomic orbital used in the combination. Thus, each overlap of two atomic orbitals must yield **two** molecular orbitals: **one bonding orbital** and **one antibonding orbital**.

3. For a simple two-electron bond (localized atomic orbitals), the mixed wave equations for the representative atomic orbitals can either be **in-phase** or **out-of-phase**. If the equations are in-phase, the amplitudes of the wave functions add to increase the electron density in the orbital overlap. If the wave equations are out-of-phase, the wave functions cancel, decreasing the electron density of the orbital overlap.

4. For mixed wave equations that are **in-phase** (amplitudes add), the result is a **bonding molecular orbital**. For mixed wave equations that are **out-of-phase** (amplitudes cancel), the result is an **antibonding molecular orbital**. The concept of antibonding orbitals is a necessary component of certain organic reaction mechanisms and properties.

1.1.2 LEWIS STRUCTURES

1. Lewis structures are pictorial representations (**not** exact theoretical explanations) of the role of all valence electrons (outer shell electrons for nonmetals) in a covalent molecule. *For Lewis structures, draw **all** valence electrons for **all** atoms in the molecule.*
2. Electrons will usually exist as pairs: electrons in a bond are **bonding pairs** and are shown as a line. Thus, one line equals one bonding electron pair.
3. Multiple covalent bonds represent additional bonding electron pairs between two atoms. Each line represents a **separate** set of orbital overlaps (molecular orbital) and thus a separate electron-bonding pair:
   two lines equal a double bond (x=x).
   three lines equal a triple bond (x≡x).
4. *Electrons not used in bonding are **lone pairs***; these are shown as a pair of dots. Occasionally, non-octet atoms will have an unpaired electron, shown as a single dot.
5. ***Central atoms** in a molecule are bonded to **more than one** other atom. **Outside atoms** are bonded to **only one** other atom.* Simple covalent molecules and polyatomic ions often consist of one central atom surrounded by outside atoms bonded only to the central atom. The central atom is most often (but not always) the one with the lower electronegativity. Atoms that take a maximum of one bond can never be central atoms.
6. More complex molecules with more than one central atom can be interpreted with a knowledge of normal bonding numbers and/or with additional information about bonding patterns. Lewis structures for these molecules are put together by considering each central atom in sequence.
7. Excluding metal complexes, only a few elements are found in the large majority of organic molecules; these are **C, H, O, N, P, S, and halogens (F, Cl, Br, I)**. Lewis structures for **neutral** organic compounds can be put together based on simple **normal neutral bonding rules**, as defined in **Section 1.2**.
8. The large majority of organic molecules will follow the normal neutral bonding rules. Molecules and molecular fragments which do not follow these rules can be understood by applying **exception rules**. The exception rules describe bonding processes which alter the normal bond number, most often producing a formal charge on a bonding atom.

1.2 BONDING RULES FOR NONMETALS IN COVALENT MOLECULES

1.2.1 GENERAL CONCEPT OF NORMAL NEUTRAL BOND NUMBERS

1. Most molecules can be constructed using what are termed (arbitrary terminology) **normal neutral bonding rules** for most nonmetal elements as described below.
2. If each atom in the molecule follows these rules, formal charges on all atoms in the molecule will be **zero** (i.e., the atoms are neutral).
3. Normal bonding rules for hydrogen (H), plus the row-2 elements (C, N, O, F), are based on the concept that *each atom has a tendency to form a sufficient number of covalent bonds to achieve a complete outer shell set of **s** and **p** orbitals.*

    For H (shell $n = 1$), this equals two electrons; for the row-2 elements ($n = 2$), this equals eight electrons, an **"octet."**
4. The specific bond numbers are determined from the fact that *each* covalent bond adds **one** additional electron to the atom's bonding outer valence shell. Therefore, for **neutral** atoms which follow the **octet** rule: *the number of covalent bonds an atom will make will be equal to the number of additional valence electrons the atom needs to complete the outer valence shell (total to equal 2 or 8 electrons).*

    To calculate the number of electrons in any neutral atom's bonding outer valence shell:

    > **# electrons in bonding valence shell**
    > **= [# of atom's original valence electrons]**
    > **+ [# of electrons provided from bonds]**

5. **Normal neutral bonding rules** assign **one** electron from **each** of the bonding atoms to form the shared electron pair in the covalent bond. The "left over" valence electrons, the ones **not** used in bonding, are written in pairs and designated as **lone pairs.** *The number of nonbonding electrons found in lone pairs is equal to the number of original valence electrons minus the number of electrons used in covalent bonding.*

    To calculate the number of electrons for neutral atoms found in lone pairs:

    > **# unused (left over) electrons for lone pairs**
    > **= [# of atom's original valence electrons]**
    > **− [# of electrons used in bonding]**

6. There is one important **restriction** to normal neutral octet bonding: the total **maximum** number of covalent bonds possible for a **neutral** atom is limited to the total number of valence

electrons: a neutral atom cannot share valence electrons that it does not have. *Thus, bonding rules for **neutral** Be and B are based on the number of outer shell electrons available for forming covalent bonds.*

1.2.2 SUMMARY TABLE FOR NORMAL NEUTRAL BOND NUMBERS IN COVALENT MOLECULES

Note that the results apply to neutral atoms; exceptions will occur for non-neutral atoms.

1.2.2.1 FOR H, B, C, N, O, F (H PLUS ROW 2)

Atom/# of Valence e⁻	# of e⁻ Needed to Fill Shell	# of Covalent Bonds	# e⁻ Left Over/# of Lone Pairs
H· (1 valence e⁻)	**1** ($1 + 1 = 2$)	**1** covalent bond	0 e⁻ left over = **0** lone pairs
·B· (3 valence e⁻)	**5** ($5 + 3 = 8$)	Maximum of **3** bonds (restriction rule)	0 e⁻ left over = **0** lone pairs
·C· (4 valence e⁻)	**4** ($4 + 4 = 8$)	**4** covalent bonds	0 e⁻ left over = **0** lone pairs
·N· (5 valence e⁻)	**3** ($3 + 5 = 8$)	**3** covalent bonds	2 e⁻ left over = **1** lone pair
·O: (6 valence e⁻)	**2** ($2 + 6 = 8$)	**2** covalent bonds	4 e⁻ left over = **2** lone pairs
:F: (7 valence e⁻)	**1** ($1 + 7 = 8$)	**1** covalent bond	6 e⁻ left over = **3** lone pairs

1.2.2.2 COMMON BOND NUMBERS FOR CERTAIN ELEMENTS IN ROWS 3–6 (SI, P, S, CL, BR, I)

Bonding rules for the nonmetals in **rows 3–6** (Si, P, S, Cl, Br, I) are **often** based on the **octet** rule. These atoms may behave similar to their column mates to form normal-filled s and p subshells. Elements in **rows 3–6** can also have *expanded valence shells: the use of empty outer shell (highest shell number which has electrons in the s and p subshells) **d-orbitals** to accept extra bonding electrons.* These atoms can then have more than eight electrons in the valence shell and can have **more** bonds than normal for an octet. The maximum number of bonds is still restricted to the number of available valence electrons.

Atom/# of Valence e⁻	# of e⁻ Needed to Fill Shell	# of Octet Covalent Bonds	# of Bonds for Expanded Valence Shell
·Si· (**4** valence e⁻)	**4** ($4 + 4 = 8$)	**4** covalent bonds	Not applicable
·P· (**5** valence e⁻)	**3** ($3 + 5 = 8$)	**3** covalent bonds	**5** covalent bonds
·S: (**6** valence e⁻)	**2** ($2 + 6 = 8$)	**2** covalent bonds	**4** covalent bonds or **6** covalent bonds
Cl, Br, I (7 valence e⁻)	**1** ($1 + 7 = 8$)	**1** covalent bond	**3**, **5**, or **7** covalent bonds

1.3 CONSTRUCTING LEWIS STRUCTURES FOR COVALENT MOLECULES: MOLECULES WHICH DO NOT REQUIRE EXCEPTIONS TO NORMAL NEUTRAL BONDING RULES

1.3.1 LEWIS STRUCTURE CONCEPTS

1. Atoms connected in covalent molecules are of two types:
   a. **Outside atoms:** *These are atoms which are bonded to only one other atom.* All atoms which take **only one bond must** be outside atoms and are termed *required outside atoms.* Atoms which can take two or more bonds **can** be outside atoms if all of their bonds are to only one other atom through a double or triple bond.
   b. **Central atoms:** *These are atoms bonded to two or more other atoms.* All *possible central atoms* **must** be capable of taking **two or more** bonds.
2. All valence electrons for all atoms must be shown in a Lewis structure. Each covalent bond, shown by a line, represents one bonding electron pair. Multiple bonds are shown by a double line for a double bond and a triple line for a triple bond. Nonbonding electrons are shown by dots as lone pairs.
3. Lewis structures are formed in such a way as to determine at least one valid bonding structure for a particular molecular formula. Sometimes, only one Lewis structure is possible for a molecular formula. In other cases, especially for larger molecules, there may be many possible bonding patterns. *Different bonding patterns (shown as Lewis structures) for the same molecular formula are said to be related as* **constitutional isomers.**
4. Guidelines for forming Lewis structures depend on the types of atoms involved and the number of possible central atoms. **The guidelines in this section apply to Lewis structures for smaller molecules containing 1–5 central atoms.**

1.3.2 MOLECULES CONTAINING H, BE, B, C, N, O, F THAT FOLLOW NORMAL NEUTRAL BONDING RULES

1. H plus the row-2 elements (H, Be, B, C, N, O, F) cannot have expanded valence shells. If no exceptions need to be used, the normal neutral bond numbers for each of these elements can be used directly to form a valid molecule.
2. A valid Lewis structure can be determined using the following procedure. **An example for the molecule CF_3N is shown.**
   Step (1) For a neutral covalent molecule, identify the normal neutral bonding numbers for each of the atoms in the molecular formula.
   Step (2) Identify all possible central atoms and required outside atoms.

For the example:

1 carbon;   each C = 4 bonds   (possible) central atom
3 fluorines;   each F = 1 bond   (required) outside atoms
1 nitrogen;   each N = 3 bonds   (possible) central atom

Step (3) If there is more than one central atom, connect the central atoms with a single bond (to start with). If there are three or more central atoms, a valid structure can **usually** be found if they are connected in any order.

For the example: C—N

Step (4) A valid Lewis structure will be **any** atom-bonding pattern which will form a molecule with correct bond numbers for **each** atom. Begin connecting all the required outside atoms to the possible central atoms using single bonds. Try to find a connection pattern such that all atoms have the correct number of bonds. This is like putting a puzzle together.

For the example (arbitrary selection of pattern):

$$F—C—N—F$$
$$\quad\;\; |$$
$$\quad\;\; F$$

Check the bonding at   each F has one bond = correct number
this point: C has only three bonds, needs four
                         N has only two bonds, needs three

Step (5) If there are not enough outside atoms to give all central atoms the correct numbers of bonds, this means that double or triple bonds will be required. Use these types of bonds to increase the number of bonds to deficient atoms; **rearrangements of the initial pattern may be necessary to accommodate all atoms.** Continue by trial and error until all bonding requirements are satisfied.

One solution for the example to give the carbon and nitrogen one extra bond apiece to form a **double** bond between them:

$$F—C=N—F$$
$$\quad\;\; |$$
$$\quad\;\; F$$

Check the bond numbers: F = 1 bond each = **correct #**
                           C = 4 bonds = **correct #**
                           N = 3 bonds = **correct #**

Step (6) To complete the Lewis structure, add the required lone electron pairs to each atom. These numbers are equal to the number of electrons left over after normal bonding and were shown in the summary for row-2 atoms. The method of calculation was:

To calculate the number of electrons found in lone pairs for neutral atoms:

 # unused (left over) electrons for lone pairs

= [# of atom's original valence electrons]

− [# of electrons used in bonding]

The number of electrons used for bonding is based on <u>one</u> electron for each bond formed.

# of lone pairs for F = [7 e⁻] − [1 e⁻ used in bonding] = 6 e⁻ left over = 3 lone pairs
# of lone pairs for C = [4 e⁻] − [4 e⁻ used in bonding] = 0 e⁻ left over = 0 lone pairs
# of lone pairs for N = [5 e⁻] − [3 e⁻ used in bonding] = 2 e⁻ left over = 1 lone pair

A valid Lewis structure for this molecule is

3. Note that an initial bonding pattern for the outside and central atoms from **Step (4)** may need to be rearranged if double or triple bonds are required. For example, assume that the initial bonding pattern from **Step (4)** was

$$F—\underset{\underset{F}{|}}{\overset{\overset{F}{|}}{C}}—N$$

A double bond could not be used between the central atoms: C would have five bonds and **N** would have two bonds. A triple bond would give **N** the correct number of three bonds but would give C six bonds. To solve the problem, a shift of one of the F atoms from C to **N**, as shown in the example, is necessary before the double bond can be written correctly.

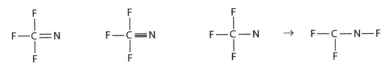

incorrect result   **incorrect result**   **incorrect pattern**   **correct pattern**

4. **Additional example:** Find **three** constitutional isomers for the formula **C₃H₂F₂O₂**.
   Step (1) and **Step (2):** Determine bond numbers and designate possible central atoms or required outside atoms:

3 carbons;	each C = 4 bonds	(possible) central atoms
2 oxygens;	each O = 2 bonds	(possible) central atoms
2 hydrogens;	each H = 1 bond	(required) outside atoms
3 fluorines;	each F = 1 bond	(required) outside atoms

Step (3) Connect the central atoms with a single bond; in a molecule with five possible central atoms, the central atoms can usually be connected in any order. Select, for example, the symmetrical pattern:

$$\text{O}-\text{C}-\text{C}-\text{C}-\text{O}$$

Step (4) Connect all the required outside atoms to the possible central atoms using single bonds. A quick informal inspection suggests that there are not enough outside atoms for all atoms to have the correct number of bonds. Arrange the outside atoms to allow for double and triple bonds. Select, for example, the continued symmetrical pattern:

$$\begin{array}{ccccc} & \text{H} & & \text{H} & \\ & | & & | & \\ \text{F}-\text{O}-\text{C}-\text{C}-\text{C}-\text{O}-\text{F} \end{array}$$

The bonding at this point:

each F has one bond = correct number
each H has one bond = correct number
each O has two bonds = correct number
two C have only three bonds, need four
one C has only two bonds, needs four

Step (5) The outside atoms were previously arranged to plan for two double bonds; rearrangements might have been necessary. The solution for the pattern in **Step (4)** is the symmetrical double-bond pattern shown; all atoms have the correct number of bonds.

$$\begin{array}{ccccc} & \text{H} & & \text{H} & \\ & | & & | & \\ \text{F}-\text{O}-\text{C}=\text{C}=\text{C}-\text{O}-\text{F} \end{array}$$

Step (6) Add the required lone electron pairs to each atom. Carbon and fluorine were calculated in the previous example; since normal neutral bond numbers were used for these atoms, the calculation need not be repeated. Apply the calculation method to hydrogen and oxygen:

unused (left over) electrons for lone pairs

= [# of atom's original valence electrons]

− [# of electrons used in bonding]

# of lone pairs for H = [1 e⁻] − [1 e⁻ used in bonding] = 0 e⁻ left over = 0 lone pairs

# of lone pairs for O = [6 e⁻] − [2 e⁻ used in bonding] = 4 e⁻ left over = 2 lone pairs

The complete Lewis structure is

$$\begin{array}{ccccc} & \text{H} & & \text{H} & \\ & | & & | & \\ \text{:F}-\ddot{\text{O}}-\text{C}=\text{C}=\text{C}-\ddot{\text{O}}-\ddot{\text{F}}\text{:} \end{array}$$

Constitutional isomers are different bonding patterns (shown as Lewis structures) for the same molecular formula. Another isomer could be formed by simply exchanging a hydrogen on a carbon with a fluorine on an oxygen. A more interesting variation would be the use of one triple bond instead of the two double bonds; note that this is not the only possible structure with a triple bond. Constitutional isomer #2:

The original central atom connection pattern can be rearranged in a variety of ways. Possible central atoms need not be in a continuous sequence, and in fact some need not be central atoms at all as shown by the double-bonded oxygen in constitutional isomer #3. As will be demonstrated in Section 1.6, all constitutional isomers of the same molecular formula must contain the same number of pi-bonds if ring patterns are specifically excluded.

Constitutional isomer #3:

1.3.3 LEWIS STRUCTURES OF MOLECULES CONTAINING ATOMS FROM ROWS 3 TO 6 (SI, P, S, CL, BR, I)

1. Atoms from rows 3 to 6 can have expanded valence shells; they thus can have a **variable** number of bonds.
   a. *If the bond numbers of all* **row** *3–6 elements are specified in the problem molecule, use the identical* **steps (1) through (6)** *described in the previous section. In* **Step (1)**, *use the bond numbers specified.*
   b. This method is useful because the row 3–6 elements very often show **octet** bonding. *These elements are always octet bonded when they act as outside atoms.*
2. **Examples:** Find valid Lewis structures for the following molecules, using the additional bonding information specified.
   a. C_2Cl_4 Cl shows octet bonding (1 bond).
   b. CS_2 S shows octet bonding (2 bonds).
   c. $COBr_2$ Br shows octet bonding (1 bond).

   Solutions: By following **steps (1) through (6)**, it is found that the only valid Lewis structures for these molecules are as shown. All bond numbers for all atoms are correct. In the examples, Cl and Br are required outside atoms (they take only one bond). The S atoms could have been central atoms but result in outside atoms for CS_2

a. $:\ddot{C}l-C=C-\ddot{C}l:$ b. $:\ddot{S}=C=\ddot{S}:$ c. $:\ddot{B}r-C=\ddot{O}:$

with structures showing:

a. (with $:\ddot{C}l:$ and $:\ddot{C}l:$ below the carbons)

c. (with $:\ddot{B}r:$ below the carbon)

1.4 CONSTRUCTING LEWIS STRUCTURES FOR COVALENT MOLECULES: MOLECULES WHICH REQUIRE EXCEPTIONS TO NORMAL NEUTRAL BONDING RULES

1.4.1 EXCEPTIONS FOR ROW-2 ELEMENTS: B, C, N, O

1. F and H have no exceptions; they must always take only one bond. B, C, N, and O can have exceptions to the normal neutral bonding rules. *The most common exceptions occur whenever these atoms are **not neutral** or are **free radicals**; these are described by exception rules.*

2. A formal charge of (+1) or (−1) is used to indicate that atoms are **not neutral** and therefore have bonding numbers different than the normal neutral values. The electron-counting rules previously demonstrated do not strictly apply to non-neutral atoms. One method of counting electrons to determine a formal charge:

   Formal charge on an atom

   = [# of original valence electrons]

   − [# of lone electrons + 1 / 2 (total # of bonding electrons in atom's bonding valence shell)]

3. **(General) Exception Rule #1** (can actually apply to all atoms):
   a. Exceptions to normal bonding occur when an atom has an **extra** electron completely held in its orbital (i.e., not shared with another atom). The atom can be thought of as being a (−1) ion bonded covalently to another atom. This exception can occur in neutral molecules or in polyatomic ions.
   b. The bonding difference is designated by a formal charge of (−1). *An extra electron compensates for **one** bond while allowing the atom to reach a normal **octet**.* For C, N, and O, a formal charge of (−1) indicates that the atom has **one fewer bond and consequently one additional lone pair**.
   c. Example: $(OH)^- = ^-:\ddot{O}-H$
      Oxygen has only one bond, three lone pairs and a formal charge of (−1), and achieves an octet. **To prove an octet for oxygen, calculate**

# of bonding outer shell valence electrons (oxygen)

= [6 original valence electrons]

+ [1 extra electron] + [1 additional electron from one covalent bond]

= **8 electrons**

To calculate the formal charge for oxygen:

$$\text{formal charge} = [6 \text{ original valence e}^-] - [(6 \text{ e}^- \text{ in lone pairs})$$
$$+ 1/2 \,(2 \text{ total bonding e}^- \text{ in bonding shell})]$$
$$= [6] - [7] = \mathbf{-1}$$

d.   Example: $(NH_2)^- = H - \overset{..}{\underset{..}{N}}{}^- - H$
     Nitrogen has only two bonds, two lone pairs, a formal charge of (–1), and reaches an octet.

To show that nitrogen has an octet, calculate

# of bonding outer shell valence electrons (nitrogen)

= [5 original valence electrons]

+ [1 extra electron] + [2 additional electrons from two covalent bonds]

= **8 electrons**

To calculate the formal charge for nitrogen:

$$\text{formal charge} = [5 \text{ original valence e}^-] - [(4\text{e}^- \text{ in lone pairs})$$
$$+ 1/2 \,(4 \text{ total bonding e}^- \text{ in bonding shell})]$$
$$= [5] - [6] = \mathbf{-1}$$

e.   Example: $(HCC)^- = H - C \equiv \overset{..}{C}{}^-$ Carbon has only three bonds, one lone pair, a formal charge of (–1), and achieves an octet.

4. **Exception Rule #2:**
    a. *In rare cases, an atom which normally shows octet bonding cannot achieve a full eight electrons (octet) in the outer-bonding shell.* This central atom has **one missing bond** and only **seven** bonding shell electrons; one of the electrons is unpaired. This atom is termed a **radical** (or **free radical**) because the unpaired electron is very reactive to make an additional bond.
    b. **Example:** $\cdot \overset{..}{N} = \overset{..}{\underset{..}{O}}$
        The nitrogen in (NO) has only **two** bonds and is non-octet; it has only **seven** bonding outer shell valence electrons.
    c. Free radicals often occur as **molecular fragments** rather than as stable molecules. Thus, both oxygen and carbon free radicals can exist **during** chemical reactions, when bonding changes are occurring, but generally do **not** exist as final products.

        Examples: $H - \overset{..}{\underset{..}{O}} - \overset{..}{\underset{..}{O}} \cdot$ $H - \overset{\textstyle .}{\underset{\textstyle |}{C}} - H$
        H

        The end oxygen in the peroxide radical (HOO) has only **one** bond and is non-octet; it has only **seven** bonding outer shell valence electrons. The carbon atom in the methyl radical

(H_3C) has only **three** bonds and is non-octet; it has only **seven** bonding outer shell valence electrons.

5. **(General) Exception Rule #3** involves formation of a **coordinate covalent bond** and is discussed in **Section 1.4.3**.

1.4.2 LEWIS STRUCTURES FOR POLYATOMIC IONS: USE OF EXCEPTION RULE #1

1. Polyatomic ions cannot be neutral and thus **require** use of the exception rules. Ions with a negative charge can be solved using **steps (1) through (6)** for normal neutral bonding, if **exception rule #1** is added to the **step (1)** determination of bond numbers and the determination of lone pairs.

   a. **Exception rule #1** states that an **extra** electron (indicated by a charge of (−1)) compensates for **one** bond for an atom while still allowing the atom to reach a normal octet; the extra electron in an orbital replaces the need for a bonding electron.

   b. The **sum** of **all** formal charges on **all** atoms in a polyatomic ion must equal the actual charge on the polyatomic ion. For anions (negative charge), the atom(s) that get the extra electron (formal charge of (−1)) should be the more electronegative element (very often oxygen). *Never give more than one extra electron to any one atom.* If the charge on the polyatomic ion is (−2) or (−3), distribute one extra electron to two or three different atoms.

2. The method for using **steps (1) through (6)** for polyatomic anions is

   Step (1) Additional process: Identify the specific atom(s) which must get the extra electron(s) indicated by the net charge on the polyatomic anion. Select the appropriate bond numbers which correspond to an extra electron compensating for one bond: *the value will be **one fewer** bond than the normal neutral bond number.*

Examples: $(CO_3)^{-2}$ **Given information: C is the only central atom**
$(SO_4)^{-2}$ **Given information: S is the only central atom**

Summary: C = 4 bonds; S (row 3) takes an unspecified number of bonds. **Two** oxygens from **each** of these polyatomic anions must take **one extra** electron apiece to reach a total charge of (−2) for each ion; these oxygens will take **one** extra electron and **only one** bond. The remaining oxygens will take the normal number of **two** bonds.

Steps (2) through (5): Complete these steps in the same manner as for neutral molecules, using the bond numbers specified. Example results of **steps (1) through (5)**:

Step (6): Add the correct number of lone pairs for each of the atoms using the electron count method: neutral C = no lone pairs, neutral O = two lone pairs (normal bonding).

$$\text{\# of lone pairs for S} = [6e^-] - [6e^- \text{ used in bonding}]$$
$$= 0\,e^- \text{ left over} = 0 \text{ lone pairs}$$
$$\text{\# of lone pairs for O with a } (-1) \text{ charge}$$
$$= [6e^-] + [1 \text{ extra } e^-] - [1e^- \text{ used in bonding}]$$
$$= (6+1-1) = 6e^- \text{ left over} = 3 \text{ lone pairs}$$

Example results through step 6:

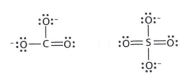

1.4.3 USE OF EXCEPTION RULE #3

1. Exceptions to normal bonding occur when an atom makes a **coordinate covalent bond**: *one atom **donates both** electrons to the covalent bond and one atom **accepts both** electrons for the covalent bond.* This bonding difference is designated by formal charges: the donor of both electrons changes its formal charge by (+1) and the acceptor of both electrons changes its formal charge by (–1).

2. The electron-counting rule for formal charge can be used:

> **Formal charge on an atom**
> = [**# of original valence electrons**]
>   –[**# of lone electrons + 1 / 2 (total # of bonding**
>   **electrons in atom's bonding valence shell)**]

3. **Donor atom description:**
   a. *Coordinate covalent bonding does **not** change the bonding valence shell electron number for the **donor** atom: atoms following octet bonding will still have **eight** outer shell bonding valence electrons.* The difference in bonding occurs because two lone pair valence electrons must now be shared: a lone electron pair becomes a bonding electron pair. The donor atom changes its formal charge by (+1) to indicate that the two electrons originally held by the atom all to itself must now be equally shared to the other bonding atom. (For counting purposes, the donor atom appears to be providing the acceptor's contribution of one electron to a normal covalent bond.)

b. **Donor** atoms are ones with lone electron pairs which can be converted into bonding electron pairs: common examples are N and O. These atoms take **one extra bond** in coordinate covalent bonding and must be designated with a formal charge of (+1).

c. For O: when formal charge is (+1) = **three** bonds and **one** lone pair instead of two bonds and **two** lone pairs. Note that oxygen still has an octet.

$$H_3O^+ = H—\overset{\overset{\displaystyle H}{|}}{\underset{..}{O}}{}^+—H \quad O \text{ Formal Charge} = [6e^-] - [2e^- + 1/2(6e^-)] = +1$$

d. For N: when formal charge is (+1) = **four** bonds and **no** lone pairs instead of **three** bonds and **one** lone pair. Note that nitrogen still has an octet.

$$NH_4^+ = H—\overset{\overset{\displaystyle H}{|}}{\underset{\underset{\displaystyle H}{|}}{N}}{}^+—H \quad N \text{ Formal charge} = [5\ e^-] - [0\ e^- + 1/2\ (8\ e^-)] = +1$$

4. Acceptor atom description:

a. *The **acceptor** atom in a coordinate covalent gains **two additional** electrons to its outer bonding valence shell, since an empty orbital accepts two bonding electrons from the donor. The acceptor atom changes its formal charge by (−1) since it gets a share of an electron pair that it did not originally have. (For counting purposes, the acceptor is getting its normal contribution of one electron for a normal covalent bond from the donor atom.)*

b. **Acceptor** atoms must be ones which have an empty p-orbital or d-orbital which can take two additional electrons; this excludes most atoms such as C, O, and N. Boron, with an empty p-orbital, is a common example of an acceptor. Acceptor atoms in a coordinate covalent bond form **one extra** bond are indicated by a formal charge of (−1).

c. For B, when formal charge is (−1) = **four** bonds instead of **three**. Note that boron now has an **octet**.

$$(BF_4)^- = {:}\underset{..}{\overset{..}{F}}—\overset{\overset{\displaystyle :\overset{..}{F}:}{|}}{\underset{\underset{\displaystyle :\overset{..}{F}:}{|}}{B}}{}^-—\overset{..}{\underset{..}{F}}: \quad B \text{ Formal charge} = [3\ e^-] - [0\ e^- + 1/2\ (8\ e^-)] = -1$$

1.5 ADDITIONAL CONCEPTS AND TECHNIQUES FOR CONSTRUCTING LEWIS STRUCTURES

1.5.1 OTHER USES OF EXCEPTION RULE #3

1. Polyatomic ions are not neutral and will **always** require the use of an exception rule. The formation of the polyatomic ions H_3O^+,

NH_4^+, and BF_4^- described in the previous section use exception rule #3.

2. Lewis structures for **neutral** compounds may also sometimes require the use of **exception rule #3**; this occurs whenever no bonding pattern can be found which will satisfy all **normal neutral** bonding rules. Although this situation is relatively rare, bond numbers and formal charges corresponding to the formation of coordinate covalent bonds can be required.

   a. **Example:** $O_3 = \ddot{\text{O}}\!=\!\text{O}^+\!-\!\ddot{\text{O}}\!:^-$ or $^-:\!\ddot{\text{O}}\!-\!\text{O}^+\!=\!\ddot{\text{O}}\!:$
      No bonding pattern with **all** oxygens taking exactly **two** bonds is possible. One oxygen is therefore given an extra bond and a formal charge of (+1) by **exception rule #3** and one oxygen is given one bond and a formal charge of (–1) by **exception rule #1**.

   b. **Example:** $N_2O = {}^-\ddot{\text{N}}\!=\!{}^+\text{N}\!=\!\ddot{\text{O}}\!:$ or $\text{N}\!\equiv\!\text{N}^+\!-\!\ddot{\text{O}}\!:^-$
      No bonding pattern with **both** nitrogens taking three bonds and oxygen taking two bonds is possible. The use of both **exception rule #3** and **exception rule #1** is required to find an acceptable bonding pattern.

3. *If a molecule is **neutral**, the sum of all formal charges for all atoms must add to **zero**.* The Lewis structures shown for the O_3 and N_2O follow this requirement.

4. The following example shows the electron donation and acceptance required for the formation of a coordinate covalent bond:

N = Donor   **B = Acceptor**   **New bond is formed**
with lone pair   with empty orbital   N has a charge = (+1)
(neutral)   (neutral)   B has a charge = (−1)

1.5.2 RESONANCE STRUCTURES

1. Often, **more than one** Lewis structure is possible for **one specific** atom connection pattern. Bonding descriptions (Lewis structures) can have the same atom connections but differ in the placement of electrons for lone pairs and multiple bonds (double and triple bonds). *Lewis structures for a specific molecule which show the exact same atom connection pattern but differ only in the placement of electrons and multiple bonds are termed **resonance structures**.*

2. **Resonance structures do not represent different molecules,** but instead represent different descriptions of the same molecule. Actual electron distribution is usually a composite of the most probable Lewis structure descriptions.

3. The examples shown on the previous page for O_3 and N_2O illustrate this idea.

   a. For O_3 = $:\ddot{O}{=}\overset{+}{O}{-}\ddot{\underset{..}{O}}:^-$ or $^-:\ddot{\underset{..}{O}}{-}\overset{+}{O}{=}\ddot{O}:$

   The two Lewis structures indicate that either end oxygen can have the negative charge. The fact that both end oxygens are found to be identical confirms the representation of the "true" molecule as a composite of the two Lewis structures.

   b. For N_2O = $^-\ddot{\underset{.}{N}}{=}{^+N}{=}\ddot{O}:$ or $\ddot{N}{\equiv}N^+{-}\ddot{\underset{..}{O}}:^-$

   The two Lewis structures differ as to the distribution of electrons within the same atom-bonding connection pattern. The structure showing the negative charge on the oxygen would contribute more to the overall picture of the molecule.

4. Resonance structures are also drawn even when a preferred completely neutral (all atoms have formal charge = 0) Lewis structure is possible. In this case, the additional structures are used to show patterns of electron distribution which provide insight into chemical behavior. *Resonance structures of a molecule are indicated by using double-headed arrows.*

Example:

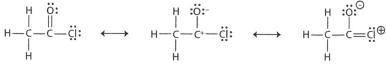

Preferred Structure:
All formal charges = 0

Resonance Structure:
Shows reactive   C $=$ O
bond.

Resonance Structure:
Shows stabilizing effect
of bonded Cl.

1.6 GUIDELINES FOR CONSTRUCTING LEWIS STRUCTURES OF LARGER MOLECULES

1.6.1 GENERAL CONCEPT

1. The process for drawing Lewis structures of larger, more complex organic molecules is based on the same principles described for general Lewis structures. *In general, larger organic molecules containing C, H, N, O, F, Cl, Br, and I will follow **normal neutral octet** bonding rules; molecules with S and P require additional information.*

2. The steps for larger molecules include additional simplifying information: an additional process to help in the identification and placement of multiple bonds (double and triple bonds); and a **Step 7** for writing one specific Lewis structure for a specific molecule based on many possible initial Lewis structure bonding patterns.

1.6.2 BONDING PATTERN TEMPLATE METHOD; PROCESS LISTED IN STEP SEQUENCE

Step (1) For a neutral covalent molecule, identify the normal neutral bonding numbers for each of the atoms in the molecular formula.

Most larger (usually organic) molecules will contain carbon atoms and usually at least some hydrogen atoms; oxygen and nitrogen are also common atoms. Phosphorus and sulfur are the only other common **potential central** atoms; bond numbers for these must be given through additional information. The elements Cl, Br, and I are almost always octet bonded (1 bond).

Step (2) Identify all possible central atoms and required outside atoms.

Carbon, nitrogen, and oxygen are **always possible central** atoms; phosphorus and sulfur, if present, will also be possible **central** atoms whether octet bonded or expanded valence shell. Hydrogen, fluorine, chlorine, bromine, and iodine will **always** be **outside** atoms unless specifically stated otherwise.

Step (3) If there is more than one central atom, connect the central atoms with a single bond (to start with). If there are three or more central atoms, a valid structure can usually be found if they are connected in any order.

Start by connecting the **possible** central atoms of **carbon, oxygen,** and **nitrogen** using **single** bonds in **any** bonding pattern that does not exceed the correct number of bonds based on normal neutral bonding rules. (A simple "straight" chain, i.e., continuous atom chain, is usually the easiest to start.) Now, add to the chain (or other pattern) any phosphorous or sulfur atoms through single bonds. *The atoms in the bonding pattern may have fewer than the required number of bonds at this stage of the solution.*

Step (4) A valid Lewis structure will be **any** atom-bonding pattern which will form a molecule with correct bond numbers for **each** atom. **Begin connecting all the required outside atoms to the possible central atoms using single bonds. Try to find a connection pattern such that all atoms have the correct number of bonds.**

Step (4A) Fluorine always takes **one** bond; chlorine, bromine, and iodine always take **one** bond (octet) in larger, organic molecules unless specifically stated otherwise. If these atoms are present, they **must** be **outside** atoms; add them through **one** bond only to any of the atoms already placed in the central atom pattern from **Steps (1) through (3)**. *Be sure that no atom so far exceeds the required bond number; some atoms will still have fewer bonds than necessary.*

Step (4B) The only atoms left (if any) will be hydrogens; these must be **outside** atoms. Add each hydrogen through **one** bond to the partial Lewis structure. *If sufficient hydrogen atoms are available, all remaining "vacant" bond positions in the molecule will* *be filled: each atom in the molecule will have the exact number of required bonds to satisfy the normal neutral bonding rules.* The result will be a **valid** Lewis structure for this formula.

Step (5A) If there are not enough outside atoms to give all central atoms the correct numbers of bonds, this means that double or triple bonds will be required. Use these types of bonds to increase the number of bonds to deficient atoms.

Step (5B) Rearrangements of the initial pattern may be necessary to accommodate all atoms. Continue by trial and error until all bonding requirements are satisfied.

Notes for Step (5A):

a. If some atoms are left with vacant bond positions (fewer than the required number of bonds), this means that *sufficient hydrogen atoms are not available.* This result indicates that the molecule must contain at least one **multiple** bond.

b. Vacant bond positions can also indicate that the atoms are bonded in a **"ring"** pattern; this type of bonding will not be specifically considered in this discussion.

c. A **multiple** bond implies a **double** or **triple** bond. The following notation will be used: *Each **double** bond counts as **one** multiple bond; each **triple** bond counts as **two** multiple bonds. A multiple bond is the same as a pi-bond (π-bond) in this notation.*

d. **The following general rule can be applied:** *The molecule will contain one multiple bond for every two vacant bonding positions. Or, stated alternatively, the molecule will contain one multiple bond for every two "missing" hydrogens.*

Notes for Step (5B):

The guidelines to this point will determine the presence and number of **total** multiple bonds, as defined by the general rule in **Step (5A)**. The initial atom-bonding pattern selected may not provide valid positions for placing double or triple bonds: *formation of multiple bonds must be accomplished without exceeding any required bond numbers.* To complete a **valid** structure, rearrange the outside bonded atoms in such a way as to put two atoms that can form a multiple bond next to each other in a sequence. It may also be necessary to rearrange the pattern of central atoms. Do this for each multiple bond required: a correct atom connection pattern will be generated.

Step (6) To complete the Lewis structure, add the required lone electron pairs to each atom.

a. These numbers are equal to the number of electrons left over after normal bonding. For B, C, N, O, F, H, the number of lone pair electrons for **normal neutral** bonding rules were shown in the bonding rule summary. The method of calculation is described as follows:

b. **To calculate the number of electrons found in lone pairs for neutral atoms:**

unused (left over) electrons for lone pairs

= [# of atom's original valence electrons]

− [# of electrons used in bonding]

Step (7) Once one possible Lewis structure has been generated from the required number of multiple bonds (Steps 5A and 5B), all other possible Lewis structures for this same formula must contain this exact same number of total multiple bonds.

a. Other possible Lewis structures may contain multiple bonds which are structured differently: for example, one triple bond can be exchanged for two double bonds or different atoms may be involved in the multiple bond.

b. This concept can be used to write additional Lewis structures if required, or to construct a more specific molecule based on additional information.

Example: $C_5H_5NOClBr$

The example is shown in step sequence:

Steps (1, 2, and 3) C—C—C—C—C—N—O

Step (4A)
$$Br—\underset{\underset{Cl}{|}}{C}—C—C—C—N—O$$

Step (4B)
$$Br—\underset{\underset{Cl}{|}}{\overset{\overset{H}{|}}{C}}—\underset{\underset{H}{|}}{\overset{\overset{H}{|}}{C}}—\underset{\underset{H}{|}}{\overset{\overset{H}{|}}{C}}—\underset{\underset{?}{|}}{\overset{\overset{?}{|}}{C}}—\underset{\underset{?}{|}}{\overset{\overset{?}{|}}{C}}—\underset{\underset{?}{|}}{N}—O—?$$

Step (5A) 6 vacant positions (marked as "?")/**2** vacant positions per multiple bond = **3** multiple bonds

Notation: one double bond = **one** multiple bond; **one** triple bond = **two** multiple bonds.

Result: A total of **three** multiple bonds can be formed for this molecular formula. The requirement can be satisfied by **three double bonds** or **one triple bond plus one double bond.**

Step (5B): The bonding pattern shown directly leads to a valid Lewis structure by using a triple bond and a double bond; complete the Lewis structure by adding the correct lone pairs to each atom:

Step (5B) plus Step (6) $:\ddot{Br}—\underset{\underset{:\ddot{Cl}:}{|}}{C}—\underset{\underset{H}{|}}{\overset{\overset{H}{|}}{C}}═\underset{\underset{H}{|}}{C}—\underset{\underset{H}{|}}{\overset{\overset{H}{|}}{C}}═C—\ddot{N}═\ddot{O}:$

Step (7): Many other possible Lewis structures can be drawn for this molecular formula; *all* must have **three multiple bonds** *(with the restriction that the **ring** pattern is **not** being used).* Different atom-bonding patterns based on the same molecular formula are termed **(constitutional) isomers.** The original central atom and outside-bonding pattern can be rearranged in a variety of ways to accommodate the required **three** multiple bonds.

Step (7) Example: Assume that a structure is required that contains **only** double bonds and **no** triple bonds. In this case, the outside atom positions cannot completely accommodate the required set of multiple bond positions. The outside atoms must be rearranged; one possible example is

Use the rearranged pattern to form the necessary double bonds, and then add the lone electron pairs to each atom. This produces the following Lewis structure:

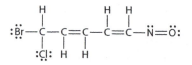

Example: Assume that a different triple bond plus double-bond structure was required; the rearrangement could be

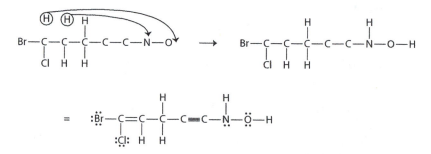

Example: Once the multiple-bond requirement is determined, major changes in the structure, including the original central atom pattern, can be easily visualized:

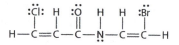

PRACTICE EXERCISES

Use the bonding rules described in this chapter to draw correct Lewis structures for the following molecules:

1. All of the following molecules will follow normal neutral bonding rules: octet number with no exceptions for C, N, O, F plus any element from rows 3 to 6; use the stated number for H and B. Each of these molecules will have only **one** bonding pattern found by following the required rules:

a. OF_2 b. HNF_2 c. N_2O_2 d. CH_2F_2 e. COF_2 f. C_2F_4
g. C_2H_6 h. CH_5N i. CH_5P j. BHClF   k. $CSBr_2$

2. The following molecules will have more than one possible Lewis structure which follows normal neutral bonding rules:

    a. C_2H_4S (S is octet bonded = 2 bonds) Find **two** Lewis structures.

    b. C_3H_6O Find at least **three** Lewis structures (This formula can actually have at least eight, all of which still satisfy all of the bonding rules.)

    c. C_2H_3NO Find at least **four** structures (many are possible).

3. Draw **five** different Lewis structures for the molecular formulas:

    a. C_4H_7NO Select any **five**.
    b. $C_5H_4F_2O_2$ Select **five** which always have at least one O double bonded to a C.

4. Draw Lewis structures for the following polyatomic ions; each will have only **one** possible structure. Use **exception rule #1 and exception rule #3**. In all cases, C and H follow normal neutral bonding rules; the exceptions and formal charges apply to oxygen or nitrogen. For (e), **both** oxygens are bonded to the same carbon.

    a. $(CH_3O)^+$ b. $(CH_5O)^+$ c. $(CH_3O)^-$ d. $(C_2H_8N)^+$ e. $(C_3H_3O_2)^-$

5. Draw **two resonance structures** for $C_2H_3N_3$. All hydrogens are bonded to carbon; the carbons are bonded to each other and all nitrogens are bonded in a sequence. The exception rules and formal charges will always apply to the nitrogens.

ANSWERS TO PRACTICE EXERCISES: LEWIS STRUCTURES

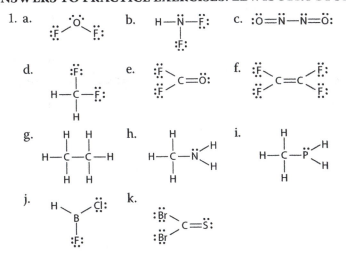

2. a.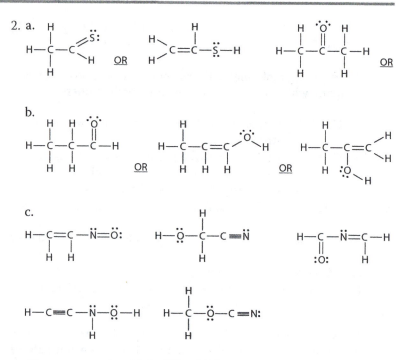

b.

c.

Other answers are possible for 2(b) and 2(c)

3. a. C_4H_7NO = <u>two</u> multiple bonds; <u>many</u> possible answers, for example,

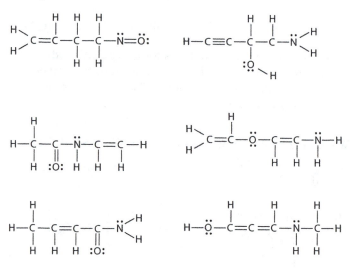

b. $C_5H_4F_2O_2$ = three multiple bonds; many possible answers
    which will have at least one C=O, for example,

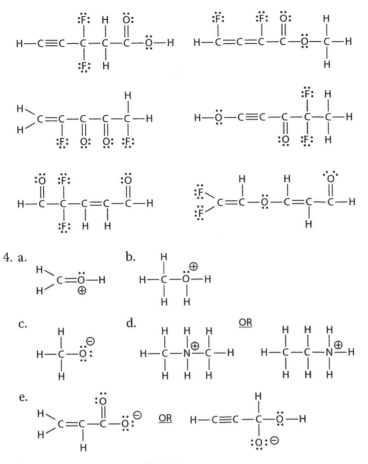

4. a.   b.

c.   d.   OR

e.

5. Resonance structures $C_2H_3N_3$:

(i) Isomer which follows example:

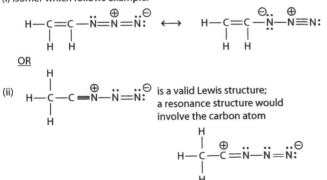

OR

(ii) is a valid Lewis structure;
a resonance structure would
involve the carbon atom

Guideline for Writing Organic Molecule Isomers and Determining Number of Rings Plus Pi-Bonds

2.1 GENERAL CONCEPT

1. **Constitutional isomers** (or structural isomers) describe the relationship between *molecules which have the same molecular formula but have different bonding patterns*. Writing correct isomer structures from organic formulas requires the same skills used for general Lewis structures. Neutral isomers should always follow the rules outlined in **Chapter 1**.

2. The term **fully "saturated"** in organic chemistry refers to a molecule that contains the **maximum** number of hydrogen atoms possible for a particular molecular formula based on carbon and other atoms. "**Unsaturated**" specifically refers to molecules that contain one or more double or triple bonds, that is, one or more pi-bonds.

3. The term "**multiple bond**" can also be used to refer to a **pi-bond**: a double bond counts as one multiple bond (one pi-bond) and a triple bond counts as two multiple bonds (two pi-bonds).

4. The general technique for writing Lewis structures for organic molecules demonstrated in Section 1.6 of Chapter 1 that **each** multiple bond (pi-bond) corresponds to **two** vacant bonding positions.
   a. Since almost all organic compounds contain **hydrogen** as a major **outside** atom, this rule can also be stated as:

each multiple bond (pi-bond) corresponds to **two "missing" hydrogen atoms**.

b. It is generally true that the presence of **each pi-bond** in a molecule results in a formula with **two fewer hydrogens than the maximum possible**. The number of pi-bonds (multiple bonds) is termed the "**degree of unsaturation.**"

5. A related concept recognizes that **two** fewer hydrogen atoms are required in any molecular formula which contains a chain of atoms in ring form. It is also generally true that **each ring** in a molecule results in a formula with **two fewer hydrogens than the maximum possible**.

6. The calculation of the total number: "**sum of rings plus pi-bonds**" (or equivalently "**sum of rings plus multiple bonds**") can be used in writing isomer structures from formulas. Two methods are described. The standard method involving the construction of a bonding-pattern template can always be used to find this number. An additional method provides direct calculation of this value from the complete molecular formula.

2.2 USE OF THE "BONDING-PATTERN TEMPLATE" METHOD TO FIND SUM OF RINGS PLUS PI-BONDS (DEGREE OF UNSATURATION)

The **sum of rings plus pi-bonds** is also sometimes abbreviated as "**degree of unsaturation**" even though the role of ring formation is misrepresented by this term. The method for constructing a sequence of central atoms and counting outside atoms is completely described in **Chapter 1**. The following summary is adapted to include the possibility of ring formation:

Steps (1)–(3): Construct a partial molecule (e.g., a continuous chain) of all central atoms (atoms which take more than one bond in organic molecules): C, N, O, P, S. Follow normal neutral bonding rules; provide octet bonding for S (two bonds) and P (three bonds) unless other information is available.

Step (4): Attach all outside atoms (single-bonding atoms) to the template (partial molecule). Outside atoms are H, F, and almost always Cl, Br, I.

Step (5): The **total** number of **rings plus pi-bonds (i.e., rings plus multiple bonds)** can be calculated as previously described: *The molecule must contain <u>one ring</u> or <u>one pi-bond</u> (multiple bond) for <u>each two</u> vacant bonding positions remaining in the bonding-pattern template.*

Step (6): Structural formulas in organic chemistry do **not** require showing lone pairs. The number of lone pairs for C, N, O, S, P, and halogens should be easily determined.

Step (7): The calculation in this procedure **does <u>not</u> distinguish between rings or pi-bonds**; potential isomers may have any number of either in any combination. Rearrangement of atoms and the resulting constitutional isomers depends on the desired molecule.

2.3 CALCULATION OF SUM OF RINGS PLUS PI-BONDS FROM THE MOLECULAR FORMULA

Molecular formulas quantitate the relationship between "missing" hydrogens and the number of rings or pi-bonds. Each ring or pi-bond represents two vacant bonding positions (or equivalently two missing hydrogens) in the bonding-pattern template. *Vacant positions (missing hydrogens or other outside atoms) can be calculated directly from a molecular formula.* Note how the general formula shows the **two** fewer hydrogens for each ring or pi-bond in the general example:

a. A **fully saturated, non-ring** hydrocarbon (alkane) has the general formula of C_nH_{2n+2}, where n = any number of carbons; # H = $2n + \underline{2}$.

b. A **one**-ring cycloalkane or a **one** double-bond alkene has a general formula of C_nH_{2n}, where n = appropriate number of carbons; # H = $2n$.

2.3.1 PROCESS FOR USE OF FORMULAS TO CALCULATE NUMBER OF RINGS PLUS PI-BONDS: MOLECULES CONTAINING <u>ONLY</u> C, H, AND O

Step (1): Multiply the number of carbons in the formula by **2**, and then add **2**:

$$2 \times (\text{# of C in formula}) + 2$$

Step (2): Subtract the actual number of hydrogens **from** the total found for [Step (1)]:

$$[2 \times (\text{# of C in formula}) + 2] - [\text{actual # of H}]$$

Step (3): Divide the total from [Step (2)] by **2** to determine the total number of rings plus pi-bonds:

$$\textbf{Sum of rings + pi-bonds} = [2 \times (\text{# of C in formula}) + 2]$$
$$- [\text{actual # of H}] / 2$$

Step (4): Treatment of oxygen in the molecule: This calculation is valid **only** for molecules containing **only carbon, hydrogen, and oxygen**. For molecules which contain oxygen: **ignore all oxygens in the calculation**; the sum of rings plus pi-bonds will still be correct. Since oxygen takes exactly two bonds, it does not change the carbon-to-hydrogen ratios described. (This result can be demonstrated by constructing a bonding-pattern template, e.g., place an oxygen between two carbons in a chain.) This **concept also applies to sulfur**, under conditions where sulfur takes **two bonds**.

2.3.2 EXPANDING THE CALCULATION METHOD TO MOLECULES WHICH INCLUDE F, CL, BR, I

Halogens (F, Cl, Br, I) generally take only one bond; they can therefore be treated as if they were hydrogens in the calculation. The corresponding process is:

Step (1): Multiply the number of carbons in the formula by **2**, and then add **2.**:

Step (2): Subtract the **total** actual number of **hydrogens plus halogens from** the total found for [Step (1)].

Step (3): The complete equation for **all** molecules containing **C, H, O, (S with two bonds), F, Cl, Br, I** would then be:

$$\text{Sum of rings + pi-bonds} = [2 \times (\text{\# of C in formula}) + 2]$$
$$- [\text{actual \# of H + F + Cl + Br + I}] / 2$$

Step (4): As before, ignore oxygen and/or S with two bonds in the calculation.

2.3.3 EXPANDING THE CALCULATION METHOD TO MOLECULES WHICH INCLUDE NITROGEN

1. Molecules which contain **nitrogen** do not follow this equation. Nitrogen takes **three** bonds and therefore changes the carbon-to-(hydrogen plus halogen) ratio. The net result is that **each nitrogen requires an additional one hydrogen (or halogen) to be present in a saturated molecular formula**. These molecules are often best analyzed using the bonding-pattern template method; however, a direct calculation can be made.

2. To expand the calculation rules to include molecules containing nitrogen:
Change Step (1) by adding "1" to the carbon and central atom term for each nitrogen in the molecule.

For molecules with **one** nitrogen:
Step (1): Multiply the number of carbons in the formula by 2, and then add **3**:

$$2 \times (\text{\# of C in formula}) + 3$$

For molecules with **two** nitrogens:
Step (2): Multiply the number of carbons in the formula by 2, and then add **4**:

$$2 \times (\text{\# of C in formula}) + 4$$

Step (3): An example of the final equation for molecules with **one nitrogen** would read:

Sum of rings + pi-bonds $= [2 \times (\text{\# of C in formula}) + 3]$
$$- [\text{actual \# of H} + \text{F} + \text{Cl} + \text{Br} + \text{I}] / 2$$

EXAMPLES

1. $C_7H_{12}O$
   a. **Bonding-pattern template method** (following general steps):

   Each ? = one vacant bonding position

   Sum of rings plus pi-bonds = 4 vacant bonding positions$/2 = \mathbf{2}$
   b. **Calculation method for $C_7H_{12}O$:**
   [Step (1)] $[2 \times (\text{\# of C}) + 2] = [(2 \times 7) + 2] = \mathbf{16}$
   [Step (2)] [actual # of H] $= \mathbf{12}$
   [Step (3)] Sum of rings plus pi-bonds $= [16] - [12]/2 = \mathbf{2}$
   [Step (4)] Note that oxygen was ignored in the calculation
   c. Molecules with the molecular formula $C_7H_{12}O$ will thus have **any** combination of rings plus pi-bonds which add to **two: two double** bonds; or **one triple** bond; or **one double** bond **plus one ring**; or **two rings**.

2. **$C_4H_2F_3NO$**
   a. **Bonding-pattern template method** (following general steps):

   ```
       H   F   ?   ?
       |   |   |   |
   H — C — C — C — C — N — O — ?        Each ? = one vacant bonding position
       |   |   |   |   |
       F   F   ?   ?   ?
   ```

   Sum of rings plus pi-bonds = 6 vacant bonding positions$/2 = \mathbf{3}$
   b. **Calculation method for $C_4H_2F_3NO$:**
   For a molecule with **one nitrogen**:
   [Step (1)] $[2 \times (\text{\# of C}) + 3] = [(2 \times 4) + 3] = \mathbf{11}$
   [Step (2)] [actual # of H + F] $= \mathbf{5}$
   [Step (3)] Sum of rings plus pi-bonds $= [11] - [5]/2 = \mathbf{3}$
   [Step (4)] Note that oxygen was ignored in the calculation
   c. Molecules with the molecular formula $C_4H_2F_3NO$ will thus have **any** combination of rings plus pi-bonds which add to **three: three double** bonds; or **one triple** bond plus **one double** bond; or **one double** bond **plus two rings**, etc.

Guideline for Complete Analysis for Central Atoms and Molecules

Bonding/Hybridization/ Geometry/Polarity

3.1 PROCESS FOR COMPLETE HYBRIDIZATION/ GEOMETRY ANALYSIS

1. Each central atom in a molecule can be described by its bond types, hybridization, geometries, and resulting polarity. The steps:

   Step (1): Determine the Lewis structure.

   Step (2): Classify the bond type for each bond in the molecule.

   Step (3): Count the number of **Electron Regions (E.R.)**.

   Step (4): Identify the hybridization of each central atom.

   Steps (5–6): Identify and then draw the E.R. geometry around each central atom in each molecule.

   Step (7): Identify the **Atom** geometry around each central atom in the molecule.

   Step (8): Determine the polarity of the molecule.

2. **The general process for central atoms with 2–4 E.R.:**

   Step (1): Draw the correct Lewis structure for each central atom in the molecule. This step specifies the number and types of **E.R.** around each central atom. Follow the procedure for Lewis structures described in **Chapter 1**.

   Step (2): Classify the bond type for each bond in the molecule according to the following identifications:

   Single Bond = **one** sigma-type bond

Double Bond = **one** sigma-type bond **plus one** pi-type bond

Triple Bond = **one** sigma-type bond **plus two** pi-type bonds

a. A **sigma bond** is an **end-to-end overlap** of orbitals which provides a spherically symmetrical electron-density distribution around the bond axis; it allows bonded atoms to rotate around the bond axis. All orbital types can engage in end-to-end overlap:

s, p, d, sp-hybridized, sp²-hybridized, sp³-hybridized, etc. Hybridized orbitals **always** form sigma-bonds, **never** pi-bonds.

b. A **pi-bond** is a **side-to-side overlap** of orbitals which produces electron-density regions distributed above and below the bond axis. The overlap is **not** spherically symmetrical and does **not** allow rotation of bonded atoms around the bond axis. Pi-bonds are usually formed through overlap of p-orbitals and sometimes d-orbitals.

Step (3): Count the number of E.R. for each central atom using the following rules:

1. Assign <u>**one** E.R.</u> to <u>**each lone electron pair**</u> (nonbonding electron pair).
2. Assign <u>**one** E.R.</u> to <u>**each sigma-bonding electron pair.**</u>
3. The results for these designations produce the following rules:

each lone electron pair	=	**one** E.R.
each single bond	=	**one** E.R. (**one** sigma bond)
each double bond	=	**one** E.R. (**only one** sigma bond, the other is a pi-bond)
each triple bond	=	**one** E.R. (**only one** sigma bond, the other two are pi-bonds)

4. Although double and triple bonds are formed by separate orbital overlaps (molecular orbitals), the separate overlaps occupy approximately the same region of space. <u>**Double and triple bonds are each considered one E.R. apiece**</u>.

Step (4): Identify the hybridization of each central atom, based on the E.R. count.

a. The pattern to follow is straightforward: **A central atom requires <u>one hybridized orbital for each E.R.</u> this requirement is independent of the number of pi-bonds needed in the molecule.** This **one-to-one ratio** occurs because central atom sigma-bonds and lone pairs (determinants of E.R. count) always use hybridized orbitals in hybridized atoms while pi-bonds always use unhybridized orbitals (p or d).

b. Hybridized orbitals are constructed by starting with the
s-orbital and then adding (in order) the proper number
of p-orbitals, then (if needed) d-orbitals, to reach the
required E.R. count. **The set of hybridized orbitals
have equivalent energies and point to the corners of
the VSEPR-determined geometrical figure**.

c. Use the summary table to match E.R. count to hybridization designation:

# E.R.	Orbitals Used For Hybridization	Orbital Notation	# of Equivalent Orbitals
2	s + p	sp	2
3	s + p + p	sp^2	3
4	s + p + p + p	sp^3	4

Steps (5 and 6): Identify and draw the E.R. geometry.
General Concept:

1. The three-dimensional (3-D) shape of a molecule, termed
the **molecular geometry**, is based on the arrangement
of all electron pairs around each central atom in a complete molecule:

a. The 3-D geometry around a central atom describes
the relative positions of each lone electron pair and
each bonding electron pair. By extension, the relative positions of the bonding electron pairs must
specify the corresponding positions of all atoms
attached to a central atom.

b. **Analysis of 3-D geometry applies only to central
atoms in a molecule**; the geometry around outside
atoms does not contribute to the overall molecular
geometry. This is because a minimum of two bonded
atoms is necessary to make positional comparisons
relevant to molecular properties.

c. **The shape of a complete molecule is a composite
of the 3-D geometry of each of the central atoms;
3-D geometry must be analyzed through one
central atom at a time**.

2. Two types of geometry analysis are used for central
atoms:

a. **Electron Region (E.R.) Geometry** describes the
arrangement of all electron pairs (bonding electron
pairs and lone electron pairs) around a central atom.

b. **Atom Geometry** (sometimes called **molecular
geometry**, used as a more narrow description)
describes the arrangement of only the bonded
atoms around a central atom. Bonding electron

pairs specify the positions of the bonded atoms; thus, lone electron pairs are considered invisible in this description. **Atom geometry must be derived from E.R. geometry**.

Step (5): Determine the E.R. geometry for each central atom as necessary; <u>each central atom is analyzed separately</u>.

Electron regions repel each other; the distribution of E.R. around a central atom is based on the fact that E.R. will occupy positions around a central atom which allow them to be as far away from each other as possible. The E.R. geometry is found by matching the E.R. count to the corresponding geometrical figure which describes the 3-D distribution of E.R. in space. **Use the summary table on the next page** to identify the geometrical figure from the E.R. count.

Summary table of Electron Region Geometries

Notation: ($\bigcirc$) = E.R. ($\bullet$) = central atom
(‹·····I) = figure line going backward (away from you)
(▬►) = figure line coming forward (toward you)

# of Electron Regions	Geometrical Figure	Drawn Figure
2	**Linear** all angles = 180°	
3	**Trigonal planar** all angles = 120°	
4	**Tetrahedral** all angles = 109.5°	

Step (6): Draw the geometry of the central atoms by placing bonded atoms and lone pairs at the corners of the correct geometrical figure, with the specific central atom at the center.

Linear, trigonal planar, and tetrahedral geometrical figures are symmetrical: all angles connecting any two corners through the central atom are identical; thus, all corner positions are equivalent. **<u>For these figures, it does not matter which corners are selected for lone pairs and which corners are selected for atoms; any arrangement will always produce equivalent final shapes</u>**.

Step (7): Determine the atom geometry (molecular geometry): <u>first draw the E.R. geometry [Step (6)]; then, find the atom geometry by "covering up" the lone pairs and identifying the figure that remains</u>.

1. **Atom Geometry** (also termed molecular geometry) describes the shape of only the bonded atoms around a central atom; the lone pairs are considered "invisible." Atom geometry must be derived from E.R. geometry; **lone pairs always influence the position of bonded atoms, even if the lone pairs are not specifically designated in the atom geometry shape**. Lone pairs or atoms can be placed at **any** corner for trigonal planar and tetrahedral geometries. **Do not rearrange atoms from the E.R. geometry**; the lone pairs are still present in the designated corners, even if they are considered invisible for the atom geometry.

2. If a central atom does not have lone pairs, then, atoms will occupy all corners of the E.R. geometrical figure: **whenever the E.R. of a central atom are all bonded atoms (no lone pairs), the atom geometry must be identical to the E.R. geometry**.

3. Atom geometry shapes include **all of the possible E.R. geometrical** figures **plus** the **additional** possibilities shown in the table below. Lone pairs (placed in arbitrary corners) are shown shaded out to indicate the shape of just the atoms. **The geometrical figure name applies to the remaining atom-only shape**.

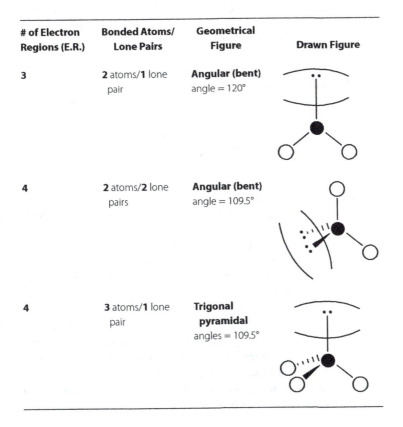

# of Electron Regions (E.R.)	Bonded Atoms/ Lone Pairs	Geometrical Figure	Drawn Figure
3	2 atoms/1 lone pair	**Angular (bent)** angle = 120°	
4	2 atoms/2 lone pairs	**Angular (bent)** angle = 109.5°	
4	3 atoms/1 lone pair	**Trigonal pyramidal** angles = 109.5°	

Step (8): Determine the polarity of the molecule.

General Concept:

1. **Molecular polarity** measures the total vector sum of all bond polarities in a complete molecule; this is termed as the molecular **dipole moment.** The dipole moment is proportional to the strength of each individual bond dipole and the direction of each bond dipole distributed around a central atom.

2. A molecule is generally polar (has a **net dipole**) when at least **one** bond in the molecule is polar (has a **dipole**) and for at least **one** central atom, all polar bonds (**dipoles**) around this central atom do **not cancel** due to a symmetrical arrangement of the bonded atoms. An exception to this can occur: a molecule will not be polar if polar central atoms themselves cancel due to symmetry.

3. The degree of polarity of a **specific bond** (termed the strength of the **dipole**) is proportional to the difference in electronegativity (Δ**EN**) between the two bonded atoms.
   An approximate, arbitrary, classification is
   (Δ**EN**) < 0.5 = weakly polar bond
   (Δ**EN**) $= 0.5–1.0$ = moderate polar bond
   (Δ**EN**) > 1.0 = strong polar bond

4. Use the arrow notation ($\longmapsto$) for polar covalent bonds; the "cross" end of the arrow is placed over the relatively positive (δ+) atom and the arrow point is placed over the relatively negative (δ–) atom of the covalent bond:
   $$\overset{\delta+}{X}—\overset{\delta-}{Y} = X—Y$$

Step (8): Process:

1. Use a table of atom electronegativities (EN) to determine the polarity of each bond in the molecule. Designate the bond polarities using the arrown notation.

2. Analyze each central atom separately; determine whether all the bond dipole arrows around a central atom show a net additive direction or whether they all exactly cancel.

General guidelines:

1. Linear, trigonal planar, and tetrahedral E.R. geometries are completely symmetrical structures. A simple generalization for **2–4 E.R. geometries is: If all corners of these figures are occupied by identical atoms, then, the central atom will not contribute to a polar molecule**. If **any** of the corners are **nonidentical** (assuming the bonds are polar), the central atom **will** contribute to a polar molecule.

2. A direct corollary is: *if any corner contains a lone pair, the arrangement around the central atom is not symmetrical and this central atom must contribute to polarity.*

3. Note that molecular dipoles can be classified in a very wide range of **net** polarity: **nonpolar** $\longleftrightarrow$ **very weakly polar** $\longleftrightarrow$ **weak** $\longleftrightarrow$ **moderate** $\longleftrightarrow$ **strongly polar**

3.2 GENERAL SUMMARIES

3.2.1 SUMMARY OF GEOMETRY/HYBRIDIZATION FOR 2–4 ELECTRON REGIONS

# Electron Regions (E.R.)	Bonded Atoms/ Lone Pairs	Hybridization	E.R. Geometry	Atom Geometry
2	**2** bonded atoms **0** lone pairs	sp	Linear	Linear
3	**3** bonded atoms **0** lone pairs	sp^2	Trigonal planar	Trigonal planar
3	**2** bonded atoms **1** lone pair	sp^2	Trigonal planar	Bent (angular)
4	**4** bonded atoms **0** lone pairs	sp^3	Tetrahedral	Tetrahedral
4	**3** bonded atoms **1** lone pair	sp^3	Tetrahedral	Trigonal pyramidal
4	**2** bonded atoms **2** lone pairs	sp^3	Tetrahedral	Bent (angular)

3.2.2 SUMMARY FOR THE GEOMETRY/HYBRIDIZATION FOR C, N, AND O

The geometry and hybridization of **carbon**, **nitrogen**, and **oxygen** as **central** atoms are important for the understanding of **organic and biological molecules**. (Normal **neutral** bonding, formal charge = 0; other geometries must be found by further E.R. analysis.)

Atom/# E.R.	Bond Types/ Lone Pairs	Sigma-Bonds/ Pi-Bonds	Hybridization	E.R. Geometry	Atom Geometry
Carbon 4 E.R.	**4** single bonds **0** lone pairs	**4** sigma-bonds **0** pi-bonds (0 lone pairs)	sp^3	Tetrahedral	Tetrahedral
Carbon 3 E.R.	**2** single bonds **1** double bond **0** lone pairs	**3** sigma-bonds **1** pi-bond (0 lone pairs)	sp^2	Trigonal planar	Trigonal planar
Carbon 2 E.R.	**2** double bonds **0** lone pairs	**2** sigma-bonds **2** pi-bonds (0 lone pairs)	sp	Linear	Linear
Carbon 2 E.R.	**1** single bond **1** triple bond **0** lone pairs	**2** sigma-bonds **2** pi-bonds (0 lone pairs)	sp	Linear	Linear
Nitrogen 4 E.R.	**3** single bonds **1** lone pair	**3** sigma-bonds **0** pi-bonds (1 lone pair)	sp^3	Tetrahedral	Trigonal pyramidal
Nitrogen 3 E.R.	**1** single bond **1** double bond **1** lone pair	**2** sigma-bonds **1** pi-bond (1 lone pair)	sp^2	Trigonal planar	Bent (angular)
Oxygen 4 E.R.	**2** single bonds **2** lone pairs	**2** sigma-bonds 0 pi-bonds (2 lone pairs)	sp^3	Tetrahedral	Bent (angular)

3.3 PRACTICE EXERCISES

1. Refer to the practice exercises for **Chapter 1**. For **problem #2(c)** only: complete analysis steps **(2)–(8)** based on the Lewis structures identified.

2. Complete the analysis steps **(2)–(8)** for the following **given** Lewis structures. **The drawings show the condensed form for Lewis structures in which bond lines between H and C or N combinations are not shown:**

$$CH_3 = H-\underset{\underset{H}{|}}{\overset{|}{C}}-H \qquad CH_2 = H-\underset{\underset{H}{|}}{\overset{|}{C}} \qquad CH = H-C \qquad NH_2 = N-\underset{\underset{H}{|}}{\overset{|}{H}}$$

a. $H_3C - CH = CH_2$ b. $H - C \equiv C - CH_3$ c. $H_2C = \underset{\underset{HC = N-H}{|}}{C} - C \equiv N$

d.

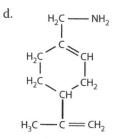

3. Complete the analysis steps **(1)** through **(8)**; use the additional information for each formula to find the **one** correct structure; then analyze each central atom in the molecule.

a. $C_3H_3F_3$	**All** fluorines are bonded to the same carbon.
b. C_4H_7N	The nitrogen is sp^3 hybridized and is bonded only to carbons.
c. $C_4H_8O_2$	**All** carbons are bonded in a connecting chain. Both oxygens are bonded to the same carbon. One oxygen uses a p-orbital to form a pi-bond; the other oxygen is sp^3 hybridized.
d. C_2H_5NO	The nitrogen is sp^2 hybridized; the oxygen is sp^3 hybridized. The carbons are bonded to each other. **No** hydrogens are bonded to nitrogen.

3.4 ANSWERS TO PRACTICE EXERCISES

1. Analysis of the Lewis structures shown in Chapter 1 answers. Other answers are possible.

   Polar

   $C^\#$ 1,2 are sp^2/trig planar (ER & atom)
   N = sp^2/trig planar-ER
   Bent-atoms

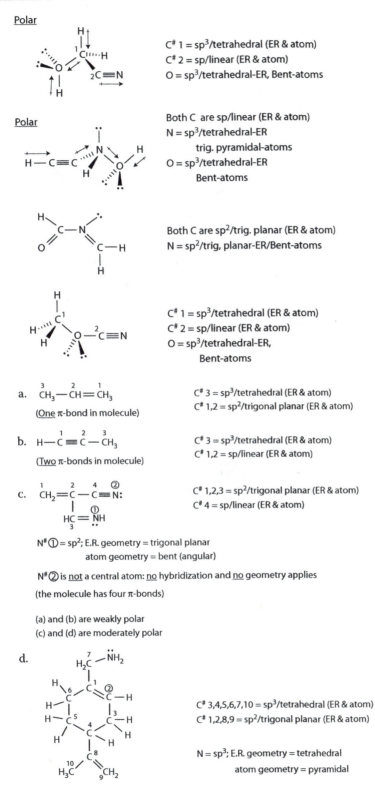

Polar

$C^{\#}$ 1 = sp^3/tetrahedral (ER & atom)
$C^{\#}$ 2 = sp/linear (ER & atom)
O = sp^3/tetrahedral-ER, Bent-atoms

Polar

Both C are sp/linear (ER & atom)
N = sp^3/tetrahedral-ER
 trig. pyramidal-atoms
O = sp^3/tetrahedral-ER
 Bent-atoms

Both C are sp^2/trig. planar (ER & atom)
N = sp^2/trig, planar-ER/Bent-atoms

$C^{\#}$ 1 = sp^3/tetrahedral (ER & atom)
$C^{\#}$ 2 = sp/linear (ER & atom)
O = sp^3/tetrahedral-ER,
 Bent-atoms

2.   a. $\overset{3}{CH_3} - \overset{2}{CH} = \overset{1}{CH_3}$

   (One π-bond in molecule)

$C^{\#}$ 3 = sp^3/tetrahedral (ER & atom)
$C^{\#}$ 1,2 = sp^2/trigonal planar (ER & atom)

  b. $H - \overset{1}{C} \equiv \overset{2}{C} - \overset{3}{CH_3}$

   (Two π-bonds in molecule)

$C^{\#}$ 3 = sp^3/tetrahedral (ER & atom)
$C^{\#}$ 1,2 = sp/linear (ER & atom)

  c. $\overset{1}{CH_2} = \overset{2}{C} - \overset{4}{C} \equiv \overset{②}{N:}$
       $\quad\quad | \quad \overset{①}{}$
       $\quad\; \underset{3}{HC} = NH$
       $\quad\quad\quad\quad ..$

$C^{\#}$ 1,2,3 = sp^2/trigonal planar (ER & atom)
$C^{\#}$ 4 = sp/linear (ER & atom)

N$^{\#}$① = sp^2; E.R. geometry = trigonal planar
         atom geometry = bent (angular)

N$^{\#}$② is not a central atom: no hybridization and no geometry applies

(the molecule has four π-bonds)

(a) and (b) are weakly polar
(c) and (d) are moderately polar

  d.

$C^{\#}$ 3,4,5,6,7,10 = sp^3/tetrahedral (ER & atom)
$C^{\#}$ 1,2,8,9 = sp^2/trigonal planar (ER & atom)

N = sp^3; E.R. geometry = tetrahedral
         atom geometry = pyramidal

(The molecule has two π-bonds)

3.

a.

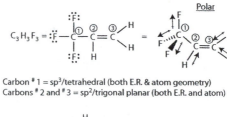

$C_3H_3F_3 =$

Carbon # 1 = sp³/tetrahedral (both E.R. & atom geometry)
Carbons # 2 and # 3 = sp²/trigonal planar (both E.R. and atom)

b.

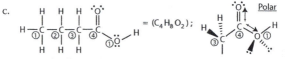

$C_4H_7N =$

(Polarity shown
for _some_ bonds)

Carbons # 1,# 2 = sp³/tetrahedral (E.R. & atom)
Carbons # 3, # 4 = sp/linear (E.R. & atom)
Nitrogen = sp³/tetrahedral _E.R._ and pyramidal _atoms_

c.

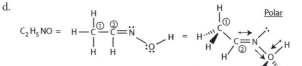

= $(C_4H_8O_2)$;

(Partial geometry &
polarity shown)

Carbons # 1, # 2, # 3 = sp³/tetrahedral (E.R. & atom)
Carbon # 4 = sp²/trigonal planar (E.R. & atom)
Oxygen # 1 = sp³/tetrahedral _E.R._; bent-_atoms_

d.

$C_2H_5NO =$

(Partial polarity
shown)

Carbon # 1 = sp³/tetrahedral (E.R. & atom)
Carbon # 2 = sp²/trigonal planar (E.R. & atom)
Nitrogen = sp²/trigonal planar _E.R._; bent-_atoms_
Oxygen = sp³/tetrahedral _E.R._; bent-_atoms_

Notation in Organic Chemistry

Guide to Writing and Using Condensed Formulas and Line Drawings

4

4.1 CONDENSED STRUCTURAL FORMULAS

Full Lewis structures can be simplified to a condensed formula by elimination of bond lines and lone pair notation. **The key requirement to any simplification is that all information about the correct structure must be retained in the condensed form.**

Simplification results from well-established electron-bonding principles of the major atoms in the molecule (C, H, O, N, plus halogens).

4.1.1 GUIDELINES FOR CONVERTING A LEWIS STRUCTURE INTO A CONDENSED STRUCTURAL FORMULA

Examples for acetone and alanine (an amino acid) will be carried through the process.

Step 1: Start with the correct complete Lewis structure.

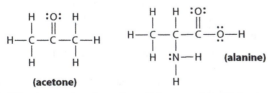

(acetone)   (alanine)

Step 2: Eliminate the dots for all lone pairs of electrons. **No information is lost**; the number of lone pairs on each atom is designated by the specific bonding arrangement (number of bonds plus any stated formal charge).

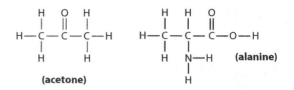

(acetone)   (alanine)

Step 3: Eliminate the bond lines between hydrogen and its bonded atoms (i.e., the structural line (−) connecting H− to its bonded atom). Replace these with the standard formula system, where the **H**s are written next to the atoms they are bonded to; indicate the number of hydrogens bonded to the same atom by a subscript. **No information is lost**; since hydrogen can take **only one bond**, the bonding pattern will be specifically designated. (Since H can **never** be in the **middle** of an atom chain, the order of writing this formula combination is not important.)

$$H_3C - \overset{\overset{O}{\|}}{C} - CH_3 \quad \text{or} \quad CH_3 - \overset{\overset{O}{\|}}{C} - CH_3 \qquad CH_3 - \underset{NH_3}{CH} - \overset{\overset{O}{\|}}{C} - OH$$

(acetone)   (alanine)

Step 4: Keep the carbon atom chain in order, and then eliminate bond lines (−) between carbon atoms. **No information is lost if the carbon sequence remains in the same order as the original Lewis structure**. The bonding pattern for the carbons is determined by its normal bonding rules; neutral carbon must always take **four** bonds; exceptions must be indicated by formal charges.

$$CH_3\overset{\overset{O}{\|}}{C}CH_3 \quad \text{(acetone)} \qquad\qquad CH_3\underset{NH_2}{CH}\overset{\overset{O}{\|}}{C} - OH \quad \text{(alanine)}$$

This level of condensation often provides the most useful description of the molecule. Further condensation is possible.

Step 5: Eliminate bond lines (−) from carbon to oxygen; place the oxygen atom symbol on the same line as the carbons, and next to the carbon (or carbon–hydrogen group) to which it is bonded. **No information is lost**, since the bonding pattern from carbon to oxygen is specified by established bonding rules.

$$CH_3COCH_3 \text{ (acetone)} \qquad CH_3\underset{NH_2}{CHCOOH} \quad \text{or} \quad CH_3\underset{NH_2}{CHCO_2H} \quad \text{(alanine)}$$

4.1.2 NOTES AND OTHER SIMPLIFICATIONS

1. Halogens (F, Cl, Br, I) almost always take **one** bond. Use the rules stated for hydrogen to apply to halogen atoms also.

$$H-\overset{\overset{\displaystyle H}{|}}{\underset{\underset{\displaystyle H}{|}}{C}}-\overset{\overset{\displaystyle Br}{|}}{\underset{\underset{\displaystyle Br}{|}}{C}}-\overset{\overset{\displaystyle H}{|}}{\underset{\underset{\displaystyle H}{|}}{C}}-O-H \ = \ CH_3CBr_2CH_2OH$$

Br cannot be part of the carbon-bonded chain because it takes only **one** bond.

2. Atom groupings (various possibilities) can be treated as a single unit by enclosing the group in parentheses.

$$CH_3-CH_2-\overset{\overset{\displaystyle }{\underset{\underset{\displaystyle CH_3}{|}}{CH}}}{}-CH_2-CH_3 \ = \ CH_3CH_2CH(\mathbf{CH_3})CH_2CH_3$$

In this case, the atom grouping {**CH$_3$**}, which is bonded to the middle-chain carbon, is placed on the same line as the carbon chain next to the carbon it is bonded to. The parentheses indicate that this atom grouping is not part of the carbon chain sequence.

3. More than one **identical** atom grouping (enclosed in parentheses) can take a numerical subscript in a condensed formula.

$$CH_3-\overset{\overset{\displaystyle H}{|}}{\underset{\underset{\displaystyle CH_3}{|}}{C}}-\overset{\overset{\displaystyle F}{|}}{\underset{\underset{\displaystyle F}{|}}{C}}-F \ = \ (\mathbf{CH_3})_2CHCHF_2$$

In this case, **two** atom groups {**CH$_3$**} are bonded to the same carbon; this is indicated by the subscript (2) after the parentheses. (Compare this to the indication of two fluorines on the other carbon.)

$$CH_3\mathbf{CH_2CH_2CH_2CH_2}CH_3 = CH_3(\mathbf{CH_2})_4CH_3$$

This example shows the simplification for writing long carbon chains, even when they are shown in the partially condensed form.

4. *Note that the examples show structural formulas in various stages (levels) of condensation. It is important to realize that* *all* *forms of* *condensed* *or* *expanded* *formulas are* *equally* *acceptable* *as long as the complete molecule can be reconstructed from the information. The level to which a Lewis structure is condensed is dictated by writing convenience, not by rigid requirements.*

4.1.3 EXPANDING CONDENSED FORMULAS: RECONSTRUCTING THE COMPLETE MOLECULE

1. Although the level to which a structural formula is condensed is arbitrary, it is always necessary to be able to reconstruct or "picture" the complete Lewis structure from any condensed formula.

If a condensed formula is valid, it must contain all information necessary to reproduce the original molecule.

2. **To expand a condensed formula to any specific level of detail, simply reverse the process steps used in condensation.**

$CH_3(CH_2)_2CH_2Br \longrightarrow CH_3CH_2CH_2CH_2Br \longrightarrow CH_3-CH_2-CH_2-CH_2-Br \longrightarrow$

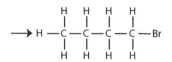

3. **To identify double and triple bonds in a formula, use the normal neutral bonding rules for each atom as a guide.**

$CH_3CHO \longrightarrow CH_3\overset{\displaystyle H}{\overset{|}{C}}-O$ (**C** has only **three** bonds, **O** has only **one** bond) $\longrightarrow$

Use a **double** bond $\longrightarrow$ $CH_3\overset{\displaystyle H}{\overset{|}{C}}=O$ $\longrightarrow$ $H-\overset{\displaystyle H}{\underset{\displaystyle H}{\overset{|}{\underset{|}{C}}}}-\overset{\displaystyle H}{\overset{|}{C}}=O$

$CH_3CHCHCH_3 \longrightarrow CH_3-CH-CH-CH_3$ (middle **C**arbons each missing **one** bond)

Use a **double** bond $\longrightarrow$ $CH_3-CH=CH-CH_3$ $\longrightarrow$ $H-\overset{\displaystyle H}{\underset{\displaystyle H}{\overset{|}{\underset{|}{C}}}}-\overset{\displaystyle H}{\overset{|}{C}}=\overset{\displaystyle H}{\overset{|}{C}}-\overset{\displaystyle H}{\underset{\displaystyle H}{\overset{|}{\underset{|}{C}}}}-H$

$CH_3CCCH_3 \longrightarrow CH_3-C-C-CH_3 \longrightarrow CH_3-C\equiv C-CH_3$

4.2 LINE DRAWINGS

Line drawings are an alternative notational system for organic structures. The key requirement is the same: ***all*** *information contained in the Lewis structure must be retained in the line drawing.* The simplification consists of **keeping the bond lines** (except for **C–H**) while **eliminating most carbon and hydrogen atom symbols: C** and **H**.

4.2.1 RULES FOR PRODUCING LINE DRAWINGS

Three examples will be carried through the process.

  Step 1: Draw a complete structural formula from the Lewis structure:

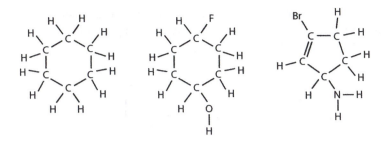

Step 2: Eliminate the atom symbol for **carbon (C)** while leaving **all** bond lines in place:

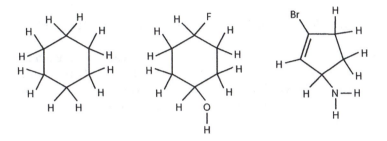

Step 3: Specifically and **only** for **hydrogens bonded to carbon (C—H)**, **eliminate both** the **atom symbol** for **hydrogen and** the **C—H bond line**. **All** other atom symbols **must** be shown; **symbols** for **all** hydrogens bonded to **any** atom other than carbon **must** be shown. **Bond lines** from hydrogen to other atoms are always optional: use the rules for condensing formulas for further simplification if desired.

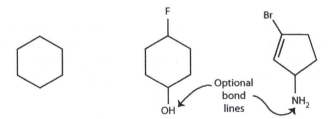

Step 4: Line drawings are standard for **ring** structures. They are also commonplace for most other organic molecules. For **non-ring** carbon chains, use "zig-zag" drawings for the chain for clarity (mimics tetrahedral carbon):

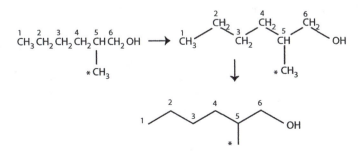

Line drawing notation can be mixed with structural formula notation to increase visual clarity. Example: to aid in seeing an isolated part of the structure or emphasize a particular part of a molecule:

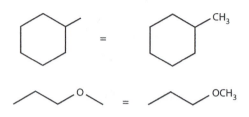

4.2.2 CONVERTING LINE DRAWINGS INTO STRUCTURAL FORMULAS

Step 1: Starting from a line drawing, replace **carbons** at **each line end** or **line vertex** (intersection of two lines) **only if no other atom is shown**. *Note: Replacing **incorrect** carbons at line ends that already show another atom is the most common error in this process!*

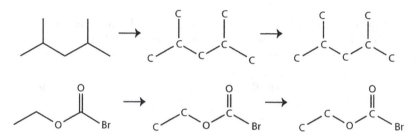

Step 2: Replace the **hydrogens (H symbols)** on **carbons**. Use the normal neutral bonding rules to determine the correct number of hydrogens on each carbon: neutral carbon (in normal compounds) **must** have **four** bonds; thus, add sufficient numbers of hydrogens to give each carbon exactly four bonds.

Conventions: For convenience, it is rarely necessary to show all C−H bond lines when expanding a line drawing; they **may** be shown to emphasize certain parts of a molecule. Exceptions to this convention are expansions of line drawings of **rings: C−H bond lines are usually shown for carbons contained in the ring**. This convention aids in visual clarity for the ring structure.

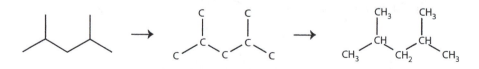

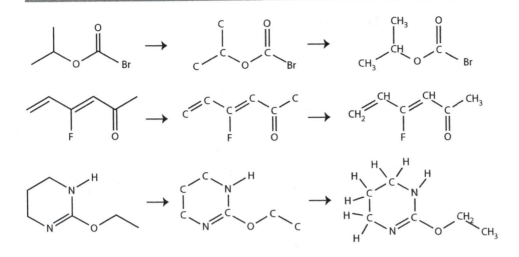

PRACTICE EXERCISES

1. Complete the following exercises; use any structure method desired.
   a. The formula C_3H_6O actually has a total of **eight** constitutional isomers when normal neutral bonding rules are applied to all atoms. Try to find all **eight**, noting that rings containing three or four atoms can be used.
   b. Find **four** constitutional isomers (one with a ring) for C_2H_5N. Use normal neutral bonding for all atoms.
   c. Find **seven** constitutional isomers (two with rings) for $C_3H_4Cl_2$. Use normal neutral bonding; Cl takes only **one** bond.
2. Expand the following condensed formulas to nearly complete structural formulas, showing lone pairs of electrons. However, for convenience: C−H bond lines are optional, keep these condensed if desired.
   a. $(CH_3)_3CCHO$
   b. $CH_3CH_2C(CH_3)_2F$
   c. $(CCl_3)_3CCBr_3$
   d. $CH_3CH_2CHCl\ COCH_3$
   e. $CH_3(CH_2)_4COOH$
   f. $(CH_3CH_2)_2CCHCOOCH_3$
   g. $Cl_2CCBrCH(CH_2Cl)_2$
   h. $CH_3CH_2OCH_2COCH_3$
3. Convert the following line drawings into full structural formulas (including lone pairs) or structural formulas to line drawings as requested. For structural formulas, show the C−H bond lines when the carbon is part of a ring, but keep the condensed CH_X simplification for non-ring carbons.

STRUCTURAL FORMULA
Corresponding Line Drawing

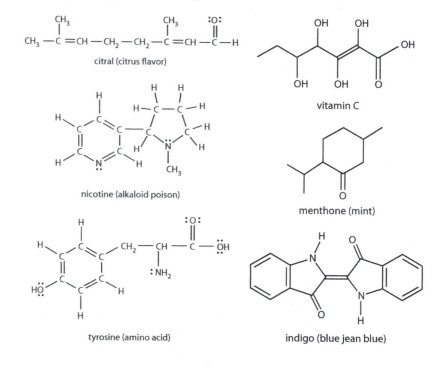

citral (citrus flavor)

vitamin C

nicotine (alkaloid poison)

menthone (mint)

tyrosine (amino acid)

indigo (blue jean blue)

ANSWERS TO PRACTICE EXERCISES:
ORGANIC FORMULA NOTATION

1. a.

b.

c.

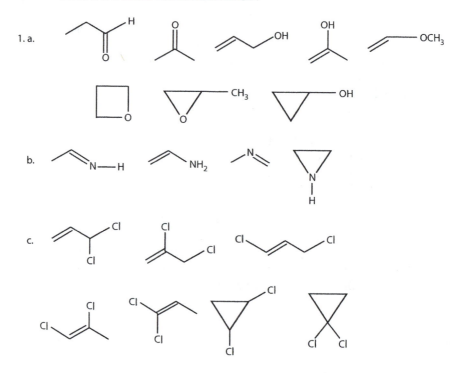

2. a. 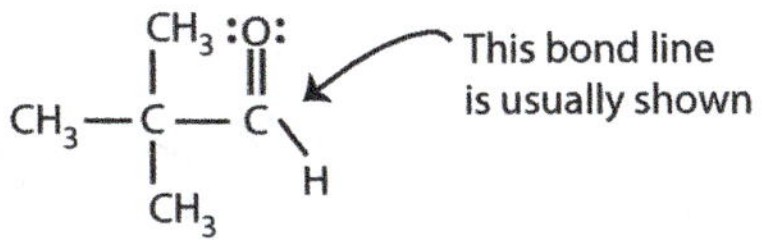
This bond line
is usually shown

b. 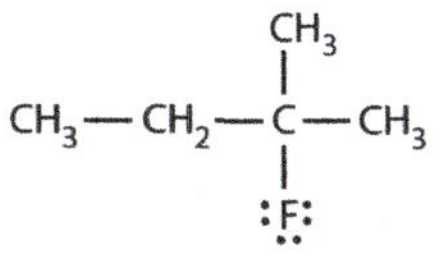

c. 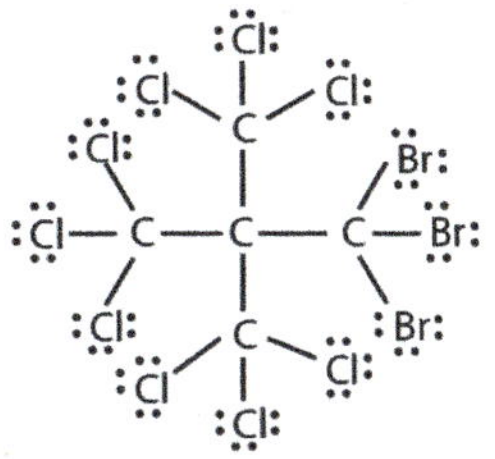

d. 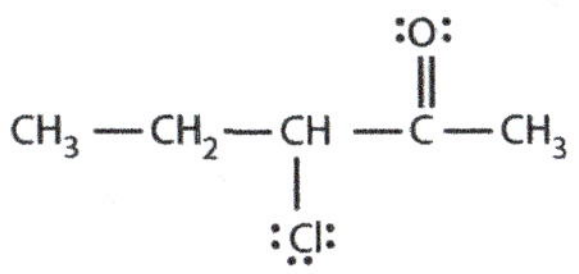

e. 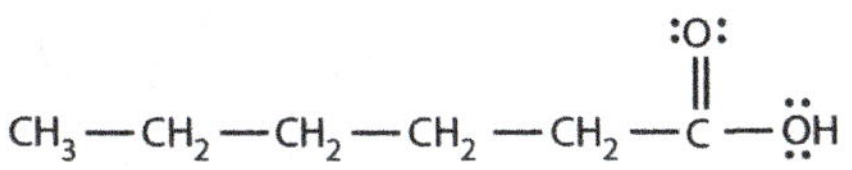

f. 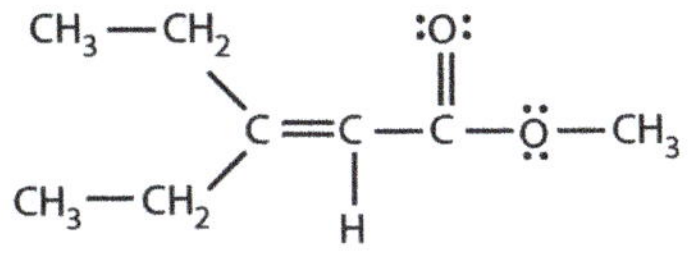

g. 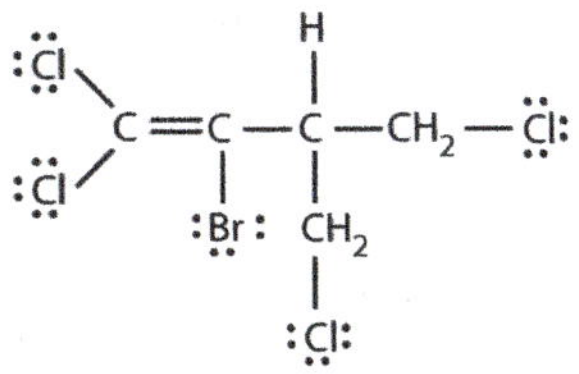

h. 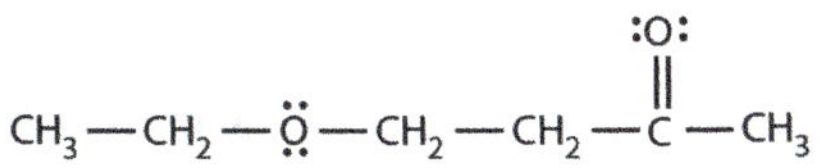

3. Convert the following line drawings into full structural formulas (including lone pairs) or structural formulas to line drawings as requested.

STRUCTURAL FORMULA

Corresponding Line Drawing

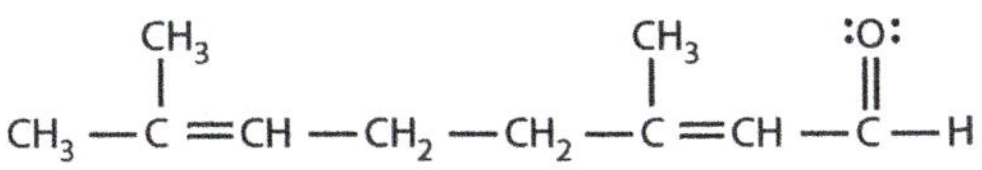

citral (citrus flavor)

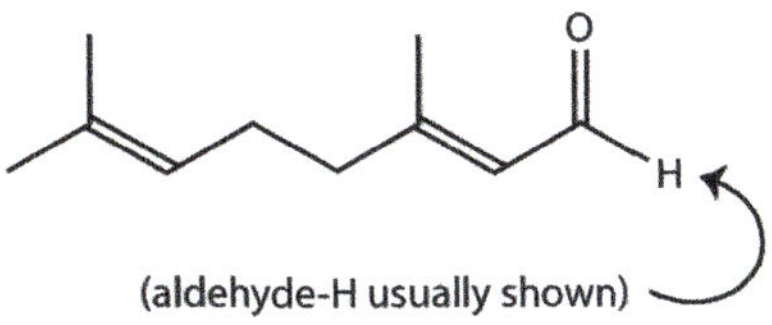
(aldehyde-H usually shown)

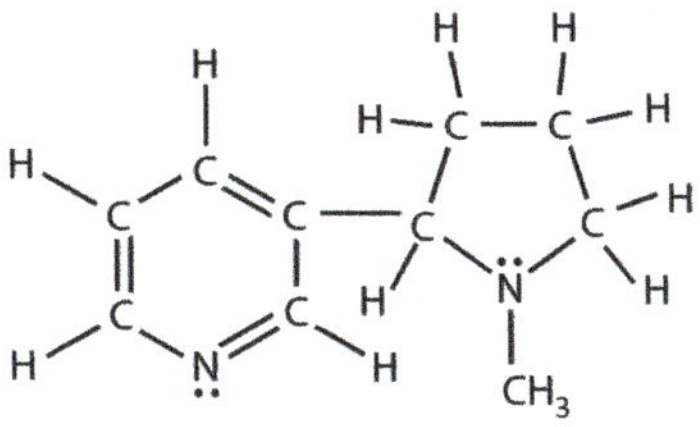

nicotine (alkaloid poison)

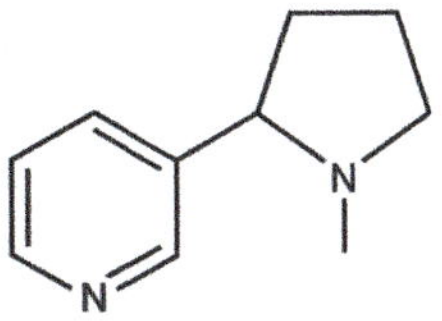

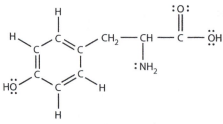

tyrosine (amino acid)

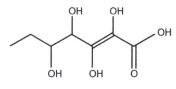

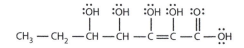

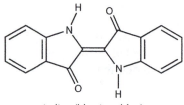

vitamin C

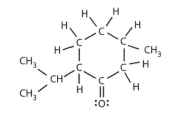

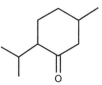

menthone (mint)

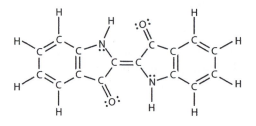

indigo (blue jean blue)

Summary Guidelines for Organic Nomenclature

5

5.1 GENERAL CONCEPTS

1. Nomenclature of organic molecules follows a systematic method (IUPAC method) based on main carbon chains or rings, attached groups, and specific prefixes and suffixes to indicate functional groups.
   a. The procedure outlined in this supplement describes a general series of **Steps** (**I–X**) with explanatory rules and information tables. This is a **summary** of the main concepts for the systematic method; *it is **not** a complete discussion of each rule and all exceptions*.
   b. Note that many organic molecules can be named through more than one system; other names are termed **common names**. Many common names, especially in biochemistry, do not follow any universal system; for example, names of sugars ("glucose"), amino acids ("glycine"), vitamins ("niacin," "pyridoxal," and "folic acid"), etc.
2. For systematic nomenclature, **a compound consists of**
   a. **A main portion:** carbon chains or rings that contain the main functional group in the molecule.
   b. **Substituents:** attached atoms or atom groupings which represent carbon groups or other functional groups connected to the main portion of the molecule.
3. For systematic nomenclature, **a complete name consists of**
   a. A **root** (base) **name** which indicates the number of carbons in the nomenclature-based main portion of the molecule.
   b. A combination of **prefix** and **ending** (suffix) combined with the root name which indicates the major functional group in the main portion.
   c. A **list** of all attached **substituents** connected to the main portion of the molecule.
   d. A listing of **numbers** to locate substituents and main functional groups within the main portion.

4. *This summary outline does **not** apply to the naming of **aromatic** compounds*; these molecules must be treated as a separate topic.
5. **Steps I–X** are not sufficient to ***directly* name carboxylic acid derivatives *(esters, amides, etc.)***. These molecules are named through an adaptation of carboxylic acid nomenclature and covered in **Part 3**.

Part 1: Description of General Process: Alkanes/Cycloalkanes/Alkyl Halides

Step I. Identify the major functional group in the molecule; this determines the main portion of the molecule.

1. The major functional group in an organic molecule identifies the carbon portion to be counted in the root name and identifies the **correct ending**.
2. For alkanes/cycloalkanes/alkyl halides, the major functional group is the alkane hydrocarbon chain or ring.

Step II. Identify the correct nomenclature-based main portion of the molecule; this determines the correct root name.

1. Alkanes/cycloalkanes/alkyl halides are molecules with ***no functional group of higher priority than alkane***. Therefore, the **main portion** of the molecule **is either**
   1a. *the longest continuous **chain** of bonded carbon atoms* for molecules that do **not** contain rings; this is termed the **main chain**.
   1b. *the **ring** with the greatest number of carbons* for molecules that contain one or more rings and no carbon chains larger than the largest ring.
   1c. *the ring or chain with the greatest number of total carbons* if the molecule contains both rings and long chains; the main portion is selected by comparing the number of carbons in the largest ring to the number of carbons in the longest chain. *If these numbers are the same, selection involves other factors.*

Step III. Determine the correct root name of the molecule based on the main portion.

1. *Count the number of carbons* in the main chain or ring identified in **Step II**.
2. *Match this carbon count to the correct root name* based on the table below.

3. *If the main portion of the molecule is a ring, add the prefix "cyclo"* to indicate a ring of the specific number of carbons.

The list below shows the root names for counting carbons from 1 to 12:

# C	Root Name	# C	Root Name	# C	Root Name
1 C	meth	5	pent	9	non
2 C	eth	6	hex	10	dec
3 C	prop	7	hept	11	undec
4 C	But	8	oct	12	dodec

Step IV. Identify the correct ending to indicate the main (highest priority) functional group in the molecule.

1. Match the highest-priority functional group to the corresponding **ending**.
2. The correct ending for alkanes/cycloalkanes/alkyl halides corresponds to the ending for the alkane major functional group: **-ane**

Step V. Name the key part of the main portion by adding the appropriate ending to the root name; this is the base name or main name.

1. The name of the *main portion of the molecule contains an ending* to indicate the highest-priority functional group; this ending is combined with the root name to form the **main (base) name**.

Functional Group	Main Name = Root + Ending	Multiples of the Same Group: Examples of Endings
Alkane	root + **-ane**	not applicable

Step VI. Identify all substituents

1. The main portion of the molecule is indicated by the main name or base name; *__all__ other atoms or atom groups attached to the main chain or ring must be named as substituents*.
2. Substituents can include **halogen** atoms. Substituents may also be carbon chains or rings that are smaller than the main chain or ring. A general term for these are (carbon) **branches** off a chain/ring. Carbon branches can occur attached to a main chain/ring or to further substituent chains/rings.
3. *Each* substituent is identified by a *specific name; match* the substituent to its name using the table shown. Depending on the types of functional groups present, almost any functional group may be a substituent in certain molecules. *Only the highest-priority functional group is used in the main name; __all__ other lower-priority functional groups must be substituents.*

Table for Step VI: Substituent Names for Alkanes/Cycloalkanes/Alkyl Halides

Atom or Atom Grouping as a Substituent	Substituent Name
—F	fluoro
—Cl	chloro
—Br	bromo
—I	iodo
alkyl: **non**-main sequential carbon chain: $CH_3(CH_2)_x$—	root + -yl
specific branched substituent: $\begin{array}{c}CH_3CH-\\|\\CH_3\end{array}$	isopropyl
specific branched substituent: $\begin{array}{c}CH_3CHCH_2-\\|\\CH_3\end{array}$	isobutyl
specific branched substituent: $\begin{array}{c}CH_3CH_2CH-\\|\\CH_3\end{array}$	secondary-butyl (sec-butyl)
specific branched substituent: $(CH_3)_3C$—	tertiary-butyl (tert-butyl or t-butyl)

Many branched groups exist that are not specifically included in this table: *For __all__ general branched carbon groups as substituents: See Step X.*

Step VII. Number the main chain or main ring based on the principle of <u>first point of difference</u>.

1. Functional groups and substituents must have a location specified relative to the main portion of the molecule; all carbons in the main chain or ring are numbered.
2. *__Alkanes__ and __cycloalkanes__*, as the main portion if with low-priority functional groups, are not numbered to locate the main functional group, but are numbered to locate substitutents.
   Rules for carbon chains (alkanes):
   2a. The main chain of carbon atoms must be numbered sequentially *such that the #1 carbon starts at one end*. Numbering of the carbon chain can begin from one of the two ends and therefore be numbered sequentially in two ways.
   Example (2a): If the main chain of a simple alkane is written horizontally, an example is

#1	#2	#3	#4	#5	#6		#6	#5	#4	#3	#2	#1
CH_3	CH_2	CH_2	CH_2	CH_2	CH_3	or	CH_3	CH_2	CH_2	CH_2	CH_2	CH_3

   left-to-right   right-to-left

   2b. The main chain is always numbered in the direction that produces *the __lowest__ number for the carbon that bears substituent groups __at the first point of difference__*.

3. ***Rules for carbon rings (cycloalkanes):***
   3a. A carbon ring must be numbered ***sequentially***. A ring has no beginning or end carbon; therefore, **any** of the ring carbons can **potentially** be designated as carbon **#1**.
   3b. The main ring is always numbered such that ***a carbon bearing a substituent is carbon #1***. For more than one substituent, number in the direction that produces the ***lowest number for the carbon that bears substituent groups at the first point of difference***.
4. ***Explanation of the concept of first point of difference:***
   The **first point** of **difference** means that of all possible ways to correctly number a chain or ring sequentially:
   4a. select the direction which gives the **lowest** number to the carbon that bears the **first** substituent to be reached along the chain or ring in the sequential counting process.
   4b. select the direction which gives the **lowest** number to the carbon that bears the **first tie-breaking** substituent to be reached along the chain or ring in the sequential counting process.

Example for (4a):

#1   #2   #3   #4   #5   #6   #6   #5   #4   #3   #2   #1
CH_2—CH_2—CH_2—CH—CH—CH_3 or CH_2—CH_2—CH_2—CH—CH—CH_3
| | | | | |
Br   Br   Br   Br   Br   Br

left-to-right   **right-to-left**

first —Br comes at carbon **#1**   first —Br comes at carbon **#2**

The correct numbering sequence is **left-to-right** to produce the first—Br at carbon #1. Since #1 is lower than #2 at the first point of difference, the location numbers for the other two bromines do not affect the selection.

Example for (4b):

#1   #2   #3   #4   #5   #6   #6   #5   #4   #3   #2   #1
CH_2—CH_2—CH_2—CH—CH_2—CH_2 or CH_2—CH_2—CH_2—CH—CH_2—CH_2
| | | | | |
Br   Br   Br   Br   Br   Br

left-to-right   **right-to-left**

first —Br comes at carbon **#1** ←—same—→ first —Br comes at carbon **#1**
second —Br comes at carbon **#4**   second —Br comes at carbon **#3**

The first —Br comes at carbon #1 for both directions and cannot be used in the selection. The first **tie-breaking** —Br is used: the correct numbering sequence is **right-to-left** to produce the **second** —Br at carbon **#3** as compared to the **second** —Br at carbon **#4**. The carbon numbers are **not** additive; selection is based only on the **first** tie-breaking substituent.

Step VIII. Complete the main name (base name) by using the carbon numbers to locate all groups of the same highest-priority functional groups.

1. **Step VIII** applies to molecules with higher-priority functional groups and is described in **Part 2**. The main name of *alkanes/ cycloalkanes/alkyl halides*, as the main portion if with low-priority functional groups, does not contain numbers to locate the main functional group.

Step IX. Finish the complete name of the molecule by naming and locating all substituents attached to the main portion; use Step X for complicated branched groups.

1. Use the list of substituents identified and matched to a name in **Step VI**.
2. *Each substituent is located by the carbon number* on the main chain or ring to which it is attached, based on **Step VII**.
3. Each substituent group must show the corresponding carbon location number directly in front of the substituent name, separated by a hyphen; different substituents are separated by a hyphen. All substituents are listed in alphabetical order (with some exceptions for certain prefixes) *in front of the main name:*

    # - substituent (A)- # - substituent (B)- # - substituent (C)main name

4. Multiple examples of the **same** substituent can be indicated by using the prefix-type labels "**di**" = 2, "**tri**" = 3, "**tetra**" = 4, "**penta**" = 5, etc. Place the numbers for the multiple examples together, separated by commas; if one specific carbon has two or more of the same substituent, the number must be repeated for each attached substituent.

    #, #, # - trisubstituent (A)- # - substituent (B)main name

5. Substituent numbers are **not** stated for
   a. carbon chains of one or two carbons: numbering cannot lead to a difference.
   b. rings with only one substituent: must automatically be #1; the #1 is not stated.

Step X. Use the techniques described in Steps VI through IX to name more complicated branched substituents.

1. Some branched substituents may have specific common names. *All branched substituent groups can be named by extending the concepts used for main chains and substituents:* Any branched substituent group can be considered to be composed of a *main branching* chain to which further groups are attached as substituents, that is, "branches on branches."
2. Identify the *connecting carbon* of the substituent carbon group.
3a. Identify the longest sequential carbon chain in the **substituent** carbon group *beginning with the connecting carbon*: this is the main substituent chain.

or
3b. Identify the connecting ring carbon in the **substituent** ring.
4. Name the ***main substituent chain as root + yl***; name the ***substituent ring as cyclo + root + yl***. **All substituent** chains or rings must take a **-yl** ending even if it is the basis of further substituents or branches.
5. Number the ***main* substituent chain or ring such that the *connecting carbon is #1***. Do **not** apply the rule of first point of difference to a substituent carbon chain or ring!
6. Indicate all ***further substituents on the main substituent chain or ring*** by following **Step** IX for the **substituent** main chain or ring. The name of the complete branched substituent group is enclosed in parentheses to avoid confusion:

#-(# - **substituent a, # - substituent b-main substituent chainyl)-main name**

5.2 EXAMPLE FOR AN ALKANE/ALKYL HALIDE

Molecule:

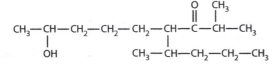

Step I. Highest-priority functional group = alkane.
Step II. For an alkane, the main portion is the longest continuous chain of bonded carbon atoms, this is shown in **bold face**. Note that the main chain is not written horizontally in this molecule:

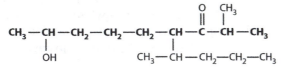

Step III. Total number of carbons in main chain = **10**; root name = **dec**.
Step IV. Correct ending for an **alkane** = **-ane**.
Step V. The correct base name is root + ending = dec + -ane = **decane**.
Step VI. Substituents are all atoms or atom groupings attached to the main chain; these are shown in **boldface**: (**a**) is a complicated branched substituent; (**b**), (**c**), and (**d**) are one-carbon substituent groups: root + ending = **meth + -yl = methyl**; (**e**) is **bromo**;

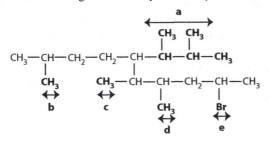

Step VII. The main chain can be numbered in one of the two ways:

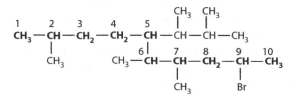

left-to-right: first substituent comes at C **#2**, second substituent comes at C **#5**; <u>or</u>

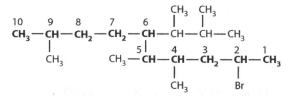

Right-to-left: First substituent comes at C **#2**, second substituent comes at C **#4**: By principle of first point to difference, the correct numbering is **right-to-left**.

Step VIII. Alkanes have no functional group number; the base name is **decane**.

Step IX. Substituents are named as **2-bromo; 4,5,9-trimethyl**. **Step X** is needed to complete the name of the complicated branched substituent.

Step X. The longest continuous chain of the complicated substituent group must begin with the connecting carbon as carbon number #1; the result is three carbons: name = root + -yl = **prop** + **-yl** = **propyl**. Attached to carbons #1 and #2 of the propyl substituent chain are **methyl** groups. The complete substituent name is (**1,2-dimethylpropyl**)

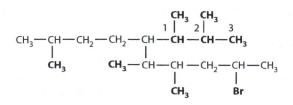

Step IX can be completed with the name of the complicated branched group. The (1,2-dimethylpropyl) is attached to the main chain at carbon **#6 = 6-(1,2-dimethylpropyl)**. Write all the substituents in alphabetical order in front of the main name: (The prefixes di- and tri- are not used for alphabetical order.) **2-bromo-4,5,9-trimethyl-6-(1,2-dimethylpropyl) -decane**

Part 2: Description of Complete General Process: Molecules with Higher-Priority Functional Groups

Step I. Identify the major functional group in the molecule; this determines the main portion of the molecule.

1. The major functional group in an organic molecule identifies the carbon portion to be counted in the root name and identifies the correct ending. _**The major functional group for naming purposes is the highest-priority functional group as defined by the following sequence**_:

Highest priority ←———————————————————————→ Lowest priority

Carboxylic > Aldehydes > Ketones > Alcohols > Alkynes > Alkenes > Alkanes/
acids and derivatives   (Aromatics)   Alkyl
(esters; amides; acid chlo-   Halides;
rides; anhydrides)   Alkoxy;
                                                                            Nitro;
                                                                            Amino;
                                                                            Epoxy

Step II. Identify the correct nomenclature-based main portion of the molecule; this determines the correct root name.

1. For molecules with **_no functional group of higher priority than alkane_**: follow the process described in **Part 1**.
2. For molecules with **_any_ _functional group of higher priority than alkane_**, (alkenes ——————→ carboxylic acid derivatives), the **main portion** of the molecule **is** either
   2a. **_the longest continuous chain of bonded carbon atoms which contains the highest-priority functional group_**; this is the **main chain**;
   2b. or **_the ring which contains the highest-priority functional group_**. Size of the ring is **not** considered unless two or more rings have the same highest-priority functional group.
3. If **_more than one_** of the same highest-priority functional group is present, select the longest chain (or ring) containing all of them **_if possible_**.

Step III. Determine the correct root name of the molecule based on the main portion.

1. **_Count the number of carbons_** in the main chain or ring identified in **Step II**.
2. **_Match this carbon count to the correct root name_** based on the table shown in **Part 1**.
3. **_If the main portion of the molecule is a ring, add the prefix "cyclo"_** to indicate a ring of the specific number of carbons.

Step IV. Identify the correct ending to indicate the main (highest-priority) functional group in the molecule.

1. Match the highest-priority functional group to the corresponding **ending** using the table shown.

Step IV and Step V Table: Endings for Nomenclature-Based Functional Group Main Portions

Functional group		Main Name = Root + Ending	Multiples of the Same Group: Examples of Endings
Alkane		root + **-ane**	not applicable
Alkene	$R_2C = CR_2$	root + **-ene**	root + -adiene (2); root + -atriene (3)
Alkyne	$RC \equiv CR$	root + **-yne**	root + -adiyne (2); root + -atriyne (3)
Alcohol	$R_3C - OH$	root + **-anol**	root + -anediol (2); root + -anetriol (3)
Ketone	$R_3C - \overset{\overset{O}{\|\|}}{C} - CR_3$	root + **-anone**	root + -anedione (2); root + -anetrione (3)
Aldehyde	$R_3C - \overset{\overset{O}{\|\|}}{C} - H$	root + **-anal**	root + -anedial (2)
Carboxylic acid	$R_3C - \overset{\overset{O}{\|\|}}{C} - OH$	root + **-anoic acid**	root + -anedioic acid (2)

Ending Combinations **Examples** for double-bond combinations:

Alkene + alcohol	root + **-enol**	**or** root + **-en-(#)-ol**
Alkene + aldehyde	root + **-enal**	
Alkene + ketone	root + **-enone**	**or** root + **-en-(#)-one**
Alkene + carboxylic acid	root + **-enoic acid**	

Step V. Name the key part of the main portion by adding the appropriate ending to the root name; this is the base name or main name.

1. The name of the ***main portion of the molecule contains an ending*** to indicate the highest-priority functional group; it is combined with the root name to form the **main (base) name**.
2. ***Except*** for additional functional groups combined with alkenes and alkynes, ***each main portion root name can have only <u>one possible</u> ending applied to the highest-priority functional group.*** Multiple ending combinations for **different** functional groups (except for double and triple bonds) are **not** allowed.
3. ***More than one group*** of the **same** highest-priority functional group is indicated by using the prefix-type labels "**di**" = 2, "**tri**" = 3, "**tetra**" = 4, "**penta**" = 5, etc.
4. ***Combinations of double or triple bonds with higher-priority functional groups*** are shown by a single combination ending;

this is the only exception to rule (**2**). The ending combination may be split to insert a number (indicated in the table as #).

5. Endings for **aromatic** compounds and carboxylic acid **derivatives** (esters, amides, etc.) are not included. These molecules must be treated as a separate topic.

Step VI. Identify all substituents

1. The main portion of the molecule is indicated by the main name or base name; **_all_ other atoms or atom groups attached to the main chain or ring must be named as substituents**.
2. Substituents can include halogen atoms or carbon branches or ring combinations as described in **Part 1**.
3. **_Each_** substituent is identified by a **_specific name; match_** the substituent to its name using the **table** shown in **Part 1** and the **additional** **table** shown in **Part 2**. Depending on the types of functional groups present, almost any functional group may be a substituent in certain molecules. **_Only the highest-priority functional group is used in the main name; all other lower-priority functional groups must be substituents_**.

Table of Additional Substituent Names

Atom or Atom Grouping as a Substituent		Substituent Name
		epoxy
$-NO_2$		nitro
$-NH_2$		amino
alkoxy:	$-O(CH_2)_xCH_3$	root + -oxy
alkenyl:	$R_2C=CR-$	root + -enyl
alkynyl:	$RC\equiv C-$	root + -ynyl
alcohol:	$-OH$	hydroxy
keto:	$=O$	oxo

Step VII. Number the main chain or main ring based on the principle of first point of difference.

1. Functional groups and substituents must have a location specified relative to the main portion of the molecule; all carbons in the main chain or ring are numbered.

2a. and 2b. **The rules for alkane carbon chains are described in Part 1.**

2c. For molecules with **_any_** functional group of higher priority than alkane, (alkenes ———→ carboxylic acid derivatives): the main chain is always numbered in the direction that produces **_the lowest number for the carbon that bears the highest-priority functional group_**.

2d. ***If more than one*** of the same highest-priority functional group is present in the main chain, the main chain is always numbered in the direction ***that produces the <u>lowest</u> number for the carbon that bears a highest-priority functional group <u>at the first point of difference</u>***.

3a. and 3b. **The rules for cycloalkane carbon rings are described in Part 1.**

3c. For molecules with ***<u>any</u>*** functional group of higher priority than cycloalkane, (alkenes ————→ carboxylic acid derivatives): the main ring is always numbered ***beginning with the carbon that bears the highest-priority functional group; this carbon must be #1***. If the molecule has more than one of the same highest-priority functional group, number in the direction that produces the ***<u>lowest</u> number for the carbon that bears a highest-priority functional group <u>at the first point of difference</u>***.

4. ***The concept of first point of difference as applied to functional groups is identical to that described in Part 1.***

The **<u>first point of difference</u>** means that of all possible ways to correctly number a chain or ring sequentially:

4a. select the direction which gives the **<u>lowest</u>** number to the carbon which bears the **<u>first</u>** functional group to be reached along the chain or ring in the sequential counting process.

4b. select the direction which gives the **<u>lowest</u>** number to the carbon which bears the **<u>first tie-breaking</u>** functional group to be reached along the chain or ring in the sequential counting process.

Example for (4a):

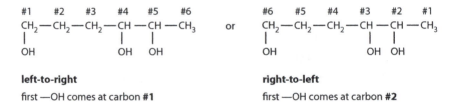

left-to-right

first —OH comes at carbon **#1**

right-to-left

first —OH comes at carbon **#2**

The correct numbering sequence is left-to-right to produce the first —OH at carbon #1. Since #1 is lower than #2 at the first point of difference, the location numbers for the other two alcohols do not affect the selection.

<u>Step</u> VIII. Complete the <u>main</u> name (base name) by using the carbon numbers to locate all groups of the same highest-priority functional groups.

1. The main name of *alkanes/cycloalkanes/alkyl halides* does not contain numbers to locate the main functional group.

2. For molecules with ***<u>any</u>*** functional group of higher priority than alkane, (alkenes ————→ carboxylic acid derivatives): the highest-priority functional group in the main name must be ***located by a carbon number based on Step VI***. If more than one of the same highest-priority functional group is present on the main

chain or ring, ***each functional group must be located by a numbered carbon.***

3. To complete the main name: the ***number*** of the carbon which bears the highest-priority functional group ***is placed directly in front of the main name (root plus ending):*** (#) – (root + ending).

    For more than one required number, __all__ ***numbers*** of the carbons which bear the highest-priority functional group ***are placed directly in front of the main name (root plus ending):*** (#), (#)......– (root + ending).

4. Although all highest-priority functional groups must be located, in some cases, the numbers may not be explicitly stated in the main name. Main functional group numbers are __not__ stated for

    a. Carbon chains of one or two carbons: numbering cannot lead to a difference.

    b. Rings with only one highest-priority functional group: must automatically be #1; the #1 is not stated.

    c. Chains with carboxylic acids or aldehydes: functional group bonding requirements force the carbon to be at a chain end: must be automatically #1; the #1 is not stated.

Step IX. Finish the complete name of the molecule by naming and locating all substituents attached to the main portion; use Step X for complicated branched groups.

1. Use the list of substituents identified and matched to a name in **Step VI.**

2. ***Each substituent is located by the carbon number*** on the main chain or ring to which it is attached, based on **Step VII.**

3–4. Complete the name as described in **Part 1.**

Step X. Use the techniques described in Steps VI through IX to name more complicated branched substituents.

1–6. Complete substituent names as described in **Part 1.**

5.3 EXAMPLES FOR A MOLECULE WITH HIGHER-PRIORITY FUNCTIONAL GROUPS

Example 1 Molecule:

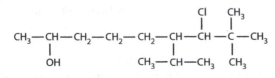

Step I. Highest-priority functional group = **alcohol.**
Step II. For an alcohol, the main portion is the longest continuous chain of bonded carbon atoms **containing the alcohol carbon**; this is shown in **boldface.**

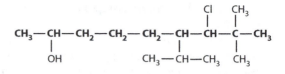

Step III. Total number of carbons in main chain = **9**; root name = **non**.

Step IV and **Step V.** Correct ending for an **alcohol** = **-anol**; the correct base name is root + ending = non + -anol = **nonanol**.

Step VI. Substituents are all atoms or atom groupings attached to the main chain; these are shown in **boldface: (a)** = **chloro**; **(b)** are one carbon group substituents: **meth + yl = methyl**; and **(c)** is **isopropyl** (specific named branched group); **Step X** may be used for an alternate name.

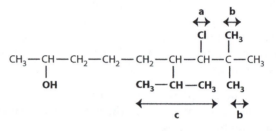

Step VII. The main chain can be numbered in one of the two ways:

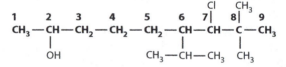

left-to-right: main functional group (alcohol) comes at C #**2**

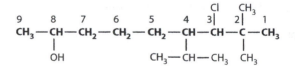

right-to-left: main functional group (alcohol) comes at C #**8**

The one alcohol provides a difference in the numbering sequence: the main functional group **must** come at the **lower** numbered carbon: the correct numbering is **left-to-right** (**bold face**).

Step VIII. Alcohols require a location number to complete the main name. Place the number of the alcohol carbon directly in front of the main name: **2-nonanol**.

Step IX. Substituents are named as (**Step X** is not required) **6-isopropyl; 7-chloro; 8,8-dimethyl**

Complete the name: substituents are listed in alphabetical order (prefix **di-** is not counted); the number for the alcohol **must** come **directly in front of the base name**.

7-chloro-6-isopropyl-8,8-dimethyl-2-nonanol

Example 2 Molecule:

$$CH_3-CH-CH_2-CH_2-CH_2-CH-\overset{\overset{O}{\|}}{C}-\overset{\overset{CH_3}{|}}{CH}-CH_3$$

with OH on the second carbon and $CH_3-CH-CH_2-CH_2-CH_3$ branch.

Step I. Highest-priority functional group = **ketone**.

Step II. For a ketone, the main portion is the longest continuous chain of bonded carbon atoms **containing the ketone carbon**; this is shown in **boldface**. With another high-priority functional group, the longest chain is selected to include both the ketone and alcohol **if possible**. Note that the longer possible 10-carbon chain which does not contain the ketone cannot be accepted as the main chain.

$$\mathbf{CH_3-CH-CH_2-CH_2-CH_2-CH}-\overset{\overset{O}{\|}}{\mathbf{C}}-\overset{\overset{CH_3}{|}}{\mathbf{CH}}-\mathbf{CH_3}$$

with OH and $CH_3-CH-CH_2-CH_2-CH_3$ branch.

Step III. Total number of carbons in main chain = **9**; root name = **non**.

Step IV and **Step V.** Correct ending for a **ketone** = **-anone**; the correct base name is root + ending = non + -anone = **nonanone**.

Step VI. Substituents are all atoms or atom groupings attached to the main chain; these are shown in **boldface: (a)** = **methyl**; **(b)** is an alcohol group which is **not** the main functional group; its name as a substituent is **hydroxy**; and **(c)** is a complicated branched substituent

$$CH_3-CH-CH_2-CH_2-CH_2-CH-\overset{\overset{O}{\|}}{C}-\overset{\overset{CH_3}{|}}{CH}-CH_3$$

with **OH** (a) and **c** = $CH_3-CH-CH_2-CH_2-CH_3$

Step VII. The main chain can be numbered in one of the two ways:

$$\overset{1}{CH_3}-\overset{2}{CH}-\overset{3}{CH_2}-\overset{4}{CH_2}-\overset{5}{CH_2}-\overset{6}{CH}-\overset{\overset{O}{\|}}{\overset{7}{C}}-\overset{\overset{CH_3}{|}}{\overset{8}{CH}}-\overset{9}{CH_3}$$

with OH and $CH_3-CH-CH_2-CH_2-CH_3$ branch.

left-to-right: main functional group (ketone) comes at C **#7**

$$\overset{9}{CH_3}-\overset{8}{CH}-\overset{7}{CH_2}-\overset{6}{CH_2}-\overset{5}{CH_2}-\overset{4}{CH}-\overset{\overset{O}{\|}}{\overset{3}{C}}-\overset{\overset{CH_3}{|}}{\overset{2}{CH}}-\overset{1}{CH_3}$$

with OH and $CH_3-CH-CH_2-CH_2-CH_3$ branch.

right-to-left: main functional group (ketone) comes at C **#3**

The one ketone provides a difference in the numbering sequence: the main functional group **must** come at the **lower** numbered carbon: the correct numbering is **right-to-left** (**boldface**).

Step VIII. Ketones require a location number to complete the main name. Place the number of the ketone carbon directly in front of the main name: **3-nonanone**.

Step IX. Substituents are named as **2-methyl** and **8-hydroxy**. **Step X** is needed to complete the name of the complicated branched substituent.

Step X. The longest continuous chain of the complicated substituent group **must begin with the connecting carbon as carbon number #1**; the result is a four-carbon chain: name = root + -yl = **but** + **-yl** = **butyl**.

Although all five carbons of this group are sequentially connected, they cannot all be placed in a main substituent chain if the connecting carbon begins the chain as carbon #1. The remaining carbon is attached to carbon #1 and is a **methyl** group. The complete substituent name is (**1-methylbutyl**)

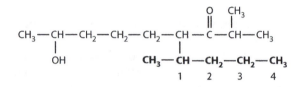

Step IX. The (1-methylbutyl) is attached to the main chain at carbon #**4**:
= **4-(1-methylbutyl)**. The complete name is
8-hydroxy-2-methyl-4-(1-methylbutyl)-3-nonanone

Example 3 Molecule: (**example for a combined** ending)

$$CH_3-CH-CH_2-CH_2-CH_2-CH-CH=CH_2$$
$$||$$
$$OHCH_2-CH_3$$

Step I. Highest-priority functional group = **alcohol**.

Step II. For an alcohol, the main portion **must** contain the alcohol carbon. With another high-priority functional group, the longest chain is selected to include both the alcohol and double bond **if possible**. The main chain is shown in **boldface**.

$$CH_3-CH-CH_2-CH_2-CH_2-CH-CH=CH_2$$
$$||$$
$$OHCH_2-CH_3$$

Step III. Total number of carbons in main chain = **8**; root name = **oct**.

Step IV and **Step V.** Correct ending for an **alcohol** combined with one **double bond** is **-enol**; or **-en- # -ol**. **Step VIII** will show that a number is needed between the two endings: the correct base name is root + ending = oct + -en-#-ol = **octen-#-ol**.

Step VI. The only substituent **(a)** = **eth** + **-yl** = **ethyl**;

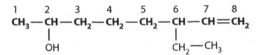

Step VII. The main chain can be numbered in one of the two ways:

$$8 \quad 7 \quad 6 \quad 5 \quad 4 \quad 3 \quad 2 \quad 1$$
$$CH_3 - CH - CH_2 - CH_2 - CH_2 - CH - CH = CH_2$$
$$\quad\quad\; | \quad\quad\quad\quad\quad\quad\quad\quad\quad |$$
$$\quad\quad OH \quad\quad\quad\quad\quad\quad CH_2 - CH_3$$

right-to-left: main functional group (alcohol) comes at C **#7**
alkene comes at C **#1**

(The number for an alkene is the lower of the two numbers for the double-bonded carbons when numbered sequentially.)
The other numbering possibility is

$$1 \quad 2 \quad 3 \quad 4 \quad 5 \quad 6 \quad 7 \quad 8$$
$$CH_3 - CH - CH_2 - CH_2 - CH_2 - CH - CH = CH_2$$
$$\quad\quad\; | \quad\quad\quad\quad\quad\quad\quad\quad\quad |$$
$$\quad\quad OH \quad\quad\quad\quad\quad\quad CH_2 - CH_3$$

left-to-right: main functional group (alcohol) comes at C #2
alkene comes at C #7:

As the highest-priority functional group, the alcohol **must** be located on the **lower** numbered carbon, despite the very low number possible for the alkene: the correct numbering is **left-to-right** (boldface).

Step VIII. Both the alcohol and the alkene require a location number to complete the main name. The number for the alkene is placed directly in front of the main name. The number for the alcohol is placed **between** the parts of the combined ending, directly in front of the **-ol** suffix: **7-octen-2-ol**.

Step IX. The substituent is named as **6-ethyl**. (**Step X** is not needed.) Complete name is **6-ethyl-7-octen-2-ol**.

Part 3: General Process for Esters and Amides

General Concept: Carboxylic acid derivatives (esters, amides, acid chlorides, acid anhydrides, and nitriles) take names based on the carboxylic acid from which they are derived.

1. For naming purposes, an ester is considered to be composed of an acid piece and an alcohol piece.

2. For naming purposes, an amide is considered to be composed of an acid piece and an amine piece.

5.4 ADAPTED PROCESS FOR NOMENCLATURE OF ESTERS

An ester is considered to be composed of an acid piece and an alcohol piece. The alcohol piece is considered to be composed of the alcohol oxygen (with the two single bonds) and the carbon (CR_3) group connected to this single-bonded oxygen.

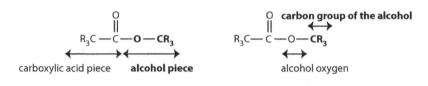

1. ***Identify the carboxylic acid*** from which the ester is derived: that is, reconstruct the carboxylic acid piece by (conceptually) replacing the alcohol piece with —**OH**.
2. Use **Steps** I–X ***to name*** the carboxylic acid determined in **(1)**.
3. ***Adapt the name*** of the carboxylic acid from **(2)** to an **ester** name by changing the ending "**-ic acid**" to "**-ate**."
4. The complete name of an ester must include a description of the alcohol piece. ***Name the carbon group of the alcohol as a substituent*** and place this name at the very beginning of the complete ester name, separated by a space with no hyphen. ***The alcohol oxygen is not specifically named; it is indicated through the named ester functional group.***

Example:

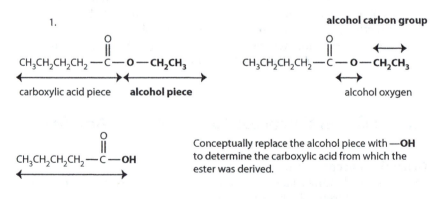

2. **Steps** I–X show that the carboxylic acid from which the ester is derived is **pentanoic acid**.
3. Adapted ester name: change "ic acid" to ate = **pentanoate**.

4. The **carbon group** of the alcohol piece named as a <u>**substituent**</u> is **ethyl**. The alcohol oxygen is <u>**not**</u> named specifically: eth**oxy** is <u>**not**</u> correct.

> Complete name = **ethyl pentanoate**.

5.5 ADAPTED PROCESS FOR NOMENCLATURE OF AMIDES

An amide is considered to be composed of an acid piece and an amine piece. The amine nitrogen may have hydrogen atoms or carbon groups connected.

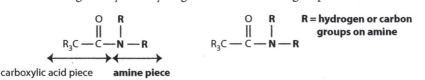

1. ***Identify the carboxylic acid*** from which the amide is derived: that is, reconstruct the carboxylic acid piece by (conceptually) replacing the amine piece with —**OH**.
2. Use <u>**Steps I–X**</u> *to name* the carboxylic acid determined in **(1)**.
3. ***Adapt the name*** of the carboxylic acid from **(2)** to an <u>**amide**</u> name by changing the ending "**-ic acid**" to "**-amide**," or "**-oic acid**" to "**-amide**." Note that the vowel "o" from the acid name is always dropped.
4. ***Name any carbon groups on the nitrogen as substituents***; the location is indicated by a capital "N-," instead of a number. Substituent groups on nitrogen must be placed at the very beginning of the complete amide name.

Example:

1.

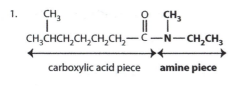

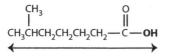

Conceptually replace the alcohol piece with —**OH** to determine the carboxylic acid from which the amide was derived.

2. <u>**Steps I–X**</u> show that the carboxylic acid from which the amide is derived is **6-methylheptanoic acid**.
3. Adapted amide name: change the ending "-oic acid" to "-amide." Note that the vowel "o" from the acid name is dropped in this case = **6-methylheptanamide**.
4. The **carbon groups** of the nitrogen are named as **N-ethyl** and **N-methyl**:

Complete name = **N-ethyl-N-methyl- 6-methylheptanamide**.

ADDITIONAL PRACTICE PROBLEMS

The following molecules shown in line-drawing form will provide a challenge. Determine the correct name for each.

1. Molecules for Part 1:

(a) 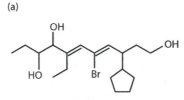   (b)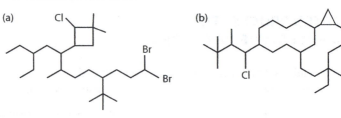

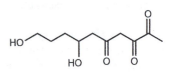

2. Molecules for Part 2:

(a)   (b)

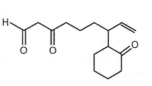

(c)

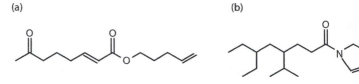

3. Molecules for Part 3:

(a)   (b)

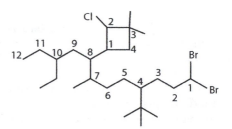

ANSWERS TO ADDITIONAL PRACTICE PROBLEMS

1.(a)

Longest chain = 12 carbons: main chain: dodec + -ane = **dodecane**.
Number from right-to-left to place the first substituent at carbon #1.
Non-ring substituents: 1-bromo; 1-bromo; 4-tert-butyl: 7-methyl; 10-ethyl.

Ring substituent at carbon #8: 4 carbons = cyclo + but + -y = cyclo-
  butyl: number the connecting carbon #1 and follow the first-
  point-of difference procedure.

Final name: list substituents in alphabetical order (the tert-prefix
  is not counted):

**1,1-dibromo-4-tert-butyl-8-(2-chloro-3,3-dimethylcyclobu-
  tyl)-10-ethyl-7-methyldodecane.**

1.(b)

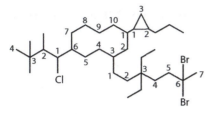

Longest chain = 7 carbons; largest ring = 10 carbons

main ring: cyclo + dec + -ane = **cyclodecane**

Lowest numbering of the ring places substituents at ring carbons
  1, 3, and 6.

Step X for complicated substituents must be used for each;
  connecting carbons must be #1.

1-(2-propylcyclopropyl)

3-(6,6-bibromo-3,3-diethylheptyl)

6-(1-chloro-2,3,3-trimethylbutyl)

**Final name: 3-(6,6-bibromo-3,3-diethylheptyl)-6-(1-chloro-2,3,
  3-trimethylbutyl)-1-(2-propylcyclopropyl)-cyclodecane**

2.(a)

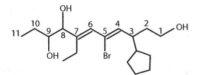

Highest-priority functional group = alcohol. Select the main chain
  to include all alcohol carbons; number from right-to-left; the
  first alcohol group comes at carbon #1.

Double bond plus alcohol results in the ending: -en-#-ol.

Two double bonds plus three alcohols: -adiene-#-triol.

Main chain = 11 carbons; double bonds are at carbons 4 and 6; alcohols are at carbons 1, 8, and 9. **Main chain = 4,6-undecadien-1,
  7, 8-triol.**

Substituents: 3-cyclopentyl; 5-bromo; 7-ethyl.

**Final name: 5-bromo-3-cyclopentyl-7-ethyl-4,6-undecadien-1,
  7, 8-triol**

2.(b)

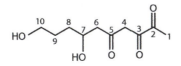

Highest-priority functional group = ketone. Select the main chain to include all ketone carbons; number from right-to-left; the first ketone group comes at carbon #2.

Three ketones: -anetrione.

Main chain = 10 carbons; ketone carbons are at 2, 3, and 5.

Main chain = 2,3,5-decanetrione

Substituents: alcohol groups at 7 and 10 must be substituents = hydroxy.

Final name: 7,10-dihydroxy-2, 3, 5-decanetrione

2.(c)

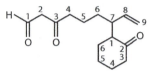

Highest-priority functional group = aldehyde; number from left-to-right to place the aldehyde carbon #1. Double bond plus aldehyde ending: -enal

Main chain = 9 carbons; the double bond is carbon 8; **Main chain = 8-nonenal**.

Substituents: ketone group at chain carbon 3 must be a substituent = oxo.

Step X for complicated substituents must be used for the ring: 2-oxocyclohexyl.

Final name: 3-oxo-7-(2-oxocyclohexyl)-8-nonenal

3.(a) The molecule is converted into a condensed structure.

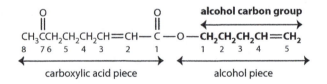

Highest-priority functional group = ester.

Replace the ester alcohol piece with —OH and name the carboxylic acid from which the ester is derived: 7-oxo-2-octenoic acid.

Change the carboxylic acid ending to an ester ending: change "-ic acid" to "-ate": **7-oxo-2-octenoate**.

Name the carbon group of the alcohol piece as a substituent: 4-pentenyl.

Final name: 4-pentenyl is placed first in the name with a full space followed by the carboxylic ester main name: **4-pentenyl 7-oxo-2-octenoate**

3.(b) The molecule is converted into a condensed structure.

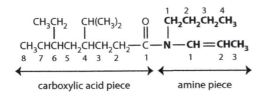

Highest-priority functional group = amide.

Replace the amide nitrogen piece with —OH and name the carboxylic acid from which the amide is derived: 6-ethyl-4-isopropyloctanoic acid.

Change the carboxylic acid ending to an amide ending: change "-oic acid" to "-amide": **6-ethyl-4-isopropyloctanamide**.

Name the carbon groups on the nitrogen as substituents; locate them with a capital "N" and place them first in the name.

Final name: N-butyl,N-(1-propenyl)-6-ethyl-4-isopropyloctanamide

Guidelines for Analysis of Intermolecular Forces for Organic Molecules

6.1 INTERPARTICLE AND INTERMOLECULAR FORCES FOR INDIVIDUAL PURE COMPOUNDS: FORCES BETWEEN MOLECULES OF THE SAME COMPOUND

6.1.1 INTERPARTICLE FORCES FOR NONMOLECULAR COMPOUNDS

1. Certain types of compounds do not exist in molecular form. Instead, large amounts of matter exist as extended 3-D arrays of particles (atoms or ions) held together by the strong forces of **ionic** or **covalent** bonding.

   a. **Ionic** compounds are extended 3-D arrays of ions held together by **ionic forces**: forces of electrostatic attraction. Ionic forces are extremely strong; all ionic compounds are solids at 25°C and have very high melting points. Examples are most (metal) + (non-metal) combinations.

   b. **Network covalent** compounds are extended 3-D arrays of atoms held together by **covalent bonds**. The structure can be completely covalent; examples are diamond (carbon solid), quartz, or glass (SiO_2). The structure may also be a "hybrid" of large covalent sections partially held together by other intermolecular forces; an example is graphite, another form of solid carbon.

   c. **Metals** in the elemental form are extended 3-D arrays of metal atoms held together by **metallic bonds**. The metallic bond is formed as a result of overlap of many individual metal atomic orbitals to produce very extended molecular orbitals: "delocalized" orbitals. The delocalized orbitals allow electrons to move freely through the metal mass; the electrons are **not** bound to individual molecular orbitals.

6.1.2 INTERMOLECULAR FORCES FOR MOLECULAR COMPOUNDS

1. Molecular compounds are those that exist in discrete molecular form.
   a. Individual molecules are formed through covalent bonding between the atoms in one molecule. However, ***discrete individual molecules cannot be covalently bonded to each other*** to assemble into larger amounts of matter.
   b. Large amounts of **molecular** matter must be held together through non-covalent bonding forces termed ***intermolecular forces***: **intermolecular forces are nonbonding attractive forces <u>between</u> separate individual molecules**.
2. **Intermolecular forces are (permanent) dipole force; hydrogen bonding; and dispersion forces (induced dipole).**
3. The **dipole force** (or permanent dipole force) is the electrostatic force of attraction between oppositely charged ends of permanent molecular dipoles. The strength of the dipole force is proportional to the degree of polarity of the entire molecule: the size of the permanent molecular dipole. **The size of the permanent molecular dipole is proportional to**
   a. The **electronegativity difference (Δ E.N.)** between bonded atoms. Very generally, the greater the electronegativity difference between atoms in a covalent bond, the larger the potential dipole (related to the degree of charge separation). For practical experience, ranges of Δ E.N. can be thought of as leading to certain approximate dipole strengths:
   Δ E.N. ≤ 0.4 …usually **weak** dipoles
   Δ E.N. of 0.5–1.0…usually leads to **moderate** dipoles
   Δ E.N. > 1.0…usually leads to **strong** dipoles
   b. The **degree of asymmetry of the central atoms** involved in polar bonds. Individual bond dipoles are added together as vectors; **symmetric** central atoms (and molecules) tend to have dipoles that completely or mostly cancel each other out.
   c. Molecular dipoles can range from

   very weak ↔ weak ↔ moderate ↔ strong ↔ very strong

   Very weak dipoles will provide only very weak dipole intermolecular attractive forces; *the strength of the dipole attractive forces will increase with the strength of the molecular dipole.*
4. **Hydrogen bonding** is a special form of the strong dipole force. The hydrogen "bond" is **not** a covalent bond but a force of attraction between very specific positive and negative dipole ends.
   a. The **positive** end of the attractive force is specifically hydrogen covalently bonded to either oxygen, nitrogen, or fluorine:

   $$\delta+ \; \delta- \qquad \delta+ \; \delta- \qquad \delta+ \; \delta-$$
   $$H—O \qquad\quad H—N \qquad\quad H—F$$

b. The **negative** end of the hydrogen bond can be the lone pairs of oxygen, nitrogen, or fluorine, **regardless of whether these atoms are bonded to a hydrogen**:

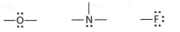

5. **Hydrogen bonding** is stronger than other dipole forces due to the small sizes and high relative charge concentrations of hydrogen and the row-2 elements (O, N, F). Hydrogen bonding can occur in a variety of ways between molecules of the **same** compound. The positive end is very specific but the **negative end can be more variable**.
   Examples: The dashed line (||||||||||||) indicates hydrogen bonding.

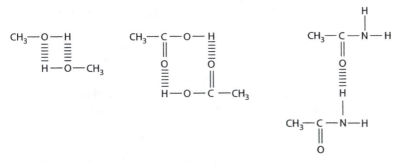

6. **Dispersion forces (induced dipole)** do not involve a permanent dipole, but instead are based on a **temporary** or **induced** dipole.
   a. Electrons in orbitals (electron density) are not in fixed positions, but are described as probability distributions. In response to a neighboring charge separation, a symmetrical electron density across a nonpolar molecule can distort, providing a **temporary** positive and negative molecular end: an **induced dipole**.
   b. The induced dipole can then cause a distortion of electron density in another neighboring molecule, etc.; the induced dipoles are temporarily held together by the attractive forces. Although the temporary dipoles rapidly decompose back to a symmetrical structure, new induced dipoles are in the process of constantly forming and breaking; the **net** result is an attractive force.

7. Although dispersion forces are the weakest of the intermolecular forces **per individual attraction**, the total force can be large due to the **quantity** of the induced dipole attractions. They are present in all molecules, and are the major attractive force in nonpolar or weakly polar molecules.

8. *Total dispersion force between molecules of the same compound is proportional to the size and number of the atoms in the molecule, proportional to molar mass (MM).*
   a. **Size** (very generally) **of the atom** is related to the concept of **polarizability**: electron densities are more easily distorted (larger induced dipole) as the number of electrons and the

distance to the outer shell increases. Both of these criteria relate generally to the atom size (atomic number).

b. **Total size (MM) of the molecule** (for molecules containing similar atoms) affects dispersion forces because the larger the molecule (number of total atoms), the greater (generally) the number of possible locations for forming induced dipole attractive forces.

c. Both criteria indicate that, as a general rule, *for most molecular comparisons, the total strength of the molecular dispersion forces is proportional to the MM of the compound.*

6.1.3 GUIDELINE FOR COMPARING TOTAL STRENGTH OF INTERMOLECULAR FORCES IN INDIVIDUAL COMPOUNDS

1. Boiling point and melting point temperatures for compounds indicate the energy required to break the attractive forces between molecules. These temperatures are thus proportional to the **total** strength of intermolecular forces: *a **higher** melting or boiling temperature corresponds to a **greater total** intermolecular force.*

2. Other factors can influence total intermolecular attractive force, for example, 3-D molecular shape, especially in the solid; large differences in polarizability (e.g., F vs. I).

3. To compare **total intermolecular force** (or boiling/melting temperatures) for different compounds:

    (1) *Determine the **force types** for each compound and identify the **dominant force-type**.*

    a. The **dominant force-type** is the **strongest** force-type based on

Hydrogen bonding (**H-bonding**) > **dipole force** > **dispersion force**

    b. Additionally, identify the **number of sites** in the molecule for hydrogen bonding or dipole forces and distinguish among the **relative strength ranges** for dipole forces.

    (2) *For molecules of approximately the same size (measured by MM): The compound with the **greater** total intermolecular forces will be the one with the **stronger dominant force-type** or the greater number of sites of attraction.*

    (3) *For molecules with similar force types/number of sites: The compound with the **greater** total intermolecular forces will be the molecule with the **larger MM**.*

    (4) **For comparing molecules with different force types and greatly different MM**: If the two trends are in agreement, use the previous rules. If the two trends lead to opposite conclusions, experience is necessary to estimate the net effect.

Comparison Tables

1. **Comparing Boiling Point to Different MM in Molecules with Same Force Types**

Compound	MM	Bond Example	Bond ΔE.N.	Force Types	Boiling Point (°C)
C_4H_{10}	58	C—H	0.4	**Dispersion** Very weak dipole	0
C_5H_{12}	72	C—H	0.4	**Dispersion** Very weak dipole	36
$C_{10}H_{22}$	142	C—H	0.4	**Dispersion** Very weak dipole	174
CH_3—OH	32	C—O O—H	1.0 1.4	Strong dipole H-bonding **dispersion**	64
C_2H_5—OH	46	C—O O—H	1.0 1.4	Strong dipole H-bonding **dispersion**	78
C_3H_7—OH	60	C—O O—H	1.0 1.4	Strong dipole H-bonding **dispersion**	97

2. **Comparing Boiling Point to Different Force Types in Molecules with Approximately the Same MM**

Compound	MM	Bond Example	Bond ΔE.N.	Force Types	Boiling Point (°C)
C_4H_{10}	58	C—H	0.4	**Very weak dipole** dispersion	0
CH_3—O—CH_2CH_3	60	C—O	1.0	**Moderate dipole** dispersion	8
$CH_3CH_2CH_2$—OH	60	O—O O—H	1.0 1.4	**Strong dipole** **H-bonding** dispersion	97
HO—CH_2CH_2—OH	62	C—O O—H	1.0 1.4	**Strong dipole** **H-bonding: 2 *sites*** dispersion	198

6.2 INTERPARTICLE AND INTERMOLECULAR FORCES FOR COMBINATIONS OF COMPOUNDS: SOLUTIONS AND SOLUBILITY

6.2.1 SOLUBILITY AND SOLUTION TERMS

1. **Solutions** are homogeneous mixtures of compounds; mixing must occur at the molecular level. General terminology describes the **solute** as the compound being dissolved and the **solvent** as the compound acting as the dissolver.

2. The term **solubility** refers to a measure of a specific amount of solute which is dissolved in a particular amount of solvent or solution. This measure is expressed as a concentration unit, for example, molarity (moles solute/liter solution), grams solute/mL solution, etc. A **saturated solution** describes the **maximum solubility** of a solute in a solvent.

6.2.2 SOLUBILITY REQUIREMENTS

1. Molecules (or particles) of a pure solvent and a pure solute are each held together by existing compound attractive forces. Formation of a solution requires molecular intermixing: some of these attractive forces must be broken to "expand" the solute and solvent compounds to allow interpenetration. **Solution formation involves**
   a. **Breaking** of solute–solute interparticle (intermolecular) forces, requiring an input of energy into the system.
   b. **Breaking** of some solvent–solvent intermolecular forces, requiring an input of energy into the system.
   c. **Formation** of solute–solvent interparticle (intermolecular) attractive forces. This represents an energy release and must compensate the system for the energy input necessary to break the existing solute–solute and solvent–solvent interparticle forces.
2. In the optimum circumstance, the **sum** of energies of solute–solvent attractive forces (formed) will be **greater** than the **sum** of energies of solute–solute and solvent–solvent attractive forces (broken). This leads to a **negative** value for ΔH (solution).
3. An **exothermic** process (**negative** ΔH {solution}) is the most favorable energy balance for solution formation. However, a spontaneous **endothermic** solution process (**positive** Δ H {solution}) is possible due to the effects of **entropy** (**maximization** of **randomness**).
4. *The key to achieving optimum solubility is maximization of the solute–solvent interparticle (intermolecular) attractive forces.* When this occurs, the energy balance for solvation will be most favorable. **This requirement forms the basis of solubility analysis: *solute–solvent attractive forces, and therefore solubility, will be maximized whenever the force types of each are equivalent or as similar as possible, often phrased as "like dissolves like."***
5. Matching of force types does **not** guarantee solubility. This process only ensures that the solute will have the best possible chance of dissolving in a properly selected solvent. The wide variety of interparticle or intermolecular forces, and the effect of entropy, produces variable maximum solubilities of solutes in solvents, even under optimum force-type matching.

6.2.3 INTERPARTICLE (INTERMOLECULAR) FORCES BETWEEN SOLUTE AND SOLVENT

1. The force types for solute–solvent interaction are the same as for single compounds with, however, the possibilities for additional combinations:
   a. **Dipole–dipole**
   b. **Hydrogen bonding**
   c. **Induced dipole–induced dipole (dispersion)**
   d. **Dipole–induced dipole**
   e. **Ion–dipole**
2. Particles in solutions may not have equivalent force types; closely similar force types may represent the best "match." Examples:
   a. Solvation of **ions** in water (**strong dipole/hydrogen bonding**) is possible because ion–strong dipole is the closest match possible. Ion-induced dipole (i.e., solvent has mainly dispersion forces) would **not** be expected to provide good solubility; thus, ionic compounds are **not** soluble in non-polar, dispersion forces-based hydrocarbons.
   b. Solvation of non-polar, **dispersion**-based compounds would be higher in a **weakly polar** solvent than in a strongly polar solvent. The combination of dispersion–weak dipole is a closer match than dispersion–strong dipole.

Examples:

Good Force Match: dispersion–weak dipole; example result: nonpolar oil or grease dissolves in gasoline or CH_3Cl.

Poor Force Match: dispersion–strong dipole; example result: oil or grease does **not** dissolve in water.

6.2.4 ANALYSIS OF SOLUBILITY FOR POSSIBLE SOLUTE/SOLVENT COMBINATIONS

1. **To analyze a solute/solvent combination**: Apply the general rule: *Solubility (and total attractive force) will be maximized by the closest match of solute–solvent interparticle (intermolecular) forces.*
2. Two types of comparison problems are common:
   (1) Deciding which solvent would be best for dissolving a particular solute.
   (2) Deciding which solute would have the greater solubility in a particular solvent.
3. **Both problems require the same analysis:**
   Step 1: *Determine the major force types present in the pure solute*; that is, identify the forces between particles or molecules of the specific solute compound or possible solute compounds.

Step 2: *Determine the major force types present in the pure solvent*; that is, identify the forces between molecules of the specific solvent compound or possible solvent compounds.

Step 3: ***Compare the force types of all possible combinations of solute/solvent***. The closest force-type match will represent the best solvent for a specific solute (for problem type 1) or the solute with higher solubility in a specific solvent (problem type 2).

4. Example: List the order of solubility from **most** soluble to **least** soluble for the following solutes in water as a solvent: C_3H_7—OH; C_4H_9—OH; C_5H_{11}—OH; and C_6H_{13}—OH.

Analysis:

Summary for the solvent: Water has strong dipole/hydrogen bonding with a very small contribution from dispersion forces.

Summary for each solute: The **—OH** portion of each molecule matches water through strong dipole/hydrogen bonding. The **C_xH_y** (hydrocarbon) portion of each molecule represents dispersion forces and is a very poor match with water (very small dispersion force contribution).

Result: As the **dispersion** (hydrocarbon) portion of the solute molecule **increases in proportion to** the **strong dipole/ hydrogen bonding** (—OH) portion, the **solubility** in **water decreases**. The hydrocarbon portion is indicated by the number of carbons and hydrogens in the molecule while [—OH] portion remains constant at one per molecule. **The following order is the result; the solubility values are shown as verification:**

Solute Molecule	Maximum Solubility in Water
C_3H_7—OH	Soluble in all concentrations
C_4H_9—OH	1.1 Molar
C_5H_{11}—OH	0.30 Molar
C_6H_{13}—OH	0.056 Molar

PRACTICE EXERCISES

1. **Analyzing individual compounds; boiling/melting points:** Select the **one** compound from the following pairs that has the **stronger total** interparticle or intermolecular force, and thus would be expected to have the **higher** boiling point or melting point. **Lewis structures will be required to analyze the force types; some Lewis structures are indicated by condensed drawings**.

   a. PCl_3 or $MgCl_2$

   b. CH_3NH_2 or CH_3F

c. CH_3Br or CH_3Cl

d. CH_3CH_2—OH or CH_3—O—CH_3

e. BH_3 or HF

f. H_2NNH_2 or PH_3

g. $CH_3CH_2CH_3$ or CH_3Cl

h. C_6H_6 or $C_{10}H_8$

i.
$$\underset{CH_3CH_2\overset{\textstyle O}{\overset{\|}{C}}-OCH_3}{} \quad or \quad \underset{CH_3CH_2CH_2\overset{\textstyle O}{\overset{\|}{C}}-OH}{}$$

j.
$$\underset{HO-CH_2\overset{\textstyle OH}{\overset{|}{C}H}CH_2-OH}{} \quad or \quad Cl-CH_2CH_2-Cl$$

2. **Analyzing solubility:**

Which of the two compounds listed below would be expected to be more soluble in water? Which of the two compounds would be expected to be more soluble in hexane (C_6H_{14})?

a. PCl_3 or $MgCl_2$

b. CH_3NH_2 or CH_3F

c. CH_3CH_2—OH or CH_3—O—CH_3

d. H_2NNH_2 or PH_3

e. $CH_3CH_2CH_3$ or CH_3Cl

f.
$$\underset{CH_3CH_2\overset{\textstyle O}{\overset{\|}{C}}-OCH_3}{} \quad or \quad \underset{CH_3CH_2CH_2\overset{\textstyle O}{\overset{\|}{C}}-OH}{}$$

g.
$$\underset{HO-CH_2\overset{\textstyle OH}{\overset{|}{C}H}CH_2-OH}{} \quad or \quad Cl-CH_2CH_2-Cl$$

ANSWERS TO PRACTICE EXERCISES

1. **Total intermolecular force/boiling point/melting point analysis**.

   a. PCl_3 or $MgCl_2$ **$MgCl_2$ would have the higher melting point**.

   PCl$_3$ has a polar P–Cl bond. The molecule has pyramidal geometry and thus is polar, with a polarity estimated to be moderate. $MgCl_2$ is ionic; interparticle forces are ionic bonds. Ionic bonds are much stronger than moderate dipole force.

   b. CH_3NH_2 or CH_3F **CH_3NH_2 would have the higher boiling point**.

   CH_3NH_2 must have N–H bonds and can therefore form hydrogen bonds. CH_3F has all atoms bonded to carbon; no H–F bonds indicates no hydrogen bonding. Both molecules are polar, but hydrogen bonding is the strongest intermolecular force-type. Since the MM of the two molecules is about the same, select the molecule with the strongest intermolecular force-type.

   c. CH_3Br or CH_3Cl **CH_3Br would have the higher boiling point**.

   Both molecules have weak-to-moderate dipoles and thus have similar force types. The deciding factor will be the total size of the dispersion force, which is proportional to MM. Select the larger molecule (largest MM).

d. CH_3CH_2-OH or CH_3-O-CH_3 **CH_3CH_2-OH would have the higher boiling point**.

The two molecules have identical MM. Select the compound with the stronger force-type. CH_3CH_2-OH has strong dipole/hydrogen bonding. CH_3-O-CH_3 has a moderate dipole and no hydrogen bonding.

e. BH_3 or HF   **HF would have the higher boiling point**.

HF can be a hydrogen bond; BH_3 cannot. Select the stronger force-type. (HF also has a slightly higher MM.)

f. H_2NNH_2 or PH_3 **H_2NNH_2 would have the higher boiling point**.

The two molecules have approximately the same MM. H_2NNH_2 can be a hydrogen bond; PH_3 cannot. Select the stronger force-type.

g. $CH_3CH_2CH_3$ or CH_3Cl **CH_3Cl would have the higher boiling point**.

The two molecules have approximately the same MM. The C–Cl bond provides a larger bond dipole than C–H bonds. CH_3Cl would have a stronger dipole force than $CH_3CH_2CH_3$; select the stronger force-type.

h. C_6H_6 or $C_{10}H_8$ **$C_{10}H_8$ would have the higher boiling point**.

Both molecules have essentially the same force types: very weak dipole and a predominant dispersion force. Select the molecule with the greater total dispersion force, which is proportional to MM.

i.
$$CH_3CH_2\overset{\overset{\displaystyle O}{\|}}{C}-OCH_3 \quad \text{or} \quad CH_3CH_2CH_2\overset{\overset{\displaystyle O}{\|}}{C}-OH$$

$CH_3CH_2CH_2\overset{\overset{\displaystyle O}{\|}}{C}-OH$ **would have the higher boiling or melting point.**

The molecules have the same MM; select the molecule with the stronger force-type. Both molecules have strong dipoles; however, the molecule selected can also participate in hydrogen bonding.

j.
$$HO-CH_2\overset{\overset{\displaystyle OH}{|}}{C}HCH_2-OH \quad \text{or} \quad Cl-CH_2CH_2-Cl$$

$HO-CH_2\overset{\overset{\displaystyle OH}{|}}{C}HCH_2-OH$ **would have the higher boiling point.**

The molecules have approximately the same MM; select the molecule with the stronger force-type. Both molecules have a dipole; however, the molecule selected has a much stronger dipole and can participate in hydrogen bonding.

2. **Intermolecular and interparticle forces for the solute compounds were analyzed for problem #1; these are used to solve problem #2.**

Solvent
intermolecular forces: **water = strong dipole/hydrogen bonding**
hexane = very weak dipole/dispersion
predominant

a. **PCl_3** would be the more soluble of the two compounds in **hexane**; the weak dipole matches hexane. **$MgCl_2$** would be the more soluble of the two compounds in **water**; ionic force matches water.

b. **CH_3NH_2** would be the more soluble of the two compounds in **water**; hydrogen bonding matches water. **CH_3F** would be the more soluble of the two compounds in **hexane**; despite the C–F bond, dispersion forces are predominant and dispersion matches hexane.

c. **CH_3CH_2—OH** would be the more soluble of the two in **water**; hydrogen bonding matches water. **CH_3—O—CH_3** would be the more soluble of the two in **hexane**; moderate dipole/dispersion matches hexane the best.

d. **H_2NNH_2** would be the more soluble of the two in **water**; hydrogen bonding matches water. **PH_3** would be the more soluble of the two in **hexane**; weak dipole/dispersion dominant matches hexane.

e. **$CH_3CH_2CH_3$** would be the more soluble of the two in **hexane**; very weak dipole/dispersion matches hexane. **CH_3Cl** would be the more soluble of the two in **water**. Although this compound is not very soluble in water, the dipole provided by the C–Cl bond would be a better match for water than the very weak dipole of $CH_3CH_2CH_3$.

f. $CH_3CH_2CH_2\overset{\overset{\displaystyle O}{\|}}{C}$—OH would be the more soluble of the two in **water**; hydrogen bonding provides the difference compared to the other molecule.

$CH_3CH_2\overset{\overset{\displaystyle O}{\|}}{C}$—$OCH_3$ would be the more soluble of the two in hexane. Although this compound would also be soluble in water due to its strong dipole, it would be more soluble in hexane than the previous molecule. This is because it cannot participate in hydrogen bonding; hydrogen bonding is a bad match for hexane.

g. HO—$CH_2\overset{\overset{\displaystyle OH}{|}}{C}HCH_2$—OH would be the more soluble of the two in **water** due to the extensive sites for hydrogen bonding. Cl—CH_2CH_2—Cl would be the more soluble of the two in **hexane**; moderate dipole/dispersion is a better match for hexane than the extensive hydrogen bonding of the other compound.

Alkane and Cycloalkane Conformations

7.1 CONFORMERS

Conformers (conformational isomers) are 3-D structures representing different 3-D arrangements of atoms within the **same molecule** (i.e., exact same bonding pattern).

Different conformers represent the same molecule because conformers are interconvertible by free rotation around single/sigma (σ) bonds. Interconversion between conformational structures does **not** require bond breaking and reformation; thus conformers are **not** isomers.

For alkane-related molecules, conformers are often identified by using a perspective that relates atom/group spatial positions through a **torsion angle**.

7.2 PROCESS FOR RECOGNITION AND IDENTIFICATION OF ALKANE CONFORMERS THROUGH THE USE OF TORSION ANGLE

Step 1: Select **any** specific C−C bond (or other central atom) in the molecule; this is the **C−C bond axis**. Each C−C bond in a molecule can be used as a bond axis, depending on the type of 3-D information to be viewed.

Step 2: Sight exactly straight down this C−C bond axis from the "front" carbon to the "back" carbon. Torsion angle results do **not** depend on which carbon is chosen as "front" or "back."

Step 3: Each of the axis carbons will have **three remaining** atoms or atom groups attached: the other three bonds to carbon not including the C−C axis bond. Select **one** of the three atom/groups on the front carbon and **one** of the three atom/groups on the back carbon. Visualize these attached groups as being connected to the axis carbons through elongated "stick-model" bonds. These stick-model bond lines will be the torsion angle reference lines.

Step 4: Sighting directly down the original C—C bond axis from front to back foreshortens this dimension and leaves an **apparent** two-dimensional representation. Using this perspective, identify the **apparent** angle formed between the selected atom/group bond line connected to the front carbon and the selected atom/group bond line on the back carbon (**Step 3**). This apparent angle is the <u>**torsion**</u> angle (or dihedral angle) between these two atom/groups.

7.3 DRAWING AND USING NEWMAN PROJECTIONS

1. For alkane-type molecules, conformational structures and torsion angles can be represented by **Newman projections**: these are two-dimensional representations of the selected C—C bond axis and all torsional angles between any one-to-one combination of an atom/group on the front carbon to an atom/group on the back carbon.

2. Process for Drawing Newman Projections

   Step 1: Sight exactly straight down the C—C bond axis from the "front" carbon to the "back" carbon (**Step 2 for torsion angle**). Draw a small solid circle to represent the front carbon, and then draw an artificially large open circle to represent the back carbon. This artificial scaling allows the back carbon to be "observed" even though directly behind the front carbon.

   Step 2: The remaining three attached atoms/groups on both the front and back carbons now appear in the two dimensions of the paper surface. Connect these groups to the axis carbons with the elongated bond lines; the pattern will be a <u>**symmetrical**</u> (bond lines separated by 120°) "Y" shape on each axis carbon circle. (The 120° bond line separation is caused by the flattening of the tetrahedral bond angles into two dimensions.)

   Step 3: Any set torsion angle combinations between group **X** on the front carbon and group **Y** on the back carbon can be represented by rotating around the C—C bond sigma-bond axis. Some common torsion angles with terminology and the corresponding conformations are

   Eclipsed X/Y torsion angle = 0° gauche: X/Y torsion angle = 60°.

   Eclipsed conformation: torsion angle = 0° for all adjacent groups on axis carbons.

   Staggered conformation: torsion angle = **60°** for all adjacent groups on axis carbons.

   Anti: specific **X/Y torsion angle = 180°** found in the **staggered** conformation.

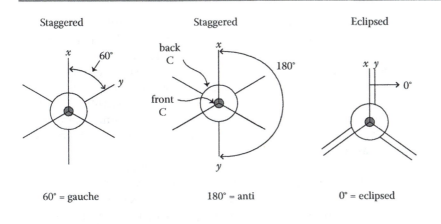

Staggered   Staggered   Eclipsed

60° = gauche   180° = anti   0° = eclipsed

7.4 POTENTIAL ENERGIES/STABILITIES OF ALKANE CONFORMATIONS

7.4.1 POTENTIAL ENERGY AND TORSIONAL STRAIN

Potential energies of the conformations produced by rotation around the C—C bond sigma-bond axis are dependent on the torsion angles due to **torsional strain**.

1. **Torsional strain** is the increase in potential energy (destabilization) caused by increasing electron–electron repulsion of closely approaching bonding electron pairs for the compared attached X and Y groups on axis carbons.
2. Torsional strain and potential energy increase as bonding electron pairs of X and Y come closer together: minimum separation = 0° torsional angle. Torsional strain and potential energy decrease with increasing torsion angle. For general conformations:
3. **Eclipsed conformation:** Torsion angles for **all** attached groups on the axis carbons are 0°. The minimum torsion angle conformation has the highest potential energy and is least stable.
4. **Staggered conformation:** Torsion angles for **all** attached groups on the axis carbons are 60°. This represents the maximum possible **average** torsion angle and represents the lowest potential energy and most stable conformation.

7.4.2 POTENTIAL ENERGY AND VAN DER WAAL'S REPULSION (STERIC REPULSION)

Potential energies of the conformations produced by rotation around the C—C bond sigma-bond axis are also dependent on **Van der Waal's (steric) repulsion**.

1. **Van der Waal's (steric) repulsion** is electron repulsion between electron shells of closely approaching atomic radii for the compared attached X and Y groups on axis carbons.
2. Van der Waal's (steric) repulsion and potential energy increase as electron shells of X and Y come closer together: minimum separation $= 0°$ torsional angle. Repulsion and potential energy decrease with increasing torsion angle: maximum separation for any specific X and Y comparison $= 180°$; the X and Y groups are **anti**.
3. For any specific X and Y torsion angle, Van der Waal's (steric) repulsion and potential energy increase with the 3-D size ("steric bulk") of the compared X and Y groups involved.
4. The 3-D size of an attached group may be reflected in atomic radius, for example:

$$I(133\,pm) > Br\,(114\,pm) > Cl(99\,pm) > F(71\,pm)$$

5. The 3-D size of an attached group may be reflected in the number, size, and molecular geometry of the bonded atoms in a grouping, for example:

—C(CH$_3$)$_3$	>	—CH(CH$_3$)$_2$	>	—CH$_2$CH$_3$	>	—CH$_3$
tert-butyl		**isopropyl**		**ethyl**		**methyl**

7.4.3 GENERAL CONCLUSIONS FOR CONFORMER STABILITY

1. Torsional angle relationships between the **largest** groups on **each** attached axis carbon dominate the net potential energy of conformations due to both torsion strain and steric repulsion.
2. Conformations become more stable with net increasing separation between the larger attached groups to each axis carbon:

most stable ◄————————————————————► **least stable**

larger groups **anti** (180°) > larger groups **gauche** (60°) > larger groups **eclipsed** (0°).

7.4.4 EXAMPLE

Follow the process shown below to identify the most stable and least stable conformations of 2,3-dimethylpentane when viewed down the **C#3—C#4** bond axis; C#3 and C#4 are determined based on correct nomenclature numbering rules.

1. Draw the structure of 2,3-dimethylpentane.
2. Draw any staggered conformation of the molecule using the **C#3—C#4** bond axis by drawing standard tetrahedral carbons for **C#3** and **C#4**.

3. Convert this conformation into a Newman projection with C#3 in front and C#4 at the back.
4. Determine and draw Newman projections for the **<u>most</u>** stable conformation of this molecule when viewed down the **C#3—C#4** bond axis.
5. Determine and draw Newman projections for the **<u>least</u>** stable conformation of this molecule when viewed down the **C#3—C#4** bond axis.

1. 2,3-dimethylpentane is shown as a partially condensed structure:

$$CH_3CH_2CH(CH_3)CH(CH_3)CH_3$$
#5   #4 #3   #2   #1

2. The critical part of this problem is the correct determination of the groups attached to C#3 and C#4 excluding the **C#3—C#4** bond axis. Draw tetrahedral carbons with 3-D wedges and dashes; draw a staggered conformation:

C#3 has an attached isopropyl group: —CH(CH₃)CH₃   CH₃CH₂CH(CH₃)**CH(CH₃)CH₃**
C#3 has an attached methyl group: (CH₃)—   CH₃CH₂CH(**CH₃**)CH(CH₃)CH₃
C#3 has an attached H—   CH₃CH₂**CH**(CH₃)CH(CH₃)CH₃

C#4 has an attached methyl group: CH₃—   **CH₃**CH₂CH(CH₃)CH(CH₃)CH₃
C#4 has **two** attached H—   CH₃**CH₂**CH(CH₃)CH(CH₃)CH₃

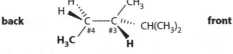

3. Reproduce the specific conformational drawing as a Newman projection.
4. Keep one axis carbon fixed (e.g., the front carbon) and rotate around the back carbon to find the conformation in which the largest group on each carbon is **anti**; this is the **most** stable conformation.
5. Rotate around the back carbon to find the conformation in which the largest group on each carbon is **eclipsed**; this is the **least** stable conformation.

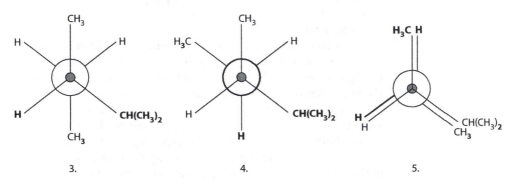

3.   4.   5.

7.5 CYCLOHEXANE CONFORMATIONS

Two distinct conformations of cyclohexane are the "boat" and "chair" conformations:

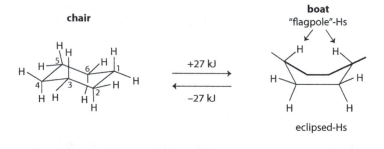

The torsion strain of eclipsed hydrogens coupled with Van der Waal's repulsion of the flagpole hydrogens makes the boat conformation less stable by about 27 kJ/mole; cyclohexane exists predominantly in the more stable chair conformation.

7.5.1 DESCRIPTION OF THE CHAIR CONFORMATION OF CYCLOHEXANE

Tetrahedral carbon bond angles are close to the optimum 109.5°; the "zig-zag" structure mimics the tetrahedral carbon.

1. With the ring viewed "edge-on" carbons #1 and #4 are in the plane of the paper.
2. With the ring viewed "edge-on" carbons #2 and #3 are coming out of the plane **toward** the viewer.
3. With the ring viewed "edge-on" carbons #5 and #6 are going **backward** away from the viewer.

Each carbon in the six-membered ring is bonded to two ring carbons, leaving two additional bonding positions available for hydrogens or other substituents. These two substituent positions are not energetically equivalent in a specific conformational structure.

1. With the ring viewed "edge-on" the **axial** position on each ring carbon will show a bonding direction either "straight-up" or "straight-down."
2. With the ring viewed "edge-on" the **equatorial** position on each ring carbon will show a bonding direction either slightly down or slightly up with the predominant direction mostly "out-to-the-side," that is, away from the ring center.
3. Axial and equatorial positions in the cyclohexane chair conformation are not permanent: these positions are interchanged upon rotation around multiple ring carbon sigma bonds during a conformational change ("ring-flip"). A ring-flip converts all axial positions into equatorial and all equatorial positions to axial.

7.5.2 DRAWING THE CHAIR CONFORMATION

Step 1: Draw the edge-on view for the alternating zig-zag line-drawing structure of the chair carbon ring.

Step 2: Draw the six **axial** positions, one on each ring carbon:

1. If the two ring C−C bond lines in the line drawing are slanting **upward** to reach the ring carbon vertex with the axial position to be drawn, the **axial** position is drawn **straight up**.

2. If the two ring C−C bond lines in the line drawing are slanting **downward** to reach the ring carbon vertex with the axial position to be drawn, the **axial** position is drawn **straight down**.

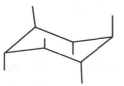

Step 3: Draw the six **equatorial** positions, one on each ring carbon; equatorial positions must be on the opposite face of the ring from the axial position. Ring face refers to the top or "up" face versus the bottom or "down" face of the ring when the ring is viewed edge-on.

1. If the **axial** position from step 2 is drawn straight **up**, the **equatorial** position must be drawn **downward** and out to the side to mimic the direction away from the ring center.

2. If the **axial** position from step 2 is drawn straight **down**, the **equatorial** position must be drawn **upward** and out to the side to mimic the direction away from the ring center.

The 3-D view of the equatorial positions will match the corresponding orientation of the ring carbons to which they are attached:

1. The edge-on view of carbons #1 and #4 is in the plane of the paper; the equatorial position bond lines on C#1 and C#4 is in the plane of the paper.

2. The edge-on view of carbons #2 and #3 is coming out of the plane **toward** the viewer; the equatorial position bond lines on C#2 and C#3 are coming out of the plane **toward** the viewer.

3. The edge-on view of carbons #5 and #6 is going **backward** away from the viewer; the equatorial position bond lines on C#5 and C#6 are going **backward** away from the viewer.

4. The 3-D orientation for the equatorial positions is shown in some cases; however, it is usual to draw the ring with simple lines implying the direction.

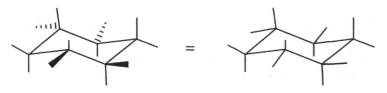

7.5.3 WORKING WITH THE CHAIR CONFORMATION

Each ring carbon in the cyclohexane chair conformation will have four atoms/atom groups attached: two ring carbons, one axial group, and one equatorial group. Relative 3-D directions for these groups can be visualized by drawing a double Newman projection. The chair drawn below numbers each ring carbon **1 → 6**. The remaining positions are marked: **ax5 = axial** position on ring carbon **#5**; **eq2 = equatorial** position on carbon **#2**, etc.

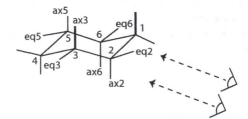

Select the following Newman projection axes:

$$C\#2(\text{front}) \longrightarrow C\#3(\text{back})$$
$$C\#6(\text{front}) \longrightarrow C\#5(\text{back})$$

The combined Newman projection is

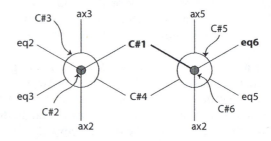

1. Each axis carbon in a Newman projection has three groups attached excluding the connected axis carbon: an equatorial group, an axial group, and a **non-axis** ring carbon group. For example, C#2 (front) has attached groups: ax2, eq2, and C#1; the fourth bond is to axis C#3 (back).
2. The chair conformation is symmetrical: all selected axes combinations will show the same torsion angle relationships:
   a. Axial groups on axes carbons will have a torsion angle **anti** (180°).
   b. Equatorial groups on axes carbons will have a torsion angle **gauche** (60°).
   c. The **non-axis** ring carbon groups on axes carbons will have a torsion angle **gauche** to each other.
   d. A **non-axis** ring carbon group on one axis carbon will be **anti** to the equatorial group on the other axis carbon.

7.5.4 PROCESS FOR SOLVING CHAIR CONFORMATION PROBLEMS

Example: Draw two methyl groups on the ring shown which are **anti** to the fluorine substituent.

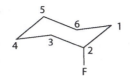

Step 1: **Determine to which carbons on the ring the methyl groups must be bonded.**
Torsion angle relationships for the chair conformation, as for any Newman projection, apply only to groups on axis carbons; axis carbons by definition must be directly connected.

1. **F** is a substituent on **C#2**: C#2 must be one of the axis carbons.
2. The other axis carbon must be directly connected to C#2: C#3 can be selected as the other axis carbon forming the axis **C#2(front)** ⟶ **C#3(back)**; identical results would occur for C#3(front) ⟶ C#2(back).
3. C#1 can also be selected as an alternative back axis carbon; this forms the axis **C#2(front)** ⟶ **C#1(back)**; (or C#1 (front) ⟶ C#2(back).
4. Methyl groups must be placed on C#3 and C#1.
5. Note that despite the more regular shape of the chair conformation, no torsion angle relationship exists between the F and any group on C#4, C#5, or C#6 since no connected axis can be drawn between these carbons and the C#2 with the F group.

Step 2: **Determine which specific position(s) on the required carbon(s) will be anti to the F on C#2.**

1. The F on C#2 is axial. The generalized Newman projection of the chair conformation shows that axial groups on axes carbons are **anti** (180°).
2. The **axial** positions on C#3 and C#1 are **anti** the F on C#2.

Step 3: **Complete the problem:**
Methyl groups are placed on the axial positions on C#3 and C#1:

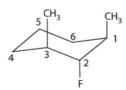

7.5.5 ADDITIONAL EXAMPLES

Example: Draw two ethyl groups on the ring shown which are **gauche** to the chlorine substituent.

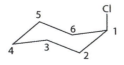

Step 1: **Cl** is a substituent on **C#1**: The two axes must be **C#1**
(front) ——→ **C#2(back)** and **C#1(front)** ——→ **C#6(back)**;
ethyl groups must be placed on C#2 and C#6.

Step 2: The Cl on C#1 is axial. The Newman projection of the chair
conformation shows that an axial group on one axis carbon is
gauche to the equatorial position on the other axis carbon. The
same axial position is also gauche to the non-axis ring carbon
group; however, ethyl substituents cannot be inserted in place
of a ring carbon. The equatorial positions on C#2 and C#6 are
gauche to the Cl on C#1.

Step 3: Ethyl groups are placed on the equatorial positions on C#1
and C#6:

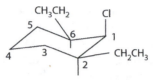

Example: Draw two bromines on the ring shown above which are
anti to **ring C#1**.

Step 1: The <u>group</u> for torsion angle comparison is the **non**-axis ring
carbon **C#1**; C#1 cannot be part of a torsion angle axis. As the
considered group, C#1 is bonded to C#2 and C#6; C#2 and C#6
can be the front carbons starting the two axes:
C#2(front) ——→ C#3(back) and C#6(front) ——→ C#5(back);
bromines must be placed on C#3 and C#5.

Step 2: The Newman projection of the chair conformation shows
that the non-axis ring carbon on one axis carbon is anti to the
equatorial position on the other axis carbon. The equatorial
positions on C#3 and C#5 are anti to the group C#1 for the axes
determined in step 1.

Step 3: Bromines are placed on the equatorial positions on C#3
and C#5:

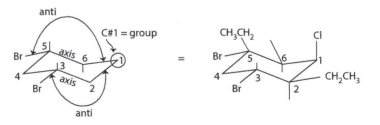

Summary Guide to Thermodynamic Concepts for Organic Chemistry

8

Part 1: Bond Energy, Enthalpy, and Potential Energy Diagrams

8.1 CONCEPT OF ENERGY

1. Energy is the capacity to do work and/or to produce heat. It is composed of two major classifications: **potential energy (PE)** and **kinetic energy (kE)**. The metric units of energy are $(kg\text{-}m^2/sec^2) =$ one **Joule (J)**; one kiloJoule **(kJ)** = 1000 J; (4.184 Joules = 1 calorie).

2. **PE** is the capacity to do work/produce heat based on the relative positions or arrangements of matter; *__PE is stored energy__*. The PE of matter is released when the matter, in its specific position or arrangement, can respond to a force.

   Molecules and compounds store **chemical/electrical PE** through the arrangements of electrons and nuclei relative to each other in chemical covalent or ionic bonds. Chemical PE is released through the response to the electromagnetic force; during a chemical reaction, atoms and electrons rearrange to form new combinations and new bonding/positional relationships.

3. **kE** is the energy of __motion__ of matter. If stored (potential) energy is released, it appears as kE.

   a. kE includes both the production of work (mechanical energy represented by motion of bulk matter) and the production of heat: Heat is thermal energy and is a measure of the kE of motion of individual atoms, molecules, or ions in bulk matter.

    b. Temperature (in K) is a measure of the average kE of atoms and molecules.

8.2 CONSERVATION OF ENERGY: FIRST LAW OF THERMODYNAMICS

1. The **law of conservation of energy** (a form of the **first law of thermodynamics**) states that energy can neither be created or destroyed during any process in the universe: ***total energy of the universe is constant***.
2. **Forms** of energy are interconvertible; a major relationship between kinetic and PE is the interconversion of the two forms:

$$\textbf{Potential Energy} \xrightleftharpoons{\hspace{3cm}} \textbf{Kinetic Energy}$$

3. The law of conservation of energy (first law of thermodynamics) requires that any change in total potential energy (ΔPE) must be accompanied by an equal and opposite change in the kinetic/radiant energy (ΔkE): $\Delta PE = -\Delta kE$. This is necessary if the total energy of the universe is to remain constant: $\Delta PE + \Delta kE = 0$.

8.3 CHEMICAL BOND ENERGETICS

8.3.1 PE AND CHEMICAL BONDS

1. The **PE** of an **electron** in an atom is based on the electromagnetic force of attraction to the positively charged nucleus: **a greater strength of attractive force is equated to a lower PE**. The force of attraction is proportional to the size of the positive and negative charges and inversely proportional to the distance between the charges.
2. The strength of the attractive force and the stability of an electron are inversely proportional to the distance of the electron from the nucleus: **the closer the electron to the nucleus, the greater the attractive force and thus the lower the PE**.
3. The **PE** of a (**covalent**) chemical **bond** is based on the arrangements of the bonding electrons to the nuclei of the two bonded atoms. Bonding energy is derived through the electromagnetic force of attraction between both nuclei and the bonding pair of electrons.

    a. The ***sum total* of attractive forces for *both* electrons to *both* bonding nuclei is always greater for an electron pair** *in a covalent bond than for the same electrons in sepa-rated atoms*. This increase in the attractive force is the basis for the strength of the covalent bond.

b. It is therefore, always true that the ___total PE of atoms in a covalent bond is always less than the ___total PE of the same atoms as separated atoms__. A summary of the relationships:

Bonded Atoms: **Separated Atoms:**

greater attractive force   ←————— less ————→   zero attractive force

lower total potential energy   ←——————————→ **higher total potential energy**

Atoms more stable   ←——————————————→   **Atoms less stable**

8.3.2 BOND ENERGIES AND CHANGES

1. ___Total PE always decreases when a covalent bond is formed__ from the corresponding separated atoms. The energy released by the decrease in potential energy (−ΔPE) is converted into an equal increase in kinetic energy (+ΔkE). The release of energy usually appears as work, heat, or light.

2. Conversely, ___total PE must always increase when a covalent bond is broken__ to reform the original separated atoms. The required increase in potential energy (+ΔPE) must be supplied by an equal decrease in kinetic energy (−ΔkE). This means that energy must be supplied from an outside source to break a chemical bond.

3. A specific PE associated with a specific bond, a **bond** PE, can be identified based on the following criteria:

    a. The bond PE of **non**bonded separated atoms is defined as **zero** (bond PE = 0).

    b. PE always **decreases** when a bond is formed; thus, the **change in bond PE, ΔPE (bond formation)**, must be **nega-tive** for bond formation: ___a stronger bond equates to a larger numerical value for the negative number and thus equates to a lower PE.__

    c. Conversely, the amount of energy required (to be supplied) to **break** a specific covalent bond must be equal in value, but opposite in sign, to the amount of energy released when the bond is formed: +ΔPE **(bond breaking)** = −ΔPE **(bond formation)**. The value [+ΔPE **(bond breaking)**] is termed as the **bond dissociation energy (BDE)**; this value is always **positive** and is equal to the amount of **added** energy (as kE) required to break a specific covalent bond = **strength** of the bond.

4. This concept can be displayed through a **PE diagram**. The diagram, read from **left-to-right** in these cases, describes the changes in potential energy (**ΔPE**).

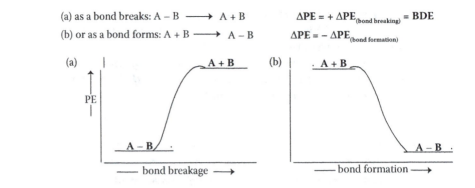

8.4 ENERGY AND CHEMICAL REACTIONS

8.4.1 REACTIONS AND BOND CHANGES

1. Chemical reactions occur through rearrangement of atoms in molecules and compounds and thus **always involve bond changes: certain bonds in the reactants are broken in order to form new bonds in the products.**

2. ***Bonding changes in chemical reactions always involve a corresponding change in PE.***
   a. The total bond PE of a complete molecule is the sum of bond PEs of each of the individual bonds comprising the molecule; the total bond PE of all reactants or all products in a chemical reaction is equal to the sum of the bond PEs of all reactant or product molecules:

   bond PE (reactants) = sum of bond PE

   (each bond in reactant molecule)

   bond PE (products) = sum of bond PE

   (each bond in product molecule)

   b. The PE **change** in a chemical reaction, **ΔPE (reaction)**, is measured by the **difference** between the PE sum of all the reactant molecules and the PE sum of all product molecules. A summary of bond energy and PE relationships for molecules:
   Molecules with

total stronger bonds ⟵⟶	**total weaker bonds**
lower bond PE ⟵⟶	**higher bond PE**
higher BDE values ⟵⟶	**lower BDE values**
more stable ⟵⟶	**less stable**

3. Under a certain restricted set of conditions, the bond PE change in a reaction is defined by the term **enthalpy change**, symbol **ΔH**:

$$\Delta PE \text{ (reaction)} = \Delta H \text{ (reaction)}$$

Enthalpy (ΔH) can be viewed as the change in potential energy (ΔPE) of a process <u>as</u> <u>measured</u> <u>by</u> heat transfer to the surroundings at constant pressure. It is important to recognize, however, that enthalpy (ΔH) is a measure of PE change, with equivalent meaning for signs:

$-\Delta PE$ **(reaction)** $= -\Delta H$ **(reaction)**

= potential energy decreases; energy released

$+\Delta PE$ **(reaction)** $= +\Delta H$ **(reaction)**

= potential energy increases; energy required

For a PE decrease, the energy released can be measured as heat (kE); for a PE increase, the required energy can be supplied as heat (kE). The following terms describe the relationship between chemical reactions and the sign of enthalpy change:

$-\Delta H$ **(reaction)**: reaction is termed **exothermic** ("heat out")

$+\Delta H$ **(reaction)**: reaction is termed **endothermic** ("heat in")

8.4.2 CALCULATING REACTION ENTHALPY VALUES USING BOND ENERGIES

1. The calculation equation for enthalpy based on the use of **BDE** is

ΔH (reaction) = [total bond BDE (reactants)]

− [total bond BDE (products)]

2. The summation of all bonds in all reactants and products for large molecules is impractical. The enthalpy (PE) change in a chemical reaction will be based only on the actual bonds that change; bonds that are not involved in the reaction will not affect the net PE change. Therefore, a more useful form of the equation is

ΔH (reaction) = [Sum of BDE for bonds broken]

− [Sum BDE for bonds formed]

3. Both equational formats for the subtraction sequences ("reactants **minus** products"; "bonds broken **minus** bonds formed") are required to produce the correct sign for enthalpy based on the positive values used for BDEs.
4. To calculate the bond PE (enthalpy) change for a chemical reaction based on a given balanced equation:

Step (1): Draw **Lewis structures** of each reactant and product molecule to identify all necessary bonds.

Step (2a) (Shorter method): Identify **all** the bonds **broken** and **all** the bonds **formed**; be certain to include the correct numbers of each bond based on the coefficients in the balanced equation.

<u>Alternate</u>: **Step (2b) (Can be used when bond changes are difficult to identify):** Identify <u>**all**</u> bonds in <u>**all**</u> reactants and

products; be certain to include all molecules based on the coefficients in the balanced equation.

Step (3): Match each identified bond with its corresponding **BDE** value from BDE tables.

Step (4): Add all the required bond BDE values for the necessary categories: [BDE of bonds broken] and [BDE of bonds formed] for step (**2a**) **or** [BDE of all reactant bonds] and [BDE of all product bonds] for step (**2b**). *__Be certain to multiply each bond BDE value by the number of each of the bond types identified__*.

Step (5): Complete the **calculation of ΔH (reaction)** using the appropriate equation; be careful to keep the subtraction sequence in the correct order.

For step (**2a**):

$$\text{ΔH (reaction)} = [\textbf{Sum of BDE for bonds broken}]$$
$$- [\textbf{Sum BDE for bonds formed}]$$

For step (2b):

$$\text{ΔH (reaction)} = [\textbf{total bond BDE (reactants)}]$$
$$- [\textbf{total bond BDE (products)}]$$

5. **Example:**
Calculate the value for **ΔH (reaction)** for the following balanced equation; state whether the reaction is exothermic or endothermic:

$$CH_4 + O_2 \longrightarrow CH_2O + H_2O$$

Answer based on the process using step (2a):

(**1**)   (Lone pairs are not shown in Lewis structures.)

(**2a**) **Bonds broken:** Compare CH_4 with CH_2O: **two** of the reactant C—H bonds from CH_4 must be broken; the other two are retained in CH_2O and thus do not change. The reactant O=O double bond must be broken.
Bonds formed: The product CH_2O has a newly formed C=O bond; the product H_2O has **two** newly formed O—H bonds.

(**3**)   Use BDE tables:

Bonds broken: C—H = **415 kJ**;   O=O = **494 kJ**

Bonds formed: C=O = **750 kJ**;   O—H = **460 kJ**

(4)   **Bonds Broken**   **Bonds Formed**

Bonds broken: $2 \times C{-}H = 2 \times 415 = 830$ kJ $1 \times C{=}O = 1 \times 750 = 750$ kJ

Bonds formed: $1 \times O{=}O = 1 \times 494 = \underline{494}$ kJ $2 \times O{-}H = 2 \times 460 = \underline{920}$ kJ

                            1324 kJ   **1670 kJ**

(5) The equation used for calculations based on bonds broken and bonds formed is

$$\Delta H \text{ (reaction)} = [\text{Sum of BDE for bonds broken}]$$
$$- [\text{Sum BDE for bonds formed}]$$
$$\mathbf{\Delta H \text{ (reaction)}} = [1324\,\text{kJ}] - [1670\,\text{kJ}] = \mathbf{-346\ kJ}$$
$$= \textbf{exothermic}$$

(An equivalent answer can be obtained based on the process using step **(2b)** instead of **(2a)**.)

8.4.3 REACTION ENERGY-LEVEL DIAGRAMS

1. Changes in reaction PE are very often displayed in the form of a **reaction PE diagram**. This is a pictorial description of a reaction based on plotting the **relative potential energies** of **all** reactants and products as a function of the general concept termed **reaction progress**.
   a. **Reaction progress** is nonspecific and represents some sequential description of how the reaction changes; reaction progress is shown along the horizontal axis.
   b. **Relative potential energies** can be shown as bond PE along the vertical axis.
2. For the **simplest** form of the diagram, the **complete** set of initial **reactants** and the **complete** set of final **products** are each displayed as a separate energy "platform." Connecting the two in sequence is a smooth curve, which represents the energy changes of all the molecules for the reaction. The diagram follows the convention of balanced chemical equations: the reaction described by the equation, termed the forward reaction, is read from left-to-right; the reactants are on the left side and the products are on the right side.
3. **Examples:** Consider the complementary reactions describing the interconversion of hydrogen, oxygen, and water. It is found that 474 kiloJoules (kJ) of PE are released (**bond PE decreases**; $\Delta H_{(\text{reaction})} = (-)$) in the reaction of hydrogen gas plus oxygen gas to form water according to the balanced equation; this information is displayed in plot **(a)**. Conversely, it is true that 474 kJ of energy must be added to the reaction (**bond PE increases**; $\Delta H_{(\text{reaction})} = (+)$) to allow the formation of hydrogen gas and oxygen gas from water; this information is displayed in plot **(b)**.

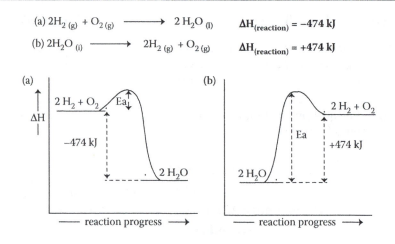

(a) $2H_{2\,(g)} + O_{2\,(g)} \longrightarrow 2\,H_2O_{(l)}$ $\Delta H_{(reaction)} = -474\text{ kJ}$

(b) $2H_2O_{(l)} \longrightarrow 2H_{2\,(g)} + O_{2\,(g)}$ $\Delta H_{(reaction)} = +474\text{ kJ}$

4. The curve that shows the change in PE as the reaction proceeds from reactants to products does not connect the two energy platforms with a straight line: all complete reactions show an **initial increase** in energy in the reaction progress; this is true regardless of the relative PE positions of reactants or products. This initial increase in energy for all reactions is termed the **activation barrier** or **activation energy (symbol = Ea)**; it represents the energy which must be initially "invested" to allow a chemical reaction to proceed.

5. **Example:**
   Draw a **very general** PE diagram based on the enthalpy change (PE change) information that was calculated for the reaction in a previous example:

   $$CH_4 + O_2 \longrightarrow CH_2O + H_2O$$

   Answer for general PE diagram:
   Follow the previous examples to draw the axes for a PE diagram; label the reactant platform [**CH$_4$ + O$_2$**]; label the product platform [**CH$_2$O + H$_2$O**]. The calculated value of $\Delta H_{(reaction)}$ is **negative**; $\Delta PE_{(reaction)}$ decreases: **reactant platform must be higher than product platform**. Include an initial increase in $\Delta PE_{(reaction)}$ to indicate the **activation energy, Ea**.

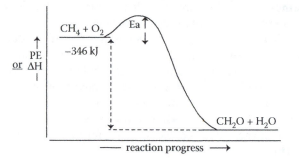

8.4.4 PE, ENTHALPY, AND REVERSIBLE REACTIONS

1. A conventional chemical reaction describes the process of converting a specific set of reactants into a specific set of products. The corresponding balanced equation, read from left-to-right, describes what is termed the **forward** **reaction**.

2. Any specific original chemical reaction can **theoretically** have an opposite direction reaction: the products of the original reaction act as the reactants in the opposite reaction; the reactants of the original reaction represent the products of the opposite reaction.

   *Based on the original left-to-right form of the balanced equation, the **original reaction is termed the forward reaction direction; the opposite reaction is termed the reverse reaction direction.***

   A **reversible reaction** is one in which both the forward reaction and the reverse reaction can occur at the same time. Reversible reactions can be written as one balanced equation, with arrows pointing in opposite directions. The **forward reaction is read from left-to-right; the reverse reaction is read from right-to-left**.

3. PE change as measured by the enthalpy change (ΔH) can also be calculated from standard enthalpies of formation ($\Delta H°f_i$):

$$\Delta H°_{(reaction)} = [\text{Sum } n_i \ \Delta H°f_i \ (\text{products})]$$
$$-[\text{Sum } n_i \ \Delta H°f_i \ (\text{reactants})]$$

The requirement for a **reversible** chemical reaction, regardless of the method of enthalpy calculation, matches that developed for bond energies:

$$\Delta H \ (\textbf{forward reaction}) = -\Delta H \ (\textbf{reverse reaction})$$

Example: For the following reversible reaction, write the forward and reverse reactions separately with energy as a product or reactant.

$$2 \ H_{2 \ (g)} + O_{2 \ (g)} \ \underset{\longleftarrow}{\longrightarrow} \ 2 \ H_2O_{(g)} \quad \Delta H \ (\text{forward reaction}) = -474 \ \text{kJ/mole-reaction}$$

$$2 \ H_{2 \ (g)} + O_{2 \ (g)} \longrightarrow 2 \ H_2O_{(g)} + \textbf{474 kJ} \quad \Delta H \ (\textbf{forward}) = (-); \text{heat is a product}$$

$$2 \ H_2O_{(g)} + \textbf{474 kJ} \longrightarrow 2 \ H_{2 \ (g)} + O_{2 \ (g)} \quad \Delta H \ (\textbf{reverse}) = (+); \text{heat is a reactant}$$

4. PE change and enthalpy change are state functions:
   A **state function** is one in which the measured values depend only on the state of the system and **not on how the state occurred**. This requires that **changes of state** (e.g., ΔH, depend only on the **initial** and **final** states and not how the change occurred).

In contrast, a **path function** depends on the method of the state change: the pathway, or how the change occurred.

8.4.5 EXAMPLE PROBLEM: READING AN ENERGY DIAGRAM

1. A PE diagram displays certain information about a reaction: if certain energy values are included, energy values for specific differences can be read or calculated. If the reaction is reversible, both the forward and reverse reactions can be read from the same diagram: the forward reaction is read from left-to-right; the reverse reaction is read from right-to-left.

   Example: Answer the following questions by reading the energy diagram shown:
   a. What is the **forward** reaction (i.e., the reaction formula equation for the **forward** reaction as shown in the diagram)?
   b. What is the value for ΔH, the PE change for this **forward** reaction?
   c. What is the value for activation energy (activation barrier) for this **forward** reaction?
   d. What is the **reverse** reaction (i.e., the reaction formula equation for the **reverse** reaction as shown in the diagram)?
   e. What is the value for ΔH, the PE change for this **reverse** reaction?
   f. What is the value for activation energy (activation barrier) for this **reverse** reaction?
   g. Which reaction is endothermic?
   h. Which reaction is spontaneous based on the energy criterion?
   i. Which set of molecules, reactants, or products represent the greater **total** (i.e., all bonds in the molecules) bond strengths?

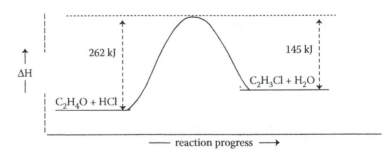

ANSWERS:
   a. **Forward** reaction: read left-to-right =
       $C_2H_4O + HCl \longrightarrow C_2H_3Cl + H_2O.$
   b. ΔH for **forward** reaction: the difference between the two energy platforms can be calculated as 262 kJ $- 145$ kJ $= 117$ kJ; the PE increases: $\Delta H = +117$ **kJ**.
   c. Activation energy (activation barrier) for **forward** reaction: complete height of the barrier is read from left-to-right: **Ea = 262 kJ**.

d. **Reverse** reaction: read right-to-left =
$C_2H_3Cl + H_2O \longrightarrow C_2H_4O + HCl.$

e. ΔH for **reverse** reaction: energy **difference** must be the same value as for the forward reaction; the PE decreases: $\Delta H = -117$ **kJ.**
    This demonstrates that ΔH for the forward and reverse reactions must be related as equal value but opposite sign.

f. Activation energy (activation barrier) for **reverse** reaction: complete height of the barrier is read from right-to-left: **Ea = 145 kJ.**
    Note that Ea does <u>not</u> show the same relationship as ΔH.

g. Endothermic reaction is the one that has a **positive** $\Delta H =$ the **forward reaction**.

h. The spontaneous direction based on the energy criterion is the one that shows a **negative** $\Delta H =$ the **reverse reaction**.

i. The lower PE represents the stronger bonds. The reactants are at the lower PE position; therefore, the reactants ($C_2H_4O + HCl$) have the greater total bond strengths.

Part 2: Entropy, Free Energy, Spontaneity, and Equilibrium

8.5 ENTROPY

8.5.1 GENERAL CONCEPTS

1. _**The entropy of a system (symbol = $S_{(system)}$) is a measure of the disorder or randomness of that system.**_ The larger the entropy value, the greater the randomness or disorder of the system; the smaller the entropy value, the lesser the randomness or disorder (or, equivalently, the smaller the entropy value, the more ordered the system).

    a. Entropy measures the statistical probabilities of specific system configurations versus all possible configurations (termed the **degrees of freedom**).

    b. _**Randomness or disorder always represents the more probable arrangement of matter.**_ Thus, it is always true that _**entropy always increases in the direction of increasing probability of the arrangement of matter**_ within a system.

2. Randomness or disorder in a system of matter is related to the number of arrangements in which the matter can exist. **A general summary of trends for probability and entropy, applied to a <u>specific</u> amount of matter, is**

    a. A specific amount of matter has more probable arrangements and is more disordered **(higher entropy) if it exists in a larger number of small, simpler molecules rather than a fewer number of more complex, larger molecules.**

Formation of large, complex molecules produces a more ordered, less random (lower entropy) system.

b. For more than one substance, a specific amount of matter has more probable arrangements and is more disordered **(higher entropy) if it exists as a mixture or solution rather than if the components are separated as pure substances.**

c. A specific amount of matter has more probable arrangements and is more disordered **(higher entropy) in proportion to the degree of independence of the molecules**: gas is more disordered (higher entropy) than liquid, which is more disordered (higher entropy) than solid.

d. A specific amount of matter has more probable arrangements and is more disordered **(higher entropy) in proportion to the amount of molecular motion, as measured by average kE**: disorder (entropy) <u>increases</u> as temperature <u>increases</u>.

8.5.2 PREDICTING ENTROPY CHANGES FOR A CHEMICAL PROCESS

1. A chemical process is a chemical reaction or physical change. The **<u>change</u>** in entropy for any process relates the degree of disorder of the reactants compared to the products.

a. The **change** in the entropy of a system is symbolized as ΔS. ΔS is a **positive** value whenever entropy **increases**: the system becomes more disordered and tends toward a more probable arrangement of matter. ΔS is a **negative** value whenever entropy **decreases**: the system becomes less disordered (i.e., more ordered) and tends toward a less probable arrangement of matter.

b. The general probability trends for a specific amount of matter can be used to analyze a process based on a **balanced** equation, phase change, or solution change. The following descriptions indicate the relationship between common processes and the corresponding direction of entropy change. ***These general concepts can be used to predict the direction of entropy change (sign of ΔS) in the absence of any calculations***.

2. Entropy changes for reactions between simple and complex molecules:

a. **Entropy increases (disorder increases)** for reactions in which a fewer number of larger or more complex molecules are broken up into a greater number of smaller, simpler molecules: ***more moles of product molecules form from fewer moles of reactant molecules***. This is especially true whenever the number of moles of **gas** increases during the reaction. **Example**: The following reaction shows 7 moles of reactants with one complex molecule decomposing to 12 moles of smaller products; in addition, the number of moles of gas molecules increase. The entropy change of the reaction is positive: entropy increases.

$$C_6H_{12}O_{6\,(s)} + 6\,O_{2\,(g)} \xrightarrow[\text{7 moles form 12 moles}]{\text{Entropy increases }(\Delta S = +)} 6\,CO_{2\,(g)} + 6\,H_2O_{\,(g)}$$

b. Conversely, **entropy decreases (disorder decreases**, i.e., **system** becomes **more ordered**) if a fewer number of larger, more complex molecules are produced by combination of smaller, simpler molecules: *__fewer moles of product molecules form from a greater number of moles of reactant molecules__*. **Example**: The reverse of the previous example: 7 moles of products with one complex molecule are formed by combination of 12 moles of simpler reactant molecules. The entropy change of the reaction is negative: entropy decreases.

$$6\,CO_{2\,(g)} + 6\,H_2O_{\,(g)} \xrightarrow[\text{12 moles form 7 moles}]{\text{Entropy decreases }(\Delta S = -)} C_6H_{12}O_{6\,(s)} + 6\,O_{2\,(g)}$$

c. *__Changes in number of moles of products and reactants can be easily read from the coefficients in the balanced equation__*.

3. Entropy changes for solution formation:
   a. **Entropy increases (disorder increases)** during the process of **solution formation** from pure substances, or during the **dilution** of a solution to a lower concentration.

Examples:

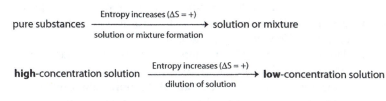

pure substances $\xrightarrow[\text{solution or mixture formation}]{\text{Entropy increases }(\Delta S = +)}$ solution or mixture

high-concentration solution $\xrightarrow[\text{dilution of solution}]{\text{Entropy increases }(\Delta S = +)}$ **low**-concentration solution

b. **Entropy decreases (disorder decreases**, i.e., the **system** becomes **more ordered**) for processes which purify or separate mixtures into pure substances, or during the concentration of a solution to higher concentration.

Example:

solution or mixture $\xrightarrow[\text{separation/purification}]{\text{Entropy decreases }(\Delta S = -)}$ pure substances

low-concentration solution $\xrightarrow[\text{solution concentration}]{\text{Entropy decreases }(\Delta S = -)}$ **high**-concentration solution

4. Entropy changes for phase changes: **Entropy increases (disorder increases)** during the phase change in the sequence solid melting to liquid and liquid evaporation to a gas; **entropy decreases (disorder decreases**, i.e., the **system** becomes **more ordered**) for condensation of a gas to a liquid and freezing of a liquid to a solid.

Examples:

$$\text{solid} \xrightarrow{\text{Entropy increases } (\Delta S = +)} \text{liquid} \xrightarrow{\text{Entropy increases } (\Delta S = +)} \text{gas}$$

$$\text{gas} \xrightarrow{\text{Entropy decreases } (\Delta S = -)} \text{liquid} \xrightarrow{\text{Entropy decreases } (\Delta S = -)} \text{solid}$$

5. Entropy change as a function of temperature: For any specific type of matter, **entropy increases (disorder increases)** as the temperature (average kE) **increases; entropy decreases (disorder decreases**, i.e., the **system** becomes **more ordered)** as the temperature (average kE) **decreases**.

Example:

$$\text{matter at } \textbf{lower} \text{ temperature} \xrightarrow{\text{Entropy increases } (\Delta S = +)} \text{matter at } \textbf{higher} \text{ temperature}$$

$$\text{matter at } \textbf{higher} \text{ temperature} \xrightarrow{\text{Entropy decreases } (\Delta S = -)} \text{matter at } \textbf{lower} \text{ temperature}$$

8.5.3 ENTROPY AND THE SECOND AND THIRD LAWS OF THERMODYNAMICS

1. The second law of thermodynamics states that the total entropy of the **universe** must increase during a spontaneous process: $\Delta S_{(universe)} > 0$ or $\Delta S_{(universe)} = (+)$. Non-spontaneous processes measured for the complete universe (system plus surroundings) do not occur; thus, the entropy of the universe is always increasing.

2. The third law of thermodynamics states that the entropy of a pure, perfectly ordered crystal is zero at a temperature of absolute zero Kelvin: $S_{(substance)}$ **at 0 K = 0**.
   a. Unlike enthalpy (H) or energy (E), entropy (S) has a true zero point. This allows the expression of the entropy of any substance in the form of an **absolute molar entropy ($S°$)** in units of Joules/Kelvin (J/K) rather than a change as used for enthalpy ($\Delta H°f$).
   b. Absolute standard molar entropies are provided under **standard conditions**: 1 atmosphere of pressure and 298 K. Since S = 0 only at 0 K, all values for absolute molar entropies at 298 K must be **positive** values (J/K); they are provided in standard tables similar to listings for $\Delta H°f$.

3. The standard entropy **change** ($\Delta S°$) for a reaction (or other process) measures the general **increase (positive** value for entropy change) or **decrease (negative** value for entropy change) in the total disorder present in all products and reactants under standard conditions. The equation for entropy change:
   a.

$\Delta S°_{(reaction)} = $ **[Sum of the absolute molar entropies of all products]**

$\qquad\qquad$ **−[Sum of the absolute molar entropies of all reactants]**

or

$\Delta S°_{(reaction)} = $ **[Sum n_i $S°_i$ (products)]−[Sum n_i $S°_i$ (reactants)]** where the terms S°_i are the **absolute** molar entropies of each species, while n_i designates the **individual** coefficients in the balanced equation.

b. Entropy change is a **state function**: $\Delta S^\circ_{(forward\ reaction)} = -\Delta S^\circ_{(reverse\ reaction)}$

c. **Example:** Predict the direction of entropy change for the following reaction and then use tables of S^0 values to calculate $\Delta S^\circ_{(reaction)}$ for the reaction

$$C_2H_{2\ (g)} + 2\,H_{2\ (g)} \longrightarrow C_2H_{6\ (g)}$$

Prediction: 3 moles of gas reactants combine to form 1 mole of gas product. This represents the direction of a greater number of smaller, simpler molecules combining to form fewer more complex molecules: entropy should decrease; $\Delta S^\circ_{(reaction)} = (-)$.

Calculation:

Reactants: $S^\circ\,(C_2H_{2\ (g)}) = 201$ J/K; $S^\circ\,(H_{2\ (g)}) = 131$ J/K
Products: $S^\circ\,(C_2H_{6\ (g)}) = 230$ J/K

$\Delta S^\circ_{(reaction)} = $ Sum of $[S^\circ(C_2H_{6(g)})] - $ Sum of $[(S^\circ(C_2H_{2(g)})) + 2(S^\circ(H_{2(g)}))]$

$\Delta S^\circ_{(reaction)} = [(230\text{ J/K})] - [(201\text{ J/K}) + 2(131\text{ J/K})]$

$\qquad = -\textbf{233 J/K}$ or $-\textbf{0.233 kJ/K}$

8.6 REACTION SPONTANEITY

8.6.1 GENERAL CONCEPTS

1. A chemical **process** is a **chemical reaction** or a **physical change** (such as a phase change: solid ⟷ liquid ⟷ gas). **Most** chemical processes are **theoretically reversible**: reactants and products may exchange roles; solids, liquids, and gases can be interconvertible.

2. It is valuable to determine the **expected direction** of reversible chemical reactions or processes. Expected direction for a chemical reaction refers to determining which molecules will be favored as being products and which will be favored as being reactants under a defined set of energy conditions. The concept used for expected process direction is **chemical spontaneity**.

3. **Spontaneity** in chemical PE concepts refers to the direction a chemical reaction proceeds as a function of several energy-related criteria, that is, whether the forward or reverse reaction is favored based on these criteria.

   a. **A spontaneous chemical process is one which, once started, will continue on its own without any additional stimulus of any type, for example, addition of required energy.**

   b. **A non-spontaneous chemical process is one which, once started, will not continue on its own; this process requires continuous input of required energy in order to proceed.**

c. Spontaneity does **not** indicate **if** the reaction will occur or **how fast** a reaction will occur (the **rate** of reaction). It answers the question: "**If** a reaction starts, what is the probable result and what is the energy change?"

4. *A spontaneous reaction need not proceed at a measurable rate.* For example, the combustion of wood or sugar (i.e., the reaction of cellulose or carbohydrate plus oxygen to yield CO_2 and water) is a highly spontaneous reaction (very large negative value for ΔH), but a block of wood or a bowl of sugar in air is stable nearly indefinitely in the absence of a spark (heat), enzymes, or biological organisms to accelerate the rate of the reaction.

5. *A non-spontaneous reaction can still occur if energy (e.g., as kE) is supplied to the reaction system from the surroundings.* Even when sufficient energy is provided, however, a "path" or sequence of bond-making and bond-breaking steps must be available for energy input.

8.6.2 ENTHALPY AND SPONTANEITY

1. PE change, as measured by the enthalpy change, is one main criterion for determining the spontaneous direction of a chemical process:

   a. ***Based on the PE criterion*** (i.e., ignoring other factors), ***all chemical processes are spontaneous in the direction of PE decrease: the direction of higher PE to lower PE***. Spontaneous processes "go down the energy hill." The equivalent statement is that a forward chemical reaction is ***spontaneous if the enthalpy change (ΔH) for the direction as written is negative***.

   b. Spontaneous chemical reactions are favored by bond PE energy release ($-\Delta H$). This is indicated by reactant molecules of higher PE being converted into product molecules of lower PE. ***The favored direction for a spontaneous reaction is to form products with stronger bonds from reactants with weaker bonds***.

2. The corollary: ***Based on the PE criterion*** (i.e., ignoring other factors), ***all chemical processes are non-spontaneous in the direction of PE increase: the direction of lower PE to higher PE***. Non-spontaneous processes "go up the energy hill." The equivalent statement is that a forward reaction is ***non-spontaneous if the enthalpy change (ΔH) is positive***.

8.6.3 ENTROPY AND SPONTANEITY

1. The **entropy** of a chemical process is the second main criterion for determining the spontaneous direction:

   a. ***Based on the entropy criterion*** (i.e., ignoring other factors), ***all chemical processes are spontaneous in the direction*** *of entropy increase*, that is, the direction of greater ran-domness or disorder: a positive value for entropy ($\Delta S = +$).

b. Spontaneous chemical reactions are favored by an increase in entropy ($+ \Delta S$). *The favored direction for a spontaneous reaction is in the direction of more ordered reactants to less ordered (more random) products*.

c. The direction of spontaneity as a function of entropy is best recognized by noting that entropy is driven by probability. Disorder or randomness (as compared to order) is the most probable arrangement for a system. *Thus, it is true that chemical systems, based on the entropy criterion, always tend toward the most probable molecular configuration*.

2. The corollary is: ***based on the entropy criterion, all chemical processes are non-spontaneous in the direction of entropy decrease*** (i.e., in the direction of lower randomness or disorder [or equivalently in the direction of greater order]): a negative value for entropy ($\Delta S = -$). The relationship between a non-spon-taneous process and a decrease in entropy can be recognized by noting that ***energy is required to produce order from disorder***.

8.6.4 REACTION SPONTANEITY: COMBINING ENTHALPY AND ENTROPY

1. Reversible reactions are those that can have both the forward and reverse reactions occurring simultaneously. A **general** reac-tion can be symbolized by the equation:

$$aA + bB \; \underset{\longleftarrow}{\longrightarrow} \; cC + dD$$

a. The uppercase letters (A, B, C, D) represent the general reac-tants and products. The lowercase letters (a, b, c, d) represent the stoichiometric coefficients of A, B, C, D, respectively in the balanced equation.

b. The **forward** reaction is read from **left-to-right** and has the **reactants A** and **B** with **products C** and **D**. The **reverse** reaction is read from **right-to-left** and has **reactants C** and **D** with **products A** and **B**.

2. It is valuable to determine the spontaneous direction of chemi-cal reactions (i.e., to determine whether the forward or reverse reaction is thermodynamically favored). Determining reaction spontaneity requires combining the functions of enthalpy and entropy, plus temperature and reactant/product concentrations; spontaneity ultimately depends on

a. The standard enthalpy ($\Delta H°$) of the reaction.

b. The standard entropy ($\Delta S°$) for the reaction.

c. The temperature (T).

d. The starting concentrations of **all** compounds, [A], [B], [C], [D], etc.

3. Based on the PE criterion, all chemical processes are **spontane-ous** in the direction of PE decrease: negative value for enthalpy:

$\Delta H^0 = (-)$. Based on the entropy criterion, all chemical processes are spontaneous in the direction of entropy increase (i.e., in the direction of greater randomness or disorder: a positive value for entropy ($\Delta S = +$)). These independent relationships between reaction spontaneity and enthalpy or entropy produce the following general conclusions:

a. A chemical process is **always spontaneous** when PE decreases ($\Delta H^0 = (-)$) and entropy increases ($\Delta S^0 = (+)$); both enthalpy and entropy show the same trend toward spontaneous.

b. A chemical process is **always non-spontaneous** (**never spontaneous**) when PE increases ($\Delta H^0 = (+)$) and entropy decreases ($\Delta S^0 = (-)$); both enthalpy and entropy show the same trend toward non-spontaneous.

c. The spontaneity of a chemical process in which the enthalpy and entropy trends oppose each other one toward spontaneous and one toward non-spontaneous (i.e., $\Delta H^0 = (-)$ with $\Delta S^0 = (-)$ or $\Delta H^0 = (+)$ with $\Delta S^0 = (+)$ must be determined by further analysis and calculation of all factors).

8.7 FREE ENERGY (ΔG)

8.7.1 GENERAL CONCEPTS

1. The second law of thermodynamics provides a **general** definition for spontaneity under **all** possible circumstances: all spontaneous processes must show an increase in the entropy of the universe, $\Delta S_{(universe)} > 0$. The thermodynamic function which describes this requirement is termed **free energy** (symbol = ΔG).

2. Free energy incorporates the values for **enthalpy (ΔH), entropy (ΔS)**, and **temperature (T)**. The dependence on product/reactant **concentration** is included in the total entropy; the general equation is $\Delta G = \Delta H - T \Delta S$.

3. The form of the equation allows the sign of ΔG to specify reaction spontaneity:

a. A forward reaction is **spontaneous** if the free energy change (ΔG) for the direction as written is **negative**: $\Delta G = (-)$. This means that the free energy of the system **decreases**: the reaction (based on the definition of free energy) is energetically "**downhill**."

b. A forward reaction is **non-spontaneous** if the free energy change (ΔG) is **positive**: $\Delta G = (+)$. This means that the free energy of the system **increases**: the reaction (based on the definition of free energy) is energetically "**uphill**."

4. Free energy, based on the state functions of enthalpy and entropy, is also a **state function**; measurements for the forward and reverse reactions must be symmetrical:

a. If the forward reaction is spontaneous ($\Delta G = (-)$), the reverse reaction must be non-spontaneous ($\Delta G = (+)$).

b. If the forward reaction is non-spontaneous ($\Delta G = (+)$), the reverse reaction must be spontaneous ($\Delta G = (-)$).

c. Based on the symmetry of calculations for ΔH and ΔS, it is always true that

$$\Delta G_{(\text{forward reaction})} = -\Delta G_{(\text{reverse reaction})}$$

5. The equation for free energy demonstrates the previously established relationships between spontaneity and the signs for enthalpy and entropy:

a. Based on the equation $\Delta G = \Delta H - T\,\Delta S$, a **negative** value for reaction enthalpy change contributes to a **negative** value for ΔG: $\Delta H = (-)$ contributes to $\Delta G = (-)$, which contributes to reaction **spontaneity**.

b. The equation $\Delta G = \Delta H - T\,\Delta S$ shows a negative sign in front of the complete term for entropy (including the temperature): $[-T\,\Delta S]$. Since temperature (T) in K is always a positive value, the $[-T\,\Delta S]$ term must be **negative** for all **positive** values of $[\Delta S]$. Based on the equation, a **positive** value for reaction entropy change contributes to a **negative** value for ΔG: $\Delta S = (+)$ contributes to $\Delta G = (-)$ which contributes to reaction **spontaneity**.

6. Spontaneity determination requires a calculation for free energy based on all possible variables of enthalpy, entropy, temperature, and concentrations.

8.7.2 STANDARD FREE ENERGY: $\Delta G°_{298}$ AND $\Delta G°$ AT VARIABLE TEMPERATURES

1. **Standard conditions** for free energy calculations are defined as a temperature of 298 K, pressure equal to 1 atmosphere, and concentrations of all components equal to exactly **1 Molar** (gas concentrations may be expressed as partial pressures). The thermodynamic function which determines spontaneity under standard conditions is termed the **standard free energy**; symbol = $\Delta G°$.

2. The equation for standard free energy includes the **standard** enthalpy ($\Delta H°$) and **standard** entropy ($\Delta S°$); the temperature (**T**) in this case is **298 K**:

$$\Delta G° = \Delta H° - T\,\Delta S°$$

a. Calculation of standard free energy can be performed using standard free energies of formation ($\Delta G°f$); this method limits the temperature to 298 K.

b. An alternative method of calculation involves the separate calculation of enthalpy and entropy, followed by use of the equation $\Delta G° = \Delta H° - T\,\Delta S°$; T = 298 K.

3. The equation for standard free energy can be adapted to use **temperature** as a variable:

$$\Delta G°_T = \Delta H° - T \, \Delta S°$$

$\Delta G°_T$ is the standard free energy under standard conditions of exactly 1 Molar concentrations and 1 atmosphere pressure, **but at a variable temperature specified by "T."** The values for standard enthalpy and standard entropy at 298 are used in the equation $\Delta H°$ means $\Delta H°_{298}$ and $\Delta S°$ means $\Delta S°_{298}$.

4. The equation $\Delta G°_T = \Delta H° - T \, \Delta S°$ shows that the **reaction standard free energy measured at different specified potential temperatures remains a strong function of temperature.** This is based on the entropy term $[-T\Delta S°]$ in which $[\Delta S°]$ itself is multiplied by **T** in the equation.

5. **Example:** Calculate $\Delta G°_{(reaction)}$ for the following reaction at 298 K using the equation for $\Delta G°_T$; $\Delta H°_{(reaction)} = -312$ kJ/mole-reaction; $\Delta S°_{(reaction)} = -0.233$ kJ/K

$$C_2H_{2\,(g)} + 2\,H_{2\,(g)} \longrightarrow C_2H_{6\,(g)}$$

$$\Delta G°_T = \Delta H° - T \, \Delta S°$$
$$\Delta G°_{298} = (-312\,kJ) - [(298\,K)(-0.233\,kJ/K)] = (-312\,kJ) - [-69.4\,kJ]]$$
$$= -242.6\,kJ/mole - rxn$$

6. The general equation for reaction free energy at **any** temperature indicates that temperature affects the value of $\Delta G°_T$, and thus the spontaneity of a reaction, through the entropy term but not the enthalpy term.

   a. These effects are especially important if the **number** of **gas**-phase molecules change. Reactions producing gases from solids or liquids generally have relatively high entropy increases. The effect of entropy is much smaller when the number of molecules of the reactant and product remain the same and/or if gases are not involved.

   b. The sign of enthalpy and entropy often shows opposite tendencies for spontaneity: $\Delta H° = (-)$ with $\Delta S° = (-)$: enthalpy favors spontaneity; entropy favors non-spontaneity. $\Delta H° = (+)$ with $\Delta S° = (+)$: entropy favors spontaneity; enthalpy favors non-spontaneity.

   c. **If enthalpy and entropy show opposite tendencies for spontaneity, the sign of $\Delta G°_{(reaction)}$ and the spontaneous direction can be changed as a function of a temperature increase or decrease.**

7. The entropy term is the only variable in the general equation for $\Delta G°_T$ that is multiplied by the temperature; the greater the value of the temperature, the greater the energy contribution of the entropy term $[-T\Delta S°]$ to the numerical value of $\Delta G°_T$. As a general trend:

a. ***High temperature favors the spontaneity direction indicated by the entropy contribution of a process***. High temperature results in a relatively **higher** numerical value for the $[-T\Delta S°]$ term compared to the $\Delta H°$ term.

b. ***Low temperature favors the spontaneity direction indicated by the enthalpy contribution of a process***. Low temperature results in a relatively **lower** numerical value for the $[-T\Delta S°]$ term compared to the $\Delta H°$ term.

c. **When the numerical value of $\Delta S°$ is large, the value of $\Delta G°_T$ can change dramatically with temperature variation**.

8. **Examples:** Consider the previous example reaction:

$$C_2H_{2\,(g)} + 2\,H_{2\,(g)} \longrightarrow C_2H_{6\,(g)}$$

$\Delta H°_{(reaction)} = -312$ kJ/mole-reaction; $\Delta S°_{(reaction)} = -233$ kJ/K

a. Based on the **signs** of $\Delta H°$ and $\Delta S°$ determine, **without calculation**, the effect of temperature on the spontaneity of the forward reaction; select from the options: the **reaction is always** spontaneous; **never** spontaneous; spontaneous at sufficiently **low** temperatures; and spontaneous at sufficiently **high** temperatures.

b. Calculate $\Delta G°_{(reaction)}$ for the reaction at 400, 800, and 1400 K using the equation for $\Delta G°_T$. From the calculation, state which direction is spontaneous at each of these temperatures. Determine if the trend in the numbers helps to confirm the answer for (a).

ANSWERS:

a. The enthalpy for the forward reaction is a negative value: contributes to spontaneity. The entropy for the forward reaction is negative (decreasing): contributes to **non**-spontaneity. **Therefore, the spontaneity depends on the temperature (i.e., on the balance between enthalpy and entropy).**

The spontaneity of the forward reaction is favored by the **enthalpy**; **low temperature** favors the spontaneity direction indicated by the **enthalpy** contribution of a reaction: **forward reaction is spontaneous at sufficiently low temperatures.**

(High temperature favors the spontaneity of the reverse [entropy-positive] reaction.)

b. $\Delta G°_T = \Delta H° - T\,\Delta S°$

$\Delta G°_{400} = (-312\text{ kJ}) - [(400\text{ K})(-0.233\text{ kJ}/\text{K})] = (-312\text{ kJ}) - [-93.2\text{ kJ}]$

$= -218.8\text{ kJ}/\text{mole} - \text{rxn}$: forward reaction spontaneous

$\Delta G°_{800} = (-312\text{ kJ}) - [(800\text{ K})(-0.233\text{ kJ}/\text{K})] = (-312\text{ kJ}) - [-186.4\text{ kJ}]$

$= -125.6\text{ kJ}/\text{mole} - \text{rxn}$: forward reaction spontaneous

$\Delta G°_{1400} = (-312\text{ kJ}) - [(1400\text{ K})(-0.233\text{ kJ}/\text{K})] = (-312\text{ kJ}) - [-326.2\text{ kJ}]$

$= +14.2\text{ kJ}/\text{mole} - \text{rxn}$: reverse reaction spontaneous

The calculations agree with the prediction: the forward reaction is spontaneous at low temperatures and becomes non-spontaneous at high temperatures.

8.7.3 NONSTANDARD FREE ENERGY (ΔG) AND CONCENTRATIONS

1. Standard free energy at variable temperatures (ΔG°_T) requires concentration conditions of exactly **1 Molar** (excluding gas concentrations). A **complete** analysis of spontaneity requires the consideration of **variable concentrations** of product and reactant species described through the **nonstandard free energy, (ΔG_T)**. This function takes into account the **variable** concentration entropy; it applies to all nonstandard conditions, *including concentrations other than 1 Molar.*

2. The concentration dependence for the nonstandard free energy is produced through the **nonstandard concentration entropy** of the reaction product/reactant mixture. Each reversible reaction has two sets of molecules: the set of molecules on the left represent reactants in the forward reaction and products in the reverse reaction; the set of molecules on the right represent products in the forward reaction and reactants in the reverse reaction. Specific concentrations of products and reactants depend on all possible conditions of the reaction mixture; however, a general trend can be observed:

   a. *Based only on the nonstandard concentration entropy* (i.e., ignoring standard enthalpy and standard entropy), *the spontaneous direction of a reversible reaction is to consume (use as reactants) the set of molecules in greater concentration to produce (form as products) the set of molecules in lower concentration.*

   b. The net spontaneity result of the nonstandard concentration entropy is to "even out" all concentrations: to decrease the higher concentrations and to increase the lower concentrations.

3. Consider the general reversible reaction shown previously:

$$aA + bB \longrightarrow\!\!\!\longleftarrow cC + dD$$

   A, B, C, D = general reactants and products; **a, b, c, d** = stoichiometric coefficients of A, B, C, D respectively in the balanced equation.

   a. Viewed from the perspective of the forward reaction, the **reactants** are **A** and **B**; the **products** are **C** and **D**. A specific ratio of products to reactants, termed the **concentration quotient** (symbol = **Q**), is defined for **any** actual values for concentrations of components at any point in the reaction. The form of the ratio expression uses the specific concentration of each species, usually as **molarity (M)**, raised to the power of its coefficient in the balanced equation

$$Q_{(forward\ reaction)} = [C]^c[D]^d / [A]^a[B]^b \propto [products]/[reactants]$$

4. The value of the **nonstandard** ΔG_T under **all** circumstances is given by a general equation:

$$\Delta G_T = \Delta G^\circ_T + RT \ln(Q)$$

a. ΔG°_T = standard free energy under **standard** conditions of pressure and **concentration** but at a variable temperature specified by "**T**." **R** = gas constant measured in units of kJ/mole-K = **0.008314 kJ/mole-K**; **T** is temperature in units of **K**. The term "**ln**" is the natural log function.

b. The natural log term (**ln**) in the equation relates the two free energy measurements through the **concentration** of all components; the value for the nonstandard free energy change for a reaction system is dependent on the **actual starting concentrations** of all reactant and product species.

c. If the concentration of reactants [reactants] is greater than the concentration of products [products], the numerical value defined by the ratio **Q** is **less than 1** and **RTln(Q) is negative**: ΔG_t for the forward reaction becomes more negative and more spontaneous. If the concentration of [products] is greater than [reactants], the numerical value defined by the ratio **Q** is **greater than 1** and **RTln(Q) is positive**: ΔG_T for the forward reaction becomes more positive and less spontaneous (the reverse reaction becomes more spontaneous).

d. Note that if all concentrations are exactly 1 M, the calculation for the ratio **Q** becomes **exactly 1**. Since **ln(1) = 0**, the expected equivalence results:

$$\Delta G_T = \Delta G^\circ_T + RT \times (0) \text{ or}$$
$$\Delta G_T = \Delta G^\circ_T \textbf{ when all concentrations} = 1\,M$$

8.8 EQUILIBRIUM

8.8.1 CONCENTRATIONS AND EQUILIBRIUM

1. The rate of a reaction is proportional to the concentrations of the molecules reacting. Viewed from the perspective of the forward reaction in the general reversible reaction

$$aA + bB \longrightarrow cC + dD$$

the rate of the **forward** reaction is proportional to the concentrations of the reactants **A** and **B**; the rate of the reverse reaction is proportional to the concentrations of the **products C** and **D**.

2. A reversible reaction can reach a state termed **equilibrium**:

a. *Equilibrium for a reversible reaction occurs whenever the forward and reverse reactions proceed at the same rate.*

b. This means that reactant molecules are forming product molecules at the same rate that product molecules are being reconverted into reactant molecules: *a system at equilibrium shows no net change in concentrations of products or reactants*.

3. Since the concentrations of all products and reactants remain in balance (no change) at equilibrium, the ratio determined by the concentration quotient, Q remains a constant value:

$$Q_{(forward\ reaction)} = [C]^c[D]^d / [A]^a[B]^b = \textbf{constant value at equilibrium}$$

a. The concentration ratio defined by Q is also termed the **equilibrium concentration ratio** when applied to reactions at equilibrium.

b. The constant **numerical value** for Q at equilibrium is termed the **equilibrium constant**, general symbol = **K** or **K**(with a subscript) to indicate a specific type of reaction. An equation that expresses the correct equilibrium concentration ratio equal to the equilibrium constant (either the symbol or its actual numerical value) is termed the **equilibrium expression**. An example for the general reversible reaction:

$$\textbf{Equilibrium expression}: K_{(forward\ reaction)} = [C]^c[D]^d / [A]^a[B]^b$$

c. **Example:** Use the general definitions to write the complete equilibrium expression for the following reversible reaction: the symbol, **K**, is used since no numerical value is given.

$$CH_4 + 2\,Cl_2 \xrightleftharpoons{\qquad} CH_2Cl_2 + 2\,HCl$$

Answer: K = $[CH_2Cl_2][HCl]^2/[CH_4][Cl_2]^2$

4. Under certain specified conditions, the numerical value of the equilibrium constant can indicate the spontaneity of the forward or reverse reactions:

If the numerical value of **K is greater than 1**, this means that **at equilibrium**, the concentration of products is higher than the concentration of reactants; products are formed from reactants: the **forward reaction is spontaneous**.

If the numerical value of **K is less than 1**, this means that **at equilibrium**, the concentration of reactants is higher than the concentration of products; reactants are formed from products: **the reverse reaction is spontaneous**.

5. A system at equilibrium shows no **net** change in concentrations of products or reactants: *for a reaction at equilibrium, neither the forward nor reverse reactions are spontaneous*. The mathematical requirement for a system at equilibrium is that $\Delta G_T = 0$.

6. The equation for nonstandard free energy ($\Delta G_T = \Delta G°_T + RT \ln(Q)$) is most often used to evaluate reaction concentrations at **equilibrium**.

a. Specifically for an **equilibrium**, substitute $\Delta G_T = 0$ and $Q \equiv K$ in the equation $0 = \Delta G°_T + RT \ln K$; **rewrite as** $\Delta G°_T = -RT \ln K$.

b. **To use $\log_{10}$:** $\Delta G°_T = -2.303\ RT \log_{10} K$; $R = 0.008314$ kJ/mole-K, T **is in** K.

7. Any mixture of exactly 1 Molar reactants and products for a reversible reaction will proceed in the spontaneous direction given by the standard enthalpy and entropy, expressed as a value for $\Delta G°_T$. The reaction will continue until the concentration of products in the spontaneous direction build up to the point where the concentration entropy exactly cancels the $\Delta G°_T$: this is the point of equilibrium; the equilibrium constant defines the corresponding concentrations.

Examples: Calculate the numerical value for the equilibrium constants, **K**, at the temperatures of 400 Kelvin, 800 Kelvin, and 1400 Kelvin for the following reaction:

$$C_2H_{2\,(g)} + 2\,H_{2\,(g)} \rightleftharpoons C_2H_{6\,(g)}$$

$\Delta G°_T = -2.303\ RT \log_{10} K$; solve for $\log_{10}$ **K:** $\log_{10}$ **K** $= \Delta G°_T / -2.303\ RT$

$\Delta G°_T$ from previous examples: $\Delta G°_{400} = -218.8$ kJ/mole-rxn

$\Delta G°_{800} = -125.6$ kJ/mole-rxn; $\Delta G°1400 = +14.2$ kJ/mole-rxn

$\log_{10}\ K_{400} = \Delta G°_T / -2.303\ RT = (-218.8\ kJ)/(-2.303)(0.008314\ kJ/mole\text{-}K)(400\ K)$

$\log_{10} K_{400} = 28.6$; K = antilog(28.6) $= 10^{28.6} = 3.98 \times 10^{28}$

$\log_{10}\ K_{800} = \Delta G°_T / -2.303\ RT = (-125.6\ kJ)/(-2.303)(0.008314\ kJ/mol\text{-}K)(800\ K)$

$\log_{10} K_{800} = 8.20$; K = antilog(8.20) $= 10^{8.20} = 1.58 \times 10^8$

$\log_{10}\ K_{1400} = \Delta G°_T / -2.303\ RT = (+14.2\ kJ)/(-2.303)(0.008314\ kJ/mol\text{-}K)(1400\ K)$

$\log_{10} K_{1400} = -0.530$; K = antilog(-0.530) $= 10^{-0.530} = 2.95 \times 10^{-1}$

8.8.2 EQUILIBRIUM SHIFTS

1. A reaction at equilibrium means that reactant molecules are forming product molecules at the same rate that product molecules are being reconverted into reactant molecules: concentrations of products or reactants cannot change. However, an **outside** change can be applied to a reaction after it has reached equilibrium. In this case, the concentrations of all reactants and products must change in order to reestablish the required equilibrium ratios.

a. The outside changes may be addition or removal of some of the moles of a reactant or product; this will then change the concentration of compound added or removed and thus **temporarily** change the [product]/[reactant] ratio.

b. The outside change may be addition (temperature increase) or removal (temperature decrease) of some heat. Heat can be considered as a product or reactant based on the sign of the enthalpy (ΔH): if ΔH is positive, heat is a reactant; if ΔH is negative, heat is a product. A change in temperature results in the change of the heat product or reactant: it will also produce a **temporary** change in the actual concentrations described by the [product]/[reactant] ratio.

2. An equilibrium principle (Le Chatelier's principle) states that _**if an additional outside change is induced on an established equilibrium reaction, the reaction will respond by attempting to counteract the outside change**_.

Examples:

a. If additional **reactants are added** to an equilibrium, the **forward reaction increases its rate** to convert some of these reactant molecules into product molecules, and thus remove some of these extra reactant molecules. This response **decreases** the concentration of the reactants, **increases** the concentration of product molecules, and thus reestablishes the original required equilibrium ratio.

b. If **products are removed** from an equilibrium, the **forward reaction increases its rate** to convert some existing reactant molecules into more product molecules to replace the ones that were removed. This response **decreases** the concentration of the reactants, **replaces** (partly) the concentration of product molecules, and thus reestablishes the original required equilibrium ratio.

c. **If additional products are added to an equilibrium, or if reactants are removed from an equilibrium, the reverse reaction increases its rate; this reestablishes the original required equilibrium ratio.**

d. The response of an equilibrium to a temperature change follows the same principle: a reaction will counteract a **temperature increase** by increasing the rate of the direction that **consumes heat** (direction of ΔH positive); a reaction will counteract a **temperature decrease** by increasing the rate of the direction that **produces heat** (direction of ΔH negative).

3. The response of an equilibrium to an outside change can be determined by analyzing the balanced equation and predicting the reaction direction which must increase its rate to offset the outside change:

If the **rate of the forward reaction must increase**, the reaction is said to **shift to the right** or to **shift toward products**.

If the **rate of the reverse reaction must increase**, the reaction is said to **shift to the left** or to **shift toward reactants**.

8.8.3 COMPREHENSIVE EXAMPLE OF ALL CONCEPTS

Consider the common reversible reaction forming ammonia:

$$N_{2\,(g)} + 3\,H_{2\,(g)} \rightleftharpoons 2\,NH_{3\,(g)}$$

a. Use the following BDE to calculate ΔH for the **forward** reaction as shown:

N≡N (triple bond): 940 kJ ; N—H : 390 kJ ; H—H : 436 kJ

b. Predict the sign of entropy change for the forward reaction **without calculation**.

c. Use the sign of ΔH plus the predicted entropy change to decide what can be determined about the spontaneity of the forward reaction; select from the options: the **reaction is always** spontaneous; **never** spontaneous; spontaneous at sufficiently **low** temperatures; and spontaneous at sufficiently **high** temperatures.

d. The actual value of $\Delta S°_{(reaction)} = -0.200$ kJ/K; use this value to calculate the numerical values for standard free energy, $\Delta G°_T$ at temperatures of 298 and 770 K; state which direction is spontaneous at each of these temperatures.

e. Calculate the numerical value for K_{298} and K_{770} for this reaction.

f. Using your answer from (a), show the equilibrium reaction equation to include **heat** (energy) as a reactant or product.

g. State which **direction** the equilibrium will **shift** for **each** of the following induced changes: (i) Hydrogen gas is removed; (ii) Nitrogen gas is added; (iii) Ammonia is removed; (iv) Ammonia is added; (v) Temperature is increased; and (vi) Temperature is decreased.

ANSWERS:

a. $\Delta H_{(reaction)} = $ [Sum n_i BDE$_i$ (bonds broken)] − [Sum n_i BDE$_i$ (bonds formed)]

$\Delta H_{(reaction)} = $ [1 × (BDE N≡N) + 3 × (BDE H—H)] − [6 × (BDE N—H)]

$\Delta H_{(reaction)} = $ [1 × (940 kJ) + 3 × (436 kJ)] − [6 × (390 kJ)] = **−92 kJ/ mole-reaction**

b. The balanced equation shows **4 moles of gas reactants** combining to form **2 moles of gas products**. The standard entropy change is predicted to be **negative**: greater number of smaller molecules combine to form a fewer number of product molecules; the system becomes more ordered (less disordered).

c. The enthalpy for the forward reaction is a negative value: contributes to spontaneity. The entropy for the forward reaction is negative (decreasing): contributes to non-spontaneity. The spontaneity of the forward reaction is favored by the **enthalpy**; **low temperature** favors the spontaneity direction indicated by the **enthalpy** contribution of a reaction: **forward reaction is spontaneous at sufficiently low temperatures**.

(High temperature favors the spontaneity of the reverse [entropy-positive] reaction.)

d. $\Delta G°_T = \Delta H° - T\Delta S°$

$\Delta G°_{(298)} = -92\,kJ - [(298\,K)(-0.200\,kJ\,/\,K)]$

$= -92\,kJ + 59.6\,kJ = \mathbf{-32.4\,kJ\,/\,mole}$

Forward reaction is spontaneous at 298 K.

$\Delta G°_{(770)} = -92\,kJ - [(770\,K)(-0.200\,kJ\,/\,K)]$

$= -92\,kJ + 154.0\,kJ = \mathbf{+62.0\,kJ\,/\,mole}$

Reverse reaction is spontaneous at 770 K.

e. $\Delta G°_T = -2.303\,RT\,\log_{10}K$ thus $\log_{10}K = \Delta G°_T / - 2.303\,RT$

$\log K_{(298)} = -32.4\,kJ\,/-2.303\,(0.008314\,kJ\,/\,K)\,(298\,K) = \mathbf{+5.68}$

$K_{(298)} = $ antilog $(5.68) = 10^{5.68} = \mathbf{4.8 \times 10^5}$

$\log K_{(770)} = +61.8\,kJ\,/-2.303\,(0.008314\,kJ\,/\,K)\,(770\,K) = \mathbf{-4.21}$

$K_{(770)} = $ antilog $(-4.21) = 10^{-4.21} = \mathbf{6.2 \times 10^{-5}}$

f. ΔH is negative; heat is a product of the forward reaction

$$N_{2\,(g)} + 3\,H_{2\,(g)} \;\longrightarrow\; 2\,NH_{3\,(g)} + heat$$

g. i.   Hydrogen gas is removed: Hydrogen gas is a reactant; the **reverse** reaction must increase its rate to replace some of the hydrogen molecules that were removed: the reaction **shifts to the left** (or **shifts toward reactants**).

   ii.   Nitrogen gas is added: Nitrogen gas is a reactant; the **forward** reaction must increase its rate to remove some of the extra nitrogen molecules that were added: the reaction **shifts to the right** (or **shifts toward products**).

   iii. Ammonia is removed: Ammonia is a product; the **forward** reaction must increase its rate to replace some of the ammonia molecules that were removed: the reaction **shifts to the right** (or **shifts toward products**).

   iv. Ammonia is added: Ammonia is a product; the **reverse** reaction must increase its rate to remove some of the extra ammonia molecules that were added: the reaction **shifts to the left** (or **shifts toward reactants**).

   v.   Temperature is increased: heat is a product; the **reverse** reaction must increase its rate to remove some of the extra heat that was added: the reaction **shifts to the left** (or **shifts toward reactants**) (i.e., the reaction shifts in the direction of the ΔH positive).

vi. Temperature is decreased: the **forward** reaction must increase its rate to replace some of the heat that was removed: the reaction **shifts to the right** (or **shifts toward products**) (i.e., the reaction shifts in the direction of the ΔH negative).

PRACTICE PROBLEMS

The following is a partial list of **approximate** bond dissociation energies (BDE):

C—H	415 kJ
C—C	345 kJ
C＝C (double bond)	615 kJ
C—Cl	325 kJ
C—I	215 kJ
C—O	360 kJ
C＝O (double bond in most molecules)	750 kJ
C＝O (for each double bond specifically contained in carbon dioxide)	805 kJ
C≡O (triple bond)	1071 kJ
O—H	460 kJ
O＝O (double bond)	494 kJ
H—H	436 kJ
H—Cl	428 kJ
H—I	295 kJ
Cl—Cl	240 kJ

1. i. Calculate the value for $\Delta H_{(reaction)}$ for each of the following balanced equations using the BDE given.
   ii. For **each** reaction, answer the following parts:
      (1) Is the reaction exothermic or endothermic?
      (2) Which set of molecules, reactants, or products, represent the greater **total** (i.e., all bonds in the molecules) bond strengths?
      (3) **Based only on the energy** criterion, which reaction, forward or reverse, is spontaneous?
   iii. Draw a **general** PE diagram using the available information.

   a. $CH_4 + Cl_2 \longrightarrow CH_3Cl + HCl$

   b. $CH_4 + HCl \longrightarrow CH_3Cl + H_2$

   c. $C_3H_8 + 5 O_2 \longrightarrow 3 CO_2 + 4 H_2O$

   d. $C_2H_6 \longrightarrow C_2H_4 + H_2$

   e. $CH_3I + H_2O \longrightarrow CH_4O + HI$

2. Use the hypothetical energy diagram shown to answer the following questions:
   a. What is the **forward** reaction (i.e., the reaction formula equation for the **forward** reaction as shown in the diagram)?

b. What is the value for ΔH, the PE change for this **forward** reaction?

c. What is the value for activation energy (activation barrier) for this **forward** reaction?

d. What is the **reverse** reaction (i.e., the reaction formula equation for the **reverse** reaction as shown in the diagram)?

e. What is the value for ΔH, the PE change for this **reverse** reaction?

f. What is the value for activation energy (activation barrier) for this **reverse** reaction?

g. Which reaction is endothermic?

h. Which reaction is spontaneous based on the energy criterion?

i. Which set of molecules, reactants, or products, represent the greater **total** (i.e., all bonds in the molecules) bond strengths?

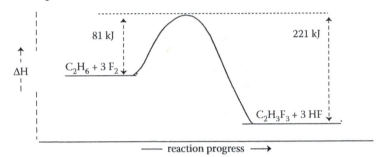

3. Consider the following reactions or physical processes. Predict the direction of entropy change for the forward process (left-to-right direction as written) (i.e., determine whether the sign of the entropy is positive [entropy increase] or negative [entropy decrease]). State whether the process as written is spontaneous or non-spontaneous **based only** on **the entropy** criterion.

a. $C_6H_{6\,(l)} + 3\,H_{2\,(g)} \longrightarrow C_6H_{12\,(l)}$

b. $CH_3OH_{\,(l)} \xrightarrow{H_2O\ solvent} CH_3OH_{\,(aq)}$

c. $C_4H_6O_{3\,(l)} \longrightarrow C_3H_6O_{\,(l)} + CO_{2\,(g)}$

d. $CH_3CH_2OH_{\,(aq)}$ at 1.0 M $\xrightarrow{\text{distillation (solution concentration)}} CH_3CH_2OH_{\,(aq)}$ at 5.0 M

4. Complete parts (i)–(iv) for **each** of the following potential reversible reactions:

i. **Without calculation,** predict the sign of the entropy change for the forward reaction.

ii. Based on the **signs** of ΔH° and ΔS° given, determine, **without calculation**, the effect of temperature on the spontaneity of the forward reaction; select from the options; the **reaction is** always spontaneous; never spontaneous; spontaneous at sufficiently low temperatures; and spontaneous at sufficiently high temperatures.

iii. Use the values for $\Delta H°$ and $\Delta S°$ given to calculate the numerical values for standard free energy, $\Delta G°_T$ at the two temperatures provided for each reaction; from the calculation, state which direction is spontaneous at each of these temperatures. Does the trend in the numbers help to confirm your answer for (ii)?

iv. Calculate the numerical value for $K_{(forward)}$ for each temperature provided.

    a. $N_2H_{4\,(l)} + 2\,H_2O_{2\,(l)} \longrightarrow N_{2\,(g)} + 4\,H_2O_{(g)}$

    $\Delta H°_{(reaction)} = -642.2$ kJ/mole-reaction; $\Delta S°_{(reaction)} = +0.6059$ kJ/K
    Calculate $\Delta G°_T$ and $K_{(forward)}$ at 200 and 1500 K.

    b. $12\,NH_{3\,(g)} + 21\,O_{2\,(g)} \longrightarrow 8\,HNO_{3\,(g)} + 4\,NO_{(g)} + 12\,H_2O_{(l)}$

    $\Delta H°_{(reaction)} = -3597.12$ kJ/mole-reaction; $\Delta S°_{(reaction)} = -2.804$ kJ/K
    Calculate $\Delta G°_T$ and $K_{(foward)}$ at 1000 and 1500 K.

    c. $HCO_2H_{(l)} \longrightarrow CO_{2\,(g)} + H_{2\,(g)}$

    $\Delta H°_{(reaction)} = +31.2$ kJ/mole-reaction; $\Delta S°_{(reaction)} = +0.2154$ kJ/K
    Calculate $\Delta G°_T$ and $K_{(forward)}$ at 100 and 500 K.

5. Complete parts (i) through (iv) for the following potential reversible reactions:

i. Use the table of BDE to calculate ΔH for the **forward** reaction as shown.

ii. Without calculation, predict the sign of entropy change for the forward reaction.

iii. Use the sign of ΔH plus the predicted entropy change to decide what can be determined about the spontaneity of the forward reaction; select from the same options as part (ii) from the previous problem.

iv. Using your answer from (a), show the equilibrium reaction equation to include **heat** (energy) as a reactant or product.

v. State which **direction** the equilibrium will **shift** for **each** of the following induced changes: (1) Hydrogen gas is removed; (2) Carbon monoxide is added; (3) Temperature is increased; and (4) Temperature is decreased.

Lewis structure for CO contains a triple bond: $:C≡O:^+$

    a. $CO_{(g)} + H_2O_{(l)} \longrightarrow CO_{2\,(g)} + H_{2\,(g)}$

    b. $CO_{(g)} + H_{2\,(g)} \longrightarrow CH_2O_{(g)}$ (formaldehyde)

    c. $CO_{(g)} + 2\,H_{2\,(g)} \longrightarrow CH_3OH_{(l)}$ (methanol)

ANSWERS TO PRACTICE PROBLEMS

1. i. **All answers are based on the process using step (2a);** lone pairs are not shown in Lewis structures.

a.

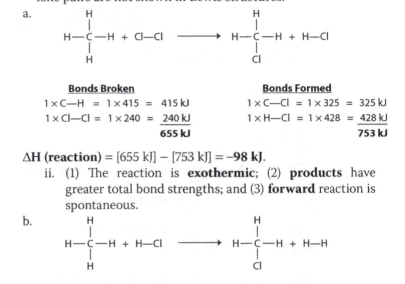

Bonds Broken	Bonds Formed
1 × C—H = 1 × 415 = 415 kJ	1 × C—Cl = 1 × 325 = 325 kJ
1 × Cl—Cl = 1 × 240 = 240 kJ	1 × H—Cl = 1 × 428 = 428 kJ
655 kJ	**753 kJ**

ΔH (reaction) = [655 kJ] – [753 kJ] = **–98 kJ**.

ii. (1) The reaction is **exothermic**; (2) **products** have greater total bond strengths; and (3) **forward** reaction is spontaneous.

b.

H—C—H + H—Cl ⟶ H—C—H + H—H
(with H on top and bottom of left C; Cl on bottom of right C)

Bonds Broken	Bonds Formed
1 × C—H = 1 × 415 = 415 kJ	1 × C—Cl = 1 × 325 = 325 kJ
1 × H—Cl = 1 × 428 = 428 kJ	1 × H—H = 1 × 436 = 428 kJ
843 kJ	**761 kJ**

ΔH (reaction) = [843 kJ] – [761 kJ] = **+82 kJ**

ii. (1) Reaction is **endothermic**; (2) **reactants** have greater total bond strengths; and (3) **reverse** reaction is spontaneous.

c.

H—C—C—C—H + 5 O=O ⟶ 3 O=C=O + 4 H—O—H
(propane structure with H on top and bottom of each C)

Bonds Broken	Bonds Formed
8 × C—H = 8 × 415 = 3320 kJ	6 × C=O = 6 × 805 = 4830 kJ
2 × C—C = 2 × 345 = 690 kJ	8 × O—H = 8 × 460 = 3680 kJ
5 × O=O = 5 × 494 = 2470 kJ	**8510 kJ**
6480 kJ	

ΔH (reaction) = [6480 kJ] – [8510 kJ] = **–2030 kJ**

b. (1) Reaction is **exothermic**; (2) **products** have greater total bond strengths; and (3) **forward** reaction is spontaneous.

d.

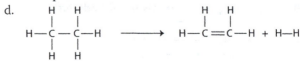

Bonds Broken	**Bonds Formed**
$2 \times C—H = 2 \times 415 = 830\,kJ$	$1 \times C{=}C = 1 \times 615 = 615\,kJ$
$1 \times C—C = 1 \times 345 = \underline{345\,kJ}$	$1 \times H—H = 1 \times 436 = \underline{436\,kJ}$
1175 kJ	**1051 kJ**

$$\Delta H\,(\textbf{reaction}) = [1175\,kJ] - [1051\,kJ] = \textbf{+124 kJ}$$

ii. (1) Reaction is **endothermic**; (2) **reactants** have greater total bond strengths; and (3) **reverse** reaction is spontaneous.

e.

$$\underset{\underset{H}{|}}{\overset{\overset{H}{|}}{H—C}}—I \ + \ H—O—H \ \longrightarrow \ \underset{\underset{H}{|}}{\overset{\overset{H}{|}}{H—C}}—O—H \ + \ H—I$$

Bonds Broken	**Bonds Formed**
$1 \times C—I = 1 \times 215 = 215\,kJ$	$1 \times C—O = 1 \times 360 = 360\,kJ$
$1 \times O—H = 1 \times 460 = \underline{460\,kJ}$	$1 \times H—I = 1 \times 295 = \underline{295\,kJ}$
675 kJ	**655 kJ**

$$\Delta H(\textbf{reaction}) = [675\,kJ] - [655\,kJ] = \textbf{+20 kJ}$$

ii. (1) Reaction is **endothermic**; (2) **reactants** have greater total bond strengths; and (3) **reverse** reaction is spontaneous.

1. iii. **Energy diagrams are general and <u>not to scale</u>.**

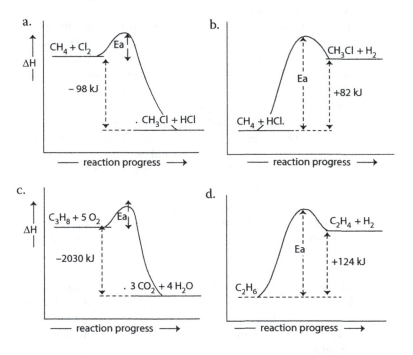

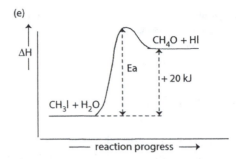

(e)

2. a. **Forward** reaction: read left-to-right =
$$C_2H_6 + 3F_2 \longrightarrow C_2H_3F_3 + 3HF$$
   b. ΔH for **forward** reaction: the difference between the two energy platforms can be calculated as $81 - 221$ kJ $= -140$ kJ; the PE decreases: $\Delta H = \mathbf{-140\ kJ}$.
   c. Activation energy (activation barrier) for **forward** reaction: complete height of the barrier is read from left-to-right: **Ea = 81 kJ**.
   d. **Reverse** reaction:
   read right-to-left = $C_2H_3F_3 + 3HF \longrightarrow C_2H_6 + 3F_2$
   e. ΔH for **reverse** reaction: energy **difference** must be the same value as for the forward reaction; the PE increases: $\Delta H = \mathbf{+140\ kJ}$.
   f. Activation energy (activation barrier) for **reverse** reaction: complete height of the barrier is read from right-to-left: **Ea = 221 kJ**.
   g. Endothermic reaction is the one which has a **positive** ΔH = the **reverse reaction**.
   h. The spontaneous direction based on the energy criterion is the one which shows a **negative** ΔH = the **forward reaction**.
   i. The lower PE represents the stronger bonds. The products are at the lower PE position; therefore, the products ($C_2H_3F_3 + 3$ HF) have the greater total bond strengths.

3. a. $C_6H_{6\ (l)} + 3\,H_{2\ (g)} \longrightarrow C_6H_{12\ (l)}$
   Entropy change is negative (decreases): greater number of smaller, simpler molecules produces fewer number of larger, more complex molecules; the total number of gas molecules decreases. Based on entropy, the reaction as written is **non-spontaneous**.
   b. $CH_3OH_{\ (l)} \xrightarrow{H_2O\ solvent} CH_3OH_{\ (aq)}$
   Entropy change is positive (increases): formation of a solution (mixture) from pure components. Based on entropy, the reaction as written is **spontaneous**.
   c. $C_4H_6O_{3\ (l)} \longrightarrow C_3H_6O_{\ (l)} + CO_{2\ (g)}$
   Entropy change is positive (increases): fewer number of larger, more complex molecules produces greater number of smaller, simpler molecules; the total number of gas molecules increases. Based on entropy, the reaction as written is **spontaneous**.

d. $CH_3CH_2OH_{(aq)}$ at 1.0 M $\longrightarrow$ $CH_3CH_2OH_{(aq)}$ at 5.0 M

Entropy change is negative (decreases): the solution becomes more concentrated: reverse of dilution. Based on entropy, the reaction as written is **non-spontaneous**.

4. a. $N_2H_{4\,(l)} + 2\,H_2O_{2\,(l)} \longrightarrow N_{2\,(g)} + 4\,H_2O_{(g)}$

  i. The balanced equation shows **3 moles of liquid reactants** breaking down to form **5 moles of gas products**. The standard entropy change is predicted to be **positive**: fewer number of larger reactant molecules break down to form a greater number of product molecules; the system becomes less ordered (more disordered).

  ii. Enthalpy is negative: contributes to spontaneity; entropy is positive: contributes to spontaneity. Both enthalpy and entropy show the same trend toward spontaneous: the **reaction is always spontaneous**.

  iii. $\Delta G°_T = \Delta H° - T\,\Delta S°$

$$\Delta G°_{(200)} = -642.2\text{ kJ} - [(200\text{ K})(0.6059\text{ kJ/K})]$$
$$= -642.2\text{ kJ} - 121.18\text{ kJ} = -763.38\text{ kJ/mole}$$

Forward reaction is spontaneous at 200 K.

$$\Delta G°_{(1500)} = -642.2\text{ kJ} - [(1500\text{ K})(0.6059\text{ kJ/K})]$$
$$= -642.2\text{ kJ} - 908.85 = -1551.05\text{ kJ/mole}$$

Forward reaction is spontaneous at 1500 K.

  iv. $\log_{10} K = \Delta G°_T / -2.303\ RT$
$\log_{10} K_{200} = (-763.38\text{ kJ})/(-2.303)(0.008314\text{ kJ/mole-K})(200\text{ K})$
$\log_{10} K_{200} = 199.35;\ K = \text{antilog}(199.35) = 10^{199.35} = 2.24 \times 10^{199}$
$\log_{10} K_{1500} = (-1551.05\text{ kJ})/(-2.303)(0.008314\text{ kJ/mole-K})(1500\text{ K})$
$\log_{10} K_{1500} = 54.00;\ K = \text{antilog}(54.00) = 10^{54} = 1.00 \times 10^{54}$

4. b. $12\,NH_{3\,(g)} + 21\,O_{2\,(g)} \longrightarrow 8\,NHO_{3\,(g)} + 4\,NO_{(g)} + 12\,H_2O_{(l)}$

  i. The balanced equation shows **33 moles of gas reactants** combining to form **24 moles total (12 gas) products**. The standard entropy change is predicted to be **negative**: greater number of smaller molecules combine to form a fewer number of product molecules; the system becomes more ordered (less disordered).

  ii. Enthalpy is negative: contributes to spontaneity; entropy is negative: contributes to **non**-spontaneity. Enthalpy and entropy trends oppose each other. **Low** temperature favors the direction indicated by the enthalpy: **reaction is spontaneous at sufficiently low temperatures**.

  iii. $\Delta G°_T = \Delta H° - T\,\Delta S°$

$$\Delta G°_{(1000)} = -3597.12\text{ kJ} - [(1000\text{ K})(-2.804\text{ kJ/K})]$$
$$= \mathbf{-3597.12\text{ kJ} + 2804.0\text{ kJ}} = -793.12\text{ kJ/mole}$$

Forward reaction is spontaneous at 1000 K.

$$\Delta G^{\circ}{}_{(1500)} = -3597.12\,kJ - [(1500\,K)(-2.804\,kJ/K)]$$
$$= -3597.12\,kJ + 4206.0\,kJ = +608.88\,kJ/mole$$

Forward reaction is non-spontaneous at 1500 K.

$\log_{10} K = \Delta G^{\circ}{}_{T}/-2.303\ RT$

$\log_{10} K_{1000} = (-793.12\ kJ)/(-2.303)(0.008314\ kJ/mole\text{-}K)(1000\ K)$

$\log_{10} K_{1000} = 41.42;\ K = \text{antilog}(41.42) = 10^{41.42} = 2.63 \times 10^{41}$

$\log_{10} K_{1500} = (+608.88\ kJ)/(-2.303)(0.008314\ kJ/mole\text{-}K)(1500\ K)$

$\log_{10} K_{1500} = -21.2;\ K = \text{antilog}(-21.2) = 10^{-21.2} = 6.31 \times 10^{-22}$

4. c. $HCO_2H_{(l)} \longrightarrow CO_{2\,(g)} + H_{2\,(g)}$

   i.   The balanced equation shows **1 mole of liquid reactants** breaking down to form **2 moles of gas products**. The standard entropy change is predicted to be **<u>positive</u>**: fewer number of larger reactant molecules break down to form a greater number of product molecules; the system becomes less ordered (more disordered).

   ii. Enthalpy is positive: contributes to **non**-spontaneity; entropy is positive: contributes to spontaneity. Enthalpy and entropy trends oppose each other. **<u>High</u>** temperature favors the direction indicated by the entropy: **reaction is spontaneous at sufficiently high temperatures**.

   iii. $\Delta G^{\circ}{}_{T} = \Delta H^{\circ} - T\,\Delta S^{\circ}$

   $$\Delta G^{\circ}{}_{(100)} = +31.2\,kJ - [(100\,K)(+0.2154\,kJ\,/\,K)]$$
   $$= +31.2\,kJ - 43.0\,kJ = +9.7\,kJ\,/\,mole$$

   Forward reaction is non-spontaneous at 100 K.

   $$DG^{\circ}{}_{(500)} = +31.2\,kJ - [(500\,K)(+0.2154\,kJ/K)]$$
   $$= +31.2\,kJ - 107.7\ kJ = -76.5\,kJ/mole$$

   Forward reaction is spontaneous at 500 K.

   $\log_{10} K = \Delta G^{\circ}{}_{T}/-2.303\ RT$

   $\log_{10} K_{100} = (+9.7\ kJ)/(-2.303)(0.008314\ kJ/mole\text{-}K)(200\ K)$

   $\log_{10} K_{100} = -5.07;\ K = \text{antilog}(-5.07) = 10^{-5.07} = 8.51 \times 10^{-6}$

   $\log_{10} K_{500} = (-76.5\ kJ)/(-2.303)(0.008314\ kJ/mole\text{-}K)(500\ K)$

   $\log_{10} K_{500} = 7.99;\ K = \text{antilog}(7.99) = 10^{7.99} = 9.77 \times 10^{7}$

5. a. $CO_{(g)} + H_2O_{(l)} \longrightarrow CO_{2\,(g)} + H_{2\,(g)}$

   i. $\Delta H_{(reaction)} = [\text{Sum } n_i\ BDE_i\ (\text{bonds broken})] - [\text{Sum } n_i\ BDE_i\ (\text{bonds formed})]$

   $\Delta H = [1 \times (BDE\ C{\equiv}O) + 2 \times (BDE\ O{-}H)] - [2 \times (BDE\ C{=}O) + 1 \times (BDE\ H{-}H)]$

   $\Delta H = [1 \times (1071\ kJ) + 2 \times (460\ kJ)] - [2 \times (805\ kJ) + 1 \times (436\ kJ)]$

   $= -55\ kJ/mol\text{-}rxn$

   ii. The balanced equation shows **1 mole of gas plus 1 mole of liquid reactants** forming **2 moles of gas products**. Owing to the difference between liquid and gas, the standard entropy change is predicted to be (slightly) **positive** (increasing): the system becomes less ordered (more disordered).

   iii. The enthalpy for the forward reaction is negative: contributes to spontaneity. The entropy for the forward reaction was predicted to be increasing: contributes to spontaneity. **The reaction is always spontaneous.**

iv. ΔH is negative; heat is a product in the forward reactions

$$CO_{(g)} + H_2O_{(l)} \;\longrightarrow\; CO_{2\,(g)} + H_{2\,(g)} + heat$$

v.   Changes: (1) Hydrogen gas is removed. Hydrogen is a product; the **forward** reaction must increase its rate to replace some of the hydrogen molecules which were removed: the reaction **shifts to the right** (or **shifts toward products**).

(2) Carbon monoxide is added. Carbon monoxide is a reactant; the **forward** reaction must increase its rate to remove some of the extra carbon monoxide molecules which were added: the reaction **shifts to the right** (or **shifts toward products**).

(3) Temperature is increased. Heat is a product; the **reverse** reaction must increase its rate to remove some of the extra heat which was added: the reaction **shifts to the left** (or **shifts toward reactants**) (i.e., the reaction shifts in the direction of the ΔH positive).

(4) Temperature is decreased. Heat is a product; the **forward** reaction must increase its rate to replace some of the heat which was removed: the reaction **shifts to the right** (or **shifts toward products**) (i.e., the reaction shifts in the direction of the ΔH negative).

5. b. $CO_{(g)} + H_{2\,(g)} \;\longrightarrow\; CH_2O_{(g)}$ (formaldehyde)

i. $\Delta H_{(reaction)} = [\text{Sum } n_i\, BDE_i \text{ (bonds broken)}] - [\text{Sum } n_i\, BDE_i \text{ (bonds formed)}]$

$\Delta H = [1 \times (BDE\ C\equiv O) + 1 \times (BDE\ H\!-\!H)] - [1 \times (BDE\ C\!=\!O + 2 \times (BDE\ C\!-\!H)]$

$\Delta H = [1 \times (1071\ kJ) + 1 \times (436\ kJ)] - [1 \times (750\ kJ) + 2 \times (415 \quad kJ)]$
$= \mathbf{-73\ kJ/mol\text{-}rxn}$

ii.   The balanced equation shows **2 moles of gas reactants** combining to form **1 mole of gas product**. The standard entropy change is predicted to be **negative** (decreasing): greater number of smaller molecules combine to form a fewer number of product molecules; the system becomes more ordered (less disordered).

iii.   Enthalpy is negative: contributes to spontaneity; entropy is negative: contributes to **non**-spontaneity. Enthalpy and entropy trends oppose each other. **Low** temperature favors the direction indicated by the enthalpy: **reaction is spontaneous at sufficiently low temperatures.**

iv ΔH is negative: heat is a product in the forward reaction

$$CO_{(g)} + H_{2\,(g)} \;\longrightarrow\; CH_2O_{(g)} + heat$$

v.   Changes: (1) Hydrogen gas is removed. Hydrogen is a reactant; the **reverse** reaction must increase its rate to replace some of the hydrogen molecules which were removed: the reaction **shifts to the left** (or **shifts toward reactants**).

(2) Carbon monoxide is added. Carbon monoxide is a reactant; the **forward** reaction must increase its rate to remove some of the extra carbon monoxide molecules which were added: the reaction **shifts to the right** (or **shifts toward products**).

(3) Temperature is increased. Heat is a product; the **reverse** reaction must increase its rate to remove some of the extra heat which was added: the reaction **shifts to the left** (or **shifts toward reactants**) (i.e., the reaction shifts in the direction of the ΔH positive).

(4) Temperature is decreased. Heat is a product; the **forward** reaction must increase its rate to replace some of the heat which was removed: the reaction **shifts to the right** (or **shifts toward products**) (i.e., the reaction shifts in the direction of the ΔH negative).

5. c. $CO_{(g)} + 2\,H_{2\,(g)} \longrightleftharpoons CH_3OH_{(l)}$ (methanol)

   i. $\Delta H_{(reaction)} = [\text{Sum } n_i\,BDE_i\,(\text{bonds broken})] - [\text{Sum } n_i\,BDE_i\,(\text{bonds formed})]$

   $\Delta H = [1 \times (BDE\ C\equiv O) + 2 \times (BDE\ H\!-\!H)]$

   $\quad -[1 \times (BDE\ C\!-\!O) + 3 \times (BDE\ C\!-\!H) + 1 \times (BDE\ O\!-\!H)]$

   $\Delta H = [1 \times (1071\ kJ) + 2 \times (436\ kJ)] - [1 \times (360\ kJ) + 3 \times (415\ kJ)$

   $\quad + 1 \times (460\ kJ)]$

   $= \mathbf{-122\ kJ/mol\text{-}rxn}$

   ii. The balanced equation shows **3 moles of gas reactants** combining to form **1 mole of liquid product**. The standard entropy change is predicted to be **negative** (decreasing): greater number of smaller molecules combine to form a fewer number of product molecules; the system becomes more ordered (less disordered).

   iii. Enthalpy is negative: contributes to spontaneity; entropy is negative: contributes to **non**-spontaneity. Enthalpy and entropy trends oppose each other. <u>Low</u> temperature favors the direction indicated by the enthalpy: **reaction is spontaneous at sufficiently low temperatures**.

   iv. ΔH is negative; heat is a product in the forward reaction

   $CO_{(g)} + 2\,H_{2\,(g)} \longrightleftharpoons CH_3OH_{(l)} + heat$

   v. Changes: The answers to changes (1), (2), (3), and (4) are identical to the reaction shown for part (b).

Guide to Kinetics and Reaction Mechanisms

9

9.1 GENERAL CONCEPTS

9.1.1 REACTION MECHANISMS

1. A **reaction mechanism** is an accepted sequence of elementary reaction steps which describe all (based on available information) bond-making and bond-breaking events characterizing the change of reactant molecules to product molecules.

2. A **reaction step (or elementary step)** is the smallest observable change in molecular bonding, an individual **bond-making**, **bond-breaking**, or **combination** event (simultaneous bond-making and bond-breaking) that can be distinguished experimentally from other such events.

3. The complete reaction mechanism may be composed of only **one** step or **many** steps depending on the overall (complete) reaction and the conditions.

4. A reaction mechanism, along with the parameters that describe it such as rate, activation energies, and intermediates (described in other sections) is a **path function**. *A path function is dependent on the "pathway" or method by which a change occurs*. Regardless of the numerical value or sign of the free energy change, a reaction can occur only if there exists an available pathway by which reactant molecules can be converted into product molecules.

5. Path functions must be distinguished from **state functions** such as ΔG, ΔH, and ΔS. These depend **only** on the initial and final states of the system: the total energies of the reactants versus the products. *State functions do **not** depend on how the reaction changes occur.*

6. All reactants and products in a complete reaction or in a single reaction step **must** exist as an independent species for some measurable amount of time. This existence is due to the presence of energy barriers blocking "instant" decomposition. The compound is considered to be in a "potential energy well" (i.e., a stable energy "valley" similar to a rock sitting in a hole) termed a **local energy minimum**. A "deep" hole represents a very stable molecule (slow to react) because

the energy barriers on each side are high. A "shallow" hole represents a **relatively** unstable molecule (faster reacting) because the energy barriers to reaction are low.

7. A **reaction intermediate** is the product of an individual reaction step that is later consumed (used as a reactant) in a subsequent step. Since it is an actual product of one reaction step and an actual reactant in a following step, an intermediate is a detectable independent species at a **local energy minimum**.

8. A **transition state** is a description of atom arrangements showing partial bonds formed or broken for the required molecular changes involved in a reaction step.

   a. The partial bonds indicate **how** the reactant atoms are rearranging to form the correct product molecules.

   b. The transition state specifically shows bonding changes at the highest energy point, **potential energy maximum**, of the reaction step. Atom bonding arrangements at the energy maximum transition state **cannot** represent an intermediate and cannot be isolated as an individual molecule.

9.1.2 GENERAL CONCEPT OF KINETICS

1. **Kinetics** is the study of rates of complete reactions by mathematical and experimental analysis to determine reaction mechanisms.

2. **Rates** of reaction are measured by quantifying **concentration changes**, either the disappearance of a reactant or the appearance of a product as a function of time (**t**):

$$\text{rate(r)} = \frac{-\Delta[\text{reactant}]}{\Delta t} \quad \text{or} \quad \text{rate(r)} = \frac{+\Delta[\text{product}]}{\Delta t}$$

3. Rates of reactions are generally proportional to the concentration of one or more of the reactants raised to a specific power (exponent). This relationship is termed the **rate expression** or rate law:

$$\text{rate(r)} = k[\text{reactant 1}]^x [\text{reactant 2}]^{y\cdots}$$

   a. The constant in the expression is called the **rate constant, k**; the numerical values of the exponents x, y, etc. are called the **reactant orders**. For example, if $x = 2$ in the above general expression, the reaction is "second-order in reactant 1."

   b. The **sum** of all the **exponents** x + y ... is called the **reaction order**. For example, if $x = 2$ and $y = 1$ in the above general expression, the reaction would be described as a "third-order reaction."

   c. **The values of k, x, y ... depend on the mechanistic steps.**

9.1.3 ADDITIONAL VARIABLES AFFECTING REACTION RATES

1. Rates are **inversely** proportional to the energy barriers that prevent a specific reaction step from occurring; this barrier is called the **activation energy**, or **activation barrier** for this step (general symbol: **Ea**).
   a. The activation energy, **Ea**, represents the energy that must be added to reactants to allow bonding changes to occur.
   b. A reaction rate will increase as the activation energy (barrier) decreases. Ea is a path function and does not depend directly on ΔG for the complete reaction. The specific Ea, most important for a complete reaction containing many steps, depends on the mechanism.
2. Rates will be shown to be directly proportional to **temperature**: *reaction **rates increase** as **temperature increases**.*
3. Reaction rates increase with the presence of a **catalyst**.

9.1.4 INFORMATION RELATIONSHIPS FOR KINETICS AND MECHANISMS

1. Kinetic experiments analyze specific functions of reactant concentration (or product) versus time: $f\{[A]\}$ **versus t**; **t** = time; **[A]** = concentration of **any** reactant (or product) molecule; $f\{\ \}$ can be certain functions.
2. The experimental information (if complete) yields the **rate expression**:

$$\text{rate} = k[\text{reactant 1}]^x[\text{reactant 2}]^y$$

3. The complete rate expression with reactant exponents, value of k (and other information) can result in the determination of an **acceptable** description of the reaction steps (mechanism). A mechanism cannot be **proven**, but can be shown to provide a correct description of the experimental facts.
4. Methods for relating information in kinetics:
   a. Experimental $f\{[A]\}$ versus time produces the rate expression, which produces a logical mechanism.
   b. A specific suggested **potential** mechanism can be used to predict the rate expression. The predicted rate expression can then be used in turn to predict the expected results of the experimental rate analysis. This is the reverse of the analysis in a.
   c. Both approaches can be combined: *A valuable technique for testing the validity of a mechanism is to compare the predicted rate expression derived from the suggested mechanism to the experimentally derived rate expression from kinetic analysis.* **The process to apply this technique will be emphasized**.

9.2 DESCRIPTION OF REACTIONS BY MECHANISMS

9.2.1 REACTION STEPS AND COMPLETE REACTIONS

1. Complete reactions *indicated by a balanced equation* can be composed of one or more **elementary reaction steps**. When **all** bonds are broken and/or made simultaneously (at least to the experimental ability to distinguish) then the complete reaction mechanism is **one step**. **A one-step reaction can have one bond making, one bond breaking, or any multiple combination if the events cannot be distinguished. Examples:**

$$N_2O_4 \longrightarrow 2NO_2 \quad \textbf{(N—N bond broken)}$$

$$O_2 \longrightarrow 2O \quad \textbf{(O=O bond broken)}$$

$$Br + Br \longrightarrow Br_2 \quad \textbf{(Br—Br bond formed)}$$

In the following example, all of the bond changes occur simultaneously or the events cannot be distinguished; it is thus considered one step:

$$CH_3Br + Cl \longrightarrow CH_3Cl + Br \quad \textbf{(C—Cl formed; C—Br broken)}$$

2. Multistep **reactions** must **involve**:
   a. **More than one** bonding change during the complete reaction, and
   b. Have bond-making and bond-breaking events which are **not simultaneous** and **can be distinguished** from each other.
   Example: More than one step is detected for the following reaction:

$$CH_4 + Cl_2 \longrightarrow CH_3Cl + HCl \text{ (balanced equation of complete reaction)}$$

Distinguishable events:

Step 1 $Cl_2 \longrightarrow 2Cl$ **Cl—Cl bond is broken**

Step 2 $Cl + CH_4 \longrightarrow CH_3 + HCl$ **C—H broken, H—Cl formed**

Step 3 $CH_3 + Cl \longrightarrow CH_3Cl$ **C—Cl bond formed**

Cl and CH_3 are **intermediates** in this mechanism, generated in a previous step and consumed in a subsequent step. They are actually detected by instruments and analytical reactions.

9.2.2 VARIABILITY OF MECHANISMS

1. Except for the simplest cases, *a mechanism cannot be determined directly from the balanced equation*. A specific reaction given by a balanced equation can potentially proceed by a number of different paths (mechanisms).

2. **General example**:

A—B + C ⟶ A—C + B **(balanced equation of complete reaction)**

a. **Mechanism #1** to accomplish this complete reaction, **one step**:

Only step A—B + C ⟶ A—C + B

Since the complete mechanism is **only one step,** it must be depicted as equivalent to the balanced equation. The **transition state** of this **one** step would have to show both the **A—B** bond breaking and the **A—C** bond forming simultaneously, for example, as: **(pictorial model)**

C------------A------------B

(forming) (breaking)

b. **Mechanism #2** to accomplish this complete reaction, **two steps**:

Step 1   A—B   ⟶   A + B   **(A—B bond breaks)**

Step 2   A + C   ⟶   A—C   **(A—C bond forms)**

_____

Add Steps: A—B + C ⟶ A—C + B

Steps 1 + 2 **must** add
to equals balanced equation

"A" is an **intermediate**: produced in Step 1, consumed in Step 2. Addition of all molecules in Steps 1 and 2, after cancellation of the intermediate "**A**," **must** equal the balanced equation.

c. **Mechanism #3** to accomplish this complete reaction, an **alternative two-step**:

Step 1   A—B + C   ⟶   C—A—B   **(A—C bond forms)**

Step 2   C—A—B   ⟶   C—A + B   **(A—B bond breaks)**

_____

Add Steps: A—B + C   ⟶   C—A + B

Steps 1 and 2 **must** add to
equal the balanced equation

"C—A—B" is an **intermediate** and is cancelled from both sides.

9.3 REACTANT CONCENTRATIONS AND EXPERIMENTAL KINETICS: DETERMINING REACTANT ORDERS AND RATE CONSTANTS

9.3.1 EXPERIMENTAL CONCEPTS

1. The **rate expression** describes the mathematical relationship between rate of the complete reaction as a function of **all possible** reactant concentrations.

$$\text{rate}(r) = k[\text{reactant 1}]^X[\text{reactant 2}]^Y$$

    a. This relationship depends on the type and sequence of reaction steps: the **mechanism**. Since the balanced equation does not reveal the mechanism, *it is also true that the rate expression cannot be determined directly from the balanced equation.*

    b. The rate expression, however, can be predicted for any reaction if the mechanism is known or suggested; this technique is described in **Section 9.4**.

2. The dependence of the reaction rate on the concentration of each reactant can be found experimentally. *The data is generated by measuring changes in reactant concentration as a function of time.*

    a. **Reaction rates are rates of concentration changes** (concentration changes per unit time). *Changes in a concentration of any specific reactant may or may not affect how fast the reaction occurs.*

    b. Mathematical analysis of **f{concentrations}** versus **time** provides the exponent (order) for each reactant (the order may be zero, equivalent to no relationship) and a calculation for the rate constant, k.

9.3.2 ZERO-ORDER REACTANTS

1. The simplest case, possible for multistep reactions, is a **rate** which is **independent** of a **specific reactant** concentration, symbolized by **[A]**. This does not mean that the reactant is unused, just that the specific molecule being analyzed has no role in the reaction steps that affect the rate of the overall (i.e., complete) reaction.

    a. The rate of the reaction is independent of **[A]**; the reaction rate expression **for this specific reactant** for any reaction would be

$$\textbf{rate(r)} = \textbf{k}[\textbf{A}]^0 ; \text{or } \textbf{rate} = \textbf{k} (\text{i.e.} = \textbf{rate constant} \times 1)$$

    (Any number raised to the zero power = 1.) The reaction is said to be **zero order** in this component.

    b. Using the general letter **A** for any specific reactant, the plot of **[A]** versus **t** (time) would be a straight line with a single (constant) slope. (**[A]** means **"concentration of A."**) This is shown in **Figure 1**.

2. Do not confuse rate with **concentration**. The **concentration** of "**A**" changes continuously with time, but the **rate of change** of the concentration remains the same (constant slope). The number of molecules of "**A**" consumed by reaction per unit time does not change as the concentration of "**A**" decreases.

3. *A kinetic analysis which yields a straight line plot of reactant concentration ([reactant]) versus time, specifically*

characterizes that reactant as being zero order in the overall rate expression.

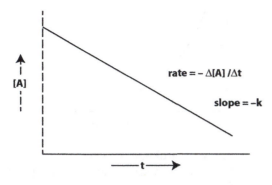

9.3.3 FIRST-ORDER REACTANTS

1. A more common situation occurs when the rate of the reaction changes as a direct function of the concentration of a specific reactant. In this case, the reaction is termed "**first order in component 'A'.**"

$$\textbf{rate(r)} = \textbf{k[A]}^1 \quad (\text{equivalent to } \textbf{rate} = \textbf{k[A]})$$

   a. In addition to chemical reactions, this general behavior is widely seen in examples such as exponential growth of population or exponential growth of invested money.
   b. Exponential decrease (exponential decay) can also occur, as is the case for a chemical reaction where "**A**" is a **reactant** being **consumed**. The relationships can be summed up as: *the more you have to start with, the faster the results (increase or decrease) multiply.*
2. To derive useful equations, consider a simple one-step reaction with "**A**" as the only reactant: (An example would be a rearrangement of bonding pattern using all the same atoms; termed an isomerization reaction.)

$$A \longrightarrow B$$

3. **Figure 2** depicts the change in **[A]** versus **time**. This is an exponential "decay" (decrease); the slope of the curve becomes continuously less negative as concentration of reactant "**A**" decreases (as the reaction proceeds).
   a. The plot shows that the **rate** at which molecules of "**A**" disappear (by reaction) **decreases** as **[A] decreases**.
   b. The molecular explanation is that the number of molecules of "**A**" which react **per unit time** decreases as the number of **available** molecules of "**A**" decrease through their conversion to molecules of "B."

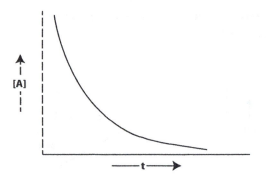

4. **To solve for a more useful equation:**
   a. Rate is measured by disappearance of "**A**" per unit time: **rate = −Δ[A]/Δt**.
   b. Rate is also equal to k[A] for a first-order reaction: **rate = k[A]**.
   c. Since both expressions give the rate, we can set them equal to each other: **−Δ[A]/Δt = k[A]**.
   d. Rearrange the variables: **−Δ[A]/[A] = kΔt**.
   e. The above equation is solved using calculus. This leads to the **first-order integrated rate equation**, which provides the function of **[A]** versus **time** that yields a **linear** equation:

$$\ln[A] = -kt + \ln[A_0]$$

   Compared to general linear equation: **y = mx + b**.
   f. This equation provides a straight-line graph shown in **Figure 3**. The **y-axis** is "**ln [A]**," the natural log (log to the base e) of the concentration of **A** at any time during the reaction; the **x-axis** is **time** (general units).
   g. Comparison to the general linear equation shows that the **slope** of the line is = **−k**, and the **y-intercept** is **ln [A_0]**. **[A_0]** in the equation is the initial concentration of "**A**" (i.e., at time = 0, before the reaction starts).
5. *A kinetic analysis which yields a straight-line plot of the natural log of reactant concentration (ln [reactant]) versus time, specifically characterizes that reactant as being first order in the overall rate expression.*

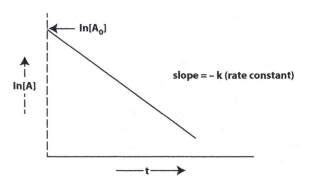

6. Do not confuse **change in [A]** (Δ**[A]**) with **percent change in [A]**. The rate equation measures the actual number of molecules (i.e., moles) of "**A**" which react as a function of time. In fact, it can be shown that the **percentage** of available "**A**" which reacts per unit time **remains constant**. The actual number of reacting "**A**" per time, however, decreases because the same percentage is multiplying successively smaller concentration numbers.

7. **Example for applying the integrated rate equation:**
   a. **Problem:** The "**half-life**" of a reactant is the time it takes for half (50%) of the reactant to be used up in a reaction. For any **starting concentration** of "**A**" ($[A_0]$), use the equation to determine the **half-life** of "**A**" if the rate constant, k, is found to be 0.0025/s. (Note: The "units" of rate constants are whatever is necessary to produce an answer with the correct units for the particular situation. In the case of a first-order equation, the units are inverse time [e.g., 1/s].)
   b. **Solution:**
      1. An alternative form of the first-order equation can be found by rearranging variables to isolate **t**: $\ln [A] - \ln [A_0] = -\mathbf{k}\mathbf{t}$.
      2. The subtraction of logs can be expressed as the log of a division (i.e., the ratio $\{[A]/[A_0]\}$: **ln** $\mathbf{([A]/[A_0])} = -\mathbf{kt})$.
      3. Use the fact that the half-life is defined as the time required for the concentration [A] to be reduced to half of the original amount ($[A_0]$); thus **[A] = (0.5) [A$_0$]**; the ratio $\{[A]/[A_0]\} = \{(0.5) [A_0]/[A_0]\} = \mathbf{0.5}$.
      4. Substitute the known rate constant and the value for the ratio $\{[A]/[A_0]\}$ into the integrated rate equation; **be careful of signs** (Units are left out for clarity.)

$$\mathbf{\ln(0.5) = -0.0025t}$$

      5. Complete the calculation: the value of ln (0.5) = −0.693; substitute (−**0.693**) into the equation for ln (0.5):

$$-0.693 = -0.0025\mathbf{t}; \quad \mathbf{t} = -0.693 / -0.0025 / s; \quad \mathbf{t = 277\,s}$$

   (A general expression is: t for half-life = 0.693/k)

8. The half-life of a first-order process is constant: decreasing the amount by 50% will always take the same amount of time (277 s in the example) regardless of the starting amount. A common example of this is radioactive decay.

9.3.4 SECOND-ORDER REACTANTS

1. The rate of reaction may also be proportional to the concentration of a reactant to the second power. The reaction is termed "**second-order in reactant A**":

$$\mathbf{rate = k[A]^2}$$

2. To derive more useful equations, consider the simplest example for analysis: a complete reaction which has only one step with only one reactant.

$$A + A \longrightarrow A_2$$

Examples: $\quad NO_2 + NO_2 \longrightarrow N_2O_4$

$$Br + Br \longrightarrow Br_2$$

3. **Figure 4** shows a plot of **[A]** versus **time** for this example. The number of "**A**" molecules decreases with reaction progress (time) and the **rate of change** in the concentration of "**A**" is not constant. As reactant "**A**" is used up, the **rate** of reaction, as measured by the slope of the line, continuously **decreases**.

4. The molecular reason for a dependence on the **square** of the concentration is based on the simple collision requirements of the reaction. A molecule of "**A**" must collide with another molecule of itself in order to achieve a bonding overlap.

   a. The rate of the reaction will be proportional to the number of collisions per unit time (frequency of collision). Although, the two molecules of **A** in the general equation are identical for this example, consider them individually. Doubling the concentration of the "**first**" **A** will double the frequency of collisions. Independently doubling the concentration of the "**second**" **A** will also double the frequency.

   b. Since **A** is actually the same molecule (or atom), generally doubling the concentration of "**A**" (**[A]**) will increase the frequency of collisions, hence the rate, by **four** times (statistically, **2 squared**).

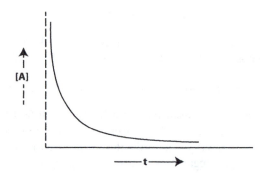

5. **To solve for the integrated rate equation,** set up the same relationships as for the first-order example:

   a. Rate is measured by disappearance of "**A**" per unit time: **rate** $= -\Delta[A]/\Delta t$.

   b. Rate is also equal to $k[A]^2$ for a second-order reaction: **rate** $= k[A]^2$.

   c. Since both expressions give the rate, we can set them equal to each other: $-\Delta[A]/\Delta t = k[A]^2$.

   d. Rearrange the variables: $-\Delta[A]/[A]^2 = k\Delta t$.

e. The equation is solved using calculus. This leads to the **second-order integrated rate equation**, which provides the function of **[A]** versus **time** that yields a **linear** equation:

$$1/[A] = kt + 1/[A_0]$$

Compared to general linear equation: $y = mx + b$.

f. This equation provides a straight-line graph shown in **Figure 5**. The **y-axis** is "**1/[A]**," an inverse function of the concentration of **A** at any time during the reaction; the **x-axis** is **time** (general units).

g. Comparison to the general linear equation shows that the **slope** of the line is $= + k$, and the **y-intercept** is $1/[A_0]$ (1/initial concentration of "**A**").

6. *A kinetic analysis which yields a straight-line plot of the inverse of reactant concentration (1/[reactant]) versus time, specifically characterizes that reactant as being second order in the overall rate expression.*

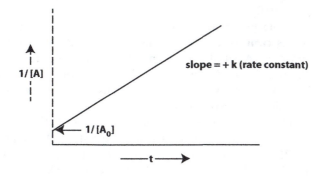

9.3.5 EXPERIMENTAL DETERMINATION OF REACTANT ORDERS IN MULTIPLE-CONCENTRATION RATE EXPRESSIONS

1. A complete reaction mechanism will most often be described by a rate expression that contains **more than one reactant** concentration.

a. The kinetic and mathematical analysis for rate functions with more than one concentration variable is complex. To solve, isolate each of the reactants in experiments, so that (usually) one of the three previous equations will apply to each reactant concentration (**[reactant]**) in turn. (The term for this is "**pseudo-order**" analysis.)

b. The experiment is carried out simply by keeping the concentrations of all reactants nearly constant, **except** the one being studied. The design usually requires a reaction in which the concentration of the studied reactant is **much smaller** than all the others.

c. The studied concentration will undergo a relatively **large percentage change** during the reaction while the others will change by only a very small percentage, that is, the others will stay **nearly constant**.

d. *The zero-order, first-order, and second-order integrated rate equations, plus the technique of isolating variables, can be used to determine the rate expression for a complete reaction mechanism.*

2. **Example:** Consider a simple one-step, two-reactant reaction:

a. The **rate expression** for this one-step reaction may include a dependence term for **both A** and **B**; none of the integrated rate equations will provide a correct mathematical relationship if the concentrations of A and B are **both measurably changing**.

b. By design, let **[A]** = 0.1 M and **[B]** = 10 M. At the completion of the reaction, **[A]** will approach **zero** but **[B]** will still be **9.9** M, a total change of only about **1%**. A similar experiment could be performed where **[B]** = 0.1 M and **[A]** = 10.0 M.

c. Example **problem**: Assume experiments show that **ln [A]** versus **time** is a straight line whenever **[B]** is held nearly constant. A similar experiment, exchanging the roles of "**A**" and "**B**," indicates that **ln [B]** versus **time** is a straight line whenever **[A]** is held nearly constant. **Determine the complete rate expression**.

d. **Solution:** If ln **[A]** versus **t** is a straight line, the reactant order for "**A**" is 1 (i.e., **first order**). The straight-line plot of ln **[B]** versus **time** shows that the reactant order for "**B**" is also 1. Combine the information to give: rate = $k[A]^1[B]^1$; that is: rate = k[A][B] (second-order **overall**).

9.4 PREDICTING RATE EXPRESSIONS FROM REACTION MECHANISMS

9.4.1 GENERAL CONCEPTS FOR RATE EXPRESSION COMPARISON; IDENTIFICATION OF THE RATE-DETERMINING STEP

1. The validity of a hypothetical mechanism can be tested against experimental results. One technique to accomplish this:

a. *Predict what the rate expression should be based on the proposed mechanistic steps for the complete reaction.*

b. *Compare the predicted rate expression to the rate expression derived from the kinetic analysis (reactants and orders).* **The results must match in a valid mechanism**.

2. Reaction mechanisms may be composed of only a single step. However, in a complete multistep reaction, one of the steps is very often much slower than the rest of the steps. (This is the molecular equivalent of a "bottleneck" in a manufacturing plant.)

The slowest chemical step is termed the **rate-determining step** (abbreviated **r.d.s.**).

   a. The r.d.s. is identified through kinetic experiments. It will be specifically marked in a mechanism description as "**slow**" or "**r.d.s.**"

   b. The slowest step is usually the one with the highest energy barrier (**Ea**), and can therefore be identified on a **potential energy (level) diagram (Section 9.4).**

3. The rate of the complete reaction depends on the rate of the **r.d.s.** plus all steps occurring **before** the **r.d.s.**

   a. Previous steps affect the **overall rate** because these are the steps that produce the intermediates required for reaction in the slow (r.d.s.) step.

   b. By implication: the overall rate of the reaction is **independent** of steps occurring **after** the **r.d.s.**

4. In less complicated reactions, it is very often true that a fast step which precedes a slow step reaches **equilibrium**. This is because the product of the fast step (an intermediate) is used as a reactant in the slow step. Because it is used slowly, the concentration of this intermediate buildup and increases the rate of the **reverse** reaction in the preceding fast step. (Recall that equilibrium is achieved when the rate of the forward and reverse reactions are equivalent.)

5. *The overall rate depends on all reactants and intermediates which participate at or before the r.d.s.; thus, the overall rate expression depends on, and provides information about, the mechanism.*

9.4.2 PREDICTING RATE EXPRESSIONS FROM MECHANISMS: REACTION STEP RATE RULE AND ONE-STEP REACTIONS

1. The rate expression for a complete reaction can be predicted based on information about specific steps in the suggested reaction mechanism. **A complete (overall) rate expression** is derived from individual rate expressions for appropriate elementary steps.

   a. The key connecting concept is (arbitrary terminology) the **reaction step rate rule**:
The rate of any __elementary__ reaction step is proportional to the concentrations of each reactant species involved in that step raised to the power of the number of molecules of the specific reactant involved in that reaction step. This power will equal the reactant __coefficient__ in the __specific reaction step__ chemical equation.

   b. The rule applies **only** to reaction steps, **not** to general balanced equations. Rate expressions for multistep reactions **must** be derived from elementary step rate expressions and the additional rules given in Section 9.4.3.

2. If the complete reaction mechanism consists of **only one (elementary) step**, then the rate expression for the overall (complete)

reaction is identical to the rate expression for the single step that describes the mechanism. In this case, *the reaction step rate rule is used directly,* nothing else is required.

Examples for one-step reactions:

(a) **one-step reaction:** $A + A + C \longrightarrow A_2C$

equivalent to: $\qquad\qquad 2A + C \longrightarrow A_2C$

rate (step) = rate (overall) = $k[A]^2[C]$

(b) **one-step reaction:** $A + B + C \longrightarrow ABC$

rate (step) = rate (overall) = $k[A][B][C]$

(c) **one-step reaction:** $NO_2 + NO_2 \longrightarrow N_2O_4$

rate (step) = rate (overall) = $k[NO_2]^2$

9.4.3 PREDICTING RATE EXPRESSIONS FROM MECHANISMS: MULTIPLE-STEP REACTION MECHANISMS

1. Rate expressions for multiple-step mechanisms can**not** be predicted directly from the **reaction step rate rule: the rule applies only to reaction steps, not to general balanced equations**.
   a. **Example:** Consider the balanced equation for the synthesis of ammonia:

   $$N_2 + 3H_2 \longrightarrow 2NH_3$$

   b. Experiments have determined that this is a **multistep** reaction; this could **not** be inferred simply from the equation.
   c. **A multistep reaction implies that the rule cannot be used directly:**

   rate (overall) is **not equal** to $k[N_2][H_2]^3$.

2. **Intermediates** are found only in reactions that have more than one step. *The overall rate expression for a multistep reaction should contain as few intermediate concentration terms as possible.*
   a. It is generally possible to eliminate **all** organic molecule intermediate concentrations; nonorganic pre-equilibrium concentrations must sometimes be included.
   b. Elimination of intermediate concentrations provides the basis for determining the overall rate expression for a multistep reaction. To determine the **overall** rate expression for a **multistep reaction, follow the method outlined**:

Step 1: *Identify the critical step in the reaction mechanism, the* **r.d.s.**

Step 2: *Use the* **reaction step rate rule** *to write the specific rate expression for the* **r.d.s:** rate (r.d.s) = k[specific reactant 1]^x[specific reactant 2]^Y, etc. for the specific step. **This expression will be valid because the rate expression generated in this case is specific only for the single step, not for the complete reaction.**

Step 3: *Determine whether* **any** *of the specific reactants found in the rate expression for the* **r.d.s.** *are intermediates.*

Step 4: *If the rate expression for the* **r.d.s.** *contains* **no** *intermediate concentrations, then the process is finished. In this case, the* **overall** *rate expression is equivalent to the* **r.d.s.** *rate expression and no further analysis is required.*

Step 5: *If the rate expression for the r.d.s contains* **one or more intermediate concentrations [intermediate]***, they must be solved and substituted into the rate expression for the* **r.d.s.**

The rate expression for the r.d.s. generated in **Step 2 may** contain terms for intermediate concentrations (abbreviated as: **[intermediate]**) because the **r.d.s.** does **not** represent a complete balanced equation.

a. An **overall** rate expression is derived from the expression for the **r.d.s.** by mathematically solving for **[intermediate]** in terms of reactant concentrations (**[reactant]**).

b. *Select the fast/equilibrium step* in the reaction mechanism that involves **both** the **intermediate** to be eliminated and the **reactants** from which it is produced; this step must precede the **r.d.s.**

c. *Set up an equilibrium equation* (or rate equivalence equation) using this **[intermediate]** and all appropriate **[reactant]**.

d. *Solve (isolate) the required [intermediate] term* as a function of the corresponding **[reactant]** terms.

e. *Substitute the solved [reactant] terms for [intermediate] into the* **r.d.s.** *rate expression* determined in **Step 2**. The **overall** rate expression will be the rate expression produced when the [reactant] terms have replaced [intermediate] in rate (r.d.s).

9.4.4 GENERAL REACTION EXAMPLES FOR MULTISTEP REACTIONS

1. Consider a reaction in which the overall balanced equation is as shown below. The reaction may proceed by a number of different mechanisms; two possible examples will be used to demonstrate a prediction of the rate expression from a knowledge of the reaction mechanism steps.

$$2A + B \longrightarrow A_2B$$

2. **Mechanism notation:** Each step in the reaction will have an associated rate constant or equilibrium constant, generally written above and/or below the arrow for each step.
   a. Rate constants are designated by a lower case "**k**." A subscript number identifies the step number to which this constant applies; a negative sign for the number indicates that the constant applies to the reverse reaction.
   b. A capital "**K**," with a number subscript, indicates an equilibrium constant for a fast/equilibrium step in the mechanism.

3. **Mechanism #1 problem: Determine the expected rate expression.**

Step 1	$A + A \xrightarrow{k_1} A_2$	slow (r.d.s.)
Step 2	$A_2 + B \xrightarrow{k_2} A_2B$	fast

Add Steps:	$2A + B \longrightarrow A_2B$	"A_2" is an **intermediate**.

Problem solving process:

a. Verify that the proposed mechanism fits the balanced equation: add the equations for all steps. (**Process is shown above.**)
b. Determine the predicted rate expression, following the process **Steps 1 through 5**. These are identified by numbers in parentheses to avoid confusion with **reaction step numbers**.
   1. *Identify r.d.s.*
      This is shown as **reaction Step 1**:
      $$A + A \longrightarrow A_2 \quad \text{slow (r.d.s.)}$$
   2. *Write rate expression for r.d.s.* using ***reaction step rate rule***; use the specific symbol for the rate constant for that **reaction step**.
      $$\text{rate (r.d.s.)} = k_1[A]^2$$
   3. *Determine whether any of the concentrations in the rate expression for r.d.s. are intermediates.*
      There are **no** [intermediate]. In this reaction, "**A**" is a reactant.
   4. *Whenever the rate expression for the r.d.s. contains **no** [intermediate], then **rate (r.d.s.) equals rate (overall)**; no further analysis is necessary:*
      $$\text{rate (overall)} = k_1[A]^2$$

4. **Mechanism #2 Problem: Determine the expected rate expression.**
   a. In this mechanism, **reaction Step 1** is a fast, reversible reaction immediately preceding the slow step; it is thus an equilibrium. This is indicated by showing both the forward (k_1)

and reverse (k_{-1}) rate constants. The equilibrium constant is defined as the forward rate divided by the reverse rate; for one **elementary** step: $K_1 = k_1/k_{-1}$

Step 1 $A + A \xrightleftharpoons[k_{-1}]{k_1} A_2$ **fast and equilibrium;** $k_1/k_{-1} \equiv K_1$

Step 2 $A_2 + B \xrightarrow{k_2} A_2B$ **slow (r.d.s.)**

Add steps: $2A + B \longrightarrow A_2B$ "A_2" is an **intermediate.**

Problem solving process:

a. Verify that the proposed mechanism fits the balanced equation: add the equations for all steps. (**Process is shown above**.)

b. Determine the predicted rate expression, following the **process Steps 1 through 5**.

   1. *Identify r.d.s.*
      This is shown as **reaction Step 2**:

      $A_2 + B \longrightarrow A_2B$ **slow (r.d.s.)**

   2. *Write the rate expression for **r.d.s.** using **reaction step rate rule***; use the specific symbol for the rate constant for that **reaction step**.
      $$\text{rate (r.d.s.)} = k_2[A_2][B]$$

   3. *Determine whether any of the concentrations in the rate expression for **r.d.s.** are intermediates.*
      "A_2" is an intermediate; the rate (r.d.s.) is not equivalent to rate (overall).

   4. **Process Step 4** does not apply in this case; go to process **Step 5**.

   5. Solve for [A_2] in terms of [reactant].

c. Completion of rate expression determination for **mechanism #2**. **Process Step 5**:

   i. The only possible equation which will provide the proper relationship is **reaction Step 1**, which contains **both** [A_2] (intermediate) and [**A**] (reactant from which it is produced). This follows mechanistic concepts: *reaction steps **before** the r.d.s are expected to influence the rate of the overall reaction and thus the rate (overall) expression.*

Step 1 $A + A \xrightleftharpoons[k_{-1}]{k_1} A_2$ **fast and equilibrium;** $k_1/k_{-1} \equiv K_1$

   ii. Set up the corresponding equilibrium expression:
       $K_1 = [A_2]/[A]^2$.

iii. Isolate the [intermediate], that is, $[A_2]$: $[A_2] = K_1[A]^2$.

iv. Complete the process by substituting the term $(K_1[A]^2)$ for the term $([A_2])$ in the rate (r.d.s), **rate (r.d.s.) = $k_2[A_2]$ [B]**; this substitution produces the rate (overall) expression:

Substitute $(K_1 [A]^2)$ for $([A_2])$: **rate (overall) = $k_2(K_1[A]^2)$ [B]**

or rewriting: **rate (overall) = $k_2K_1[A]^2[B]$**.

v. The term k_2K_1 represents a multiplication of constants; this combination can be represented by any general rate constant **k** in the expression: **rate (overall) = $k[A]^2[B]$**.

5. Compare the overall rate expression for each mechanism:

a. The rate expression for **mechanism #1** does **not** contain a term for **[B]**. The rate of the reaction does not depend on the concentration of "**B**" because this reactant is not involved in the reaction at or before the **r.d.s.**

b. **Mechanism #2** does contain **[B]** in the rate (overall) expression: the reactant "**B**" is present in the slow step. *This is a demonstration of the statement that the same balanced equation will produce different rate (overall) expressions depending on mechanism.*

Additional Example for Analyzing Mechanisms

The following three hypothetical reaction mechanisms refer to the net balanced equation developed in question #1. Complete the following questions for these mechanisms.

Reaction mechanism #1:

Step 1	CH_2O	+	NH_3	$\xrightarrow{k_1}$	CH_2OH^+	+	NH_2^-	**slow/r.d.s.**
Step 2	CH_2OH^+	+	NH_2^-	$\xrightarrow{k_2}$	$CH_2NH_2^+$	+	OH^-	**fast**
Step 3	$CH_2NH_2^+$	+	OH^-	$\xrightarrow{k_3}$	CH_2NH	+	H_2O	**fast**

1. What is the balanced equation?
2. What are the intermediates?
3. What is the predicted rate expression for the r.d.s.?
4. What is the predicted rate expression for the overall rate?
5. Kinetic analysis shows that a plot of $\ln[CH_2O]$ versus time is straight line whenever $[NH_3]$ is held approximately constant. Kinetic analysis also shows that a plot of $\ln[NH_3]$ versus time is straight line whenever $[CH_2O]$ is held approximately constant. Is mechanism #1 a possible valid mechanism for the reaction?

Reaction mechanism #2: The reaction determined from the **balanced equation** for mechanism #1 occurs in **one step**.

6. What is the predicted rate expression for the overall rate for mechanism #2?

7. Is mechanism #2 a possible valid mechanism for the reaction based on the same kinetic data described for mechanism #1?

Reaction mechanism #3:
The balanced equation and intermediates are the same as for mechanism #1.

Step 1 CH_2O + NH_3 $\underset{k_{-1}}{\overset{k_1}{\rightleftharpoons}}$ CH_2OH^+ + NH_2^- **fast/equilibrium**

Step 2 CH_2OH^+ + NH_2^- $\xrightarrow{k_2}$ $CH_2NH_2^+$ + OH^- **slow/r.d.s.**

Step 3 $CH_2NH_2^+$ + OH^- $\xrightarrow{k_3}$ CH_2NH + H_2O **fast**

8. What is the predicted rate expression for the overall rate for mechanism #3?
9. Is mechanism #3 a possible valid mechanism for the reaction based on the same kinetic data described for mechanism #1?

Answers to Additional Example for Analyzing Mechanisms
Reaction mechanism #1:

1. Balanced equation:

Step 1 CH_2O + NH_3 $\xrightarrow{k_1}$ CH_2OH^+ + NH_2^- **slow/r.d.s.**

Step 2 CH_2OH^+ + NH_2^- $\xrightarrow{k_2}$ $CH_2NH_2^+$ + OH^- **fast**

Step 3 $CH_2NH_2^+$ + OH^- $\xrightarrow{k_3}$ CH_2NH + H_2O **fast**

CH_2O + NH_3 $\longrightarrow$ CH_2NH + H_2O **balanced equation**

2. Intermediates: CH_2OH^+; NH_2^-; $CH_2NH_2^+$; OH^-.
3. Predicted rate expression for the r.d.s.: rate (r.d.s.) = $k_1[CH_2O][NH_3]$.
4. Predicted rate expression for the overall rate:

   Rate (r.d.s.) expression contains no intermediate concentrations
   Rate (r.d.s.) = rate (overall) = $k_1[CH_2O][NH_3]$

5. Kinetic analysis shows that a plot of $\ln[CH_2O]$ versus time is straight line whenever $[NH_3]$ is held approximately constant. Kinetic analysis also shows that a plot of $\ln[NH_3]$ versus time is straight line whenever $[CH_2O]$ is held approximately constant.

   The rate expression based on kinetic analysis is rate = $k[CH_2O][NH_3]$.

   The kinetic rate expression agrees with the predicted rate expression for mechanism #1; mechanism #1 is a possible valid mechanism based on kinetics.

Reaction Mechanism #2: The reaction determined from the balanced equation for mechanism #1 occurs in one step.

6. Predicted rate expression for the overall rate for mechanism #2 based on the balanced equation as one step:

$$\text{Rate (overall)} = \text{rate (one step)} = k[CH_2O][NH_3]$$

7. The predicted rate expression for mechanism #2 agrees with the kinetic data: mechanism #2 is a possible valid mechanism.

Reaction Mechanism #3:

The balanced equation and intermediates are the same as for mechanism #1.

Step 1	CH_2O	$+$	NH_3	$\underset{k_{-1}}{\overset{k_1}{\rightleftarrows}}$	CH_2OH^+	$+$	NH_2^-	**fast/equilibrium**
Step 2	CH_2OH^+	$+$	NH_2^-	$\xrightarrow{k_2}$	$CH_2NH_2^+$	$+$	OH^-	**slow/r.d.s.**
Step 3	$CH_2NH_2^+$	$+$	OH^-	$\xrightarrow{k_3}$	CH_2NH	$+$	H_2O	**fast**

8. Predicted rate expression for the overall rate for mechanism #3:

$$\text{rate (r.d.s.)} = k_2[CH_2OH^+][NH_2^-]$$

The rate (r.d.s.) expression contains intermediate concentrations: solve for the intermediate concentrations as a function of starting reactants.

$$k_1/k_{-1} = K_1 = \frac{[CH_2OH^+][NH_2^-]}{[CH_2O][NH_3]}.$$

Solve for $[CH_2OH^+][NH_2^-]$:

$$[CH_2OH^+][NH_2^-] = K_1[CH_2O][NH_3]$$

Substitute $\{[K_1[CH_2O][NH_3]\}$ for $[CH_2OH^+][NH_2^-]$ into rate (r.d.s.) expression to produce the rate (overall) expression:

$$\text{Rate (overall)} = k_2\{K_1[CH_2O][NH_3]\} = k_2K_1[CH_2O][NH_3]$$
$$k_2K_1 = \text{(general) } k$$
$$\text{Rate (overall)} = k[CH_2O][NH_3]$$

9. The predicted rate expression for mechanism #3 agrees with the kinetic data; mechanism #3 is a possible valid mechanism.

9.5 ENERGY AND CHEMICAL REACTIONS

9.5.1 BONDING CHANGES AND ACTIVATION ENERGY

1. Reaction steps always involve some combination of bond making and/or bond breaking. The net energies required or released for these combinations dictate the type of steps in the mechanism and influence the relative size of the activation barriers (Ea).
2. Generally, a reaction will "choose" (i.e., statistically follow) the path/mechanism of lowest energy barriers. An analogy: travelers follow the lowest elevation "passes" through mountain ranges. The reason for this behavior (whether for molecules or people) is that this provides the path of lowest energy investment.
3. Values for ΔG or ΔH for a complete reaction will have some influence on the energy of individual reaction steps, since they are related through bond energies. However, there is *no direct relationship between free energy of an overall reaction and the mechanism by which the overall reaction occurs.*
   a. Recall that free energy/enthalpy/entropy are **state** functions, while mechanism is a **path** function.
   b. As an example, a highly spontaneous reaction (large negative value for AG) could have a larger energy barrier (greater value for Ea) for the **r.d.s.** than, for example, a weakly spontaneous reaction or even a non-spontaneous reaction.

9.5.2 ENERGETICS OF BOND MAKING AND BOND BREAKING

1. The energy barrier (Ea) for a bond-breaking step is essentially the bond-dissociation energy (**B.D.E.**).
   a. Bond breaking can occur through **unimolecular** decomposition: the bond in a molecule is broken with **no** outside molecule participating:

$$A\!-\!B \longrightarrow A + B$$

   b. The energy to overcome the **B.D.E.** barrier is supplied by heat, visible light, UV radiation, etc. The energy is added to the internal **bond-vibration** energy of the molecule. Sufficient energy will cause the required bond to break in some statistical fraction of the reactant molecules.
   c. **Bond (heat) vibration energy is proportional to temperature.**
2. Bond breaking can also occur through **bimolecular** decomposition: the bond in a molecule is broken through the (kinetic) energy of **collision** with another molecule:

$$A\!-\!B + C \longrightarrow A + B + C$$
$$A\!-\!B + C \longrightarrow A\!-\!C + B$$

    a. The energy to overcome the **B.D.E.** barrier comes from the kinetic energy of both molecules.

    b. **Molecular kinetic energy is proportional to temperature**.

3. Bond-making steps do not require additional energy, since bond formation always decreases the total potential energy of the species involved (energy is released).

$$A + B \longrightarrow A—B$$

    a. The "barrier" in this case (termed the "diffusion barrier") simply represents the statistical probability that the appropriate molecules will collide with each other to produce a bond.

    b. The collision must occur with the correct geometry to allow orbital overlap. **All** collisions satisfying this requirement will lead to bond formation since there is **no** energy barrier.

    c. **Diffusion rates are proportional to temperature**.

9.5.3 REACTION RATES, TEMPERATURE, AND EA

1. ***The rates of all chemical reactions increase with increasing temperature.***

    a. The molecular explanation, emphasized in the last section, is based on the fact that bond-vibration energy, molecular kinetic energy, and diffusion rates all **increase** as the **temperature increases**. Temperature, therefore, increases the **rate** of bond-breaking events and bond-making events.

    b. The rate expression shows that **rate = k[reactant 1]^x [reactant 2]^Y**, etc. Excluding the complication of gases, concentrations are essentially constant with temperature; thus rate varies with temperature through the rate constant, k.

2. ***Rates of chemical reactions increase with a decreasing activation barrier, Ea.***

    a. Reaction rates are inversely proportional to the energy barrier of the slowest step (r.d.s.). Thus, **rates increase** as Ea (activation energy) **decreases; rates decrease** as **Ea increases**.

    b. As for temperature, the rate varies with Ea through the rate constant k.

3. The two relationships are summed up in the **Arrhenius equation**:

$$k = Ae^{-Ea/RT} \quad \text{or} \quad \text{(alternate form)}: \ln(k) = -Ea / RT + \ln A$$

where **T** = temperature in Kelvin; **R** = 0.008314 kJ/mole K; **e** and **ln** represent the natural log function; and **A** is a constant (called the pre-exponential factor).

    a. The mathematical directional trends for the equation follow the concepts stated:

As Ea **increases** or T **decreases**, the term (–Ea/RT) becomes a **larger negative value**, the value of k **decreases**. As Ea **decreases** or T **increases**, the term (–Ea/RT) becomes a **smaller negative value**, the value of k **increases**.

b. The constant "**A**" represents the maximum rate of the reaction, when **all** molecules possess sufficient energy to overcome the energy barrier Ea. (Mathematically, as Ea approaches zero or T approaches infinity, {–Ea/RT} approaches zero: the equation reduces to k = A or ln k = ln A.) *The value of "A" is different for every reaction* and depends on diffusion rates, geometries of molecular collisions, etc.

c. The Arrhenius equation can be used to determine the **activation barrier** (Ea) for a reaction or an r.d.s. Rate constants, k, are found from kinetic experiments described previously. The experiments are performed at a sequence of different temperatures; values of "k" as a function of "T" are generated. A plot of **ln(k)** versus **1/T** yields a straight line with **slope = –Ea/R** and a **y-intercept of ln A**.

9.5.4 RATES AND CATALYSIS

1. *A **catalyst** is an agent that increases the rate of a chemical reaction without itself being **net** consumed in the reaction.*

   a. The catalyst may act without undergoing observable change in chemical form. This is typical of **heterogeneous** catalysis: the catalyst and reactant(s) are in **different phases**. This is the dominant technology used in the chemical industry. Examples are solid metal surfaces catalyzing reactions of liquid, solution, or gaseous reactants, for example, an exhaust "catalytic converter."

   b. The catalyst may be directly involved **as a reactant** in a chemical reaction, but is **regenerated** to its initial form at the end of the reaction: *there is no **net** change in the catalyst structure or concentration.* This is typical of **homogeneous** catalysis in which catalyst and reactants are in the **same phase** (most often in solution). Examples of homogeneous catalysts are acids, bases, enzymes.

2. The concept of a **homogeneous catalyst** used as a reactant but **regenerated** at the end of the reaction can be illustrated:

Step 1 $CH_2{=}CH_2 + H^+$ **(catalyst)** $\longrightarrow$ $CH_3{-}CH_2^+$

Step 2 $CH_3{-}CH_2^+ + H_2O$ $\longrightarrow$ $CH_3{-}CH_2{-}OH_2^+$

Step 3 $CH_3{-}CH_2{-}OH_2^+$ $\longrightarrow$ $CH_3{-}CH_2{-}OH + H^+$ **(regenerated catalyst)**

Net Rxn: $CH_2{=}CH_2 + H_2O$ $\longrightarrow$ $CH_3{-}CH_2{-}OH$

a. The catalyst is **consumed** in **Step 1** to increase the rate of **Step 2**. It is then **regenerated** in **Step 3** and can be used over again: no **net** consumption.

b. Do not confuse the role of the catalyst in this type of reaction with that of an intermediate. Both can be cancelled from the net equation, ***but a catalyst can be an initial reactant (and a final product)***, *while an intermediate is always formed as a product in a reaction step and must be consumed by reaction end*. A catalyst can be (and usually is) included in an overall rate expression: ***rate usually depends on catalyst concentration***. Rates of most biochemical reactions in living systems are ultimately controlled by catalytic enzyme concentrations.

3. *The mechanistic role of the catalyst is to lower the r.d.s. activation barrier (Ea) for a reaction by providing an alternative lower energy pathway not otherwise available*; *the catalyst changes the mechanism (path function)*. **A catalyst cannot change the value of ΔG or ΔH (state functions) for the overall reaction.**

9.6 INTERPRETATION OF ENERGY LEVEL DIAGRAMS

9.6.1 GENERAL CONSTRUCTION OF ENERGY DIAGRAMS

1. Overall reaction thermodynamics, **plus** the mechanistic information developed through kinetics, are very often displayed in the form of an **energy level diagram**. This is a pictorial description of a reaction based on plotting the **relative potential energies** of **all** reactants, intermediates, products, and **activation barriers** (estimated or measured) as a function of the general concept termed **reaction progress**.

a. **Reaction progress** is nonspecific and represents some sequential description of events such as time, bonding changes or atom movements, charge distribution changes or electron movements, etc.

b. **Relative potential energies** are shown as ΔG or ΔH values along the vertical axis; reaction progress is shown along the horizontal axis.

2. The **complete** set of initial reactants, the **complete** set of final products, and every **complete** set of intermediates are each displayed as a separate energy "platform." Connecting each platform in sequence is a smooth curve, which represents the energy changes of all the molecules for the appropriate mechanistic step.

a. Each energy platform represents a **local energy minimum** (the requirement for a measurably **stable** reactant, intermediate, or product). The concept of "**local**" means that (in a two-dimensional diagram) the connecting smooth curves are higher on both sides of the platform, *indicating that energy must be added to "move" the set of molecules toward the forward or reverse reaction.*

b. The actual potential energy of the platform itself does not depend on the energy **barrier** on either side. (The height of a boulder somewhere on a mountain is not related to the depth of the hole it sits in; even a boulder in a shallow depression on the top of the mountain will resist falling for an indeterminate time.)

3. The **transition state** of a reaction step is **not** a stable species: it represents a **local energy maximum**, identified at the top of the smooth curve connecting any two platforms. (The idea of "transition" comes from the fact that at this point, the partially bonded molecular composite can "fall down" in either direction to either of the platforms.) The transition state is drawn showing the partial bonding changes.

4. The energy level diagram for a complete **multistep reaction** is put together by describing each of the steps in turn. The diagram follows the convention of chemical reaction equations: *the forward reaction is read from left to right, the reverse reaction (if it exists using the same pathway) is read from right to left.*

9.6.2 ENERGY DIAGRAM EXAMPLES

1. The energy diagrams shown below illustrate how the information can be displayed:
    a. **Figure 6a** shows an energy level diagram demonstrating **one** bond-breaking step; bond breaking is always endothermic ($+\Delta H$).
    b. **Figure 6b** shows the corresponding bond-making step, always exothermic ($-\Delta H$).
    c. **Figure 6c** is an energy diagram depicting a **one-step** reaction in which **both** bond-breaking (**A—B**) and bond-making (A—C) occur **AB + C** $\longrightarrow$ **AC + B**
        i. *The relative heights of the energy platforms for products and reactants therefore are determined by the relative bond energies of A—B (with C as a free atom) versus A—C (with B as a free atom).*
        ii. The activation barrier (Ea) for a reaction step where bond making and bond breaking occur simultaneously will depend on the relative **partial** bond energies represented by the **transition state**: how strong each bond is and to what degree is each broken or formed.
    d. **Figure 6d** shows the **same net** reaction as **Figure 6c**, but with a **different** mechanism: this mechanism depicts two reaction steps:

Step 1	AB + C	$\longrightarrow$	A + B + C
Step 2	A + B + C	$\longrightarrow$	AC + B
Net	AB + C	$\longrightarrow$	AC + B

Recall that for a complete reaction, ΔG or ΔH is determined by the initial and final states only. ***Regardless of path,*** ΔG ***must be the same for both diagrams.***

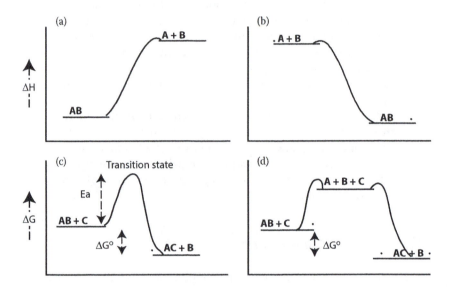

9.6.3 EXAMPLE PROBLEM: READING AN ENERGY DIAGRAM

1. **Problem:** Use the energy level diagram shown in **Figure 7** to determine the following information:
   a. What is the overall **forward** reaction?
   b. Specify **each** step and classify as **fast** or **slow** (rate-determining); try to identify a probable equilibrium step.
   c. Calculate **Ea** for the **r.d.s.**
   d. Determine the sign and value for $\Delta G°$ for the **complete forward** reaction.
   e. What is the specific $\Delta G°$ for reaction **Step 1**?
   f. If reaction **Step 1** was an equilibrium, what would be the value for the equilibrium constant?
2. **Extra problem:** With the answers to **Problem 1** as a guide (next page), use the same diagram (**Figure 7**) to answer the corresponding parts a through d for the **reverse** reaction.
   a. What is the overall **reverse** reaction?
   b. Specify **each** step and classify as **fast** or **slow** (rate-determining); try to identify a probable equilibrium step.
   c. Calculate **Ea** for the **r.d.s.**

d. Determine the sign and value for $\Delta G°$ for the **complete reverse** reaction.

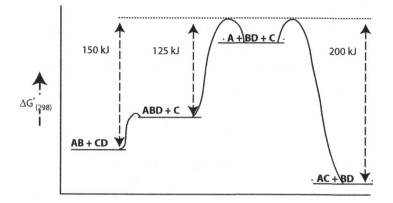

9.6.4 PROBLEM 1 ANSWER

a. Overall reaction: AB + CD $\longrightarrow$ AC + BD

b. Mechanism: steps for the **forward** reaction.

Step 1	AB + CD	$\underset{\longleftarrow}{\longrightarrow}$	ABD + C	**fast/equilibrium**
Step 2	ABD	$\longrightarrow$	A + BD	**slow**
Step 3	A + C	$\longrightarrow$	AC	**fast**

Net Reaction AB + CD $\longrightarrow$ **AC + BD**

Analysis:

Step 1 is **relatively** fast because it has a much smaller **Ea** than **Step 2** for the **forward** reaction. The **reverse of Step 1** (ABD + C $\longrightarrow$ AB + CD) has an **Ea** that is even smaller than the **forward Ea**; thus it is **probable** that this step is at equilibrium. (This conclusion cannot be proved from the diagram alone.)

Step 2 is the slow step because it has the highest activation energy (**Ea**). **Note** that the compound "**C**" was not included on either side of the equation for **Step 2**, even though it **must** appear in the diagram. *The energy level diagram depicts **total potential energy** and thus **must** include all species which appear from reactants to products.* By convention, identical species are cancelled from each side of a balanced step reaction equation: in this case "**C**" undergoes no change.

Step 3 is relatively fast because it has a much smaller Ea than Step 2 for the **forward** reaction.

Based on the energy values provided in the diagram, plus some simple calculation:

c. Ea for the **r.d.s. (forward)** is **125** kJ/mole.
d. $\Delta G°$ for the complete **forward** reaction is **−50** kJ/mole.
e. The specific **$\Delta G°$ for Step 1** is calculated as **+25** kJ/mole.
f. The equilibrium constant for Step 1 at 298 K can be calculated from:
$$\log K = \Delta G°/{-2.303RT} = +25 \text{ kJ}/{-2.303(0.008314)(298)}$$
$$= -4.38$$
$$\log K = -4.38 \quad K = 4.2 \times 10^{-5}$$

Additional Example for Reading an Energy Diagram

The potential energy diagram shown describes a hypothetical suggested mechanism for an organic reaction. Use the diagram and the suggested labeled energy values to answer the 14 questions.

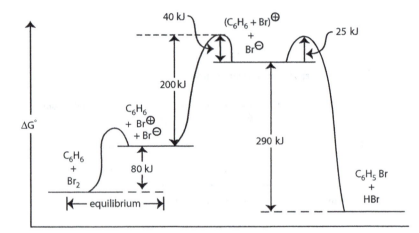

Forward reaction:

1. What is the overall **forward** reaction (balanced equation)?
2. Write the forward mechanism: Specify **each** step and classify as **fast** or **slow**; identify the r.d.s.
3. Identify all the **intermediates**.
4. Calculate **Ea** for the **forward r.d.s.**
5. Calculate $\Delta G°$ for the **complete forward** reaction.
6. Calculate $\Delta G°$ for the **forward r.d.s.** step.
7. Determine the **rate expression** for the **forward** reaction.

Reverse reaction:

8. What is the overall **reverse** reaction?
9. Write the **reverse** mechanism steps; label r.d.s. reverse.
10. Calculate $\Delta G°$ for the **reverse r.d.s.** step.
11. Calculate **Ea** for the **r.d.s.**
12. Determine $\Delta G°$ for the **complete reverse** reaction.

13. Determine the **rate expression** for the **reverse r.d.s.** step.
14. Determine the **rate expression** for the **reverse** reaction overall.

Answers to Additional Example for Reading an Energy Diagram

1. Overall forward reaction (balanced equation):

$$C_6H_6 \ + \ Br_2 \ \longrightarrow \ C_6H_5Br \ + \ HBr$$

2. Mechanism:

Step 1 $Br_2 \ \underset{k_{-1}}{\overset{k_1}{\rightleftarrows}} \ Br^+ \ + \ Br^-$ **fast/equilibrium**

Step 2 $C_6H_6 \ + \ Br^+ \ \xrightarrow{k_2} \ (C_6H_6Br)^+$ **slow/r.d.s.**

Step 3 $(C_6H_6Br)^+ \ + \ Br^- \ \xrightarrow{k_3} \ C_6H_5Br \ + \ HBr$ **fast**

Reaction steps: Each reaction step in the forward direction is read from left to right. All atoms or molecules are carried through the diagram; atoms or molecules which do not change for a particular step are eliminated from each side of the step equation: Exclude (C_6H_6) from Step 1, since it is not involved (remains unchanged) in this step. Exclude (Br^-) from Step 2, since it is not involved (remains unchanged) in this step.

Activation energies for the forward reaction are read from left to right. Assuming the PE diagram is approximately to scale, the activation energy for the first step, although not specifically marked, can be determined to be 80 kJ plus ≈20 kJ for the extra energy up to the peak. The activation energy for the third step is marked as 25 kJ. The r.d.s. (highest activation energy) is the second step with the activation energy directly marked as 200 kJ. Step 2 is labeled as slow/r.d.s. and the other two steps are labeled as fast; Step 1 is also labeled as an equilibrium.

3. Intermediates: $(C_6H_6Br)^+$; Br^+; Br^-.
4. The r.d.s. was determined to be the second step. The Ea (activation energy) was read directly from the P.E diagram = 200 kJ.
5. The $\Delta G°$ for the complete forward reaction is the energy difference for the balanced equation: the difference between the reactant (first) platform and the product (last) platform. The numerical value must be calculated from the given marked energy values: $\Delta G° = [80 + 200] - [40 + 290] = -50$ kJ/mole-reaction.
6. $\Delta G°$ for the forward r.d.s step: $\Delta G° = [200] - [40] = +160$ kJ/mole-reaction.
7. Predicted rate expression for the overall rate: rate (r.d.s.) = $k_2[C_6H_6]$ $[Br^+]$.

The rate (r.d.s.) expression contains intermediate concentrations: solve for the intermediate concentrations as a function of starting reactants.

$$k_1/k_{-1} = K_1 = \frac{[Br^+][Br^-]}{[Br_2]} \quad \text{Solve for } [Br^+]\text{: } [Br^+] = K_1[Br_2]/[Br^-]$$

Substitute $\{K_1[Br_2]/[Br^-]\}$ into rate (r.d.s.): rate $= k_2[C_6H_6]\{K_1[Br_2]/[Br^-]\}$.

Rate (overall) $= k_2K_1[C_6H_6][Br_2]/[Br^-] = k_2K_1[C_6H_6][Br_2]/[Br^-]$

$[Br^-]$ could not be eliminated from the expression.

8. Overall reverse reaction (balanced equation):

$$C_6H_5Br \quad + \quad HBr \quad \longrightarrow \quad C_6H_6 \quad + \quad Br_2$$

9. Reverse mechanism:

Step 1 $\quad C_6H_5Br \ + \ HBr \ \xrightarrow{k_1} \ (C_6H_6Br)^+ \ + \ Br^- \qquad$ **slow/r.d.s.**

Step 2 $\qquad\qquad (C_6H_6Br)^+ \ \xrightarrow{k_2} \ C_6H_6 \ + \ Br^+ \qquad\qquad$ **fast**

Step 3 $\qquad\qquad Br^+ \ + \ Br^- \ \xrightarrow{k_3} \ Br_2 \qquad\qquad\qquad$ **fast**

Activation energies for the forward reaction are read from right to left. Step 1 in the reverse direction clearly has the highest Ea and is labeled as slow/r.d.s.; the other two steps are labeled as fast.

10. $\Delta G°$ for the reverse r.d.s step is marked as +290 kJ/mole-reaction.
11. The r.d.s. was determined to be the first step. The Ea (activation energy) can be calculated as [290 + 25] = 315 kJ.
12. The $\Delta G°$ (reverse) $= -\Delta G°$ (forward) $= -(-50) = +50$ kJ/mole-reaction.
13. Rate (r.d.s./reverse) $= k[C_6H_5Br][HBr]$.
14. Predicted rate expression for the overall reverse rate: the rate (r.d.s./reverse) contains no intermediate concentrations: rate (r.d.s./reverse) = rate (overall) $= k[C_6H_5Br][HBr]$.

9.7 PRACTICE PROBLEMS

1. Consider a first-order reaction:

$$A \ \longrightarrow \ B \qquad \text{rate} = k\,[A]$$

An experiment starts with $[A_0]$ (initial concentration) equal to 1.0 M. The value for $k = 0.010$/s.
   a. What will be **[A]** after 300 s?
   b. How many seconds must the reaction proceed to reduce the **[A]** to 0.020 M?

2. Consider the following general reaction:

$$AB + CD \ \longrightarrow \ AC + BD$$

Use the **kinetic experimental** information to determine the rate expressions for the separate mechanisms #1, #2, and #3 based on the general equation shown above. *No mechanistic steps are given for any mechanism.*

Mechanism #1: A plot of [AB] versus time is a straight line whenever [CD] is held very high and nearly constant. A plot of 1/[CD] versus t is a straight line.

Mechanism #2: A plot of ln [AB] versus t is a straight line when [CD] is held nearly constant. A plot of ln [CD] versus t is a straight line when [AB] is held nearly constant.

Mechanism #3: A plot of ln [AB] versus t is a straight line with a slope of −3.5 and a y-intercept of 0.2. A plot of [CD] versus t is a straight line when [AB] is held nearly constant. Include the actual numerical value for rate constant k in the expression.

3. Use the **reaction step rate rule** to predict the rate expression for each of the following hypothetical **one-step** reactions (i.e., the rate expression if the reaction were one-step only).

a. $O_3 \longrightarrow O_2 + O$

b. $CH_3-CH_3 \longrightarrow CH_2=CH_2 + H_2$

c. $CO + O \longrightarrow CO_2$

d. $CH_3I + F \longrightarrow CH_3F + I$

e. $CH_2 + Cl + Cl \longrightarrow CH_2Cl_2$

f. $CO + Cl_2 \longrightarrow COCl_2$

g. $C_4H_6 + C_4H_6 \longrightarrow C_8H_{12}$

h. $CH_2 + H + Cl \longrightarrow CH_3Cl$

4. For the following two reactions:
   a. Write the **balanced** overall equation by adding the steps in the mechanism and eliminating intermediates.
   b. Determine the predicted rate (overall) expression following the procedure described in this chapter.

Reaction Mechanism #1:

Step 1 $C_4H_9Br \xrightarrow{k_1} C_4H_9^+ + Br^-$ **slow**

Step 2 $C_4H_9^+ + Cl^- \xrightarrow{k_2} C_4H_9Cl$ **fast**

Reaction Mechanism #2:

Step 1 $C_3H_8O + HBr \underset{k_{-1}}{\overset{k_1}{\rightleftarrows}} C_3H_9O^+ + Br^-$ **fast/equilibrium**

Step 2 $C_3H_9O^+ \xrightarrow{k_2} C_3H_7^+ + H_2O$ **slow**

Step 2 $C_3H_7^+ + Br^- \xrightarrow{k_3} C_3H_7Br$ **fast**

5. Draw an energy level diagram for **Reaction Mechanism #2** described in **practice problem 4**. Include as much detail as is possible by using the additional information: The **net overall** reaction is spontaneous based on free energy. The cation intermediate formed in **Step 2** is relatively unstable (high in potential energy). **Step 1** is an equilibrium (with relatively fast forward and reverse rates). The equilibrium constant for the **Step 1** equilibrium has a value of **less than 1**.

6. The energy level diagrams shown in **Figure 8a and b** describe **two different mechanisms** for the **same** reaction. For **each** mechanism, answer the following:
   a. What is the **overall** reaction?
   b. Specify each step and classify as **fast** or **slow** (rate-determining). Do **not** include molecules that are unchanged in the step.
   c. Calculate **Ea** for the **r.d.s.**
   d. Determine the sign and value for **ΔG°** for the **overall** reaction.
   e. What is the **rate expression** specifically for the **r.d.s.**?
   f. What is the **rate expression** for the **overall** reaction, that is, excluding intermediates, if different than e?
   g. What might the **transition state** of the **r.d.s.** look like?
   h. For **mechanism #2 only: Step 1** is an equilibrium. Calculate the equilibrium constant **for this step** from the appropriate ΔG°.

   Extra practice: Repeat the above for the **reverse reaction** for **mechanism #2**.

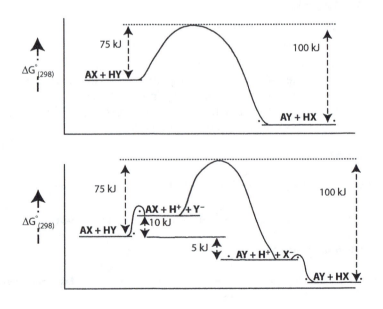

9.8 ANSWERS TO PRACTICE PROBLEMS

1. Consider a first-order reaction: **A ⟶ B   rate = k [A]**
   An experiment starts with $[A_0]$ (initial concentration) equal to 1.0 M. The value for k = 0.010/s.

a. What will be [**A**] after 300 s?

$\ln [A] = -kt + \ln [A_0]$

$\ln [A] = -(0.01/s)(300\ s) + \ln (1.0)$

$\ln [A] = -3 + 0 = -3$

[**A**] = antiln (-3) = e^{-3} = **0.0498 M**

b. How many seconds must the reaction proceed to reduce the [A] to 0.020 M?

$\ln (0.02) = -(0.01)(t) + \ln (1.0)$

$-3.91 = -(0.01)(t) + 0$

t $= -3.91/ - 0.01 = $ **391 s**

2. Consider the following general reaction: **AB + CD $\longrightarrow$ AC + BD**
Use the **kinetic experimental** information to determine the rate expressions for the separate mechanisms #1, #2, and #3 based on the general equation shown above. *No mechanistic steps are given for any mechanism.*

Mechanism #1: A plot of [AB] versus time is a straight line whenever [CD] is held very high and nearly constant. A plot of 1/[CD] versus t is a straight line.

Analysis: Whenever [CD] is constant, if [AB] versus t = straight line, reaction is zero order in [AB]. Since [AB] is zero order, the reaction rate depends **only** on [CD]. The straight line plot of 1/[CD] versus time, regardless of [AB], indicates that the reaction is second order with respect to [CD], combine the information:

$$rate = k[AB]^0[CD]^2 \quad or \quad \textbf{rate} = \textbf{k[CD]}^2$$

Mechanism #2: A plot of ln [AB] versus t is a straight line when [CD] is held nearly constant. A plot of ln [CD] versus t is a straight line when [AB] is held nearly constant.

Analysis: A straight-line plot of ln [AB] versus time (with [CD] constant) indicates that the reaction is first order in [AB]. The equivalent result for [CD] indicates the reaction is first order in [CD], combine the information:

$$rate = k[AB]^1[CD]^1 \quad or \quad \textbf{rate} = \textbf{k[AB][CD]}$$

Mechanism #3: A plot of ln [AB] versus t is a straight line with a slope of -3.5 and a y-intercept of 0.2. A plot of [CD] versus t is a straight line when [AB] is held nearly constant. Include the actual numerical value for rate constant k in the expression.

Analysis: The reaction is zero order in [CD] based on the straight line for [CD] versus time. The reaction is first order in [AB] based on the straight line for ln [AB] versus time. The integrated rate equation for first order shows that the slope $= -k$. Thus, k = $-$(slope) for the data given, or k = $-(-3.5) = 3.5/s$ (or 3.5 s^{-1}) if time is in seconds.

$$\textbf{rate} = \textbf{3.5 s}^{-1}\ \textbf{[AB]}$$

3. Use the **reaction step rate rule** to predict the rate expression for each of the following **one-step** reactions.

a. $O_3 \longrightarrow O_2 + O$ rate = k [O_3]

b. $CH_3 \!-\! CH_3 \longrightarrow CH_2 \!=\! CH_2 + H_2$ rate = k [CH_3CH_3]

c. $CO + O \longrightarrow CO_2$ rate = k [CO] [O]

d. $CH_3I + F \longrightarrow CH_3F + I$ rate = k [CH_3I] [F]

e. $CH_2 + Cl + Cl \longrightarrow CH_2Cl_2$ rate = k [CH_2] [Cl]^2

f. $CO + Cl_2 \longrightarrow COCl_2$ rate = k [CO] [Cl_2]

g. $C_4H_6 + C_4H_6 \longrightarrow C_8H_{12}$ rate = k [C_4H_6]^2

h. $CH_2 + H + Cl \longrightarrow CH_3Cl$ rate = k [CH_2] [H] [Cl]

4. For the following two reactions:
   a. Write the **balanced** overall equation by adding the steps in the mechanism and eliminating intermediates.
   b. Determine the predicted rate (overall) expression following the procedure described in this chapter.

Reaction Mechanism #1:

		$\xrightarrow{k_1}$		
Step 1	C_4H_9Br	$\longrightarrow$	$C_4H_9^+ + Br^-$	**slow**

		$\xrightarrow{k_2}$		
Step 1	$C_4H_9^+ + Cl^-$	$\longrightarrow$	C_4H_9Cl	**fast**

Equation: $C_4H_9Br + Cl^- \longrightarrow C_4H_9Cl + Br^-$

rate (r.d.s.) = $k_1[C_4H_9Br]$; C_4H_9Br is **not** an intermediate, thus
rate (overall) = rate (r.d.s.) = k[C_4H_9Br]

Reaction Mechanism #2:

		$\underset{k_{-1}}{\overset{k_1}{\rightleftharpoons}}$		
Step 1	$C_3H_8O + HBr$		$C_3H_9O^+ + Br^-$	**fast/equilibrium** $K_1 = k_1/k_{-1}$

		$\xrightarrow{k_2}$		
Step 2	$C_3H_9O^+$	$\longrightarrow$	$C_3H_7^+ + H_2O$	**slow**

		$\xrightarrow{k_3}$		
Step 3	$C_3H_7^+ + Br^-$	$\longrightarrow$	C_3H_7Br	**fast**

Equation: $C_3H_8O + HBr \longrightarrow C_3H_7Br + H_2O$

rate (r.d.s.) = $k_2 [C_3H_9O^+]$; $C_3H_9O^+$ **is an intermediate.**
Define $K_1 = [C_3H_9O^+][Br^-]/[C_3H_8O][HBr]$
$[C_3H_9O^+] = K_1[C_3H_8O][HBr]/[Br^-]$; *([Br$^-$] cannot be eliminated from expression.)*

Substitute expression for $[C_3H_9O^+]$ into rate (r.d.s.)

rate (overall) = k_2 $K_1[C_3H_8O][HBr]/[Br^-]$

5. Draw an energy level diagram for **Reaction Mechanism #2** described in **practice problem 4**. The **Reaction Mechanism #2** was

Step 1 $C_3H_8O + HBr \xrightarrow[k_{-1}]{k_1} C_3H_9O^+ + Br^-$ **fast/equilibrium**

Step 2 $C_3H_9O^+ \xrightarrow{k_2} C_3H_7^+ + H_2O$ **slow**

Step 3 $C_3H_7^+ + Br^- \xrightarrow{k_3} C_3H_7Br$ **fast**

Key additional information and conclusions:

a. **The net overall reaction is spontaneous based on free energy**: Products must be shown lower than reactants on the diagram.

b. **The cation intermediate formed in Step 2 is high in potential energy**: The energy platform for the products of Step 2 should be relatively high on the diagram.

c. **Step 1 is an equilibrium (with relatively fast forward and reverse rates); the equilibrium constant for the Step 1 equilibrium has a value of less than 1**.

The energy platforms for the products of Step 1 should be higher than the reactants; if the equilibrium constant is less than 1, $\Delta G°$ for the equilibrium step must be positive. An **equilibrium** for this step indicates that the activation energies for the forward and reverse reactions for Step 1 should be smaller than for Step 2, which is the r.d.s.

The energy level diagram shown in **Figure 9** *can thus be assembled:*

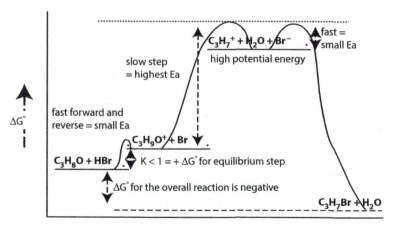

6. The energy level diagrams shown in **Figure 8a and 8b** describe two different mechanisms for the same reaction. For **each** mechanism, answer the following:

For **mechanism #1** shown in **Figure 8a**:

    a. What is the overall reaction: **AX + HY ⟶ AY + HX**
    b. Specify each step and classify as fast or slow: **One-step only**
    c. Calculate **Ea** for the r.d.s. (only step): **Ea = 75 kJ/mole as shown**
    d. Determine the sign and value for **ΔG°** for the overall reaction:

$$\Delta G° = 75 - 100 = -25 \text{ kJ/mole-reaction}$$

    e. What is the rate expression specifically for the **r.d.s.**: **rate(r.d.s.) = k[AX][HY]**
    f. What is the rate expression for the overall reaction, that is, excluding intermediates, if different than e: **same: rate (overall) = rate (r.d.s.) = k[AX][HY]**
    g. What might the transition state of the **r.d.s.** look like: **A−X and H−Y are being broken; A−Y and H−X are being formed. Transition state must show all changing bonds as partial bonds. A possible configuration (dashed lines = partial bonding):**

For **mechanism #2** shown in **Figure 8b**:

    a. What is the overall reaction? and
    b. Specify each step and classify as fast or slow:

Step 1	HY $\xrightarrow[k_{-1}]{k_1}$ H$^+$ + Y$^-$	**fast/equilibrium K$_1$ = k$_1$/k$_{-1}$**
Step 2	AX + Y$^-$ $\xrightarrow{k_2}$ AY + X$^-$	**slow/r.d.s.**
Step 3	H$^+$ + X$^-$ $\xrightarrow{k_3}$ HX	**fast**

Balanced equation   HY + AX ⟶ AY + HX

    c. Calculate **Ea** for the r.d.s.: **r.d.s. is Step 2 Ea (Step2) = 75 − 10 = 65 kJ/mole**
    d. Determine the sign and value for **ΔG°** for the overall reaction: ΔG° = 75 − 100 = − 25 kJ/mole-reaction
    e. What is the rate expression specifically for the **r.d.s.**: **rate(r.d.s.) = k$_2$[AX][Y$^-$]**
    f. What is the rate expression for the overall reaction (i.e., excluding intermediates, if different than e):

(Y^-) **is an intermediate and must be substituted for**

$$K_1 = [H^+][Y^-]/[HY] \quad [Y] = K_1[HY]/[H^+]$$

substitute into rate (r.d.s.):

rate (overall) $= k_2 \, [AX] \, \{K_1 \, [HY]/[H^+]\}$
rate (overall) $= k_2 \, K_1 \, [AX][HY]/[H^+]$

g. What might the transition state of the **r.d.s.** look like:
   Step 2 is the r.d.s. In Step 2, the A—X bond is breaking; the A—Y bond is forming.

   A representation of the transition state might be (dashed lines show partial bonds)

   Y-------A-------X

h. For mechanism #2 only: Step 1 is an equilibrium. Calculate the equilibrium constant for this step from the appropriate $\Delta G°$.
   The $\Delta G°$ specifically for Step 1 is shown as 10 kJ/mole.
   Solve using $\Delta G° = -2.303RT \log K$

   $\log K = \Delta G°/-2.303RT$
   $\quad = +10\,kJ/-2.303(0.008314\,kJ/K)(298\,K) = -1.75$
   $K = \text{antilog}(-1.75) = 1.8 \times 10^{-2}$

Review of Acid/ Base Concepts for Organic Chemistry

10

10.1 REVIEW OF GENERAL ACID/BASE CONCEPTS

10.1.1 GENERAL DEFINITIONS

1. A Bronsted **acid** is any molecule that transfers, in a chemical reaction, a hydrogen atom as an H^+ **ion** (also termed a **proton**) to another atom or molecule called a **base**.
2. A Bronsted **base** is any atom or molecule that picks up (accepts) the H^+ **ion** from the acid in a chemical reaction. Bases must have a lone pair of electrons that is donated to the H^+ ion in a coordinate covalent bond. Acid and base are complementary definitions.
3. An **acid/base reaction** is the complete description of the transfer of one or more H^+ ions from the acid to the base. An H^+ ion does not have an independent existence; the reaction involves the exchange of one covalent bond involving the hydrogen for another covalent bond. The hydrogen is transferred without its electron; thus, the species transferred is an H^+ ion (a proton).

If the acid and base are both neutral, a general equation is

$$H\text{---}Acid + :Base \longrightarrow (Acid)^- + (H\text{---}Base)^+$$

H—Acid = acid molecule with an H for transfer
:Base = neutral base showing electron lone pair
(Acid)⁻ = portion of the acid left over after the H^+ was transferred
(H—Base)⁺ = base molecule after the H^+ was accepted

If the base is shown as a negative ion (anion), a general equation is

$$H\text{---}Acid + (:Base)^- \longrightarrow (Acid)^- + H\text{---}Base$$

10.1.2 REVIEW PROCESS FOR WRITING ACID/ BASE REACTION EQUATIONS

If at least one component in the reaction is a strong acid or base, an equilibrium need not be solved. For simple quantitative reactions:

1. Identify the correct number of hydrogen atom(s) in the acid which will be transferred as H^+ ion(s).
2. Identify the base portion(s) of the base which will accept the H^+ ion(s).
   a. This portion will be OH^- if the base is a compound containing the hydroxide ion; include all hydroxide ions as H^+ acceptors.
   b. If the base is ammonia (NH_3), the complete molecule NH_3 is the base portion.
3. Start a partial **formula** equation with the acid and base reactants given. Predict the final products by transferring all required acid H^+ ions from the acid to the correct number of base portions of the base molecules. Preliminarily, **balance** the reactants by matching the number of H^+ ions to the corresponding number of base acceptors.
4. Write the **correct** formula of the specific product formed by combining the H^+ ion with the base portion: H^+ plus OH^- will form water (H_2O); H^+ plus NH_3 will form NH_4^+; and H^+ plus another negative ion (X^-) will form HX.
5. Write the **correct** formula of the other product by combining the remainder of the acid molecule (without the H^+) with the remaining part of the base molecule: often the ionic compound produced when the negative ion portion of the acid combines with the positive ion portion of the base:

$$Metal^{+x} + (Acid)^{-y} \longrightarrow (metal)_y(Acid)_{x\ (aq)\ or\ (s)}$$

$$or\ H{-\!\!-}Acid_{(aq)} + NH_{3\ (aq)} \longrightarrow (NH_4)^+(Acid)^-_{(aq)}$$

6. Complete the formula format of the equation by (a) writing in all products; (b) balancing the equation; and (c) determining the correct states of matter for all species including the aqueous solubility of each compound; use the solubility guidelines.
7. For aqueous reactions if necessary: write the **total ionic** format of the reaction equation. Write all strong acids as aqueous ions; all weak acids are written as neutral covalent aqueous (aq) molecules.
8. For aqueous reactions: write the **net ionic** equation by canceling exactly identical species from both sides of the total ionic equation.

Example: Complete the following reaction and write all three equation formats for reaction of aqueous sulfuric acid with aqueous potassium hydroxide:

1. H_2SO_4 transfers **two** H^+ ions.
2. The base portion of KOH is **OH^-**.

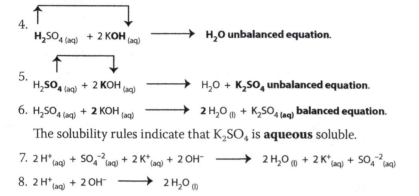

3. $H_2SO_4{}_{(aq)}$ + KOH$_{(aq)}$ ⟶

   $H_2SO_4{}_{(aq)}$ + **2** KOH$_{(aq)}$ ⟶

   two KOH are required to accept the two **H^+** ions from <u>one</u> H_2SO_4.

4. $H_2SO_4{}_{(aq)}$ + **2 KOH**$_{(aq)}$ ⟶ **H_2O unbalanced equation.**

5. **$H_2SO_4{}_{(aq)}$ + 2 KOH**$_{(aq)}$ ⟶ H_2O + **K_2SO_4 unbalanced equation.**

6. $H_2SO_4{}_{(aq)}$ + **2 KOH**$_{(aq)}$ ⟶ **2** $H_2O{}_{(l)}$ + **$K_2SO_4{}_{(aq)}$ balanced equation.**

   The solubility rules indicate that K_2SO_4 is **aqueous** soluble.

7. $2 H^+{}_{(aq)}$ + $SO_4{}^{-2}{}_{(aq)}$ + $2 K^+{}_{(aq)}$ + $2 OH^-$ ⟶ $2 H_2O{}_{(l)}$ + $2 K^+{}_{(aq)}$ + $SO_4{}^{-2}{}_{(aq)}$

8. $2 H^+{}_{(aq)}$ + $2 OH^-$ ⟶ $2 H_2O{}_{(l)}$

$SO_4{}^{-2}{}_{(aq)}$ and $2 K^+{}_{(aq)}$ are canceled from both sides of the total ionic equation.

10.1.3 CONCEPT OF CONJUGATE ACID AND CONJUGATE BASE

Acid and base are complementary definitions: an acid molecule which has lost its H^+ ion can often be considered as a possible base in the complementary **<u>reverse</u>** reaction, this species is called the **conjugate base of the acid**:

CH_3COOH +   (:Base)$^-$ ⟶ H—Base + CH_3COO^-
(acetic **acid**: transfers H^+)   (acetic acid without H^+)

$CH_3COO^-{}_{(aq)}$ = acetate ion = **<u>conjugate base</u>** of acetic acid; this ion will act as a base in the **reverse** reaction

H—Base + CH_3COO^- ⟶ CH_3COOH + (:Base)$^-$
(acetate ion **base**: accepts H^+)

A base molecule which has accepted an H^+ ion can often be considered as a possible acid in the complementary **<u>reverse</u>** reaction, this species is called the **conjugate acid of the base**

:$NH_3{}_{(aq)}$ + H—Acid ⟶ $NH_4{}^+{}_{(aq)}$ + (Acid)$^-$
(ammonia **base**: accepts H^+)   (ammonia with H^+)

$NH_4{}^+$ = ammonium ion = **<u>conjugate acid</u>** of ammonia base; this ion will act as an acid in the **reverse** reaction

$$NH_4^+ \; + \; (Acid)^- \longrightarrow \; :NH_3 \; + \; H\text{—}Acid$$
(ammonium ion **acid**: transfers H^+)

Any general reversible acid/base reaction can be identified based on the acid, base, conjugate acid, and conjugate base. The acid and base react in the forward direction; the conjugate acid and conjugate base react in the reverse direction. Example:

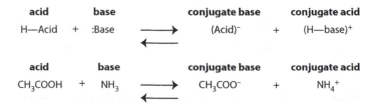

acid		base		conjugate base		conjugate acid
H—Acid	+	:Base		$(Acid)^-$	+	$(H\text{—}base)^+$

acid		base		conjugate base		conjugate acid
CH_3COOH	+	NH_3		CH_3COO^-	+	NH_4^+

An acid/base reaction can involve any combination of strong or weak acids or bases. Since the forward and reverse rates of acid/base reactions are fast, all acid/base reactions will reach the equilibrium point.

10.1.4 PROPERTIES OF ACIDS

Acids are covalent molecules; common acids show one or more hydrogen atoms covalently bonded to another atom such that the hydrogen represents the relatively positive portion (δ+) of a very polar covalent bond.

Typical inorganic (usually) strong acids used in organic reactions are HCl = hydrochloric acid; HBr = hydrobromic acid; HI = hydroiodic acid; H_2SO_4 = sulfuric acid; H_3PO_4 = phosphoric acid; $H_2Cr_2O_4$ = chromic acid; and HNO_3 = nitric acid.

Carboxylic acids (R—COOH) are generally weak acids. Other organic molecules with a polar O—H bond, such as alcohols, can behave as very weak acids. If the base is sufficiently strong, a measurable equilibrium concentration of the alkoxide can be produced.

$$R_3CO\text{—}H + (:Base)^- \longrightarrow R_3CO^- + H\text{—}Base$$
alcohol   alkoxide

The concept of a Bronsted acid is operationally defined: any molecule which transfers an H^+ during a reaction acts as an acid in that reaction. Organic molecules can transfer a hydrogen as an H^+ from a C—H bond if the base is sufficiently strong; in most cases, these molecules are extremely weak acids.

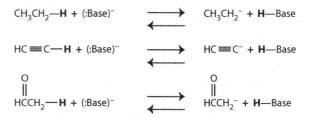

10.1.5 PROPERTIES OF BASES

A specific atom of a base molecule or ion must form a coordinate covalent bond with the H^+ ion that was transferred from the acid during a chemical reaction. Common bases contain atoms with electron lone pairs: often negatively charged oxygen or neutral oxygen or nitrogen.

Many common strong bases are ionic compounds that contain the hydroxide (OH^-) polyatomic ion: $NaOH$ (sodium hydroxide), KOH (potassium hydroxide). A corresponding class of strong alkoxide ($^-OCR_3$) bases can be formed from alcohols: $NaOCH_3$, $NaOCH_2CH_3$, $KOC(CH_3)_3$, etc.

Many other weak bases have the structure of an acid molecule that has lost one or more of its H^+ ions (the conjugate base of the acid). Examples:

CH_3COO^- = acetate
HCO_3^- = hydrogen carbonate
$H_2PO_4^-$ = dihydrogen phosphate

Organic molecules that contain neutral nitrogen or oxygen can behave as weak bases. Organic amines ($R-NH_2$) and ammonia (NH_3) are weak bases that can covalently bond to an H^+ ion through the neutral nitrogen to form the protonated nitrogen; example:

$$:NH_{3\,(aq)} + H-Acid \longrightarrow NH_4^+ + (Acid)^-$$

Neutral oxygen behaves as a weak base in alcohols and in molecules containing the carbonyl group ($C=O$); if the acid is sufficiently strong, a measurable equilibrium concentration of the protonated alcohol or carbonyl can be produced.

$$H-Acid + R_3COH \rightleftharpoons (Acid)^- + (R_3COH_2)^+$$

$$H-Acid + \overset{O}{\overset{\|}{H\text{CCH}_3}} \rightleftharpoons (Acid)^- + \overset{^+OH}{\overset{\|}{H\text{CCH}_3}}$$

Bronsted bases are also operationally defined: any molecule that accepts an H^+ during a reaction acts as a base in that reaction. Carbon can be an acceptor during the formation of a $C-H$ bond. Carbon anions, if formed, are extremely strong bases:

$$Li^+CH_3^- + H-Acid \rightleftharpoons CH_4 + Li^+(Acid)^-$$

Neutral organic molecules containing a pi-bond can accept a hydrogen as an H^+ to form a $C-H$ bond if the acid is sufficiently strong; in most cases, these molecules act as very weak bases:

$$H_2C=CH_2 + H-Acid \rightleftharpoons H_2C-CH_3^+ + (Acid)^-$$

10.2 ACID AND BASE STRENGTH

10.2.1 REFERENCE REACTIONS OF ACIDS WITH WATER

An aqueous solution of acid molecules will react with water in a reversible process with a water molecule acting as the base. This reaction is acid dissociation, also termed an ionization reaction because a covalent molecule is converted into ions by reaction with water:

$$H\text{—}Acid_{(aq)} + H_2O: \rightleftarrows (Acid)^-_{(aq)} + (H_2O\text{—}H)^+$$

$(H_2O\text{—}H)^+$ is written as H_3O^+ or more generally as $H^+_{(aq)}$

It is generally true that H_3O^+ and $H^+_{(aq)}$ can be considered as identical descriptions of the same species. Therefore, the reaction of an acid with water can also be written as

$$H\text{—}Acid \xrightarrow{\quad H_2O\ solvent\quad} H^+_{(aq)} + (Acid)^-_{(aq)}$$

10.2.2 REFERENCE REACTIONS OF BASES WITH WATER

Most bases can react with water to produce the hydroxide ion (OH^-).

Strong bases are hydroxide-containing compounds that dissolve completely in water to form approximately complete separation of the dissolved ions; in this case, water acts only as a solvent. A general example:

$$Metal(OH) \xrightarrow{\quad H_2O\ solvent\quad} Metal^+_{(aq)} + OH^-_{(aq)} \approx 100\% \text{ with high solubility}$$

Many weak bases will react with water in a reversible reaction with the water molecule acting as the **acid**:

$$:Base + H_2O \rightleftarrows OH^-_{(aq)} + (H\text{—}Base)^+_{(aq)}$$

or

$$(:Base)^- + H_2O \rightleftarrows OH^-_{(aq)} + H\text{—}Base_{(aq)}$$

10.2.3 ACID STRENGTH

Acids can very generally be defined as either **weak** acids or **strong** acids. The basis of the distinction is the equilibrium constant for the acid dissociation reaction with water. The format of this reaction showing the role of water as a base demonstrates that H_3O^+ can be considered the conjugate acid of water as a base:

$$H\text{—}Acid_{(aq)} + H_2O: \quad \longrightarrow \quad (Acid)^-_{(aq)} + H_3O^+$$

As an approximate distinction, acids that have an acid dissociation equilibrium constant with a numerical value greater than one can be considered **strong** acids:

$$H\text{—}Acid \textbf{ (strong)} \quad \xrightarrow{H_2O \text{ solvent}} \quad H^+_{(aq)} + (Acid)^-_{(aq)} \ \textbf{K}_\textbf{a} > \textbf{1}$$

$$H\text{—}Cl \textbf{ (strong)} \quad \xrightarrow{H_2O \text{ solvent}} \quad H^+_{(aq)} + Cl^-_{(aq)} \ \textbf{K}_\textbf{a} = \textbf{1} \times \textbf{10}^\textbf{7} \ (\cong 100\% \text{ reaction})$$

Acids which have an acid dissociation equilibrium constant with a numerical value less than one can be considered as **weak** acids:

$$H\text{—}Acid \textbf{ (weak)} \quad \xrightarrow{H_2O \text{ solvent}} \quad H^+_{(aq)} + (Acid)^-_{(aq)} \ \textbf{K}_\textbf{a} < \textbf{1}$$

$$CH_3COO\text{—}H \textbf{ (weak)} \quad \xrightarrow{H_2O \text{ solvent}} \quad H^+_{(aq)} + CH_3COO^-_{(aq)} \ \textbf{K}_\textbf{a} = \textbf{1.8} \times \textbf{10}^{-\textbf{5}} \ (\cong 0.5\% \text{ reaction})$$

Polyprotic acids have separate equilibrium constants for each proton transfer; examples:

$$H_2SO_{4\,(aq)} \quad \xrightarrow{H_2O \text{ solvent}} \quad H^+_{(aq)} + HSO_4^-{}_{(aq)} \ \textbf{K}_\textbf{a} = \textbf{6} \times \textbf{10}^\textbf{4}$$

$$HSO_4^-{}_{(aq)} \quad \xrightarrow{H_2O \text{ solvent}} \quad H^+_{(aq)} + SO_4^{-2}{}_{(aq)} \ \textbf{K}_\textbf{a} = \textbf{1.2} \times \textbf{10}^{-\textbf{2}}$$

10.2.4 BASE STRENGTH

The format of the base reaction showing the role of water as an acid demonstrates that OH^- can be considered as the conjugate base of water as an acid:

$$:Base + H_2O \quad \longrightarrow \quad OH^-_{(aq)} + (H\text{—}Base)^+_{(aq)}$$

Generally, **strong** bases are those with a base strength equal to or greater than hydroxide ion. **Weak** bases are those for which the reaction with water has an equilibrium constant less than one:

$$:Base \textbf{ (weak)} \quad \xrightarrow{H_2O \text{ solvent}} \quad OH^-_{(aq)} + (H\text{—}Base)^+_{(aq)} \ \textbf{K}_\textbf{b} < \textbf{1}$$

$$:NH_{3\,(aq)} \textbf{ (weak)} \quad \xrightarrow{H_2O \text{ solvent}} \quad OH^-_{(aq)} + NH_4^+{}_{(aq)} \ \textbf{K}_\textbf{b} = \textbf{1.8} \times \textbf{10}^{-\textbf{5}}$$

10.3 RELATIONSHIP BETWEEN ACID STRENGTH (pK$_a$) AND CONJUGATE BASE STRENGTH

The acid dissociation equilibrium constants for acids and bases are often listed as a derived numerical value.

The **pK$_a$** is defined as the negative logarithm of the numerical value of the K$_a$:

$$pK_a = -\log_{10}(K_a).$$

The **pK$_b$** is defined as the negative logarithm of the numerical value of the K$_b$:

$$pK_b = -\log_{10}(K_b).$$

The acid strength of an acid is inversely proportional to the base strength of its conjugate base: very strong acids have very weak conjugate bases; very weak acids have strong conjugate bases. This conclusion is based on the spontaneity of the forward and reverse reactions: the more spontaneous the forward acid reaction (stronger acid), the less spontaneous the reverse conjugate base reaction (weaker base). The converse statement is true for weaker acids compared to stronger conjugate bases.

The autoionization of water provides a relationship between the K$_a$ of an acid and the K$_b$ of the corresponding conjugate base. Neutral water will self-ionize to form H$^+$ aqueous (or equivalently H$_3$O$^+$ aqueous) and OH$^-$ aqueous; the reaction is an equilibrium

$$H_2O_{(l)} + H_2O_{(l)} \rightleftharpoons H_3O^+_{(aq)} + OH^-_{(aq)} \quad K_w = 1 \times 10^{-14}$$

or, equivalently:

$$H_2O_{(l)} \rightleftharpoons H^+_{(aq)} + OH^-_{(aq)} \quad K_w = 1 \times 10^{-14}$$

The logarithmic relationship based on pK$_w$, pK$_a$, and pK$_b$ is

$$pK_a \text{ (of an acid)} + pK_b \text{ (of the conjugate base)} = pK_w = 14$$

The relationship between acid and conjugate base is true regardless of the reaction point of view: that is, the relationship can be expressed in the form of a base and its conjugate acid. For this reason, the strength of all acid/base species is generally described as the pK$_a$ value of the acid or conjugate acid. The table in the following page lists a general comparison table of a number of acids and bases occurring in a wide variety of organic reactions.

Table of pK$_a$ Values for Common Acids and Bases

Strongest Acid		Weakest Base
↑		↑
Acid	**pK$_a$**	**Conjugate Base**
HI	−10	I$^-$
HBr	−9	Br$^-$
HCl	−7	Cl$^-$
C$_6$H$_5$SO$_3$H	−6.5	C$_6$H$_5$SO$_3^-$
H$_2$SO$_4$ (first)	−4.8	HSO$_4^-$
(CH$_3$)$_2$OH$^+$	−3.8	(CH$_3$)$_2$O
(CH$_3$)$_2$C=OH$^+$	−2.9	(CH$_3$)$_2$C=O
CH$_3$OH$_2^+$	−2.5	CH$_3$OH
H$_3$O$^+$	−1.7	H$_2$O
HNO$_3$	−1.4	NO$_3^-$
CF$_3$COOH	+0.2	CF$_3$COO$^-$
HF	+3.5	F$^-$
H$_2$CO$_3$	+3.7	HCO$_3^-$
CH$_3$COOH	+4.7	CH$_3$COO$^-$
CH$_3$COCH$_2$COCH$_3$	+9	CH$_3$COCH-COCH$_3$
NH$_4^+$	+9.2	NH$_3$
C$_6$H$_5$OH	+9.9	C$_6$H$_5$O$^-$
HCO$_3^-$	+10.2	CO$_3^{-2}$
CH$_3$NH$_3^+$	+10.6	CH$_3$NH$_2$
H$_2$O	+15.7	OH$^-$
CH$_3$CH$_2$OH	+16	CH$_3$CH$_2$O$^-$
(CH$_3$)$_3$COH	+18	(CH$_3$)$_3$CO$^-$
CH$_3$COCH$_3$	+19.2	CH$_3$COCH$_2^-$
HC≡CH	+25	HC≡C$^-$
H$_2$	+35	H$^-$
NH$_3$	+36	NH$_2^-$
H$_2$C=CH$_2$	+45	H$_2$C=CH$^-$
CH$_3$CH$_3$	+60	CH$_3$CH$_2^-$
↓		↓
Weakest Acid		**Strongest Base**

10.4 DETERMINING THE EQUILIBRIUM POSITION FOR ACID/BASE REACTIONS

The equilibrium position of an acid/base reaction describes which reaction, forward or reverse, is favored under standard concentration conditions:

acid	base		conjugate base		conjugate acid
H—Acid	+ :Base	⟶ ⟵	(Acid)$^-$	+	(H—base)$^+$

The favored (spontaneous) direction is determined by the value of $\Delta G^0_{(forward)}$ and is reflected in the corresponding equilibrium constant:

If the forward reaction is favored: $K_{(forward)} > 1$ and $K_{(reverse)} < 1$; the equilibrium favors product formation from reactants.

If the reverse reaction is favored: $K_{(forward)} < 1$ and $K_{(reverse)} > 1$; the equilibrium favors reactant formation from products.

The equilibrium position of an acid/base reaction is determined by comparing the values of K_a/pK_a for the acid and conjugate acid formed from the reacting base in the equation. The favored direction of any acid/base reaction ($K > 1$) always shows H^+ (proton) transfer from the stronger acid (lower pK_a value) to produce the weaker acid (higher pK_a value).

Example: Determine the position of the following equilibrium:

$$CH_3COOH + NH_3 \rightleftharpoons CH_3COO^- + NH_4^+$$

Step 1: Identify the acid in the forward reaction, then, use the table of pK_a values to determine the pK_a of this acid.

In the forward direction, CH_3COOH transfers the proton; it is the acid. CH_3COOH has a pK_a of 4.7.

Step 2: Identify the base in the forward reaction, then, use the table of pK_a values to determine the pK_a of the conjugate acid of this base. The pK_b of the base is not required.

In the forward direction, NH_3 accepts the proton; it is the base. NH_4^+ is the conjugate acid of NH_3 as a base; NH_4^+ has a pK_a of 9.3.

Step 3: The equilibrium is favored in the direction of H^+ (proton) transfer from the stronger acid (CH_3COOH; $pK_a = 4.7$) to produce the weaker acid (NH_4^+; $pK_a = 9.3$)

$$CH_3COOH + NH_3 \rightleftharpoons CH_3COO^- + NH_4^+ \quad \mathbf{K > 1}$$

$pK_a = 4.7$ **(stronger)** $pK_a = 9.3$ **(weaker)**

10.5 ADDITIONAL EXAMPLES

Determine the equilibrium position for the following reactions:

a. $CH_3COOH + HSO_4^- \rightleftharpoons CH_3COO^- + H_2SO_4$

b. $HC \equiv CH + OH^- \rightleftharpoons HC \equiv C^- + H_2O$

c. $HC \equiv CH + NH_2^- \rightleftharpoons HC \equiv C^- + NH_3$

ANSWERS:

a. (1) In the forward direction, CH_3COOH transfers the proton; it is the acid. CH_3COOH has a pK_a of 4.7.

(2) In the forward direction, HSO_4^- accepts the proton; it is the base. H_2SO_4 is the conjugate acid of HSO_4^- as a base; H_2SO_4 has a pK_a of −4.8.

(3) The equilibrium is favored in the direction of H^+ (proton) transfer from the stronger acid (H_2SO_4; $pK_a = -4.8$) to produce the weaker acid (CH_3COOH; $pK_a = 4.7$).

a. $CH_3COOH + HSO_4^-$ $\longrightarrow$ $CH_3COO^- + H_2SO_4$ **K < 1**

$pK_a = 4.7$ **(weaker)** $\longleftarrow$ $pK_a = -4.8$ **(stronger)**

b. (1) In the forward direction, $HC{\equiv}CH$ transfers the proton; it is the acid. $HC{\equiv}CH$ has a pK_a of 26.

(2) In the forward direction, OH^- accepts the proton; it is the base. H_2O is the conjugate acid of OH^- as a base; H_2O has a pK_a of 15.7.

(3) The equilibrium is favored in the direction of H^+ (proton) transfer from the stronger acid (H_2O; $pK_a = 15.7$) to produce the weaker acid ($HC{\equiv}CH$; $pK_a = 26$).

b. $HC{\equiv}CH + OH^-$ $\longrightarrow$ $HC{\equiv}C^- + H_2O$ **K < 1**

$pK_a = 26$ **(weaker)** $\longleftarrow$ $pK_a = 15.7$ **(stronger)**

c. (1) In the forward direction, $HC{\equiv}CH$ transfers the proton; it is the acid. $HC{\equiv}CH$ has a pK_a of 26.

(2) In the forward direction, NH_2^- accepts the proton; it is the base. NH_3 is the conjugate acid of NH_2^- as a base; NH_3 has a pK_a of 36.

(3) The equilibrium is favored in the direction of H^+ (proton) transfer from the stronger acid ($HC{\equiv}CH$; $pK_a = 26$) to produce the weaker acid (NH_3; $pK_a = 36$).

c. $HC{\equiv}CH + NH_2^-$ $\longrightarrow$ $HC{\equiv}C^- + NH_3$ **K > 1**

$pK_a = 26$ **(stronger)** $\longleftarrow$ $pK_a = 36$ **(weaker)**

Electrophiles and Nucleophiles in Organic Reaction Mechanisms

11.1 OVERVIEW OF GENERAL CONCEPTS

11.1.1 ELECTROPHILES AND NUCLEOPHILES

1. Many organic reaction mechanisms involve one or more steps that can be described as a combination of an **electrophile** with a **nucleophile**.
2. An **electrophile** ("electron-loving") is a reactant species in a complete reaction or a mechanistic step that requires additional electrons or is attracted to electron density. An electrophile may be an atom, molecule, or ion.
3. A **nucleophile** ("nuclear-" or "positive charge-loving") is a reactant species in a complete reaction or mechanistic step, which can donate electrons or is attracted to positive charge.

11.1.2 REACTION CONCEPTS

1. A nucleophile and an electrophile are complementary reactant species: a reaction induced by one species type **must** involve the complementary species type as the co-reactant. This is similar to the requirements of acid/base and oxidation/reduction reactions.
2. Although, both electrophile and nucleophile must be present in any complete step, the nature of reactant species combinations allows many organic mechanisms to be classified as **"electrophilic"** or **"nucleophilic"** (e.g., "electrophilic addition," "nucleophilic substitution," etc.). Designation of a reaction as electrophilic or nucleophilic can be based on a consideration of some selected thermodynamic or kinetic feature or may simply reflect the best description of how the spectrum of possible products is formed. For example, the reactions based on "nucleophilic addition to carbonyl," fully discussed in Chapter 18, describe one class of electrophiles (carbonyls) for which a wide variety of possible nucleophile types are available.

11.2 SPECIES IDENTIFICATION AND CHARACTERISTICS

11.2.1 ELECTROPHILES

1. *An **electrophile** can be any species that **acquires** (accepts) electrons or electron density from a nucleophile in a reaction step.* The term "species" implies either a single atom, a molecular "fragment" (i.e., at least one atom of the molecule is missing a bond), a neutral molecule, or an elemental or polyatomic ion.

2. The term "electrophile" can refer to the complete molecule, ion, or fragment that is reacting, or it can mean the **specific atom** (the "electrophilic atom") in the molecule that demonstrates the electron deficiency characteristic of the species.

   The electronic and bonding descriptions applied to an electrophile always refer to the specific atom forming the actual bond; the rest of the molecule "comes along for the ride."

   Normal neutral molecules are most often comprised of different combinations of electrophilic functional groups and nucleophilic functional groups. *For the evaluation of molecular behavior, it is important to identify the behavior of specific atom or atom combinations within a reactant molecule.*

3. *The most common electrophiles are those that accept an electron pair in a chemical reaction. Atoms that accept an electron pair to form a new bond must have an empty orbital to accommodate these electrons. Classification of general types are*

   a. **Electrophiles with an empty s-orbital or p-orbital (or d-orbital for elements other than row 2).**

   b. **Full-octet electrophiles in which an empty orbital becomes available during a reaction.**

4. The general behavior of electrophiles is as electron pair **acceptors** in a **coordinate covalent bond**, that is, they act as Lewis acids.

5. **Electrophiles with an empty s-orbital, p-orbital (or d-orbital):** An atom in a molecule or molecular fragment that has an **empty** orbital **immediately available** can accept a bonding pair of electrons to make a new coordinate covalent bond. Positive ions or non-octet neutral atoms are examples of species that have empty orbitals.

   Examples:

   H^+ : empty 1s-orbital available to accept an electron pair.

   R_3C^+ : **(R = C or H)** general form of a carbon positive ion (cation); carbon has an empty p-orbital due to a missing bond and the additional one missing electron.

   Br^+ : the bromine positive ion has only six electrons and thus has an empty p-orbital.

BF$_3$ and Cl$_3$Al : boron and aluminum are neutral but each has an empty p-orbital because each is a non-octet bonded atom

6. **Full-octet electrophiles in which an empty orbital becomes available during a reaction:**

In many cases, an electrophile (such as carbon) starts out in a reaction step with a complete bonding octet; the electrophilic atom cannot accept an electron pair in this condition due to the lack of an available orbital in which to place the electrons.

However, a full-octet atom **can** act as an electrophile if another bond to it is broken **during** the reaction step; an orbital is then made available for the new electron pair. This condition is required to produce an electrophilic full-octet row-2 atom such as carbon. *The requirement for an orbital to be made available during the electron pair transfer is that bond breaking to another bonded atom must occur simultaneously with the electron acceptance.*

The electrophilic atoms that behave in the required manner are usually the relatively positive end of a bond dipole.

Examples:

$\overset{\delta+}{R_2C} = \overset{\delta-}{O}$ Carbon ($\delta+$) can be an electrophile during reaction steps in which the pi-bond to oxygen breaks **simultaneously** with the formation of a new bond to carbon.

$\overset{\delta+}{H_3C} — \overset{\delta-}{Cl}$ Carbon ($\delta+$) can be an electrophile during reaction steps in which the bond to chlorine breaks **simultaneously** with the formation of a new bond to carbon.

11.2.2 NUCLEOPHILES

1. *A **nucleophile** can be any species that **donates** electrons or electron density to an electrophile in a reaction step.* The nucleophile may be an atom, ion, or molecule and can refer to the entire species or, more specifically, to the "nucleophilic atom."

As for electrophiles, *the electronic and bonding descriptions applied to a nucleophile always refer to the specific atom forming the actual bond.*

For the evaluation of molecular behavior, whether as a nucleophile or electrophile, it is important to identify the behavior of specific atom or atom combinations within a reactant molecule.

2. The most common nucleophiles are those that donate an electron pair to form a new bond in a chemical reaction. *Atoms that act as nucleophiles must have a free or available electron pair to donate to the electrophile in bond formation.* Classification of general types are

   a. **Nucleophiles donating a free lone pair of electrons.**
   b. **Nucleophiles donating an electron pair from a pi-bond.**

    c. **Nucleophiles donating an electron pair from a sigma-bond.**

3. The general behavior of nucleophiles is as electron pair **donors** in the formation of a **coordinate covalent bond**, that is, they act as Lewis bases.

4. **Nucleophiles donating a free lone pair of electrons**: Nucleophilic atoms that behave this way must have free lone electron pairs; the lone pair is donated to the electrophile for the coordinate covalent bond. Common nucleophiles are negative ions (such as halogens) or neutral molecules that contain atoms with available lone pairs after the formation of normal bonding (oxygen, nitrogen, sulfur, or phosphorus). The halogens (F, Cl, Br, I) are weakly nucleophilic as diatomic molecules *but are not nucleophilic when singly bonded to carbon.*
   Examples:

$$:\ddot{Br}:^- \quad ; \quad H-\ddot{O}:^- \quad ; \quad H_2\ddot{O} \quad ; \quad R_3\ddot{N} \quad ; \quad R_3C-\ddot{O}-H \quad ; \quad R-\ddot{S}:^- \quad ; \quad R\ddot{S}H$$

5. **Nucleophiles donating an electron pair from a pi-bond**: Electrons in a pi-bond are not as freely available as lone pair electrons; however, pi-bond electrons are held less tightly than sigma-bonding electrons. *An electron pair in a pi-bond can be nucleophilic (donated to form a coordinate covalent bond) during a reaction step whenever the original pi-bond is broken in the process.* **Examples** are **pi** bonds of double and triple bonds:

$$R_2C=CR_2 \quad ; \quad R_2C=O \quad ; \quad RC\equiv CR$$

6. **Nucleophiles donating an electron pair from a sigma-bond**: These nucleophiles are **not common**, but can occur in certain organic mechanisms. Donation of an electron pair from a sigma-bond is generally observed only for the highly polar metal–carbon or metal–hydrogen covalent sigma bonds.

7. Nucleophiles are usually evaluated on the basis of "**nucleophilic strength**," a measure of the influence of nucleophile chemical structure on the rate of specific reactions: a "**stronger**" nucleophile designation corresponds to a **greater rate increase** caused by the nucleophile (vs. a standard).

Strong nucleophiles can often be recognized as **strong bases**. Both are characterized by a large negative value of $\Delta G°$ for the specific electron pair donation reaction. It is often (but **not** always) true that the magnitude of the free energy change in a reaction can be directly related to the rate of the reaction. A more favorable free energy change often indicates a potentially more favorable (lower) activation energy.

Strong nucleophiles must also have an electron pair energetically **available** for donation; this requirement is independent of the ultimate free energy change. The **rate** of a nucleophile/electrophile combination will depend on the **activation energy** for electron pair donation, dependent on **electron availability** and **solvent**.

Nucleophilic strength can depend on the tightness of the solvent shell surrounding the nucleophilic atom. The strength of halogen nucleophiles follows base strength in **aprotic** solvents since all anions are solvated loosely; in this case, solvent molecules do not hinder electron donation. The strength of halogen nucleophiles is reversed in **protic** solvents (solvents with a hydrogen-bonding capable H δ+). Under these conditions, electron availability is proportional to **polarizability** since tightly bound solvent molecules cause nucleophilic electrons to be relatively unavailable. Atoms that hold electrons **tightly** (e.g., F) have a closely held valence shell and are characterized as **low polarizability**. These atoms have a tight solvent shell (due to smaller ionic radius) and demonstrate relatively slower rates of reaction: **weaker nucleophiles**. Atoms that hold electrons **loosely** (e.g., I) are characterized as **high polarizability**. These atoms have a relatively loose solvent shell (due to larger ionic radius) and will be **stronger nucleophiles**. Recall that polarizability/looseness of solvent shell for a specific family (column in the periodic table) increases with increasing atom size.

The specific atom in a nucleophilic molecule that donates the electron pair in bond formation must be the **strongest nucleophilic atom**. For polyatomic negative ions, the strongest nucleophilic atom is usually identified as the one bearing the formal negative charge.

Very generally, for a specific nucleophilic atom, the **negative** (anionic) form of the atom will be a stronger nucleophile than the **neutral** atom form:

$$\text{Br}^- > \text{Br}_2 \qquad \text{H—O}^- > \text{H}_2\text{O} \qquad \text{H}_2\text{N}^- > \text{NH}_3$$

11.3 REACTION CHARACTERISTICS

11.3.1 ELECTROPHILES AND NUCLEOPHILES IN MECHANISMS

1. Electrophile combinations with nucleophiles are specific only for single-reaction steps. Therefore, a number of different electrophiles and nucleophiles may be sequentially identified in any multistep reaction mechanism.
2. The key species that provides the classification for the mechanism type may be the electrophile or nucleophile that initiates the reaction or has the critical role in the rate-determining or product-determining step.

11.3.2 BOND FORMATION WITH ELECTROPHILES HAVING AN AVAILABLE EMPTY ORBITAL: ADDITION OF THE ELECTROPHILE TO THE NUCLEOPHILE

For reactions of **electrophiles with an immediately available empty orbital**: the most common reaction between nucleophile and electrophile is direct formation of a coordinate covalent bond between them. This is termed an **addition** reaction. The **nucleophile donates both** electrons of an electron pair to form the bond; the **electrophile accepts both** electrons forming the bond. *The nucleophile almost always donates an electron pair from a lone pair or an electron pair in a pi-bond.*

General examples: (**E** = electrophile; **Nu:** = nucleophile; *the arrow indicates the direction of electron pair transfer.*)

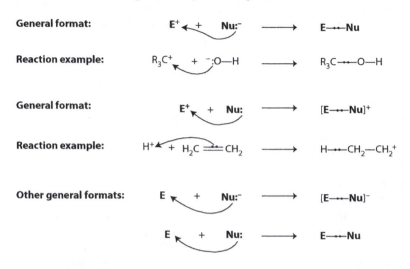

11.3.3 BOND BREAKING BETWEEN THE ELECTROPHILE AND NUCLEOPHILE BY ELIMINATION

Any bond formation step between an electrophile and nucleophile can, in principle, have a corresponding **reverse** reaction. *Bond cleavage between an electrophilic and nucleophilic portion of a molecule indicates an elimination step.* The result is formation of a **new** potential electrophile and a **new** potential nucleophile from the original combination. (The original molecule may be ionic or neutral.) The equilibrium may be viewed as

$$\text{E---Nu} \;\rightleftharpoons\; \text{E}^+ + \text{Nu:}^-$$

Examples

Bond cleavage in a neutral molecule to form a new electrophile plus nucleophile combination:

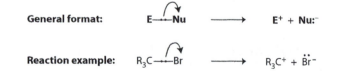

Bond cleavage in a positive ion (possible electrophile) to form a neutral molecule plus a new electrophilic atom:

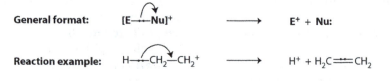

11.3.4 BOND FORMATION WITH ELECTROPHILES HAVING A FULL OCTET

1. A full-octet atom, such as carbon, acts as an electrophile if another bond to it is broken **during** the reaction step to make an orbital available for the new electron pair.

    One of the bonded atoms to this electrophilic carbon **must** be in the process of bond breaking during the electrophile/nucleophile combination as required to "open up" a position for the incoming nucleophile electron pair.

    Electrophilic carbons are usually the relatively **positive** end of a bond dipole. The bond which breaks during the reaction may be a pi-bond or a sigma-bond.

2. Reaction of a nucleophile with a **pi**-bonded carbon electrophile occurs by an **addition** mechanism. This term refers to the fact that *as the nucleophile adds to the electrophile, no portion of the electrophile molecule is removed or lost in the process of the required bond breaking.* A direct **addition** (with no loss of part of the molecule) is possible because cleavage of **only** the **pi**-portion of a double bond does not completely break the E—X bond:

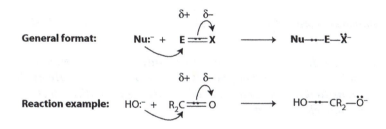

3. Reaction of a nucleophile with a carbon containing four **single** bonds (four **sigma** bonds) requires a **substitution** mechanism. This term refers to the fact that *as the nucleophile bonds to the electrophile, part of the electrophile molecule must be removed or lost in the process of bond breaking.* Since **all** bonds to the electrophilic atom are **single** bonds, the nucleophile must **substitute** at an existing bond position. The removed part may be a single atom or ion, or a molecular fragment or polyatomic ion; it is termed the **leaving group**:

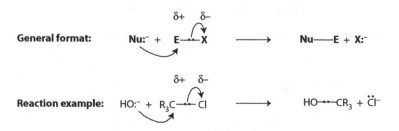

11.4 ADDITIONAL CONCEPTS

11.4.1 NUCLEOPHILES USING A SIGMA-BONDING ELECTRON PAIR

1. Electron pairs in **sigma** bonds are more tightly held than **pi**-bonding electrons.

    Donation of an electron pair from a sigma-bond is generally observed only for the highly polar metal–carbon or metal–hydrogen covalent sigma bonds, in molecules termed organometallics. In these cases, the nucleophilic electrons can be identified by viewing the molecule as an ionic compound:

$$\delta- \quad \delta+ \qquad\qquad\qquad\qquad\qquad \delta- \quad \delta+$$
$$R_3C\text{---Metal} \quad \longleftrightarrow \quad R_3C\text{:}^- \ \text{Metal}^+ \ ; \ \text{or} \ H\text{---Metal} \quad \longleftrightarrow \quad H\text{:}^- \ \text{Metal}^+$$

2. Nucleophiles of this type can add to an electrophile, most often by an addition mechanism:

$$\delta+ \quad \delta-$$
General format: $\qquad$ $Nu\text{:}^- \ \text{Metal}^+ \ + \ E\text{===}X \quad\longrightarrow\quad Nu\text{---}E\text{---}\ddot{X}^- \ \text{Metal}^+$

$$\delta+ \quad \delta-$$
Reaction example: $\qquad$ $CH_3^- \ Li^+ \ + \ R_2C\text{===}O \quad\longrightarrow\quad CH_3\text{---}CR_2\text{---}\ddot{O}^- \ Li^+$

3. *Except for reactions involving radicals* (next section) *or metal-containing compounds, nucleophiles should always be identified as being capable of donating pi-bonding electrons or lone electron pairs*.

11.4.2 SINGLE ELECTRON TRANSFER: FREE RADICAL REACTIONS

1. A free radical is a neutral atom that is missing one bond and therefore is in a non-octet bonding configuration; the atom has an odd number of valence electrons. The unpaired electron can be thought of as located in the orbital missing the bond; there are usually no empty orbitals.
2. Free **radical** molecular **electrophiles** contain one radical atom that requires only one electron to complete the valence shell and achieve normal bonding. Some example species:

$$\cdot N\text{===}\ddot{O} \quad ; \quad H_3C\cdot \quad ; \quad \text{:}\ddot{C}l\cdot \quad ; \quad \cdot\ddot{O}\text{---}\ddot{O}\text{---}H$$

3. All free radicals can be considered **electrophilic** since each must **accept** one electron into a half-filled orbital to complete covalent bond formation. In a simple radical combination, there is no practical distinction between electrophile and nucleophile.
4. Free radical electrophiles accept a net of one electron in bond formation with **nucleophiles**. The nucleophile may contribute

an **unpaired** electron; **one** electron from a **pi-bond or lone pair**; or **one** electron from a **sigma-bond**.

5. **Free radical reaction types:**

The simplest case is direct combination of an electrophile **radical** with another **radical** (either the same or different species) through each **unpaired** electron to form a covalent bond. There is usually little to distinguish the electrophile from the nucleophile:

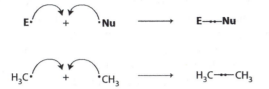

The radical electrophile can combine with a nucleophile **lone pair or pi-bond pair; only one** electron is transferred to the electrophile. In this case, the result is formation of a **new** (electrophilic) radical product. This represents an **addition** by E·

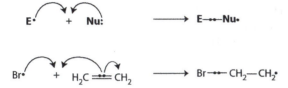

The free radical electrophile can react with a **sigma-bonding electron pair; only one** electron is transferred to the electrophile. The result is **homolytic** cleavage of the sigma-bond and formation of a **new** (electrophilic) radical product. The term **homolytic** bond breaking indicates that each of the bonded atoms retains one electron from the original bond. This represents a **substitution** by E·

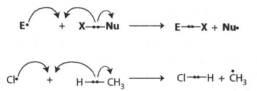

11.5 ADDITIONAL PROBLEMS

11.5.1 PRACTICE EXERCISES

1. Consider the following species for reactions of electron pair transfer. (Do not consider free radical reactions.) State whether the species will most likely act as a nucleophile or an electrophile in reaction steps. For polyatomic species, then, identify the specific nucleophilic or electrophilic atom that would be involved in bond formation. **Draw Lewis structures where necessary.**

a. Cl$^+$ b. BH$_3$ c. CH$_3$O$^-$ d. CN$^-$

e. CH$_3$CH$_2$—Br   f. H—AlH$_3$Li   g. CH$_3$CH$_2^+$ h. HCC$^-$

i. H$_3$C—MgBr   j. (CH$_3$CH$_2$)$_2$NH   k. (CH$_3$)$_2$C =CH$_2$

2. The following two compounds can act in either an electrophilic or nucleophilic capacity depending on which part of the molecule is involved in the reaction step. For **each** molecule, identify the specific atom or molecular portion that can behave as the **electrophile**. Then, repeat the analysis for the portion of the molecule that can act as the **nucleophile**.

a. H$_3$C—C(=O)—CH$_3$ b. H$_3$C—C(=O)—OH

3. For **each** of the three pairs listed, select which molecule of the pair is expected to be the stronger nucleophile.
   a. CH$_3$OH or CH$_3$O$^-$
   b. Br$^-$ or Cl$^-$
   c. H$_3$CSCH$_3$ or H$_3$COCH$_3$

11.5.2 ANSWERS TO PRACTICE EXERCISES

1. a. Cl$^+$ is positively charged, is missing a bond, and has an empty p-orbital; it will accept an electron pair as an **electrophile**.
   b. BH$_3$ is neutral, but boron has an empty p-orbital and is a non-octet bonding atom; it will accept an electron pair as an **electrophile**.
   c. CH$_3$O$^-$ The negatively charged **oxygen** will donate one of its lone electron pairs as a **nucleophile**.
   d. CN$^-$ = $^-$:C≡N The **carbon** bears a **negative** charge and has a lone pair that can be donated as a **nucleophile**. (The nitrogen lone pair is much less likely to be nucleophilic in this case.)
   e. CH$_3$CH$_2$—Br = CH$_3$CH$_2$—Br ($\delta+$, $\delta-$) The **relatively positive carbon** can act as an **electrophile if** the C—Br bond breaks during the reaction step to open up a bonding position. Note that halogens, even though they contain lone pairs, are **not** nucleophilic when singly bonded to carbon.
   f. H—AlH$_3$Li = H—AlH$_3$Li ($\delta-$, $\delta+$) Since **hydrogen** is bonded to a metal, the H—Metal electron pair can be donated and makes **H** a **nucleophile**.
   g. CH$_3$CH$_2^+$ The positively charged carbon is missing a bond, has an empty orbital, and can act as an **electrophile**.
   h. HCC$^-$ = HC≡C:$^-$ The **carbon** that bears the **negative** charge has a lone pair and can act as a nucleophile.

$$\overset{\delta-\;\;\delta+}{}$$

i. H_3C—$MgBr$ = H_3C—$MgBr$ Since **carbon** is bonded to a metal, the C—Metal electron pair can be donated and makes **C** a **nucleophile**.

j. $(CH_3CH_2)_2NH$ The **nitrogen** has a lone pair that can be donated as a **nucleophile**. (A possible answer based on the concepts discussed could be that the carbon bonded to the nitrogen might act as an electrophile if the C—N bond breaks during the reaction.)

k. $(CH_3)_2C{=}CH_2$ The **pi-bonding electrons from the double bond** can be donated as **nucleophilic** electrons; the pi-electrons would form a new sigma-bond with the corresponding electrophile in a reaction step.

2.

a.
$$\overset{\delta-}{:O:}\\ \overset{\parallel}{H_3C-C-CH_3}\\ \overset{}{\delta+}$$

b.
$$\overset{\delta-}{:O:}\\ \overset{\parallel}{H_3C-C-\overset{..}{\underset{..}{O}}H}\\ \overset{}{\delta+}$$

The **relatively positive carbon** in each of the molecules can act as an **electrophile** whenever the pi-bond to oxygen breaks in a reaction step to open up a bonding position. The **relatively negative oxygen** can donate a lone electron pair as a **nucleophile**, either with or without pi-bond cleavage. The **pi-bonding electrons** in each molecule can also act as a **nucleophile**. Molecule (b) has an additional **oxygen** bonded to the hydrogen which can also act as a **nucleophile** through its lone pair electrons.

3. a. CH_3OH or $\mathbf{CH_3O^-}$

Oxygen is the nucleophilic atom in both molecules. For an equivalent nucleophilic atom, the stronger base form is the stronger nucleophile. The negatively charged oxygen in $\mathbf{CH_3O^-}$ is a stronger nucleophile than the neutral oxygen in CH_3OH.

b. $\mathbf{Br^-}$ or Cl^-

c. $\mathbf{H_3CSCH_3}$ or H_3COCH_3

For nucleophilic atoms in the same family (column in the periodic table), the more polarizable atom will be the stronger nucleophile at least under certain conditions. Bromine holds its electrons more loosely than chlorine and is more polarizable; **Br** is a stronger nucleophile than Cl. Sulfur holds its electrons more loosely than oxygen; **S** is a stronger nucleophile than O.

Conceptual Guide to Mechanisms in Organic Chemistry

12.1 OVERVIEW FOR UNDERSTANDING REACTION MECHANISMS

12.1.1 REACTION ELECTRON CHANGES

1. A **mechanism** is a description of sequential bonding change steps that together provide a picture of how the complete reaction occurs (**see Chapter 9**).
   a. All bonding changes require a redistribution of electrons; *therefore, mechanistic descriptions are based on the principle of following the electrons*.
   b. Electron changes are often viewed as *following electron transfers from* **nucleophiles** *to* **electrophiles** (**see Chapter 11**). The nucleophile and electrophile identifications will be different for each bonding change/combination (mechanistic step) in the sequence describing the complete reaction.
2. In many cases, electrophiles or nucleophiles in sequential steps may be related as different **acid/base** forms.
   a. *Some mechanistic steps in a reaction may represent required proton transfers between an acid form and a base form of the specific molecule.*
   b. Acid/base equilibria should not obscure the key bonding changes in a reaction. *To keep track of acid/base conditions, simply add or remove protons where required.*
3. Many reaction mechanisms describe a general pathway that has the potential to form **more than one product**.
   a. In most cases, **one** product is more likely to be formed compared to the other possibilities; this is termed the **major product**.
   b. The structure of the **major product** in a reaction is determined either by thermodynamics (the **most stable** product) or by kinetics (the product formed **most rapidly**).

12.1.2 SUBSTITUTION PATTERNS ON CARBON: MECHANISTIC EFFECTS

1. Reacting carbons, especially electrophilic carbons, behave differently based on the types of atoms they carry as substituents; this is termed the **alkyl-substitution pattern**.
   a. Many mechanisms include one of the following high-energy species in the rate-determining step (as **intermediates**):

      $R_3C^{\oplus}$ (R = C or H): **electrophilic carbon cation**; carbon is missing one bond and one electron.

      $R_3C\cdot$ (R = C or H): **electrophilic carbon radical**; carbon is missing one bond but retains its unpaired electron.
   b. The overall reaction rate will be dependent upon the rate of formation of these electrophiles, which is based on the stability of the electrophilic carbon: *The faster rate of formation corresponds to the greater stability of the carbon electrophile.*
   c. The stability of the carbon electrophile is dependent upon the alkyl-substitution pattern: *Thus, the rate of a reaction, and the type of mechanism through which it proceeds, is dependent upon the alkyl-substitution pattern of the electrophilic carbon.*
2. The alkyl-substitution pattern is defined by the following designations: **in the notation, R = specifically a carbon group.**
   a. **Methyl** carbon cation or radical: carbon is bonded to **three** hydrogens and **no** other carbons. H_3C^+ or $H_3C\cdot$
   b. **Primary** carbon cation or radical: carbon is bonded to **two** hydrogens and **one** other carbon. RH_2C^+ or $RH_2C\cdot$
   c. **Secondary** carbon cation or radical: carbon is bonded to **one** hydrogen and **two** other carbons. R_2HC^+ or $R_2HC\cdot$
   d. **Tertiary** carbon cation or radical: carbon is bonded to no hydrogens and three other carbons. R_3C^+ or $R_3C\cdot$
3. Not all electrophilic carbons are equally able to bear a positive charge or a radical (unpaired electron). In general, *an electrophilic (electron deficient) carbon is more stable whenever substituted groups can provide additional electron density to it,* termed **electron donation**.
   a. Directly attached carbon groups (alkyl-substituent groups) are more efficient than substituted hydrogen atoms at donation of electron density to electrophilic carbon. This is due to the versatility (inductive and hyperconjugative effects) of the bonding electrons in the carbon group.
   b. The net result is that *a positive charge or radical on electrophilic carbon is more stable (i.e., "less unfavorable") whenever the electrophilic carbon itself is bonded to the greatest number of alkyl groups (carbon groups).*

c. **Summary:**

Least Stable Carbon Cation/Radical ——————→ Most Stable Carbon Cation/Radical

Slowest Formed ————————————————————————→ Fastest Formed

$H_3C^+/H_3C\cdot$ < $RH_2C^+/RH_2C\cdot$ < $R_2HC^+/R_2HC\cdot$ < $R_3C^+/R_3C\cdot$
(methyl)   (primary)   (secondary)   (tertiary)

12.2 FREE RADICAL HALOGENATION OF ALKANES/ALKYL GROUPS

12.2.1 REACTION CONCEPTS

1. The overall reaction is a net **substitution** of a hydrogen for a halogen in an alkane to form an alkyl halide; the other product is a hydrogen halide. The carbon being substituted can be methyl, primary, secondary, or tertiary. The halogen is usually either diatomic Cl_2 or Br_2. **Light** often is used to initiate the reaction.

   General Reaction (R = C or H): R_3C—H + X_2 ——————→ R_3C—X + HX

2. The **rate** of the reaction depends primarily on the alkyl-substitution pattern of the carbon which undergoes the halogen/hydrogen exchange.
   a. The rate-limiting step involves the breaking of the C—H bond to form a carbon radical. The ease of bond breaking follows the trends for radical stability.
   b. Resulting summary: **(R = specifically an alkyl group)**

Least Stable Radical ————————————————————————→ Most Stable Radical

Slowest Formed ————————————————————————→ Fastest Formed

$H_3C\cdot$ < $RH_2C\cdot$ < $R_2HC\cdot$ < $R_3C\cdot$
(methyl)   (primary)   (secondary)   (tertiary)

Slowest Rate of H-Exchange ——————————————→ Fastest Rate of H-Exchange

Strongest C—H Bond ——————————————————————→ Weakest C—H Bond

H_3C—H   < RH_2C—H   < R_2HC—H   < R_3C—H
(methyl)   (primary)   (secondary)   (tertiary)

   c. Hydrogens can be designated by the type of carbon they are bonded to: **Tertiary hydrogen** is a hydrogen bonded to a **tertiary** carbon; a **secondary hydrogen** is a hydrogen bonded to a **secondary** carbon; and a **primary hydrogen** is a hydrogen bonded to a **primary** carbon.
3. *The major product in a free radical reaction generally follows the relative rates of reaction for the corresponding hydrogen exchange.* This is especially true for reaction with the halogen Br_2. Exchange of H for X will occur primarily at the tertiary C—H

position if present in the molecule, followed by a secondary C—H followed by a primary C—H. **Major product formation, based on potential available** C—H

Minor Product ──────────────────────────────────────→ Major Product

H_3C—X	<	RH_2C—X	<	R_2HC—X	<	R_3C—X
(methyl)		(primary)		(secondary)		(tertiary)

12.2.2 SUMMARY OF GENERAL MECHANISM

1. The bromination of methane demonstrates the general mechanism for free radical halogenation. The overall reaction is

$$CH_4 + Br_2 \xrightarrow{\text{light or heat}} CH_3Br + HBr$$

2. A summary of the mechanistic steps are

Initiation Step: Br_2 + light or heat energy ⟶ 2 Br•

First Propagation Step: Br• + H—CH_3 ⟶ H_3C• + H—Br

Second Propagation Step: H_3C• + Br—Br ⟶ H_3C—Br + Br•

One Possible Termination Step: H_3C• + Br• ⟶ H_3C—Br

12.2.3 CONCEPTUAL DESCRIPTION OF MECHANISTIC STEPS

1. The initiation step generates the first electrophilic atom, the Br• radical:

    Br—Br + light or heat energy ⟶ 2 Br•

    The light or heat energy applied to this step is required to break the **Br—Br** bond. The bond breaks such that **each** atom of the original bond keeps **one** electron; this type of bond breaking is termed **homolytic cleavage.**

2. The next two steps of the mechanism proceed as an interdependent connected sequence termed the **propagation steps**. *Many complete sequences of the propagation steps will be completed for each initiation (or termination) step*; ratios are usually around 1000–10,000 propagations to each initiation. *The propagation steps, by themselves, will always add up to produce the balanced equation.*

3. In the **first propagation step**, the electrophilic Br• radical reacts with a (C—H) sigma-bonding electron pair in a reaction called **hydrogen abstraction**:

    Br• + H••CH_3 ⟶ H_3C• + H••Br

    a. Cleavage of the C—H bond is **homolytic** and produces a methyl radical **(H_3C•)** which behaves as the **new** electrophile in the **second** propagation step.

b. Hydrogen from the C—H bond keeps its bonding electron during the abstraction; radical combination with the bromine radical produces neutral H—Br.

4. In the **second propagation step**, the electrophilic carbon methyl radical, formed in the first propagation step, reacts with a **new** molecule of Br_2 (**not** the remaining Br• from the initiation step):

   a. The **Br—Br** bond undergoes homolytic cleavage; combination of the methyl radical with one bromine atom produces the alkyl halide.

   b. The other product of this step is a **new Br•** radical; this radical can now be used as the **electrophile** in the **first** propagation step. The result is a "closed loop" for the formation and regeneration of electrophilic bromine radicals. *A **net** reaction process which sequentially consumes and regenerates its own reactant intermediates is termed a **chain reaction**.* Once started, a chain reaction can proceed on its own through a certain number of propagation cycles without the need for further initiation steps.

5. The propagation steps in a chain reaction do not actually proceed indefinitely. They are eventually stopped by **termination step(s)**. *A reacting chain is terminated when radicals are consumed in other side reactions without generating **new** electrophilic radicals*; this generally occurs by radical recombination.

Examples:

(reverse of initiation)	Br• + Br• ⟶ Br—Br
	H_3C• + Br• ⟶ H_3C—Br
	H_3C• + H_3C• ⟶ H_3C—CH_3

12.2.4 SUMMARY: DETERMINING THE CORRECT PRODUCTS FOR FREE RADICAL HALOGENATION

1. The carbon reactants (substrates) for free radical halogenation are **alkanes** or the **alkyl portions** of higher functional group molecules. The reactive carbon/hydrogen **must** be composed of an sp^3-**carbon** to which is bonded **at least one** hydrogen.

2. The first propagation step is the product-determining step; the **Br•** (or **Cl•**) radical will favorably abstract the **weakest-bonded** hydrogen from the neutral alkane.

   a. **To find the major constitutional isomer product:** Identify the **highest alkyl-substituted** carbon in the molecule (tertiary > secondary > primary). Replace **one** hydrogen on this carbon with **one** halogen (Cl or Br from the specified reagent): this is always correct for bromination; some exceptions may occur for chlorination.

b. The other product of the reaction is the hydrogen halide, HX (this need not be shown). Note, however, that **one diatomic halogen reagent** (Br_2 or Cl_2) produces **only** a **single** substitution (**one** C—H to **one** C—X).

c. Other **minor products**, or **multiple substitutions**, are found by considering the reaction at other C—H positions in the substrate molecule.

d. Product formation occurs through an sp^2/trigonal planar carbon radical: the reaction is **not** stereospecific; no stereochemistry applies.

12.3 NUCLEOPHILIC SUBSTITUTION

12.3.1 DESCRIPTION OF THE GENERAL REACTION

1. *Nucleophilic substitution generally occurs **only** at carbons that are **sp^3/tetrahedral**.* The higher bond strength of sp^2 C—X disfavors substitution as compared to other possible reactions. Two possible general examples are shown. **Notation: R = C or H**

$$R_3C-X + Nu{:}^- \longrightarrow R_3C-Nu + X{:}^-$$

$$R_3C-X + H-Nu{:} \longrightarrow R_3C-Nu + H-X$$

2. The general forms of the reaction show a net substitution of **Nu:⁻** (**nucleophile**) for **X** (termed the **leaving group**) at sp^3-carbon, which acts as the net **electrophile**.

a. The general forms differ, however, with respect to the exact acid/base form of the nucleophile and/or leaving group. The **nucleophile** may be an **anion** (symbol = **Nu:⁻**) or a protonated **neutral** form (symbol = **H—Nu:**).

b. Bond cleavage from the carbon substrate may form an anion-leaving group (**X:⁻**), or preliminary protonation may be required to allow the leaving group to be cleaved off as a neutral molecule (**H—X**).

3. The **net electrophile** in the general reaction is the carbon connected to the **X**-leaving group, based on the relative polarity of the C—X bond:

$$\overset{\delta+}{R_3C}-\overset{\delta-}{X}$$

4. The electronic structure and potential behavior of the **leaving group** (**X:⁻** or **H—X**) **after** it has been cleaved from the electrophilic carbon is similar to the electron characteristics and behavior of the **nucleophile (Nu:⁻** or **H—Nu:**) **before** the reaction.

a. *The reaction type and reaction mechanisms are termed **nucleophilic substitution** because the net effect of the reaction is exchange of one nucleophile for another at the site of the electrophilic carbon.*

b. Very often, the **reverse** reaction, substitution of the nucleophile by the leaving group occurs simultaneously; nucleophilic substitution can be an **equilibrium**.

12.3.2 FACTORS INFLUENCING RATE AND MECHANISMS

1. Rates of nucleophilic substitution depend on
   a. **Strength** of the **nucleophile**, including the **acid/base form** of the nucleophile.
   b. The **effectiveness** of the **leaving group** (favorability of bond breaking).
   c. The electron characteristics of the **electrophilic carbon**, determined primarily by the **alkyl-substitution** pattern.
2. Nucleophilic substitution rates are directly proportional to the strength of the nucleophile in the specific solvent type. **(See Chapter 11.)**
   a. *Strong* nucleophiles are often (but not exclusively) *strong bases*: the ability to form a strong $C-Nu$ bond, measured by base strength, contributes to an increase in the nucleophilic substitution rate. This is generally true in **aprotic** solvents. *The anionic (stronger base) form of a nucleophile will always be a better nucleophile than the neutral (weaker base) form.*
   b. In protic solvents, a *strong nucleophile often contains a highly polarizable* atom: a greater availability of nucleophilic electrons due to a loose solvent shell, proportional to polarizability, contributes to an increase in the nucleophilic substitution rate.
   c. Total strength of the nucleophile is a combination of these two factors; general conclusions and summary: *strong bases are always good nucleophiles; many weak bases* **can** *be good nucleophiles, especially in protic solvents, if they contain a very polarizable nucleophilic atom.*
   Good Nucleophiles: Br^-; I^-; HS^-; RS^-; NC^-; N_3^-; HO^-; RO^-
   Moderate-to-Weak Nucleophiles: Cl^-; F^-; RCO_2^-; $:NH_3$; H_2O:; RÖH
3. Rates of substitution depend on the **effectiveness of the leaving group**. This is a general measure of the tendency of the $C-X$ bond to break such that the **X** group leaves with **both** bonding electrons, termed **heterolytic cleavage**. A greater leaving-group effectiveness contributes to a faster rate for nucleophilic substitution.
   a. *Leaving-group effectiveness is proportional to the stability of the released anion or neutral molecule and inversely proportional to the $C-X$ bond strength. These factors,* **measured approximately by base strength of the released X group**, provide the summary:

Poorest Leaving Group ⟶ **Best Leaving Group**

RO^-; HO^- < F^- < Cl^- < Br^- < H_2O; ROH < I^- < $C_7H_8SO_3^-$; $CF_3SO_3^-$

b. Note that good leaving groups, that is, **weak** bases, are the **conjugate** bases of **strong** acids. Also note that **RO⁻** and **HO⁻** almost always must leave as **neutral ROH and H$_2$O** rather than as the anions.

12.3.3 NUCLEOPHILIC SUBSTITUTION: GENERAL MECHANISTIC CONCEPTS

1. Identification of **leaving-group form**:
   a. For reactions of **alkyl halides**, the leaving-group **X** is a halide **anion: F⁻, Cl⁻, Br⁻, I⁻**. *Halides always leave as the anion, and **not** a neutral HX.* The sulfonate-leaving groups, for example, toluene sulfonate anion ($C_7H_8SO_3^-$) abbreviated as (⁻OTs), follow the reacting principles of the halides.

   $$R_3C-X + Nu:^- \longrightarrow R_3C-Nu + X:^-$$

   b. For **ethers** and **alcohols**, the leaving groups are **neutral ROH and H$_2$O**. *The anions **ro⁻** and **ho⁻** are such strong bases (poor leaving groups) that they almost always must be protonated by an acid **(ha)** before bond cleavage during the substitution.* The leaving group is then the neutral molecule

   $$R_3C-O-R + H-A \;\rightleftharpoons\; R_3C-\overset{\overset{\textstyle H}{|}}{O^+}-R + A^-$$

   $$R_3C-\overset{\overset{\textstyle H}{|}}{O^+}-R + Nu:^- \longrightarrow R_3C-Nu + H-O-R$$

2. The **rate-determining step** in nucleophilic substitution **always** includes bond cleavage of the **X** group from the electrophilic carbon. The complete reaction proceeds through one of the two mechanisms; identification of mechanism type is determined from alkyl-substitution pattern of the electrophilic carbon:
   a. **Tertiary alkyl halides** and **tertiary plus secondary alcohols** proceed through a mechanism characterized by a **unimolecular** rate-determining breaking of the C—X bond; the nucleophile is not involved in the **r.d.s.** This mechanism is abbreviated as SN1.

   SN1: r.d.s. For Alkyl Halides: $R_3C-X \longrightarrow R_3C^+ + X^-$

   SN1: r.d.s. For Alcohols: $R_3C-^+OH_2 \longrightarrow R_3C^+ + H_2O$

   b. **Primary plus secondary alkyl halides** and **primary alcohols** proceed through a mechanism characterized by a **bimolecular** rate-determining breaking of the C—X bond. *The nucleophile **is involved in the r.d.s.**;* it is in the process of forming a bond to the electrophilic carbon simultaneous with the cleavage of the C—X bond. This mechanism is abbreviated as **SN2**.

SN2: r.d.s. For Alkyl Halides: $R_3C—X + Nu:^- \longrightarrow R_3C—Nu + X:^-$

SN2: r.d.s. For Alcohols: $R_3C—^+OH_2 + Nu:^- \longrightarrow R_3C—Nu + H_2O$

3. *The selection of a mechanistic path by the* **electrophilic carbon** *is based on the relative stabilities of this carbon in the intermediate or transition state of the mechanism rate-determining step.*

   a. *The carbon cation,* R_3C^+, *will be more stable, and have the fastest rate of formation, whenever akyl substitution is the highest.* The carbon cation mechanism, **SN1**, will thus be favored for **tertiary** (halide) and **tertiary/secondary** (alcohol) substitution patterns and disfavored for primary alkyl substitution.

   b. The transition state for the bimolecular reaction requires a geometric arrangement of the leaving group directly opposite the incoming nucleophile:

   $$Nu\text{------------}R_3C\text{------------}X$$

   Steric hinderance caused by large alkyl groups, **R**, decreases the stability of this transition state. The bimolecular mechanism, **SN2**, will thus be favored for the **least** alkyl-substituted electrophilic carbons: **primary/secondary** (halide) and **primary** (alcohol). SN2 is disfavored for **tertiary** electrophilic carbons.

   c. Slight differences in energies cause secondary carbon electrophiles to select different mechanisms depending on functional group: *secondary alcohols react through SN1; secondary alkyl halides react through SN2.*

   d. **Rate Summary:**

Slowest For SN2 ⟶ **Fastest For SN2**

$R_3C—X$ (tertiary)	<	$R_2HC—X$ (secondary)	<	$RH_2C—X$ (primary)	<	$H_3C—X$ (methyl)

Slowest For SN1 ⟶ **Fastest For SN1**

$H_3C—X$ (methyl)	<	$RH_2C—X$ (primary)	<	$R_2HC—X$ (secondary)	<	$R_3C—X$ (tertiary)

12.3.4 GENERAL SUMMARIES (SELECTED FORMATS):
        NOTATION R = ALKYL GROUP

1. **SN2** for **alkyl halides** consists of only one step; carbon electrophiles are primary (RH_2CX) and secondary (R_2HCX): **X = F, CI, Br, I (or OTs)**:

Step r.d.s. (Only Step):

The single step includes simultaneous bond breaking and bond formation. The general geometry of the transition state (dashed lines [------] = partial bonding):

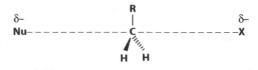

2. **SN2 for alcohols** requires **two** steps and applies **only** to **primary** carbon electrophiles (RH$_2$COH).

a. **Step 1** is a required preliminary equilibrium **protonation** of the **hydroxyl oxygen** to convert the potential **(OH⁻)** leaving group into the much better neutral **H$_2$O** leaving group. A **strong** acid acts as the electrophile (e.g., as **H⁺**); the hydroxyl oxygen is the nucleophile for this step.

b. **Step 2** is the **bimolecular** simultaneous exchange of the nucleophile and leaving group: The nucleophile transfers an electron pair to the electrophilic carbon at the same time that the oxygen leaves with the original C—O bonding electrons.

The bimolecular transition state for this step is similar to that of an alkyl halide SN2, with leaving group = ------**OH$_2$**.

c. In the example shown, the nucleophile and acid are derived from the same molecule (examples: HCl, HBr). **HNu** acts as the strong acid and **Nu:⁻** acts as the nucleophile.

Step 1   RH$_2$C—ÖH + H--Nu ⟶ RH$_2$C—⁺OH$_2$ + Nu:⁻   fast/equilibrium

Step 2   RH$_2$C--⁺OH$_2$ + Nu:⁻ ⟶ RH$_2$C—Nu + H$_2$Ö   slow/r.d.s.

3. **SN1** for **alkyl halides** applies only to **tertiary** electrophilic carbons (R$_3$CX).

a. **Step 1** is the **unimolecular** rate-determining decomposition of the alkyl halide to the carbon cation plus the halide ion. Heterolytic bond cleavage transfers the C—X bonding electron pair to the leaving group (halide).

b. **Step 2:** The reaction is then completed by bond formation with the nucleophile in a separate step. The carbon cation acts as the **new electrophile** in this step and accepts an electron pair; the nucleophile donates an electron pair. (The example shows an anionic nucleophile form.) **X = F, Cl, Br, I, (or OTs)**:

Step 1   R$_3$C--X ⟶ R$_3$C⁺ + X⁻   slow/r.d.s.

Step 2   R$_3$C⁺ + Nu:⁻ ⟶ R$_3$C--Nu   fast

4. **SN1 for alcohols** requires three steps and applies to secondary (R_2CHOH) and tertiary (R_3COH) electrophilic carbons.
   a. **Step 1** is the preliminary equilibrium protonation by the strong acid to form a good leaving group from the hydroxyl (OH), identical to the first step in SN2.
   b. **Step 2** is the unimolecular decomposition of the protonated alcohol formed in **Step 1**; the result is the formation of a carbon cation as the new electrophile. The electron pair of the original C—O bond stays with the oxygen, producing neutral water as the other product.
   c. The carbon cation electrophile formed in Step 2 accepts an electron pair from a nucleophile lone pair to complete the bonding in **Step 3**.
   d. In this example, the nucleophile and acid are derived from the same molecule: **HNu** acts as the strong acid and **Nu:⁻** acts as the nucleophile.

Step 1 R_3C—$\overset{..}{O}H$ + H—•• Nu ⟶ R_3C—⁺OH_2 + Nu:⁻ **fast/equilibrium**
⟵

Step 2 R_3C—•—⁺OH_2 ⟶ R_3C^+ + $H_2\overset{..}{O}$ **slow/r.d.s.**

Step 3 R_3C^+ + Nu:⁻ ⟶ R_3C—•—Nu   **fast**

12.3.5 SUMMARY: DETERMINATION OF THE CORRECT PRODUCT FOR NUCLEOPHILIC SUBSTITUTION

1. Identify the **nucleophilic** reagent molecule and the **specific** nucleophilic **atom** (the bonding atom); identify whether the nucleophile will substitute as an anion or neutral molecule.
2. Identify the **electrophilic** carbon **with the leaving-group X**.
3. Note the acid/base nature of the nucleophile and leaving group. Alcohols will substitute **only** in the presence of a **strong** acid.
4. Identify the **mechanism** based on the identity of **X** and the substitution pattern on the electrophilic carbon (primary, secondary, or tertiary).
   a. For primary alcohols; primary and secondary alkyl halides/sulfonates: **SN2**.
   b. For secondary and tertiary alcohols; tertiary alkyl halides/sulfonates: **SN1**.
5. Identify the correct **constitutional** isomer: directly exchange the nucleophile and leaving group, bonding the correct nucleophilic atom to the electrophilic carbon. Complete any proton transfers required to achieve normal, neutral bonding in the final products: **Add or remove protons as necessary**.

6. Identify the correct **stereochemistry**:
   a. Determine if stereoisomers are possible from the structure of the organic molecule.
   b. If **no** stereoisomers are possible, the problem is complete; show only the correct constitutional isomer from step (5).
   c. If stereoisomers are possible, use the mechanism determined from step (4):
      i. **SN1 is not stereospecific**; it produces approximately 50% (±10%) of each of the two possible stereoisomers of a stereoisomer pair: show only the correct constitutional isomer.
      ii. **SN2 is stereospecific**; it produces 100% inversion of stereoconfiguration: show the correct resulting stereoisomer.

12.4 ELIMINATION REACTIONS OF ALCOHOLS AND ALKYL HALIDES

12.4.1 GENERAL REACTION CONCEPTS

1. An **elimination** reaction proceeds through removal of two atoms or atom groupings from neighboring carbons to form a double bond.
   a. **Each** of the neighboring carbons breaks **one sigma** bond, to be replaced by a mutual **pi- bond** between them.
   b. In most cases, **one** of the removed atoms is **hydrogen**; the **other** is an atom or atom grouping that can be **usually** identified as a typical **leaving group**.

General Reaction: $R_2C—CR_2 \longrightarrow R_2C\!=\!CR_2 + HX$ (in various forms)
$||$
HX

2. The identities for the possible **X**-leaving groups, in most cases, are the same as for nucleophilic substitution: **F^-, Cl^-, Br^-, I^-, H_2O, HOR (or OTs)**.
   a. If the leaving group is a **halogen**, the substrate for the elimination reaction is thus identified as an **alkyl halide**; halogens always leave as the **anion, and not** as HX.
   b. If the leaving group is **water**, the elimination substrate is an **alcohol**; (OH) may never leave as the anion but only as neutral H_2O. A leaving group of HOR, where R = carbon group, corresponds to an ether substrate (a less common reaction).
3. The other atom removed in an elimination for alkyl halides and alcohols is hydrogen, specifically as **H^+**: hydrogen is removed **without** its bonding electron.
   a. Removal of H^+ from carbon allows the carbon to keep both of the original bonding electrons from the C—H bond. These electrons are ultimately transferred to the neighboring **electrophilic** carbon to form the carbon–carbon **pi-bond**.

   b. **Acid/base** conditions for elimination differ significantly for alcohols versus alkyl halides; these reactions are described separately.

4. The general mechanisms for elimination are classified based on the **number of molecules in the rate-determining step**:

   a. If the **r.d.s.** for elimination is **unimolecular**, the mechanism is **E1**.

   b. If the **r.d.s.** for elimination is **bimolecular**, the mechanism is **E2**.

12.4.2 IDENTIFICATION OF MAJOR ISOMER ALKENE PRODUCT IN ELIMINATIONS (NON-REARRANGEMENT)

1. Elimination reactions may undergo **rearrangements**. However, **in the absence of rearrangements**, the determination of the correct **major** product requires an analysis of the **substitution pattern** on the potential double bond.

2. The formation of an alkene through an elimination most often provides the opportunity for **more than one** possible product.

   a. For a non-rearrangement reaction, elimination occurs by removal of the leaving group plus removal of an H (as H$^+$) from a **neighboring** carbon. *In most molecules, the carbon with the leaving group can have more than one (nonidentical) neighboring carbon.*

   b. The following shows a generalized example in which the two neighboring carbons produce two distinct products: *(Group A is different than Group B.)*

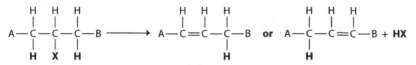

3. The possible product alkenes are classified on the basis of the specific alkyl-substitution pattern to the **sp^2-carbons** participating in the newly formed **double bond**. **Each** sp^2-carbon of the double bond will have **two** bonded substituents; thus, a double bond will have a total of **four** possible bonded atoms or atom groupings. The classification is

   a. A double bond substituted with **one** alkyl group and **three** hydrogens is **mono-substituted**: Example: $CH_3CH=CH_2$.

   b. A double bond substituted with **two** alkyl groups and **two** hydrogens is **disubstituted**:
Examples: $(CH_3)_2C=CH_2$ or $CH_3CH=CHCH_3$.

   c. A double bond substituted with **three** alkyl groups and **one** hydrogen is **trisubstituted**: Example: $(CH_3)_2C=CHCH_3$.

   d. A double bond substituted with **four** alkyl groups and **no** hydrogens is **tetra-substituted**: Example: $(CH_3)_2C=C(CH_3)_2$.

   e. An **unsubstituted** double bond refers specifically to ethene: Example: $H_2C=CH_2$.

4. *The major constitutional isomer product in a non-rearrangement elimination reaction, whether through E1 or E2, is almost always the alkene that contains the most highly alkyl-substituted double bond.*

   a. The sp^2-carbons of a double bond are slightly more electrophilic than sp^3-carbons due to the greater s-character of the sp^2-hybrid orbitals. The sp^2-carbons are thus stabilized by electron donation from alkyl-substituent groups.

   b. Elimination mechanisms show that the formation of the major product is controlled by the **rate** of the reaction. The greater stability of the more highly alkyl-substituted double bond produces a corresponding greater stability of the **transition state** leading to that double bond; the rate of the reaction, based on transition state stability, favors the more stable double bond.

   c. **Exceptions** can occur due to the geometrical constraints of the E2 transition state.

   d. **Summary:**

Least Stable Double Bond ——————————————→ **Most Stable Double Bond**

Double Bond Formed Slowest ————————————→ **Double Bond Formed Fastest**

Minor Alkene Product ————————————————————→ **Major Alkene Product**

mono-substituted   <   **di**-substituted   <   **tri**-substituted   <   **tetra**-substituted

5. Restricted rotation around double bonds in alkenes will produce potential **stereoisomer** products in elimination reactions.

   a. Larger alkyl groups influence the final stereoisomer through their 3-D arrangement: steric interaction between bulky substituents raises the energy and destabilizes the transition state.

   b. In most cases, *the **major** alkene stereoisomer will be the **most stable** stereoisomer based on the arrangement of the larger groups on each sp^2-carbon: **larger groups will be trans-** to each other. Exceptions due to the E-2 transition state can occur.*

12.4.3 SUMMARY: DETERMINATION OF MAJOR ALKENE PRODUCT FOR ELIMINATION (NON-REARRANGEMENT)

1. *Identify the carbon with the **leaving group**.* Halogens will leave under many acid/base conditions; alcohols will react only in a strong acid. *In the absence of rearrangement, the product double bond **must** include the carbon with the leaving group.*

2. *Identify **all** neighboring carbons to the carbon with the leaving group;* neighboring carbons **must** have at least **one** hydrogen. In the absence of rearrangement, *the double bond **must** include a neighboring carbon with at least one hydrogen.*

3. Select **one** of the neighboring carbons with at least one hydrogen.
   a. *Remove the leaving group from its carbon and a hydrogen from the selected neighboring carbon; replace these bonds with a **double bond** between these two carbons.*
   b. *If **stereoisomerism** is possible, select the alkene product with the **larger** groups on **each** of the two carbons in the **trans**-configuration.* Distinction between stereoisomers may not always be necessary.
4. If the carbon bearing the leaving group has **only one** neighboring carbon, the analysis is complete: **only one** constitutional isomer is possible.
5. If the carbon bearing the leaving group has **more than one** neighboring carbon:
   a. The number of **nonidentical** neighboring carbons can range from 1 to 3. *Repeat **Step 3** for **each** possible neighboring carbon:* draw the correct constitutional isomer (and stereoisomer) that would result from removal of the leaving group and a hydrogen from **each** possible neighboring carbon.
   b. Note that **identical** carbon groups as neighboring carbons will produce **equivalent** constitutional isomers. The total number of possible products will match only the total number of **nonidentical** neighboring carbons.
   c. *Analyze the substitution pattern for each of the products drawn.* **Select as the major constitutional isomer the alkene with the most highly substituted double bond: this is equivalent to removing a hydrogen from the most highly alkyl-substituted neighboring carbon.** If stereoisomers are possible, and are to be distinguished, *select the stereoisomer of this constitutional isomer which has the larger groups on each carbon trans to each other.* **Example (stereoisomers not shown):**

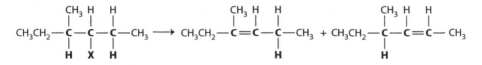

Reactant: **major product:**   **minor product:**
two neighboring carbons   **trisubstituted**   **disubstituted**

12.4.4 MECHANISM SUMMARY: E2 FOR PRIMARY, SECONDARY, AND TERTIARY ALKYL HALIDES

1. Elimination in **alkyl halides** requires removal of the halide as a negative ion **(X:⁻)** and the removal of a hydrogen **as H⁺**; conditions are generally basic or neutral.
   a. *The type of **rate-determining step** in the mechanism, bimolecular **(E2)** or unimolecular **(E1)**, depends on the strength of the base used to remove the H⁺ from the neighboring carbon, and the **alkyl-substitution** pattern of the carbon bearing the halogen leaving group.*

2. *The bimolecular r.d.s. mechanism (E2) will always occur under strong base conditions; this applies to primary, secondary, and tertiary halides.*

3. In the presence of a strong base such as **NaOCH$_3$**, the complete reaction is described as a single step; all bonds are formed and broken simultaneously:

Only Step: R$_2$C—CR$_2$ + Na$^+$ $^-$OCH$_3$ $\longrightarrow$ R$_2$C=CR$_2$ + HOCH$_3$ + Na$^+$ x$^-$
r.d.s. $\quad\quad$ | $\quad$ |
R = C or H $\quad$ **H** $\quad$ **X**

4. The strong base removal of the hydrogen as **H$^+$** leaves its bonding carbon with the C—H bonding electrons: this would form a carbon negative ion (anion). Carbon anions are very **unstable**. Strong base thus forces the removal of the **X:$^-$** leaving group simultaneously to avoid placing a negative charge on one carbon. **Strong base favors the bimolecular (E2) transition state.**

5. The rate-determining transition state (of this single step) shows the required bonding changes and geometry: (dashed lines [--------] = partial bonding).

   a. The C—X bond is being broken, with electron transfer from the carbon to the leaving X ion. The carbon becomes electrophilic.

   b. A C—H bond on the neighboring carbon is broken, with electron transfer from the hydrogen to the neighboring carbon. This carbon becomes a nucleophile; the C—H bonding electrons become the nucleophilic electrons.

   c. The **H$^+$** (proton) from the C—H bond is accepting a lone pair of electrons from the strong base anion, $^-$**OCH$_3$**.

   d. The pi-bond between the two carbons forms by electron transfer from the nucleophilic carbon to the electrophilic carbon, using the original C—H bonding electrons. The geometry of this transition state is termed anti-periplanar.

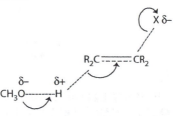

12.4.5 MECHANISM SUMMARY: E1 FOR TERTIARY ALKYL HALIDES

1. The **unimolecular** mechanism (**E1**) for **alkyl halides** can generally occur **only** with **tertiary** alkyl halides and only in the **absence** of a strong base, that is, only under a **weak base or neutral conditions**.

2. The reaction occurs in two steps: **R = C or H; :Base = weak base/solvent**

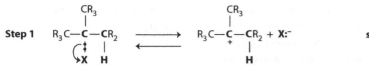

Step 1 $R_3C-\overset{CR_3}{\underset{H}{\underset{|}{C}}}-CR_2 \rightleftharpoons R_3C-\overset{CR_3}{\underset{H}{\underset{|}{\overset{+}{C}}}}-CR_2 + X:^-$ **slow/r.d.s.**

Step 2 $R_3C-\overset{CR_3}{\underset{H}{\underset{|}{\overset{+}{C}}}}-CR_2 + :Base \rightleftharpoons R_3C-\overset{CR_3}{\underset{|}{C}}\!\doteq\!\!CR_2 + (H\!\cdot\!\cdot\!Base)^+$ **fast**

3. **Step 1** is the **unimolecular** rate-determining cleavage of the C—X bond; no other molecules are involved in this reaction step.
   a. The bonding electrons are transferred from carbon to the leaving **X** atom/ion; the resulting carbon is now an **electrophile**.
   b. The rate of the **r.d.s.** step, and thus the rate of the complete reaction, is **inversely** proportional to the C—X bond strength, that is, **directly** proportional to leaving-group effectiveness: $I^- > Br^- > Cl^- > F^-$.
4. **Step 2** is the removal of the proton from a neighboring carbon. Under weak base/neutral conditions, the base in the reaction is usually a neutral **weak** base.
   a. Electron transfer occurs from the weak base lone pair to the proton. The carbon of the original C—H bond accepts **both** of the original C—H bonding electrons and transfers them to the electrophilic carbon formed in **Step 1**; the result is a new C—C **pi**-bond to complete the double bond.
   b. The **weak** base in this step is often the **solvent**, such as water or alcohol. Elimination under these **E1** conditions, along with the corresponding related **SN1** nucleophilic substitution, is termed **solvolysis**.

12.4.6 MECHANISM SUMMARY: E2 FOR PRIMARY ALCOHOLS

1. The leaving group for **alcohols** must be a neutral water molecule; **alcohols** require a preliminary protonation step by a strong acid. The **bimolecular r.d.s.** (E2) mechanism requires two steps and applies only to **primary** alcohols.
2. **:Base** = a general weak base (neutral or anionic not specified in the example):

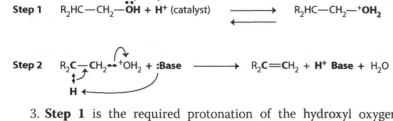

Step 1 $R_2HC-CH_2-\overset{..}{O}H + H^+$ (catalyst) $\rightleftharpoons R_2HC-CH_2-^+OH_2$ **fast/ equilibirium**

Step 2 $R_2C-CH_2\cdot\cdot^+OH_2 + :Base \longrightarrow R_2C=CH_2 + H^+ Base + H_2O$ **slow r.d.s.**

3. **Step 1** is the required protonation of the hydroxyl oxygen by a strong acid, such as H_2SO_4; this species is most often

catalytic. Electron transfer is from a hydroxyl lone electron pair (**nucleophile**) to the proton (**electrophile**).

4. **Step 2** is the **bimolecular** rate-determining cleavage of the C—O bond, which is **simultaneous** with the proton removal from the neighboring carbon by a **weak** base.

   a.   Proton removal must be **simultaneous** with the loss of the leaving group because **primary** carbon cations are **not** stable enough to form at a competitive rate.

   b.   The leaving group receives **both** of the original C—O bonding electrons during bond cleavage; the resulting carbon is **electrophilic**.

   c.   Electron transfer occurs from the **base** lone pair to the **proton**; the base may be the anion of the strong acid (e.g., HSO_4^-) or a water molecule. The original bonding electrons from the C—H bond remain with the carbon and are transferred to the electrophilic carbon to form the new **pi**-bond, completing the **double** bond.

   d.   Note that the general form [H^+ Base] is meant to show that H^+ is consumed in **Step 1** and regenerated in **Step 2**: it behaves as a catalyst.

5. The **r.d.s.** bimolecular **transition state** shows the bonding changes and required geometry: (dashed lines [−−−−−−] = partial bonding):

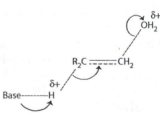

12.4.7 MECHANISM SUMMARY: E1 FOR SECONDARY AND TERTIARY ALCOHOLS

1. **Secondary** and **tertiary alcohols** will require the preliminary protonation step by a strong acid to form the neutral water leaving group. The **unimolecular r.d.s.** (**E1**) mechanism requires three steps.

2. **:Base** = a general weak base (neutral or anionic not specified in the example); **R = C or H, alcohol must be secondary or tertiary**.

Step 1 $R_2HC—CR_2—OH$ + H^+ (catalyst) ⇌ $R_2HC—CR_2—^+OH_2$ **fast/ equilibrium**

Step 2 $R_2C—CR_2—^+OH_2$ (with H below) ⇌ $R_2C—^+CR_2$ + $H_2\overset{..}{\underset{..}{O}}$ (with H below)   **slow/r.d.s.**

Step 3 $R_2C—^+CR_2$ + :Base (with H below) ⇌ $R_2C=CR_2$ + H^+ Base   **fast**

3. **Step 1** is the required protonation of the hydroxyl oxygen by the often-catalytic strong acid, for example, H_2SO_4. Electron transfer is from a hydroxyl lone electron pair **(nucleophile)** to the proton **(electrophile)**.

4. **Step 2** is the **unimolecular** rate-determining cleavage of the $C—O$ bond; no other molecule is involved in this step.

   a. Proton removal does **not** occur simultaneously with loss of the leaving group because **secondary** and **tertiary** carbon cations **are** stable enough to form with this preferred mechanism.

   b. The leaving group receives **both** of the original $C—O$ bonding electrons during bond cleavage; the resulting carbon is **electrophilic**.

5. **Step 3** completes the formation of the alkene through **pi**-bond formation.

   a. Electron transfer occurs from the **base** (e.g., HSO_4^- or H_2O) lone pair to the **proton**. The original bonding electrons from the $C—H$ bond remain with the carbon and are transferred to the electrophilic carbon to form the new **pi**-bond, completing the **double** bond.

   b. Note that the general form **[H$^+$ Base]** is meant to show that H$^+$ is consumed in **Step 1** and regenerated in **Step 3**: it behaves as a catalyst.

12.5 IDENTIFICATION OF ELIMINATION VERSUS SUBSTITUTION REACTIONS

12.5.1 REACTION ANALYSIS FOR ALCOHOLS

1. *Elimination and nucleophilic substitution reactions in alcohols proceed through common intermediates*. The presence of a common mechanistic intermediate causes different product types to be formed from the same starting compound.

   a. The bimolecular reactions (**SN2** and **E2**) both occur through the common protonated alcohol intermediate: **R$_3$C—$^+$OH$_2$**.

   b. The unimolecular reactions (**SN1** and **E1)** both proceed through the common carbon cation intermediate: **R$_3$C$^+$**.

   c. *Elimination (alkene formation) and nucleophilic substitution (e.g., alkyl halide formation) can occur simultaneously in reactions of alcohols.*

2. *The determination of the correct major product for an alcohol reaction depends on the conditions used.*

   a. *Elimination is favored over substitution at higher temperatures; substitution is favored over elimination at lower temperatures*. This is due to the positive value for entropy change found for elimination reactions: $\Delta S > 0$ for elimination reaction.

   b. *Elimination is favored in the presence of a poor nucleophile; substitution is favored in the presence of a good nucleophile*. Substitution requires the transfer of nucleophilic electrons to an electrophilic carbon; the stronger the nucleophile, the faster the rate (and the more favored the reaction) for nucleophilic substitution.

c. Strong acid is required for both reaction types, since the alcohol hydroxyl (as water) must always be a leaving group. *Strong acids with **good** nucleophile conjugate base anions, or conditions with added good nucleophiles, are generally selected for desired **substitution**. Strong acids with very **poor** nucleophilic conjugate anions (HSO_4^- and $H_2PO_4^{-2}$) are selected for desired **elimination**.*

3. **Summary: ($R = C$ or H)**

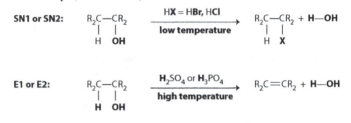

12.5.2 REACTION ANALYSIS FOR ALKYL HALIDES

1. ***Elimination and nucleophilic substitution** reactions in **alkyl halides** proceed through a **common intermediate** for the **unimolecular** reactions (**SN1 and E1**):*

a. The common intermediate is the carbon cation R_3C^+.

b. The carbon cation can undergo elimination (alkene formation) or nucleophilic substitution (e.g., formation of an alcohol or ether product from water or alcohol). The solvent (e.g., water or alcohol) often acts as both the nucleophile or weak base (**solvolysis**).

c. The unimolecular set of reactions for alkyl halides require conditions of **relatively stable** carbon cation formation and **no strong** base: *SN1 and E1 will occur simultaneously for tertiary alkyl halides under weak base/neutral conditions.*

d. *Elimination can always be favored over substitution at higher temperatures; SN1 (substitution) can rarely be favored for alkyl halides and has limited usefulness.*

e. **Summary Example: ($R = C$ or H; but alkyl halide is usually tertiary) ROH = water or alcohol:**

SN1 or E1: Formation of R_3C^+ intermediate:

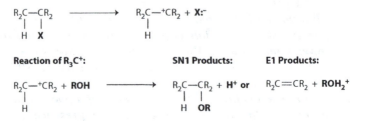

Both products can be formed; SN1 products cannot be favored.

2. The **bimolecular** mechanisms, E2 and SN2, do not proceed through a common intermediate for **alkyl halides**. However, *both products can be formed simultaneously whenever conditions for substitution (good nucleophile) and elimination (strong base) overlap*.

   a. Conditions will overlap when strong bases are used as nucleophiles in SN2; many good nucleophiles are also strong bases.

   b. *Elimination can always be favored over substitution under optimum E2 conditions of strong base and high temperature.*

3. *To favor substitution for reactions of alkyl halides, always select the best SN2 conditions possible. SN2 is optimized by*

   a. **Low** temperature.

   b. An **unhindered** (small substituent groups) **primary** or **secondary** substrate (alkyl halide).

   c. A **good** nucleophile that is also a **weak** base, **if possible**.

4. Many good nucleophiles are also weak bases: **Br⁻, Cl⁻, I⁻, CN⁻, N₃⁻**, etc. Reactions with these nucleophiles will lead to predominant substitution as the major product with good SN2 substrates.

5. For strong base nucleophiles such as **OH⁻, R₃CO⁻**, etc., the SN2 substrate **must** be primary or unhindered secondary to favor the substitution product. *Reaction of tertiary or hindered secondary alkyl halides with strong base nucleophiles will lead to an elimination (E2) product, and not the substitution (SN2) product.*

6. **Summary: R = C or H**

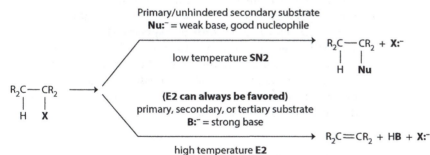

12.6 ELECTROPHILIC ADDITION TO PI-BONDS OF ALKENES AND ALKYNES

12.6.1 GENERAL REACTION CONCEPTS

1. Electrophilic addition to alkenes and alkynes involves the addition of one electrophile (**E⁺**) and one nucleophile (**Z:⁻**), one each to the carbons of a double or triple bond. In the following notation, the electrophile is shown as (**E⁺**), although in some reactions it is a radical (**E·**). The nucleophile is shown as (**Z:⁻**)

rather than (Nu:$^-$) to emphasize that the reaction type is controlled by the electrophile and not the nucleophile. **R = C or H**;

$$R_2C{=\!\!=\!\!=}CR_2 + \overset{\delta+\ \ \delta-}{E{-}Z} \longrightarrow \underset{E\quad Z}{R_2C{-}CR_2}$$

a. Each of the original double-bonded carbons forms a new **sigma** bond at the expense of breaking the mutual **pi**-bond between them: *the new sigma bonds are formed from the bonding electrons made available by cleavage of the pi-bond.*

b. The general form of the **E—Z** electrophile/nucleophile combination is widely variable: **E—Z** may be a single molecule (e.g., HCl) or each part may be derived from a separate set of molecules (e.g., H$^+$(aq)/H$_2$O).

c. The **rate-determining step** in the general reaction mechanism is the addition of the electrophile (**E$^+$**) to the nucleophilic electrons of the pi-bond; this provides the designation as **electrophilic addition**. The addition of **Z:$^-$** occurs **after** the **r.d.s.** and does not influence the overall rate.

2. Identification of reaction products and determination of the major product for electrophilic addition are based on **regiochemistry** and **stereospecificity**.

a. **Regiochemistry** refers to whether one specific constitutional isomer is favored under conditions where more than one product is expected; this will occur for addition of an **unsymmetrical E—Z** reagent to an **unsymmetrical** alkene. Regiochemistry of the major product is determined by the alkyl-substitution pattern on the original double-bonded carbons.

b. **Stereospecificity** refers to the mechanistic features of a reaction which will cause **only one specific stereoisomer** of a potential stereoisomeric pair to be formed. *Stereospecificity characterizes a **mechanism, and not** the **results** of a reaction.* Stereoisomers are not always possible for general reactions; a specific stereoisomer cannot be formed if the reaction does not lead to stereoisomers.

12.6.2 MECHANISM SUMMARY

1. Electrophilic addition proceeds through two major steps; an extra proton transfer step may also be required. **E$^+$ = electrophile; Z:$^-$ = nucleophile (general symbols)**:

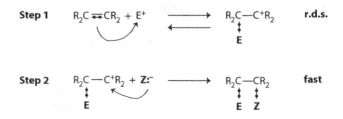

2. **Step 1** is addition of the electrophile to the nucleophile double bond. Although this first step is an equilibrium, the forward reaction is the slow **rate-determining** step.

   a. The form of the electrophile is generally shown as E^+. The actual form depends on the reaction and conditions; in many cases, the electrophile will exist as the neutral $E-Z$. The first step will then be of the general form

   Step 1 $R_2C = CR_2 + \overset{\delta+\;\;\delta-}{E\cdots Z}$ $\longrightarrow$ $R_2C-C^+R_2 + Z{:}^-$ **r.d.s.**
   $\qquad\qquad\qquad\qquad\qquad\qquad\quad\;\;\; \overset{|}{E}$

   b. The **pi**-bonding electrons of the alkene are **nucleophilic** due to the lower bond strength of pi-bonds and the extended orbital positions of the p-orbital overlaps above and below the atom plane. The pi-electron pair of the double bond (**donor**) is transferred to the electrophile (**acceptor**) in coordinate bond formation.

   c. Electron transfer of the pi-electrons forms a new sigma bond between one of the original double-bonded carbons and the electrophile. *The carbon bonded to the electrophile has four sigma bonds and thus has no formal charge.* The other carbon of the original double bond forms no new $E-C$ bond, has only three bonds, and has lost one of its pi-bonding electrons. *The positive charge carried from the electrophile goes to the carbon that does not bond to the electrophile.*

3. The electrophile may also be a radical $E\cdot$. In this case, **single** electron transfer occurs; one carbon forms the new $E-C$ bond, and the other carbon bears the free electron radical. (*The original $E\cdot$ is formed in an **initiation** step, e.g., with a **peroxide**.*)

   Step 1 $R_2C = CR_2 + E\cdot$ $\longrightarrow$ $R_2C-R_2C\cdot$ **r.d.s.**
   $\qquad\qquad\qquad\qquad\qquad\qquad\qquad\quad \overset{|}{E}$

4. **Step 2** of the complete mechanism is the combination of the carbon cation or radical with the nucleophile $Z{:}$; the carbon cation or radical formed in the first step is the electrophile. *This step is fast and does **not** affect the rate of the **overall** reaction, but may influence the structure of the major product.*

   Step 2 $R_2C-C^+R_2 + Z{:}^-$ $\longrightarrow$ R_2C-CR_2 **fast**
   $\qquad\qquad\;\; \overset{|}{E}\qquad\qquad\qquad\qquad\qquad\;\; \overset{|}{E}\;\;\overset{|}{Z}$

   a. Electron transfer from the nucleophile to the electrophile forms a new sigma bond to complete the addition molecule. The final result is replacement of the $C-C$ pi-bond with a new sigma bond for **each** of the original double-bonded carbons.

   b. During **Step 2, Z:** may exist as the anion, $Z{:}^-$. In this case, the reaction is complete in two steps. The nucleophile may also exist as a **neutral** molecule; often, this will be of the acid/base form H—Z. In this case, an additional proton transfer step is required. *(The final steps in the reaction of BH_3 with alkenes involve a rearrangement and is more complicated.)*

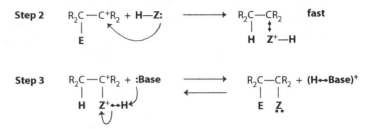

Step 2 $R_2C-C^+R_2 + H-Z:$ $\longrightarrow$ R_2C-CR_2 **fast**

Step 3 $R_2C-C^+R_2 + :Base$ $\rightleftharpoons$ $R_2C-CR_2 + (H\leftrightarrow Base)^+$

c. For the first step involving a **radical** electrophile, **Step 2** is the second step of the **chain propagation**; this step regenerates the original reactant radical electrophile, which can begin a new **Step 1** addition.

Step 2 $R_2C-R_2C\cdot + E\leftrightarrow Z$ $\rightleftharpoons$ $R_2C-CR_2 + E\cdot$ **fast**

12.6.3 REGIOCHEMISTRY IN ELECTROPHILIC ADDITION

1. Identification of the **major constitutional** isomer (**regiochemistry**) of the complete electrophilic addition to an alkene depends solely on **Step 1**, the addition of the electrophile to the pi-bonding electrons.
2. **In** the **absence of rearrangement**, the **two** options for the major constitutional isomer are determined by the **two** possible double-bonded carbons that can transfer the pi-bonding electrons to the electrophile for new sigma-bond formation (i.e., which one gets the E^+). **Example: R = <u>specifically</u> C (alkyl group)**

Step 1 $H_2C=CR_2 + E^+$ $\rightleftharpoons$ $H_2C-C^+R_2$ **r.d.s**
$\qquad\qquad\qquad\qquad\qquad\qquad\qquad\qquad\qquad\ \ \ |$
$\qquad\qquad\qquad\qquad\qquad\qquad\qquad\qquad\qquad\ \ E$

OR

Step 1 $H_2C=CR_2 + E^+$ $\rightleftharpoons$ $H_2C^+-CR_2$ **r.d.s**
$\qquad\qquad\qquad\qquad\qquad\qquad\qquad\qquad\qquad\qquad\ \ |$
$\qquad\qquad\qquad\qquad\qquad\qquad\qquad\qquad\qquad\qquad\ E$

3. The selection of the **major** product from these two potential reactions depends on the alkyl-substitution pattern of each of the two original carbons of the double bond. *The electrophile (E^+ or $E\cdot$) **always adds to the specific carbon of the double bond which results in the <u>most stable carbon cation</u> or <u>radical intermediate</u>.***
   a. Alkyl groups are more effective at stabilizing positive charge or radical character than hydrogens. *Thus, the electrophile (E^+ or $E\cdot$) **always adds to the specific carbon of the double bond that <u>produces</u> a tertiary carbon cation or radical***

intermediate (most stable) favored over a secondary carbon cation intermediate (next most stable) favored over a primary carbon cation intermediate (least stable).

b. As demonstrated in the general reaction, the **carbon that bonds to the electrophile does not bear the positive charge or radical**. *The electrophile (E+ or E•) must add to the specific carbon of the double bond that has the fewest number of bonded alkyl groups (greater number of hydrogens).* This ensures that the **other** carbon of the double bond, which **gets the positive charge** or radical, will have the **greater number of alkyl groups bonded**. This carbon will then get the nucleophile in the completion of the reaction in Step 2.

c. The correct addition pattern that will lead to the **major constitutional** isomer for the **example above** (final product is shown): (R = **specifically** C (alkyl group))

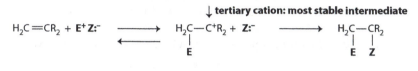

12.6.4 STEREOCHEMISTRY IN ELECTROPHILIC ADDITION

1. Stereochemistry for electrophilic addition to alkenes is determined by the **stereospecificity** of the rate-determining **Step 1** in the mechanism: the addition of the electrophile to the pi-electrons of the double bond. A complete discussion of the concepts and practice contained in this summary can be found in Chapter 13.

2. The formation of **possible** stereoisomers for alkene addition occurs because of intermediate or transition state geometries for **Step 1**; these possible geometries produce three **stereospecificity** outcomes for the complete mechanism and final product.

a. **Both** adding groups (**E** and **Z**) can be added to the **same face** of the double bond; this is termed **syn**-stereospecificity.

b. **Each** of the adding groups (**E** and **Z**) can be added to **opposite faces** of the double bond; this is termed **anti**-stereospecificity.

c. The adding groups (**E** and **Z**) can be added to **either** the **same face or opposite faces** of the double bond, usually with approximately 50%/50% probability. In this case, **no** stereospecificity applies to the mechanism or final product.

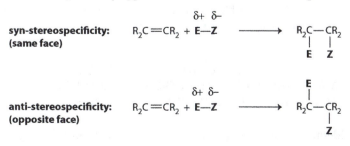

3. ***Stereospecificity characterizes a mechanism, and not the results of a reaction***. The characteristic stereospecificity of a mechanism always operates on the reactant double bond. Whether the final products demonstrate stereochemistry depends on whether the specific alkene reaction can produce stereoisomers.

4. Some examples of transition state or intermediate geometries and the related product outcomes:

   a. **The addition intermediate is based on a free carbon cation or radical:**

$$R_2C - R_2C^+ \quad \text{or} \quad R_2C - R_2C\bullet$$
$$\underset{E}{|} \qquad\qquad \underset{E}{|}$$

   In this case, the geometry of the cation or radical carbon is trigonal planar, and no stereospecificity is possible. The planar carbon of the electrophilic intermediate is symmetrical and addition of the nucleophile (**Z:**) can be to **either face** of the cation or radical carbon.

   b. **The addition intermediate is a bridged cation complex:**

$$R_2C - CR_2 \qquad \textbf{example =} \qquad R_2C - CR_2$$
$$\underset{E^+}{\backslash\,/} \qquad\qquad\qquad\qquad \underset{Br^+}{\backslash\,/}$$

   In this case, the reaction must occur with **anti**-stereo-specificity (**opposite faces**) since one face of the double bond (with the E) is inaccessible to the nucleophile (**Z:**).

   c. **The electrophile/nucleophile reagent remains bonded as E—Z after electrophilic addition:**

   Example: E—Z = **H₂B—H**

$$\overset{\delta+ \quad \delta+}{R_2C - CR_2}$$
$$|$$
$$H_2B - H$$
$$\delta-$$

   In this case, the nucleophile reacting in **Step 2** must add to the **same** face of the double bond as the electrophile to which it remained bonded in **Step 1**. The result is **syn**-stereospecificity.

   d. **The electrophilic addition occurs from a catalyst surface:**

   Example: E—Z = **H—H**

$$R_2C - CR_2$$
$$H \quad | \quad \,$$
$$\quad\;\; H$$
$$\text{............... catalyst surface}$$

   In this case, both hydrogens must be added to the **same** face of the double bond that is complexed to the catalyst surface. The result is **syn**-stereospecificity.

12.6.5 SUMMARY: DETERMINATION OF THE CORRECT MAJOR PRODUCT FOR ELECTROPHILIC ADDITION TO ALKENES (EXCLUDING REARRANGEMENTS)

1. Identify the alkene (sp²) carbons of the double bond.

2. Identify the addition electrophile (**E**) and the addition nucleophile (**Z:**) from the E—Z reagent. **Use the summary table in Section VI.G.**

3. In place of the **pi** portion of the double bond, add (E) to the **least alkyl**-substituted carbon of the original double bond. (Equivalently, this is the carbon with the **most hydrogens**.)

4. Add **(Z:)** to the **most highly alkyl**-substituted carbon of the original double bond.

5. Determine if the product molecule can have stereoisomers. If **no** stereochemistry is possible, the constitutional isomer derived in (**1**) through (**4**) is correct as written.

6. If stereoisomers are possible, determine if the specific mechanism for the E—Z reagent is stereospecific. (**See summary table.**) If the mechanism is **not stereospecific**, the constitutional isomer as written is correct, *since one specific stereoisomer cannot be exclusively formed by a non-stereospecific mechanism*.

7. If stereoisomers are possible, **and** if the specific mechanism is stereospecific, write the correct **stereoisomer** based on **syn** or **anti** addition as applicable.

12.6.6 SPECIFIC EXAMPLES

1. Addition of **HBr** to 2-methylpropene. The reaction is classified as two steps; the electrophile is one molecule as E—Z and **Z:⁻** adds as an anion.

   No stereoisomers are formed; *the intermediate of Step 1 is trigonal planar.*

Step 1 $H_2C{=}C(CH_3)_2$ + **H—Br**   ⟶ ⟵ $H_2C{-}C^+(CH_3)_2$ + **Br:⁻**   **r.d.s.**
                                                      |
                                                      H

Step 2 $H_2C{-}C^+(CH_3)_2$ + **Br:⁻**   ⟶ $H_2C{-}C(CH_3)_2$ **fast**
                  | | |
                  H   **H   Br**

2. Addition of **H⁺/H₂O** to 2-methylpropene. This reaction requires three steps; a **third** step is needed to remove a proton from the nucleophile **neutral** water molecule used in the second step. The electrophile/nucleophile combination **H⁺/H₂O** consists of separate species. *No stereoisomers are formed*; *the intermediate of Step 1 is trigonal planar.*

Step 1 $H_2C{=}C(CH_3)_2$ + **(H⁺₍aq₎)**   ⟶ ⟵ $H_2C{-}C^+(CH_3)_2$ **r.d.s.**
                              (catalyst) |
                                                               H

Step 2 $H_2C{-}C^+(CH_3)_2$ + **H₂O:**   ⟶ $H_2C{-}C(CH_3)_2$ **fast**
                  | | |
                  H   **H   ⁺OH₂**

Step 3 $H_2C{-}C^+(CH_3)_2$ + **H₂O:**   ⟶ ⟵ $H_2C{-}C(CH_3)_2$ + **H₃O⁺ (= H⁺₍aq₎)**   **fast**
                  | | **(:Base)** | |
                  H   ⁺OH₂   **H   OH**

3. **Free radical** addition of **HBr** to 2-methylpropene; in this case, *Br is the electrophile as* Br•. *The addition of **Br first as the electrophile followed by H as the nucleophile** produces the opposite regiochemistry as compared to the ionic reaction shown in number* **(1)**. *No stereoisomers are formed; the intermediate of **Step 1 propagation is trigonal planar**.*

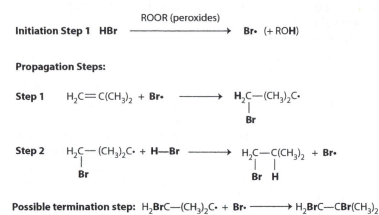

Possible termination step: H₂BrC—(CH₃)₂C• + Br• ⟶ H₂BrC—CBr(CH₃)₂

12.6.7 GENERAL SUMMARY

TABLE   E − Z Reagents for Electrophilic Addition

E − Z Reagent	E⁺	Z:⁻	Stereospecificity	Comments
H − Cl, Br, I	**H⁺**	Cl⁻/Br⁻/I⁻	None	Trigonal planar cation intermediate
H⁺/H₂O	**H⁺**	H₂O	None	Trigonal planar cation
Net result:	**H**	**OH**		intermediate; fastest rate to form tertiary C—OH
1. H₂SO₄	H⁺	H₂O	None	**Sequence** of reactions to
2. H₂O				increase the rate of alcohol
Net result:	**H**	**OH**		formation
H⁺/R₃COH	H⁺	R₃COH	None	Trigonal planar cation
Net result:	**H**	**OCR₃**		intermediate; **R₃C = alkyl C;** forms **ethers**
Br₂/Cl₂	**Br⁺/Cl⁺**	Br⁻/Cl⁻	Anti	Bridged cation intermediate
Br₂ or Cl₂/H₂O	Br⁺/Cl⁺	H₂O	Anti	Bridged cation intermediate;
Net result:	**Br/Cl**	**OH**		**bromo or chloro-alcohol** formed
Br₂ or Cl₂/R₃COH	Br⁺/Cl⁺	R₃COH	Anti	Bridged cation intermediate;
Net result:	**Br/Cl**	**OCR₃**		**bromo or chloro-ether** formed
1. B₂H₆	BH₃	H	Syn	Reaction **sequence** forms
2. H₂O₂				C − OH from C − BH₂; H is
Net result:	**OH**	**H**		transferred from BH₃ by a **rearrangement**

(Continued)

TABLE (*Continued*)   E – Z Reagents for Electrophilic Addition

E – Z Reagent	E$^+$	Z:$^-$	Stereospecificity	Comments
1. OsO$_4$ 2. ROOH	**OH**	**OH**	Syn	Reaction **sequence** forms syn **diol** through oxidation
H$_2$ **with Pt, or Pd or Ni etc. catalyst**	H	H	Syn	Syn addition due to **catalyst** surface
RCOOH $\parallel$ O		**oxygen**	Syn	Reagent is a **peroxy acid;** product is an **epoxide**
O$_3$		**oxygen**	None	Complete cleavage of the double bond to form **aldehydes** and **ketones**
1. Hg(OOCCH$_3$)$_2$/H$_2$O or R$_3$COH 2. NaBH$_4$/ROH **Net result:**	**H**	**OH/OCR$_3$**	None	Reaction **sequence** forms the **same** products as H$^+$/**H$_2$O or R$_3$COH** except **no rearrangements occur;** Hg is E$^+$, replaced by H with NaBH$_4$

Guide to Stereochemistry Concepts and Analysis of Reaction Stereochemistry as Applied to Electrophilic Addition

13

Part 1: Conceptual Guide to Stereochemistry

13.1 CONCEPTS OF STEREOCHEMISTRY AND CHIRALITY

1. **Stereoisomers:** <u>Different molecules</u> of the same constitutional isomer distinguished by different 3-D arrangements of atoms which are not interconvertible by rotation around single (sigma) bonds.
2. **Enantiomers:** <u>Stereoisomers</u> specifically related as <u>non</u>identical mirror images
   a. **Enantiomers** usually require the presence of one or more **chiral** carbons.
   b. **Chiral Carbon:** sp^3/tetrahedral carbon specifically with **four different** groups attached.
   c. **One chiral** carbon in a molecule must produce only one pair of enantiomers.

3. **Diastereomers:** <u>All other</u> stereoisomer relationships
   a. **Two** or more chiral carbons can produce both enantiomers and diastereomers.
   b. Chiral carbons are not required for certain molecular diastereomers.

13.2 ABSOLUTE CONFIGURATION OF CHIRAL CARBONS

The 3-D geometries of chiral carbons in enantiomers can be specified by a defined system which views the orientation of the four attached groups around the chiral carbon. The terms are **R** configuration (rectus/right) and **S** configuration (sinister/left).

13.2.1 PRIORITY SYSTEM FOR ATTACHED ATOM GROUPS

The priority system for attached atom groupings is based on atomic number: the higher rank applies to the higher atomic number of attached atoms in atom groups at the first point of difference.

13.2.2 PROCESS TO APPLY THE PRIORITY SYSTEM TO THE FOUR DIFFERENT GROUPS ATTACHED TO CHIRAL CARBONS

Step 1. Start at the first atom of the group attached directly to the chiral carbon: the higher atomic number has the higher priority.
Step 2. If the directly connected atoms of two or more different groups are the same atom this is a **tie**. To break the tie at the first point of difference: begin comparing, one at a time, **all** atoms attached to the equivalent connecting atoms until a difference is found.
   a. Always begin by comparing the higher atomic number atoms on each equivalent connecting atom first.
   b. Be sure to compare all atoms attached to the equivalent connecting atoms to find a difference before moving down the substituent chain.
   c. Comparisons are required to maintain a one-to-one correspondence: a double bond to an atom is counted as two individual bonds to two of these atoms; a triple bond to an atom is counted as three individual bonds to three of these atoms.

Example 1: Consider the chiral carbon drawn below:

The four groups attached to the chiral carbon can be identified; the bond line indicates the specific atom for each group directly attached to the chiral carbon:

$$-\text{H} \qquad -\text{OCH}_3 \qquad -\text{CH}_3 \qquad -\text{Br}$$

Step #1 is applied: the atomic number of the directly connecting atom for each attached group is shown above the atom.

$$\overset{1}{-\text{H}} \qquad \overset{8}{-\text{OCH}_3} \qquad \overset{6}{-\text{CH}_3} \qquad \overset{35}{-\text{Br}}$$

There are no ties from step #1; step #2 is not required. The order of the groups from highest to lowest is

$$\underset{\#1}{\overset{35}{-\text{Br}}} \longrightarrow \underset{\#2}{\overset{8}{-\text{OCH}_3}} \longrightarrow \underset{\#3}{\overset{6}{-\text{CH}_3}} \longrightarrow \underset{\#4}{\overset{1}{-\text{H}}}$$

Example 2: Consider the chiral carbon drawn below:

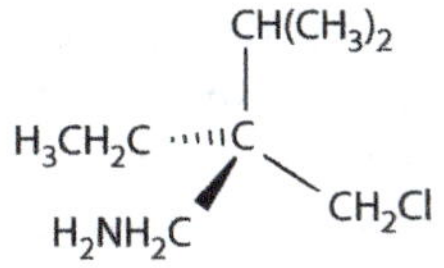

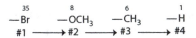

The four groups attached to the chiral carbon are identified with the bond line indicating the specific atom of each group directly attached to the chiral carbon:

$$-\text{CH}_2\text{CH}_3 \qquad -\text{CH}_2\text{NH}_2 \qquad -\text{CH}_2\text{Cl} \qquad -\text{CH(CH}_3)_2$$

Step #1 is applied: the atomic number of the directly connecting atom for each attached group is shown above the atom.

$$-\overset{6}{\text{C}}\text{H}_2\text{CH}_3 \qquad -\overset{6}{\text{C}}\text{H}_2\text{NH}_2 \qquad -\overset{6}{\text{C}}\text{H}_2\text{Cl} \qquad -\overset{6}{\text{C}}\text{H(CH}_3)_2$$

All directly connected atoms are ties from step #1; step #2 will be required for all four groups. Compare, one at a time, all atoms attached to the equivalent connecting atoms until a difference is found; compare the **highest** atomic number atoms on each equivalent connecting atom first. The continuing result is

$$-\text{CH}_2\overset{6}{\text{C}}\text{H}_3 \qquad -\text{CH}_2\overset{7}{\text{N}}\text{H}_2 \qquad -\text{CH}_2\overset{17}{\text{Cl}} \qquad -\text{CH}(\text{CH}_3)-\overset{6}{\text{C}}\text{H}_3 \text{ (isopropyl group expanded)}$$

Be sure to compare all atoms attached to the equivalent connecting atoms before moving down the substituent chain. To break the tie between the ethyl group and the isopropyl group now, compare the **second**-highest atoms bonded to the connecting carbons. For the ethyl group, this is a hydrogen (atomic number 1); for the isopropyl group, this second-highest atom is another carbon (atomic number 6):

$$-\text{C}\overset{1}{\text{H}}_2\overset{6}{\text{C}}\text{H}_3 \qquad -\text{CH}_2\overset{7}{\text{N}}\text{H}_2 \qquad -\text{CH}_2\overset{17}{\text{Cl}} \qquad -\text{CH}(\overset{6}{\text{C}}\text{H}_3)-\overset{6}{\text{C}}\text{H}_3 \text{ (isopropyl group expanded)}$$

The order of the groups from highest to lowest is

$$\underset{\#1}{-\text{CH}_2\text{Cl}} \longrightarrow \underset{\#2}{-\text{CH}_2\text{NH}_2} \longrightarrow \underset{\#3}{-\text{CH(CH}_3)_2} \longrightarrow \underset{\#4}{-\text{CH}_2\text{CH}_3}$$

13.2.3 PROCESS TO DETERMINE ABSOLUTE CONFIGURATION OF CHIRAL CARBONS

Step 1. Determine the priorities 1 ⟶ 4 for the four different groups attached to the chiral carbon using the priority system based on higher atomic number at the first point of difference.

1 = highest priority   ⟶   4 = lowest priority

Step 2. Select the specific **[chiral C——Group #4]** bond axis in which the chiral carbon is in front and the connecting atom of group #4 (lowest priority) is at the back.

Step 3. Sight directly down this axis and visualize or draw a Newman-type projection showing the orientation of groups #1, #2, and #3 on the **chiral carbon**. Ignore any atom/groups further attached to the group #4 axis atom.

Step 4. Connect the groups #1, #2, #3 attached to the chiral (front) carbon with a circular arrow in the specific order:

#1 ⟶ #2 ⟶ #3

Step 5. If the circular arrow rotates clockwise = **R** configuration.

If the circular arrow rotates counterclockwise = **S** configuration.

Completion of absolute configuration for example 1: The original chiral carbon is redrawn on the left; the group priority number determined in the previous section is shown next to each attached group. Group #4 is the hydrogen atom: rotate the original drawing such that the hydrogen is directly behind the chiral carbon; this is shown on the drawing to the right. Sighting down the [chiral C——Group #4] bond axis with the chiral carbon in front results in a Newmann-type viewpoint. The three remaining higher-priority groups produce the Y-pattern; the hydrogen (group #4) is behind the chiral carbon and cannot be seen. The circular arrow connecting groups #1 → #2 → #3 is counterclockwise; the absolute configuration is determined to be (S).

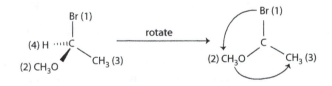

Completion of absolute configuration for example 2: The original chiral carbon is drawn on the left with the determined priority values from the previous section. In this case, group #4 is the ethyl group; the drawing on the right shows it rotated to a position directly behind the chiral carbon. The corresponding Newmann-type viewpoint for the three remaining higher-priority groups indicates that the circular arrow connecting groups #1 → #2 → #3 is clockwise; the absolute configuration is determined to be (**R**).

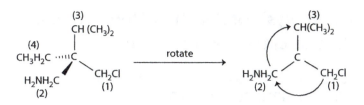

Since group #4 was already pointing mostly backward in the original drawing, both of these examples required a relatively straightforward rotation to produce the required viewpoint. Although, other examples will require a greater degree of 3-D mental gymnastics, experience will generate simplifying techniques.

13.3 MOLECULAR RELATIONSHIPS: SUMMARY EXAMPLE WITH MOLECULAR FORMULA: C_8H_{16}

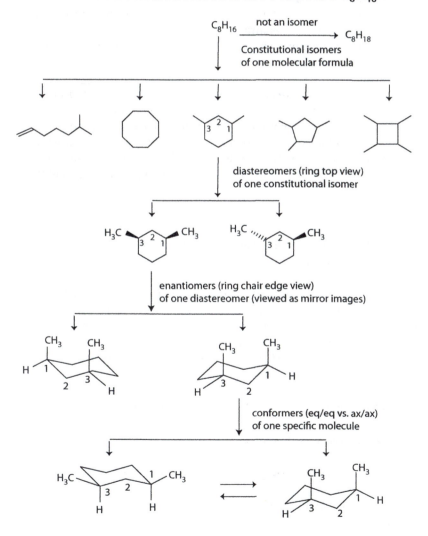

Part 2: Analysis of Reaction Stereochemistry Applied to Electrophilic Addition

13.4 STEREOCHEMISTRY CONCEPTS FOR ELECTROPHILIC ADDITION

13.4.1 GENERAL ANALYSIS FOR REACTIONS

1. Stereochemical analysis of organic molecules for reactions applies **only** to those carbons that undergo bonding changes during the reaction. Regardless of the number of stereogenic centers in a reactant molecule, chiral carbons and/or asymmetric double bonds that do not change are not analyzed.
2. Reaction stereochemical analysis describes the changes in stereogenic centers:
   a. The number and type of stereoisomers produced (diastereomers/enantiomers).
   b. The determination of how chiral carbons retain or change configuration.
3. Reactions in this first course will always involve non-chiral reagents. In any reaction in which a chiral carbon in a product molecule is formed from a non-chiral carbon from a reactant molecule, a ***non-chiral reagent must produce an exact 50% R/50% S enantiomer pair***. A **chiral** reagent is required to form a product with enantiomeric selectivity (one enantiomer favored from an enantiomeric pair) starting from a non-chiral reactant.

13.4.2 STEREOCHEMISTRY CONCEPTS APPLIED TO ALKENE ELECTROPHILIC ADDITION

Trigonal planar sp^2 alkene carbons during electrophilic addition reactions are converted into tetrahedral sp^3 carbons (i.e., four groups attached): chiral (sp^3) carbons may be produced from non-chiral (sp^2) carbons during the reaction. For any specific double bond (i.e., two sp^2 carbons reacting), three options are possible for the general reagent E—Nu (**E** = electrophile; **Nu** = nucleophile):

Option 1 alkene + E—Nu reagent ⟶ no chiral carbons formed

Option 2 alkene + E—Nu reagent ⟶ one chiral carbon formed

Option 3 alkene + E—Nu reagent ⟶ two chiral carbons formed

Each possibility leads to a specific analysis requirement and product result.

13.4.3 ANALYSIS REQUIREMENTS FOR ELECTROPHILIC ADDITION POSSIBILITIES

Option 1: alkene + E—Nu reagent ───────────→ **no** chiral carbons formed

Option 1 Analysis: Stereochemical analysis for reactions that do not produce chiral carbons is not required. If no chiral carbons are formed, no stereoisomers can be formed in the reaction. (Note again that analysis applies only to carbons that change during the reaction.) Only the correct **constitutional** isomer can be identified in this case.

Option 1 Example:

$$H_2C{=}CH_2 \xrightarrow{\quad HBr \quad} BrH_2CCH_3$$

No chiral carbons produced: no stereoisomers to describe

Option 2: alkene + E—Nu reagent ───────────→ **one** chiral carbon formed

Option 2 Analysis: A non-chiral reagent cannot produce one enantiomer favored over the other: **two stereoisomers as an enantiomer pair** must be produced in an exact *50% R/50% S enantiomer mixture* (a racemic mixture). The description of the products can show both enantiomers. An alternative is to show the product as the correct constitutional isomer with no stereochemistry indicated since the racemic (50/50) mixture must always occur.

Option 2 Example:

$$CH_3CH_2CH{=}CH_2 \xrightarrow{\quad HBr \quad} CH_3{-}CH_2{-}^*CHBr{-}CH_3$$

One chiral carbon (*) produced: one pair of enantiomers formed

Option 3: alkene + E—Nu reagent ───────────→ **two** chiral carbons formed

Option 3 Analysis: In this case, a **maximum** of 2^2 or 4 stereoisomers may be produced as *two diastereomers each with its enantiomer*. Unlike enantiomers, diastereomers do **not** have identical physical properties. Thus, in certain cases, **only one diastereomer** (with its enantiomer or meso) of the two possible diastereomers will form. Further analysis is required to determine which specific stereoisomers of the four possibilities will be formed. *Only the specific diastereomer formed in the reaction will be shown as the product.*

Option 3 Example:

Two chiral carbons (*) produced: in this case, four stereoisomers as two diastereomers each with its enantiomer are formed.

13.4.4 STEREOSPECIFICITY

Determining conditions which identify the specific number and type of stereoisomers produced in an electrophilic addition reaction requires the concept of **mechanism stereospecificity**. Mechanism stereospecificity results in **reaction stereospecificity**.

1. If starting from a specific stereoisomer reactant (or a restricted set of stereoisomers as reactants), a **stereospecific** reaction (based on mechanism) *will form a specific stereoisomer product* or a restricted set of stereoisomer products.
2. Even if starting from a specific stereoisomer or a restricted set of stereoisomers as reactants, a **non-stereospecific** reaction will form **all possible** enantiomer and diastereomer products (or meso products). Enantiomer pairs will be exactly 50%/50%; diastereomers will usually be formed in **approximately** equal ratios.
3. The designation of *stereospecificity applies to the mechanism*, **not** the reactant molecule. Stereospecificity occurs through specific mechanistic, intermediate, or transition state requirements which restrict the 3-D arrangement of atoms in the reacting molecules during the process of being converted into product molecules. These requirements exist for all reactions proceeding through the mechanism regardless of the stereochemical characteristics of the reactant molecule.
4. *Observation (detection) of stereospecificity in any specific reaction requires that the product molecule displays stereochemistry* (i.e., the molecule must be able to exist as diastereomers or enantiomer pairs). If the product constitutional isomer does not have stereoisomers, then, stereospecificity cannot affect the reaction result.

13.4.5 STEREOSPECIFIC REACTION SUMMARY

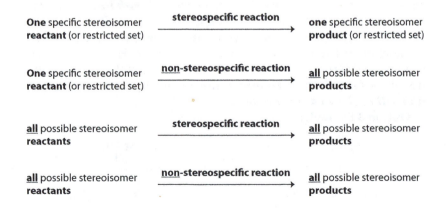

Note that **only** the combination of a specific stereoisomer reactant with a stereospecific reaction will produce a restricted set of stereoisomer products; all other combinations will produce all possible stereoisomers.

13.5 STEREOSPECIFICITY AND REACTION ANALYSIS FOR ELECTROPHILIC ADDITION

13.5.1 STEREOSPECIFICITY OPTIONS FOR ELECTROPHILIC ADDITION MECHANISMS

Double-bond <u>face</u>: View the sp^2/trigonal planar double-bond carbons edge-on; groups attached to these carbons are viewed as pointing forward and backward with the perpendicular p-orbital pi-bond overlap pointing up and down. The **double-bond face** can be observed as the "top"/"up" face versus the "down"/"bottom" face.

There is no permanent distinction between the "top"/"up" face versus the "down"/"bottom" face; the molecule can, of course, be viewed upside down. The two faces indications are used for comparing relationships to each other.

Viewpoint for distinguishing the two faces in a double bond:

Three stereospecificity results for electrophilic addition mechanisms can be observed:

1. **E** and **Nu** can add <u>only</u> to the **same** face of the double bond: the mechanism is labeled as **stereospecific <u>syn</u>**.
2. **E** and **Nu** can add <u>only</u> to **opposite** faces of the double bond: the mechanism is labeled as **stereospecific <u>anti</u>**.
3. **E** and **Nu** can add to <u>either</u> the same face or opposite faces of the double bond with approximately equal probability: the mechanism is labeled as <u>**non**</u>-**stereospecific**.

13.5.2 REACTION CONDITIONS THAT REQUIRE ANALYSIS FOR DIASTEREOMERS

Requirement 1: Diastereomer products in electrophilic addition are produced only when the starting reactant alkene contains a reacting asymmetric double bond.

Requirement 2: Diastereomer products in electrophilic addition are produced only when two chiral carbons are formed.

Requirement 3: One specific diastereomer product in electrophilic addition can be produced only when starting with a specific stereoisomer reactant alkene (E or Z/cis or trans) reacting in a stereospecific reaction/mechanism.

Example: Requirements **1** and **3** are met but not Requirement **2**:

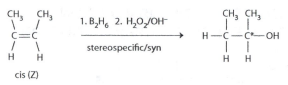

one chiral carbon (*): 50%R/50%S

Example: Requirements **1** and **2** are met but not Requirement **3**:

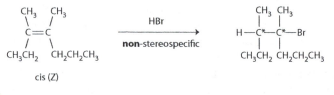

* * = chiral = [R, R]; [S, S]; [R, S]; [S, R] all formed

13.5.3 EXAMPLE: ALL REQUIREMENTS 1, 2, AND 3 ARE MET

Result: <u>one</u> specific diastereomer product formed with its enantiomer.

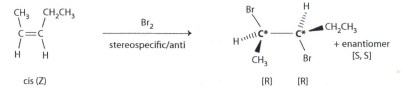

[R, S] and [S, R] not formed

13.6 CONCEPTS OF RING STEREOCHEMISTRY

13.6.1 CYCLOALKENE DIASTEROMERS

Cycloalkenes (double bonds in rings) up to about eight carbons represent a specific **stereoisomer reactant**:

1. Bonding of atoms into rings restricts full rotation around some C—C bond combinations.
2. The ring carbons attached to the sp^2 alkene carbons must be **<u>cis</u>** to each other due to the restricted cumulative bond lengths in rings below about eight carbons.
3. Cycloalkenes which indicate a specific stereoisomer reactant must always be evaluated for formation of diastereomers.
4. Diastereomers in cycloalkenes occur as a **cis/trans** relationship for substituents on the ring.

13.6.2 VIEWING RING DIASTEREOMERS

1. **Ring <u>face</u>** can be observed as the "top"/"up" face versus the "down"/"bottom" face when viewing the ring from edge-on (along the plane of the ring).

2. **Ring Diastereomers:** stereoisomers based on the relationship between two (or more) substituents on a ring are viewed as being on the same face of the ring versus being on opposite faces on the ring.
3. Stereoisomer relationships are independent of conformations.

13.6.3 STEREOSPECIFICITY OPTIONS VIEWED IN CYCLOALKENES

A **cis/trans** configuration for diastereomers formed in electrophilic addition of cycloalkenes allows convenient visualization of stereospecificity options.

1. **E** and **Nu** can add <u>only</u> to the **same** face of the double bond: the mechanism is labeled as **stereospecific <u>syn</u>**. Only **one** diastereomer (with its enantiomer) is formed: E and Nu are cis on the ring; E and Nu trans does not form.

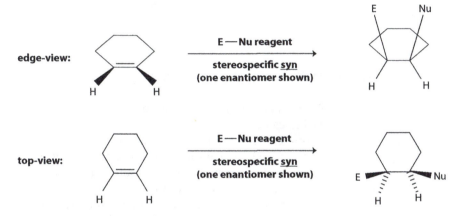

2. **E** and **Nu** can add <u>only</u> to **opposite** faces of the double bond: the mechanism is labeled as **stereospecific <u>anti</u>**. Only **one** diastereomer (with its enantiomer) is formed: E and Nu are trans on the ring; E and Nu cis does not form.

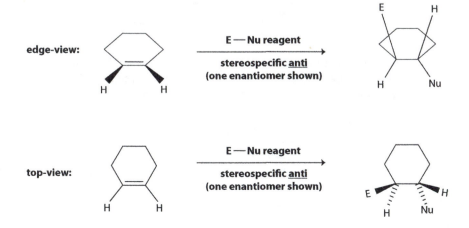

3. **E** and **Nu** can add to **either** the same face or opposite faces of the double bond with approximately equal probability: the mechanism is labeled as **non-stereospecific**. **All** stereoisomers form even when starting with a specific stereoisomer reactant. **Both** diastereomers (each with its enantiomer) are formed: E and Nu are **trans** on the ring with about 50% probability and E and Nu are **cis** on the ring with about 50% probability.

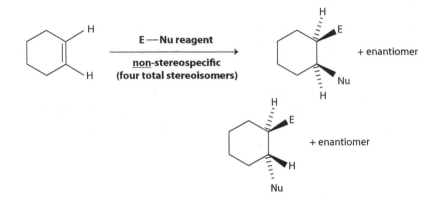

13.6.4 NOTATION FOR PRODUCT IDENTIFICATION

1. If **no** stereoisomers are formed, only the correct constitutional isomer can be indicated.
2. Formation of **one** chiral carbon forms enantiomers only. Since they must be exactly 50/50, one specific enantiomer cannot be formed: both enantiomers may be shown or alternatively, only the correct constitutional isomer needs to be indicated.
3. If a specific diastereomer is formed in a reaction, indicate **only** this diastereomer (as either enantiomer). This applies to the formation of **two** chiral carbons from a specific stereoisomer reactant with a stereospecific reaction mechanism.

4. For **non**-stereospecific reactions, either show both diastereomers or alternatively, indicate that all stereoisomers are formed by showing only the correct constitutional isomer:

A Process for Calculation of Product Distribution through Relative Rate Analysis

Examples for Free Radical Halogenation

14.1 METHODS FOR PRODUCT PERCENT AND RATIO CALCULATIONS

Organic reactions often produce a distribution of related constitutional or stereoisomer products. The determining factor for this product distribution may be free energy (thermodynamic control) or relative rates of formation (kinetic control). For those reactions under kinetic control, certain types of rate information can be used to calculate predicted isomer ratios or percentages. The processes described can be applied to three problem types using free radical halogenation as the example reaction.

1. *Problem Type: Determine <u>all</u> constitutional isomer products for free radical halogenation of one specific alkane reactant and calculate the isomer percentage and ratio for each isomer product from <u>given</u> relative rate data.*

   Example 1: Determine the number of possible constitutional isomer products for the ***monobromination*** of pentane and calculate isomer percentages and product ratios. Assume that the (**hypothetical**) **relative** rate data is

Hydrogen Type	Relative Rate per H
primary C—H :	1
secondary C—H :	20
tertiary C—H :	100

Process:

Step **(1)** First determine all possible constitutional isomer products which can be formed by exchange of one hydrogen for one halogen (bromine in this case) on the starting alkane. **Be careful to identify equivalent hydrogens on any specific carbon <u>and</u> equivalent carbons in an alkane that, when halogenated, yield the <u>same</u> constitutional isomer.** *Isomer determination in this problem type is <u>not</u> simply based on distinguishing primary versus secondary versus tertiary hydrogens.*

Example 1; Analysis for Step (1): Owing to symmetry, only three constitutional isomers are possible for **mono**bromination:

Consider the pentane carbons labeled 1 through 5:

$$\overset{5}{CH_3}\overset{4}{CH_2}\overset{3}{CH_2}\overset{2}{CH_2}\overset{1}{CH_3}$$

a. Since carbons #1 and #5 are symmetrical, exchange of Br for H at carbons #1 or #5 produces **1-bromopentane**: CH_3CH_2 $CH_2CH_2CH_2\textbf{Br} = \textbf{Br}CH_2CH_2CH_2CH_2CH_3$.

b. Since carbons #2 and #4 are symmetrical, exchange of Br for H at carbons #2 or #4 produces **2-bromopentane**: CH_3CH_2 $CH_2CH\textbf{Br}CH_3 = CH_3CH\textbf{Br}CH_2CH_2CH_3$.

c. Carbon #3 is unique; exchange of Br for H at carbon #3 produces **3-bromopentane**: $CH_3CH_2CH\textbf{Br}CH_2CH_3$.

Step **(2)**

a. For **each** possible isomer product: determine the **total** number of equivalent hydrogens which can be exchanged on the reactant alkane to produce the specific isomer product. This is found from

total number of exchangeable hydrogens per product specific product = **(number of equivalent carbons)** × **(number of equivalent hydrogens per equivalent carbon)**

b. Then, identify whether these hydrogens are primary, secondary, or tertiary (or allylic or benzylic).

Example 1; Analysis for Step (2):

a. The number of hydrogens which can be exchanged to produce 1-bromopentane is based on two *equivalent* carbon positions (C#1 and C#5), each of which has **three** exchangeable *equivalent* hydrogens = **6** total hydrogens. These are **primary** hydrogens.

b. The number of hydrogens that can be exchanged to produce 2-bromopentane is based on **two** *equivalent* carbon positions (C#4 and C#2), each of which has **two** exchangeable *equivalent* hydrogens = **4** total hydrogens. These are **secondary** hydrogens.

c. The number of hydrogens that can be exchanged to produce 3-bromopentane is based on **one** unique carbon position (C#3), with **two** exchangeable *equivalent* hydrogens = **2** hydrogens. These are **secondary** hydrogens.

Note that two __different__ isomers can be formed by the reaction of secondary hydrogens!

Step **(3)** Use the relative rate data **given** to assign a relative rate to each hydrogen type producing a specific constitutional isomer. Then, use this rate per hydrogen to calculate the relative rate for each constitutional isomer through the relationship

relative rate per isomer product	=	**relative rate of exchanged H (per H)**	×	**# of equivalent H that can produce the specific product**

Example 1; Analysis for Step (3):

d. The number of hydrogens that can be exchanged to produce 1-bromopentane was = **6** total primary hydrogens at a relative rate = **1 per H**.
Relative rate for 1-bromopentane = (primary rel. rate of **1**/H) × (**6** H) = **6**.

e. The number of hydrogens that can be exchanged to produce 2-bromopentane was = **4** total secondary hydrogens at a relative rate = **20 per H**.
Relative rate for 2-bromopentane = (secondary rel. rate of **20**/H) × (**4** H) = **80**.

f. The number of hydrogens that can be exchanged to produce 3-bromopentane was = **2** total secondary hydrogens at a relative rate = **20 per H**.
Relative rate for 3-bromopentane = (secondary rel. rate of **20**/H) × (**2** H) = **40**.

Step **(4)** To complete the **percent** calculations, **sum** the total relative rates for all products and determine the percentages of **each** isomer product from its relative rate compared to the **total** for all isomer products.

Example 1; Analysis for Step (4):

Relative rate for 1-bromopentane = (primary rel. rate of **1**/H) × (**6** H)   =   6

Relative rate for 2-bromopentane = (secondary rel. rate of **20**/H) × (**4** H)   = 80

Relative rate for 3-bromopentane = (secondary rel. rate of
20/H) × (2 H)   = **40**
Total for all isomer products   =126
percentage for 1-bromopentane = **6**/126 × 100% = **4.8%**
percentage for 2-bromopentane = **80**/126 × 100% = **63.5%**
percentage for 3-bromopentane = **40**/126 × 100% = **31.7%**

Step **(5)** To determine the **ratio** of products: express the **relative rate data** from step **(4)** as a ratio of fastest formed (greatest %) to slowest formed (least %). Then, divide by the smallest (slowest rate) number to express the ratio based on the **slowest** rate equal to **one**.

Example 1; Analysis for Step (5):
The ratio of fastest to slowest:
2-B2-bromopentane/3-bromopentane/1-bromopentane = **80/40/6**
Division by the smallest number = **13.3/6.7/1**

2. *Problem Type: For free radical halogenation, calculate the comparison rate ratio for a number of <u>different</u> alkane reactants from <u>given</u> relative rate data.*

Example 2: Calculate the rate ratio for the ***<u>monobromination</u>*** of propane, $CH_3CH_2CH_3$, versus 2-methylpropane, $(CH_3)_3CH$. **Assume that the relative rate data is the same as for <u>Example 1</u>.**

Process: If the halogenation reaction is <u>bromination</u>, use the methods described in Example 1 to calculate the relative reaction rate for products formed ***through exchange of the <u>most reactive</u> hydrogen type in each alkane molecule to be compared***. Theoretically, to compare the rate of reaction for several ***<u>different</u>*** reactant alkanes, the **total** rate for formation of **all possible** products for each reactant could be calculated. Practically, however, if the reaction is bromination, a good approximation for the rate comparison can be found by calculations based on the most reactive hydrogen type in the molecule. This is because rate differences for primary versus secondary versus tertiary (or allylic or benzylic) hydrogens are so large for a bromine reagent that contributions from the less reactive hydrogens in a reactant alkane can be ignored for approximate calculations.

Example 2; Analysis:
a.   The most reactive position on propane is the secondary carbon. Since this is a unique carbon, only one product can be formed from exchange of secondary hydrogens; 2-bromopropane: $CH_3CH\mathbf{Br}CH_3$; the **two secondary** hydrogens have a relative rate = **20 per H**.
b.   The most reactive position on 2-methylpropane is the tertiary carbon. Since this is a unique carbon, only one product can be formed from exchange of the tertiary hydrogen: 2-bromo-2-methylpropane (t-butyl bromide): $(CH_3)_3C\mathbf{Br}$; the **one tertiary** hydrogen has a relative rate = **100 per H**.

Use the same procedure as for **Example 1** to calculate the rate ratio; the equation for **rate per H** as a function of **reactant rate**, adapted from **Example 1** is

relative rate of an alkane reactant	=	**relative rate of exchanged H (per H)**	×	**# of similar reacting H in the alkane reactant.**

Relative rate for 2-bromopropane formed from the reactant **propane:** = (secondary rel. rate of **20**/H) × (**2** H) = **40**.
Relative rate for t-butyl bromide formed from the reactant **2-methylpropane:** = (tertiary rel. rate of **100**/H) × (**1** H) = **100**. The alkane **_reactant_** rate ratio is expressed as fastest/slowest: 2-methylpropane/propane = **100/40 = 2.5/1**.

3. *Problem Type: For free radical **bromination**, calculate the relative rate per H for the most reactive position comparing a number of **different** alkane reactants; the data are derived from **experimental** **reaction times**.*

Example 3: For free radical bromination, the following experiment was conducted: Under identical concentration and light conditions as measured by consumption of Br_2, ethane completely reacts in 240 min and cyclobutane completely reacts in 8 min as measured by consumption of Br_2. Calculate the relative rate for secondary hydrogens compared to primary hydrogens using this ethane versus cyclobutane experimental reaction time comparison. (**Rates from this experiment need not match any given rate data**.)

Process:
Step **(1)** Determine the rate ratio for the reactants, expressed as fastest/slowest; then, set the rate of the slowest = 1. *The rate of reaction is **inversely** proportional to the time it takes for the alkane reactant to completely react.*

Example 3; Analysis for Step (1): The rate of reaction for ethane is **_inversely_** proportional to its reaction time: **relative** rate of ethane = 1/240 = **0.004166**. The rate of reaction for cyclobutane is **_inversely_** proportional to its reaction time: **relative** rate of cyclobutane = 1/8 = **0.125**. The **_rate ratio_** (cyclobutane/ethane) = 0.125/0.004166 = **30/1**. The *faster* reactant, **_cyclobutane_**, reacts 30 times faster than ethane.
Step **(2)** Determine the **most reactive** hydrogens in each reactant, the number of these hydrogens, and their classification; the approximation described in Example 2 is applied. This calculation is **not** restricted to the formation of only one specific constitutional isomer product. **All of the most reactive hydrogen types are counted**.

Example 3; Analysis for Step (2):
The most reactive hydrogens in ethane are primary; there are **6** primary H.
The most reactive hydrogens in cyclobutane are secondary; there are **8** secondary H.
Step **(3)** In Example 2, the equation used was

relative rate of an alkane reactant	=	relative rate of exchanged H (per H)	×	# of similar reacting H in the alkane reactant.

Solve this equation for *relative rate per H* and complete the calculations:

$$\text{relative rate of exchanged H (per H)} = \frac{\text{relative rate of an alkane reactant H}}{\text{# of similar reacting H in the alkane reactant}}$$

Example 3; Analysis for Step (3): The rate ratio (cyclobutane/ethane) = **30/1**.
Relative rate for ethane = **1**
Relative rate for primary = (relative rate for ethane of **1**)/(**6** primary H per ethane) = **0.167**
Relative rate for cyclobutane = **30**
Relative rate for secondary = (relative rate for cyclobutane of **30**)/(**8** secondary H per cyclobutane) = **3.75**
For this example experiment:
ratio: secondary/primary = 3.75/0.167 = **22/1**

14.2 ADDITIONAL PRACTICE PROBLEM

For the reaction **1, 2-dimethylcyclopentane + Br₂** $\longrightarrow$?
Determine the number of possible constitutional isomer products for *monobromination* and calculate isomer percentages and product ratios. Assume that the **(hypothetical)** relative rate data is

Hydrogen Type	Relative Rate per H
primary C—H :	1
secondary C—H :	15
tertiary C—H :	28

14.3 PRACTICE PROBLEM ANSWER (FOLLOWING STEPS SIMILAR TO EXAMPLE 1)

1. Owing to symmetry, only **four** different constitutional isomers can be formed from **mono**bromination of 1,2-dimethylcyclopentane

(i.e., ignoring stereoisomers). These are indicated in the line drawings; equivalent hydrogens which lead to the same isomer match the labeled letter for the corresponding isomer product shown.

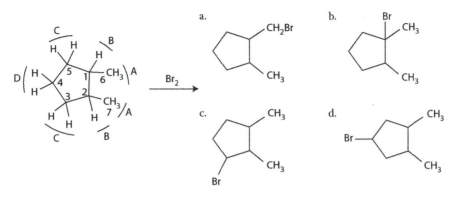

2. a. The number of hydrogens that can be exchanged to produce **A** is based on **two** *equivalent* carbon positions (C#6
and C#7), each of which has **three** exchangeable *equiva-*
lent hydrogens = **6** total hydrogens. These are **primary**
hydrogens.

   b. The number of hydrogens that can be exchanged to produce **B** is based on **two** *equivalent* carbon positions (C#1
and C#2), each of which has **one** exchangeable *equiva-*
lent hydrogen = **2** total hydrogens. These are **tertiary**
hydrogens.

   c. The number of hydrogens that can be exchanged to produce
C is based on **two** *equivalent* carbon positions (C#3 and
C#5), each of which has **two** exchangeable *equivalent* hydro-
gens = **4** total hydrogens. These are **secondary** hydrogens.

   d. The number of hydrogens that can be exchanged to produce
D is based on **one** unique carbon position (C#4), with **two**
exchangeable *equivalent* hydrogens = **2** hydrogens. These
are **secondary** hydrogens.

3 and 4:

relative rate for **A** = (primary rel. rate of **1**/H) × (**6** H)   =   **6**
relative rate for **B** = (tertiary rel. rate of **28**/H) × (**2** H)   = **56**
relative rate for **C** = (secondary rel. rate of **15**/H) × (**4** H)   = **60**
relative rate for **D** = (secondary rel. rate of **15**/H) × (**2** H)   = **30**
total for all isomer products   = **152**

percentage for **A** = **6**/152 × 100% = **3.95%**
percentage for **B** = **56**/152 × 100% = **36.8%**
percentage for **C** = **60**/152 × 100% = **39.5%**
percentage for **D** = **30**/152 × 100% = **19.7%**

5. **The ratio of fastest to slowest:**
C/B/D/A = **60/56/30/6**
Division by the smallest number = **10/9.3/5/1**

Process to Identify and Solve the Reactions for Organic I

15.1 REACTION FLOW DIAGRAM

A. The reaction flow diagram can be used as an aid for the identification of the correct reaction type for a specific starting **organic reactant**. All possible reactions are not covered, but all major reaction **types** for **Organic I** are included.

B. Included in this chapter are the reproductions of the summary sections from **Chapter 12** on determination of the correct product once the reaction type has been identified. *These are referenced in the flow diagram*.

C. **General process:**

1. a. Identify the **functional group** in the reactant organic molecule.
   b. Identify the other **reagent(s)**.

2. Combine the information from **step (1)** to determine the **reaction type**:
   a. Substitution
   b. Elimination
   c. Addition
   d. Acid/base
   e. Other reactions such as molecular cleavage, rearrangements, etc.

3. For **substitution** or **elimination** reaction types determined from **step (2)**, identify the **molecularity** of the **mechanism**; this may be required to clarify the reaction type and final product:
   a. Unimolecular: **SN1** or **E1**
   b. Bimolecular: **SN2** or **E2**

4. Use **steps (1)**, **(2)**, and **(3)** to determine important **mechanistic variables** that affect the **regiochemistry** and

stereochemistry for the major product; identification of all steps for the mechanism may not be necessary. **Examples:**
a. Alkyl substitution patterns (primary/secondary/tertiary)
b. Alkyl substitution on sp²-carbons of alkenes
c. Thermodynamic versus kinetic control
d. Intermediate or transition state geometries for stereospecificity
5. Use the listed summaries as a guide

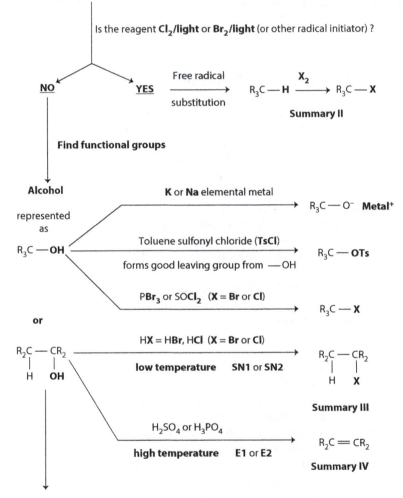

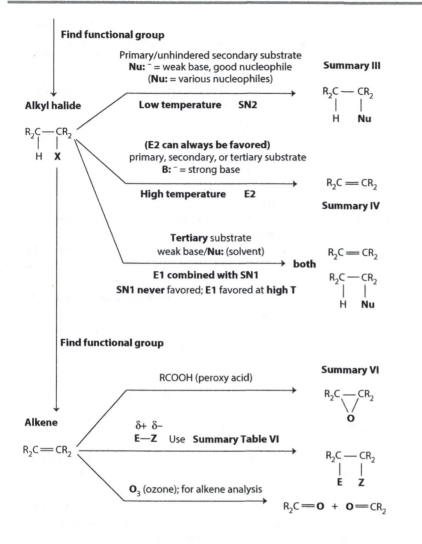

15.2 SUMMARIES FROM CHAPTER 12

A. **Summary II: Determining the Correct Products for Free Radical Halogenation**

1. The carbon reactants (substrates) for free radical halogenation are **alkanes** or the **alkyl-portions** of higher functional group molecules. The reactive carbon/hydrogen **must** be composed of an sp^3-**carbon** to which is bonded **at least one** hydrogen.

2. The first propagation step is the product-determining step; the **Br•(or Cl•)** radical will favorably abstract the **weakest-bonded** hydrogen from the neutral alkane.

   a. **To find the major constitutional isomer product:** Identify the **highest alkyl-substituted** carbon in the molecule (tertiary > secondary > primary). Replace **one** hydrogen on this carbon with **one** halogen (Cl or Br from the specified reagent): this is always correct for bromination; some exceptions may occur for chlorination.

b. The other product of the reaction is the hydrogen halide, HX; (this need not be shown). Note, however, that **one diatomic** halogen reagent (Br_2 or Cl_2) produces **only** a **single** substitution (**One** C—H to **one** C—X).

c. Other **minor products**, or **multiple substitutions**, are found by considering the reaction at other C—H positions in the substrate molecule.

d. Product formation occurs through an sp^2/trigonal planar carbon radical: the reaction is <u>not</u> stereospecific; no stereochemistry applies.

B. **Summary III: Determination of the Correct Product for Nucleophilic Substitution**

1. Identify the **nucleophilic** reagent molecule and the **specific** nucleophilic **atom** (the bonding atom); identify whether the nucleophile will substitute as an anion or neutral molecule.

2. Identify the **electrophilic** carbon **with the leaving group**, **X**.

3. Note the acid/base nature of the nucleophile and leaving group. Alcohols will substitute **only** in the presence of **a strong** acid.

4. Identify the **mechanism** based on the identity of **X** and the substitution pattern on the electrophilic carbon (primary, secondary, or tertiary).

    a. For primary alcohols; primary and secondary alkyl halides/sulfonates: **SN2**.

    b. For secondary and tertiary alcohols; tertiary alkyl halides/sulfonates: **SN1**.

5. Identify the correct **constitutional** isomer: directly exchange the nucleophile and leaving group, bonding the correct nucleophilic atom to the electrophilic carbon. Complete any proton transfers required to achieve normal, neutral bonding in the final products: **add or remove protons as necessary**.

6. Identify the correct **stereochemistry**:

    a. Determine if stereoisomers are possible from the structure of the organic molecule.

    b. If **no** stereoisomers are possible: the problem is complete; show only the correct constitutional isomer from step (5).

    c. If stereoisomers are possible, use the mechanism determined from step (4):

        1. **SN1 is** <u>not</u> **stereospecific**; it produces approximately 50% (±10%) of each of the two possible stereoisomers of a stereoisomer pair: show only the correct constitutional isomer.

        2. **SN2 is stereospecific**; it produces 100% inversion of stereoconfiguration: show the correct resulting stereoisomer.

C. **Summary IV: Determination of Major Alkene Product for Elimination (Non-Rearrangement)**

1. *Identify the carbon with the **leaving group***. Halogens will leave under many acid/base conditions; alcohols will react

only in a strong acid. ***In the absence of rearrangement***, *the product double bond **must** include the carbon with the leaving group.*

2. Identify **all** *neighboring carbons to the carbon with the leaving group*; neighboring carbons **must** have at least **one** hydrogen. In the absence of rearrangement, *the double bond **must** include a neighboring carbon with at least one hydrogen.*

3. Select **one** of the neighboring carbons with at least one hydrogen.

    a. *Remove the leaving group from its carbon and a hydrogen from the selected neighboring carbon; replace these bonds with a **double bond** between these two carbons.*

    b. *If **stereoisomerism** is possible, select the alkene product with the **larger** groups on **each** of the two carbons in the **trans**-configuration.* Distinction between stereoisomers may not always be necessary.

4. If the carbon bearing the leaving group has **only one** neighboring carbon, the analysis is complete: **only one** constitutional isomer is possible.

5. If the carbon bearing the leaving group has **more than one** neighboring carbon:

    a. The number of **nonidentical** neighboring carbons can range from 1 to 3. *Repeat **Step 3** for **each** possible neighboring carbon:* draw the correct constitutional isomer (and stereoisomer) that would result from removal of the leaving group and a hydrogen from **each** possible neighboring carbon.

    b. Note that **identical** carbon groups as neighboring carbons will produce **equivalent** constitutional isomers. The total number of possible products will match only the total number of **nonidentical** neighboring carbons.

    c. *Analyze the substitution pattern for each of the products drawn.* **Select as the major constitutional isomer the alkene with the most highly substituted double bond: this is equivalent to removing a hydrogen from the most highly alkyl-substituted neighboring carbon.** If stereoisomers are possible, and are to be distinguished, *select the stereoisomer of this constitutional isomer which has the larger groups on each carbon trans to each other.* **Example (stereoisomers not shown):**

EXAMPLE:

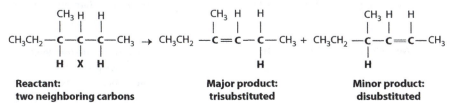

| Reactant: | Major product: | Minor product: |
| two neighboring carbons | trisubstituted | disubstituted |

D. Summary VI: Determination of the Correct Major Product for Electrophilic Addition to Alkenes (Excluding Rearrangements)

1. Identify the alkene (sp^2) carbons of the double bond.
2. Identify the addition electrophile (**E**) and the addition nucleophile (**Z:**) from the **E–Z** reagent. **Use the summary table provided below.**
3. In place of the **pi**-portion of the double bond, add **(E)** to the **least alkyl**-substituted carbon of the original double bond. (Equivalently, this is the carbon with the **most hydrogens.**)
4. Add **(Z:)** to the **most highly alkyl**-substituted carbon of the original double bond.
5. Determine if the product molecule can have stereoisomers. If **no** stereochemistry is possible, the constitutional isomer derived in **(1)** through **(4)** is correct as written.
6. If stereoisomers are possible, determine if the specific mechanism for the **E–Z** reagent is stereospecific. (**See summary table.**) If the mechanism is **not stereospecific**, the constitutional isomer as written is correct, *since one specific stereoisomer cannot be exclusively formed by a non-stereospecific mechanism.*
7. If stereoisomers are possible, **and** if the specific mechanism is stereospecific, write the correct **stereoisomer** based on **syn** or **anti** addition as applicable.

Table for E—Z Reagents for Electrophilic Addition

E—Z Reagent	E$^+$	Z:$^-$	Stereo-Specificity	Comments
H—Cl, Br, and I	**H$^+$**	Cl$^-$/Br$^-$/I$^-$	None	Trigonal planar cation intermediate
H$^+$/H$_2$O	H$^+$	H$_2$O	None	Trigonal planar cation
Net result:	**H**	**OH**		intermediate; fastest rate to form tertiary C—OH
1. H$_2$SO$_4$ 2. H$_2$O	H$^+$	H$_2$O	None	**Sequence** of reactions to increase
Net result:	**H**	**OH**		the rate of alcohol formation
H$^+$/R$_3$COH	H$^+$	R$_3$COH	None	Trigonal planar cation intermediate;
Net result:	**H**	**OCR$_3$**		**R$_3$C = alkyl C**; forms **ethers**
Br$_2$/Cl$_2$	**Br$^+$/Cr$^+$**	Br/Cl$^-$	Anti	Bridged cation intermediate
Br$_2$ or Cl$_2$/H$_2$O	Br$^+$/Cl$^+$	H$_2$O	Anti	Bridged cation intermediate; **bromo**
Net result:	**Br/Cl**	**OH**		**or chloro-alcohol** formed
Br$_2$ or Cl$_2$/R$_3$COH	Br$^+$/Cl$^+$	R$_3$COH	Anti	Bridged cation intermediate;
Net result:	**Br/Cl**	**OCR$_3$**		**bromo or chloro-ether** formed
1. B$_2$H$_6$	BH$_3$	H	Syn	Reaction **sequence** forms C—OH
2. H$_2$O$_2$				from C—BH$_2$; H is transferred
Net result:	**OH**	**H**		from BH$_3$ by a **rearrangement**
1. OsO$_4$	**OH**	**OH**	Syn	Reaction **sequence** forms syn
2. ROOH				**diol** through oxidation

(Continued)

Table for E—Z Reagents for Electrophilic Addition (*Continued*)

E—Z Reagent	E⁺	Z:⁻	Stereo-Specificity	Comments
H_2 with Pt, or Pd or Ni etc. **catalyst**	**H**	**H**	Syn	Syn addition due to **catalyst** surface
RCOOH $\overset{\parallel}{\underset{O}{}}$		**Oxygen**	Syn	Reagent is a **peroxy acid;** product is an **epoxide**
O_3		**Oxygen**	None	Complete cleavage of double bond to form **aldehydes** and **ketones**
1. Hg(OOCCH₃)₂/H₂O or R₃COH 2. NaBH₄/ROH **Net result:** **H** **OH/OCR₃**			None	Reaction **sequence** forms the **same products** as **H + /H₂O** or **R₃COH** except <u>**no**</u> **rearrangements occur**

15.3 REACTION EXAMPLES

Compound "**A**" shown is separately reacted with nine reagent combinations. Since this molecule contains multiple functional groups, the purpose of this exercise is to use functional group and reagent analysis to identify the major/most probable reaction and then determine the final product. Compound "**A**": 3-bromomethyl-4-penten-2-ol:

(Compound A) $H_2C = CH - CH - CH - CH_3$ with OH above the third carbon and CH_2Br below the third carbon $\xrightarrow[\text{Examples 1–9}]{\text{Reagents:}}$

The stereochemistry around the chiral carbons is not specified; the alkene is symmetric.

1. Compound A $\xrightarrow{\text{Br}_2/\text{light or NBS}}$?

Analysis:

a. Br₂/**light** is a somewhat shorthand notation for **free radical halogenation**. Diatomic halogen can be combined with a variety of other free radical initiations; other reagents such as N-bromosuccinimide (NBS) can be employed under conditions such as this example where electrophilic addition may compete.

b. Free radical halogenation is one of the few reactions that occur at the sp³ **C—H** functional group of an alkyl group. A competing reaction for Br₂ will be electrophilic addition to the alkene: reaction conditions can be selected to favor substitution or the special reagent, NBS, can be used. Br₂ will not react at the alcohol halide.

c. Follow Summary II: directly replace the tertiary (and allylic) (sp³) **C—H** with Br. Free radical halogenation is not stereospecific; no stereochemistry applies.

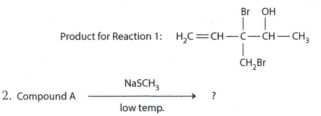

Product for Reaction 1: $H_2C = CH - \underset{\underset{CH_2Br}{|}}{\overset{\overset{Br}{|}}{C}} - \underset{}{\overset{\overset{OH}{|}}{CH}} - CH_3$

2. Compound A $\xrightarrow[\text{low temp.}]{\text{NaSCH}_3}$?

Analysis:

a. $^-SCH_3$ is a very good nucleophile: Na^+ is **not** electrophilic and has no role in the reaction: select **nucleophilic substitution**. The —OH is a very poor leaving group in the absence of a strong acid: select **SN** at the alkyl halide with bromide as the leaving group.

b. Follow Summary III: directly replace the Br with the nucleophile bonded through the nucleophilic sulfur.

c. The electrophilic C—Br carbon is primary: mechanism is **SN2**. SN2 is stereospecific but the primary carbon is not chiral; no stereochemistry applies.

Product for Reaction 2: $H_2C = CH - CH - \underset{\underset{CH_2SCH_3}{|}}{\overset{\overset{OH}{|}}{CH}} - CH_3$

3. Compound A $\xrightarrow[\text{high temp.}]{H_2SO_4 \text{ (conc.)}}$?

Analysis:

a. H^+ is a strong electrophile but HSO_4^- is a very poor nucleophile: nucleophilic substitution at either the alkyl halide or the alcohol is not favored. Electrophilic addition across the alkene is a possibility, but this is not favored especially at high temperature: select **elimination**.

b. The —OH is converted into a very good leaving group in the presence of a strong acid: select **elimination** at the alcohol.

c. Follow Summary IV: the favored constitutional isomer is trisubstituted (and conjugated).

d. The C—OH carbon is secondary: the mechanism is **E1**. E1 is not stereospecific; the stereochemistry of the tertiary carbon also has not been specified: no stereochemistry applies.

Product for Reaction 3: $H_2C = CH - CH - \underset{\underset{CH_3}{|}}{CH} - CH_3$

4. Compound A $\xrightarrow[\text{2. H}_2O_2/OH^-]{\text{1. B}_2H_6}$?

Analysis:

a. The combination of B_2H_6 (BH_3) followed by peroxide and base is specific for **electrophilic addition** across the alkene.

b. Follow Summary VI: the favored constitutional isomer is formed by a net addition of the —H to the most substituted carbon and the net —OH to the least substituted carbon.

c. The reaction occurs through a stereospecific (syn) mechanism; however, the alkene is not asymmetrical and neither sp^3 carbon formed in the reaction is chiral: no stereochemistry applies.

Product for Reaction 4:
$$H_2C-CH-CH-CH-CH_3$$
with OH above the fourth carbon, and HO, H, CH$_2$Br below the first three carbons.

5. Compound A $\xrightarrow{SOCl_2}$?

Analysis:

a. The special reagent SOCl$_2$ is specific for **substitution** of an alcohol functional group by chlorine. This reagent does not react with alkyl halides.

b. The constitutional isomer is formed by direct exchange of —OH by —Cl.

c. The reaction is stereospecific; however, the stereochemistry of the reacting chiral carbon is not specified: no stereochemistry can be indicated.

Product for Reaction 5:
$$H_2C=CH-CH-CH-CH_3$$
with Cl above the third carbon, and CH$_2$Br below.

6. Compound A $\xrightarrow[\text{high temp.}]{NaOCH_2CH_3}$?

(For simplicity for this example, ignore the acid/base reaction with the alcohol.)

Analysis:

a. NaOCH$_2$CH$_3$ is both a strong base and a good nucleophile: nucleophilic substitution at the alkyl halide will compete with elimination. The —OH is a very poor leaving group in the absence of a strong acid; elimination or substitution at the alcohol is not favored. High temperature favors alkyl halide **elimination**.

b. Follow Summary IV: only one constitutional isomer is possible.

c. The C—Br carbon is primary: the mechanism is **E2**. E2 is stereospecific; however, the alkene formed is not asymmetric: no stereochemistry applies.

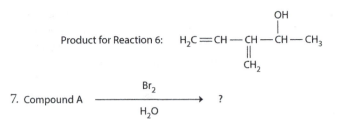

Product for Reaction 6: $H_2C\!=\!CH\!-\!CH\!-\!CH\!-\!CH_3$ with OH above the third carbon and CH$_2$ below the third carbon

7. Compound A $\xrightarrow[\text{H}_2\text{O}]{\text{Br}_2}$?

Analysis:

a. The combination of Br_2/H_2O is an electrophile/nucleophile combination used for **electrophilic addition** across the alkene. The net $E^+ = Br^+$; the net Z^- is ^-OH.

b. Follow Summary VI: the favored constitutional isomer is formed by a net addition of the —Br to the least substituted carbon and the net —OH to the most substituted carbon.

c. The reaction occurs through a stereospecific (anti) mechanism, however, the alkene is not asymmetrical; only one chiral sp³ carbon is formed in the reaction that must be 50%R/50%S: no stereochemistry can be indicated.

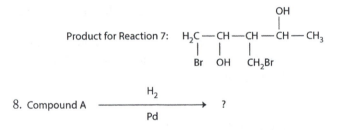

Product for Reaction 7: $H_2C\!-\!CH\!-\!CH\!-\!CH\!-\!CH_3$ with OH above the fourth carbon and Br, OH, CH$_2$Br below

8. Compound A $\xrightarrow[\text{Pd}]{\text{H}_2}$?

Analysis:

a. The combination of H_2/Pd catalyst is specific for reduction of the alkene through an **electrophilic addition** across the alkene.

b. Follow Summary VI: An —H atom adds to each alkene carbon.

c. The reaction occurs through a stereospecific mechanism (syn due to the presence of the catalyst surface); however, neither sp³ carbon formed in the reaction is chiral: no stereochemistry applies.

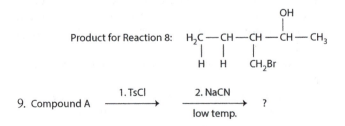

Product for Reaction 8: $H_2C\!-\!CH\!-\!CH\!-\!CH\!-\!CH_3$ with OH above the fourth carbon and H, H, CH$_2$Br below

9. Compound A $\xrightarrow[]{\text{1. TsCl}}$ $\xrightarrow[\text{low temp.}]{\text{2. NaCN}}$?

Analysis:

a. Toluenesulfonyl chloride (TsCl) is a special reagent which reacts with the alcohol oxygen to convert the −**OH** into −**OTs**, a very good leaving group under neutral or basic conditions. ⁻CN is a very good nucleophile but a relatively weak base; Na⁺ is not electrophilic and has no role in the reaction: under low-temperature conditions, select **nucleophilic substitution** at the **C−OTs** position.

b. Follow Summary III: directly replace the OTs with the nucleophile bonded through the nucleophilic carbon of the cyanide anion.

c. The electrophilic c—OTs carbon is secondary: the mechanism is **SN2**. SN2 is stereospecific; however, the stereochemistry around the reacting carbon was not specified: no stereochemistry can be indicated.

$$\text{Product for Reaction 9:} \quad H_2C=CH-CH-\overset{\displaystyle CN}{\underset{\displaystyle CH_2Br}{CH}}-CH_3$$

15.4 ADDITIONAL PRACTICE PROBLEMS

Complete the following reactions with reagents 1–7 and compound "**B**" shown; the specific stereochemistry of the molecule is indicated. Combine a functional group with the reagent to identify the major/most probable reaction and determine the final product, indicating all appropriate stereochemistry. Compound **B**:

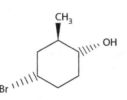

1. Compound **B** with HBr (low temperature)
2. Compound **B** with PBr₃
3. Compound **B** with Br₂/light
4. Compound **B** with NaSH (low temperature)
5. Compound **B** with H₂SO₄ (conc.) (high temperature)
6. Compound **B** with NaOCH₂CH₃ (high temperature)
7. Compound **B** with TsCl followed by NaN₃; pick kinetic major

Complete the following reactions with the reagents 8–12 and compound "**C**" shown; the stereochemistry of the one chiral carbon in the molecule is not specified. Reactions at ring carbons may require stereo chemistry analysis. Combine a functional group with the reagent to identify the major/most probable reaction and determine the final product, indicating all appropriate stereochemistry. Compound **C**:

8. Compound **C** with $NaOCH_2CH_3$ (high temperature)
9. Compound **C** with 1. OsO_4 2. $ROOH/{}^-OH$
10. Compound **C** with Br_2/CH_3OH
11. Compound **C** with $NaCN$
12. Compound **C** with H^+/H_2O

15.5 ADDITIONAL PRACTICE PROBLEMS: ANSWERS

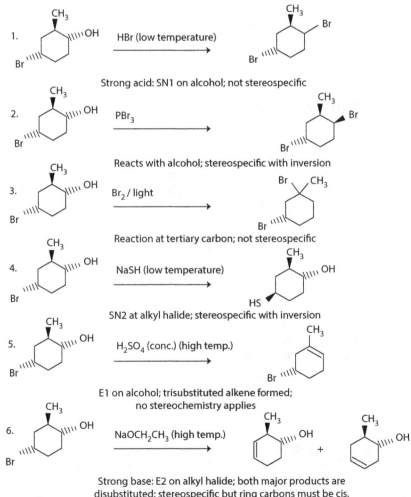

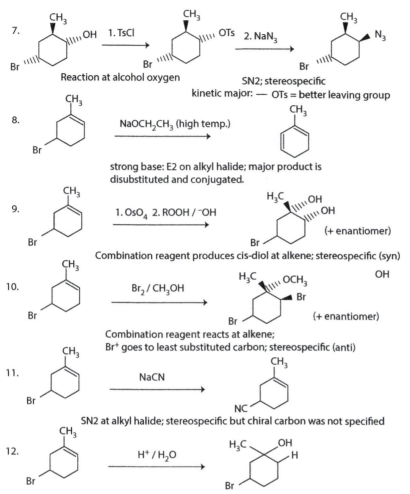

7. 1. TsCl   Reaction at alcohol oxygen   2. NaN₃   SN2; stereospecific
kinetic major: — OTs = better leaving group

8. NaOCH₂CH₃ (high temp.)
strong base; E2 on alkyl halide; major product is
disubstituted and conjugated.

9. 1. OsO₄ 2. ROOH / ⁻OH   (+ enantiomer)
Combination reagent produces cis-diol at alkene; stereospecific (syn)

10. Br₂ / CH₃OH   (+ enantiomer)
Combination reagent reacts at alkene;
Br⁺ goes to least substituted carbon; stereospecific (anti)

11. NaCN
SN2 at alkyl halide; stereospecific but chiral carbon was not specified

12. H⁺ / H₂O
Reaction at alkene; H⁺ goes to least substituted carbon;
not stereospecific

Electrophilic Addition and Addition/Elimination to Conjugated Double Bond and Aromatic Systems

16

16.1 GENERAL CONCEPTS OF PI-BONDING SYSTEMS

16.1.1 PROPERTIES OF NONAROMATIC CONJUGATED SYSTEMS

1. A system of **conjugated** double bonds is characterized by a continuous chain of sp^2 carbons (or other atoms) in which all carbons (or other atoms) participate in an extended pi-bonding molecular orbital.
   a. For most organic molecules, each atom contributes **one** p-orbital to the extended molecular orbital.
   b. Generally, each neutral carbon contributes **one** electron to the extended pi-bond. Anionic carbons or atoms with lone pairs (O and N) can contribute **two** electrons.
2. The additional molecular orbital overlap in conjugated double bonds provides extra stabilization energy compared to isolated double bonds. Conjugated double bonds are lower in potential energy and are energetically favored over isolated double bonds.
3. The conventional Lewis structure pattern for a continuous chain of sp^2 carbons participating in an extended molecular overlap will show a *sequence of alternating single and double bonds:* (R = C or H)

$$R_2C{=}CR{-}CR{=}CR_2 \quad \text{or} \quad R_2C{=}CR{-}CR{=}CR{-}CR{=}CR_2, \quad \text{etc.}$$

4. All atoms in the chain need not be carbon; pi-bonding oxygen and pi-bonding nitrogen are common, especially in biomolecules:

$$R_2C\!\!=\!\!CR\!-\!N\!\!=\!\!CR_2 \quad \text{or} \quad R_2C\!\!=\!\!CR\!-\!CR\!\!=\!\!O, \quad \text{etc.}$$

5. Conventional Lewis structures can indicate the additional pi-bonding overlap through the use of **resonance structures**. Resonance structures are valuable for depicting electron distribution in carbon cations and radicals. Resonance structures can also be used for **neutral** compounds to show the role of electron distribution in specific atom reactivities based on the role of the extended pi-electron overlap.

$$R_2C\!\!=\!\!CR\!-\!CR\!\!=\!\!\overset{..}{\underset{..}{O}} \quad \longleftrightarrow \quad R_2C^+\!\!-\!\!CR\!\!=\!\!CR\!-\!\overset{..}{\underset{..}{O}}{:}^-$$

16.1.2 PROPERTIES OF AROMATIC CONJUGATED SYSTEMS

1. A special type of conjugated system occurs in certain carbon-containing **rings** with either alternating single and double bonds and/or additional atoms with free lone pairs. This type of conjugated system is termed **aromatic**.
2. Characteristics of aromatic systems are:
   a. **All *atoms that contribute a *p-orbital to the* extended *pi-molecular orbital must be in* an *unbroken ring**, with no intervening sp³-carbon or other noncontributing atom.
   b. ***The specific number of pi-electrons in the extended pi-molecular orbital must be exactly equal to one of the values in the number series: 2, 6, 10, 14, 18, ...;*** this series in abbreviated as [**4n + 2**], where **n** = any integer.
   c. The potential energy of the molecular orbital for aromatic pi-electron overlap is **much lower** than the normal pi-bonding orbitals. ***On average, aromatic pi-bonds are much more stable than the corresponding isolated double bonds***.
3. The most common aromatic-ring form is based on benzene: a six-carbon ring with alternating single and double bonds. All carbons in the ring are sp² and contribute one p-orbital and one pi-electron each; the number of pi-electrons in the ring = **6**. The six-carbon double bond conjugation is shown by the **resonance structures**:

4. Aromatic systems containing **nitrogen, oxygen**, and **sulfur** are also very common and are termed **heterocycles**.
   a. Neutral oxygen and sulfur cannot form pi-bonds as ring members; they can contribute **two electrons in a p-orbital** through use of a **lone** electron pair (**A** and **B**) The result is often a five-membered ring with two standard double bonds (contributing two pi-electrons each) plus the p-orbital lone pair; the number of pi-electrons = **6**.

b. Nitrogen, like oxygen and sulfur, can also contribute a lone pair in a p-orbital (**C** and **D**); nitrogen in a ring may also supply a p-orbital with one electron through a pi-bond, similar to the role of a carbon atom (**E**).

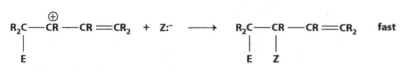

16.2 ELECTROPHILIC ADDITION TO CONJUGATED DOUBLE BONDS IN POLYENES

16.2.1 GENERAL REACTION

1. Electrophilic addition to a **polyene (diene, triene,** etc.) proceeds through the same mechanistic steps as addition to a mono-alkene.

   a. The **first step** is electrophilic addition of the electrophile (E^+) to one of the sp^2 carbons of the double bonded conjugated system. The new E—C sigma bond is formed by transfer of the pi-electrons from the broken pi-bond.
   An example for a general diene (the symbol E^+ = electrophile, although the electrophile could be a radical):

$$R_2C = CR-CR = CR_2 \; + \; E^+ \; \underset{\longleftarrow}{\longrightarrow} \; \underset{\underset{E}{|}}{R_2C-C^+R}-CR = CR_2 \quad \text{slow/r.d.s.}$$

   b. The **second step** is addition of the nucleophile ($Z:^-$) to the electrophilic carbon cation or radical generated in the first step. The new C—Z bond is formed by transfer of the $Z:^-$ nucleophilic electrons to carbon. The result is formation of two new sigma bonds to the sp^2 carbons at the expense of the original pi-bond between them.
   One possible product for a general diene:

$$\underset{\underset{E}{|}}{R_2C-\overset{\oplus}{C}R}-CR = CR_2 \; + \; Z:^- \; \longrightarrow \; \underset{\underset{E}{|}\;\;\underset{Z}{|}}{R_2C-CR}-CR = CR_2 \quad \text{fast}$$

2. The general forms of the electrophile/nucleophile combinations are the same as found for electrophilic addition to simple alkenes. The table for **E/Z** reagents is reproduced in Section 16.2.7.

3. **Regiochemistry** (determination of the correct **constitutional** isomer) for **E/Z** addition to polyenes is more complicated due to conjugation and the greater number of possible double bonds. However, *regiochemistry will still depend on the structure of the most stable electrophilic addition product formed in the rate-determining first step.*

4. **Stereochemistry** (identification of specific **stereoisomers**) follows the stereospecificity of the reaction with a particular E/Z combination. *Stereospecificity for addition to conjugated polyenes follows the same general relationship established for simple alkenes*.

16.2.2 REGIOCHEMISTRY FOR ELECTROPHILIC ADDITION TO CONJUGATED SYSTEMS: CONCEPT OF ALLYLIC CARBON CATIONS AND RADICALS

1. **Regiochemistry** for addition to a conjugated system depends on the structure of the electrophile formed in the first **electrophile addition** step.
   a. Multiple double bonds and possible rearrangements provide a number of potential constitutional isomer products. *The electrophile will always add to the conjugated system to form the most stable carbon cation or radical*.
   b. For conjugated systems, a general rule can be formulated that identifies the more energy favorable position for electrophile attachment: *the electrophile generally bonds to one of the two end sp^2 carbons in a conjugated carbon sequence; electrophile addition to an inside carbon most often results in a less stable carbon cation or radical intermediate*.
   c. An example for a general **diene:**

$$R_2C=CR-CR=CR_2 \;+\; E^+ \longrightarrow \;\; R_2C-C^+R-CR=CR_2 \quad \text{slow/r.d.s.}$$
$$\underset{E}{|}$$

2. Note that the pi-electron pair is transferred to the addition electrophile to form the new C—E bond. The carbon involved in the C—E bond is an end sp^2-carbon of the conjugated system; *the neighboring inside carbon now bears the positive charge*.
3. The regiochemistry of the intermediate cation or radical product resulting from the use of this rule often does **not** conform to the general rules employed for simple mono-alkenes.
   a. The alkyl-substitution pattern on the carbon bearing the positive charge (or radical) is not the only factor to consider, and is often not the deciding factor.
   b. This difference is due to the formation of a specific type of stable carbon cation or radical from a conjugated system termed an **allylic cation** (or **radical**).
   c. *An allylic carbon cation specifically has a positive charge on a carbon directly connected to an sp^2-carbon of a double bond*; the potential energy of this cation is lowered by electron donation from the neighboring double bond.
4. Resonance structures demonstrate the stabilizing contribution of the pi-electrons from the original conjugated double bond. The general diene example was

Step 1: R_2C=CR—CR=CR_2 + E^+ $\rightleftarrows$ R_2C—C^+R—CR=CR_2
                                                                          |
                                                                          E

Resonance Structures of the cation intermediate formed:

R_2C—C^+R⌒CR=CR_2 $\longleftrightarrow$ R_2C—CR=CR—C^+R_2
    | |
    E   E

a. The additional stabilization provided by pi-electrons of the neighboring double bond occurs by electron donation through an extended p-orbital overlap between the sp^2 carbons of the double bond and the p-orbital of the sp^2/ trigonal planar carbon cation:

R_2C—CR—CR—CR_2 $\longleftrightarrow$ R_2C—CR—CR—CR_2
    | |
    E   E

b. The result is the equivalent of spreading out the positive charge over two of the carbons in the sequence; this is termed **resonance stabilization**.

c. Equivalent resonance forms can be drawn for a radical intermediate:

R_2C—C·R⌒CR=CR_2 $\longleftrightarrow$ R_2C—CR=CR—R_2C·
    | |
    E   E

5. The general rule for electrophile addition to an **end** carbon is based on the fact that addition to an **inside** carbon will **not** generate an **allylic** cation (or radical) intermediate:

Step 1: R_2C=CR—CR=CR_2 + E^+ $\rightleftarrows$ R_2C^+—CR—CR=CR_2
                                                                          |
                                                                          E

No resonance structures can be written.

The addition step has produced an sp^3-carbon (four single bonds) bonded to the electrophile (**E**). This intervening sp^3-carbon cannot participate in an extended p-orbital overlap; the neighboring double bond is isolated and prevented from pi-electron transfer.

16.2.3 TECHNIQUES FOR DRAWING CORRECT RESONANCE STRUCTURES FOR CONJUGATED SYSTEMS

1. Identification of correct resonance structures in reactions involving conjugated systems is required for regiochemistry analysis. Allylic cations (or radicals) are generally formed in the first step of certain reactions by

a. Addition of an electrophile to an end carbon of a conjugated system:

Step 1: $R_2C{=}CR{-}CR{=}CR_2$ + E^+ $\longrightarrow$ $R_2C{-}C^+R{-}CR{=}CR_2$
$|$
E

b. Cleavage of $C{-}X$ bond from alkyl halides or alcohols; X is thus a standard leaving group:

Step 1 (or 2): $R_2C{-}CR{=}CR_2$ $\longrightarrow$ $R_2C^+{-}CR{=}CR_2 + X{:}^-$
$|$
X

c. The general format of the two cation examples shown are slightly different; ***the key feature of an allylic cation is a positively charged sp²-carbon directly connected to an sp2-carbon of a double bond***.

2. A general method for resonance structure analysis:

(1) ***Write the correct form of the allylic cation or radical intermediate*** derived from one of the reactions for their formation **(general examples shown above)**.

(2) For intermediates in which the cation or radical is conjugated to only **one** remaining double bond: ***Identify the key three connected sp² carbons in the complete molecule that represent the conjugated p-orbital pi-electron system***. These will be the positively charged (or radical) carbon plus the two carbons of the conjugated double bond: **(In the example, an arbitrary numbering scheme is shown.)**

#1#2#3#4$$#2#3#4
$R_2C{-}C^+R{-}CR{=}CR_2$ $=$ $R_2C^+{-}CR{=}CR_2$
$|$
E

(3) ***Rearrange the double bond*** by shifting the pi-electrons from the double bond between **C#3** and **C#4** to form a double bond between **C#2** and **C#3**:

#2#3#4$$#2#3#4
$R_2C^+{-}CR{=}CR_2$ $\longleftrightarrow$ $R_2C{=}CR{-}C^+R_2$

a. In the resonance structure generated, C#4 now has only three single bonds and is without its pi-electron; it therefore bears the positive charge. Carbon #2 now has four bonds and has an octet; it is neutral.

b. The electrons for the formation of this resonance structure have been transferred through the extended p-orbital overlap; ***electrons can never be transferred outside this orbital overlap***.

(4) ***The central carbon (C#3) of the three carbon sequence can never bear the positive charge***; this is because it must always be double bonded to one of the other carbons (C#2 or C#4). ***Check to make sure the positive charge has jumped over the central carbon***.

   a. ***Comparing resonance structures in this method will never show adjacent carbons bearing a positive charge.***

   b. Think of the central carbon (**C#3** in the example) as the **hinge** and the pi-electrons as the **gate**: swing the electron pair from left to right or right to left *but do not detach them from the central hinge*:

$$R_2C^+\!\!-\!\!CR =\!\!= CR_2 \longleftrightarrow R_2C^+\!\!-\!\!CR\!\!-\!\!CR_2 \longleftrightarrow R_2C^+\!\!=\!\!CR\!\!-\!\!C^+R_2$$

   c. Resonance structures for allylic **radicals** are produced using the same method. In these cases, **only one of the pi-electrons is transferred** during each shift. The transferred electron pairs up with the original radical to reform a double bond. The carbon that lost the pi-bond keeps its p-orbital electron and is now the new radical:

$$R_2\overset{\bullet}{C}\!\!-\!\!CR =\!\!= CR_2 \longleftrightarrow R_2\overset{\bullet}{C}\!\!-\!\!CR\!\!-\!\!\overset{\bullet}{C}R_2 \longleftrightarrow R_2C =\!\!= CR\!\!-\!\!R_2\overset{\bullet}{C} \cdot$$

(5) Allylic carbon cations can be formed from more extensive conjugated systems, for example by electrophile addition to an end carbon of a triene, tetraene, etc., **or to an aromatic ring**; these will produce additional resonance structures. ***Write further resonance structures for more extensive pi-bonding molecular intermediates by following steps (1) through (4) sequentially down the conjugated p-orbital sequence.***

   a. Each resonance structure formed using steps (**1**) through (**4**) then serves as the new starting point for the next structure.

   b. Always begin at one end of the conjugated sequence and move sequentially through the sp^2 carbons. Avoid trying to write new resonance structures by jumping over several carbons at once; the position of the positive charge will get lost.

   c. **Example:** Identify all resonance structures for

          #2   #3   #4   #5   #6   #7   #8
        $R_2C^+\!\!-\!\!CH =\!\!= CH\!\!-\!\!CH =\!\!= CH\!\!-\!\!CH =\!\!= CHR$

Resonance Structure: **Process To Form Next Resonance Structure:**
                                         (Indicated by elctron arrows)

(Starting molecule):

    #2   #3   #4   #5   #6   #7   #8
$R_2C^+\!\!-\!\!CH =\!\!= CH\!\!-\!\!CH =\!\!= CH\!\!-\!\!CH =\!\!= CHR$

                           shift the 3, 4 pi-electrons to form ⟶

    #2   #3   #4   #5   #6   #7   #8
$R_2C =\!\!= CH\!\!-\!\!C^+H\!\!-\!\!CH =\!\!= CH\!\!-\!\!CH =\!\!= CHR$

                           shift the 5, 6 pi-electrons to form ⟶

    #2   #3   #4   #5   #6   #7   #8
$R_2C =\!\!= CH\!\!-\!\!CH =\!\!= CH\!\!-\!\!C^+H\!\!-\!\!CH =\!\!= CHR$

                           shift the 7, 8 pi-electrons to form ⟶

    #2   #3   #4   #5   #6   #7   #8
$R_2C =\!\!= CH\!\!-\!\!CH =\!\!= CH\!\!-\!\!CH =\!\!= CH\!\!-\!\!C^+HR$

                           end of sequence

16.2.4 REGIOCHEMISTRY FOR ELECTROPHILIC ADDITION TO CONJUGATED SYSTEMS: CONSTITUTIONAL ISOMERS BASED ON RESONANCE STRUCTURES

1. In general, ***the number of potential constitutional isomer products possible in an electrophilic addition reaction is equal to the number of nonidentical carbons in a conjugated sequence which will bear some portion of the positive charge (or radical).***

   a. The carbons that bear a portion of the positive charge are determined from the resonance structure analysis. Each resonance structure shows one carbon with a full positive charge; each of these carbons are thus indicated as bearing some portion of the positive charge of the complete intermediate molecule.

   b. Each of the positively charged carbons in the sum of all the resonance structures will be electrophilic toward the nucleophile in **Step 2** of the reaction; the nucleophile can bond to any of these carbons to complete the reaction. Potential bonding of the nucleophile to each of these carbons, in turn, will produce a different constitutional isomer product for each partially positive **nonidentical** carbon.

2. **Potential constitutional isomer products are found by the following:**

   (1) Determine all possible intermediate products formed by addition of the electrophile to **each of the two ends** of the conjugated sp^2-carbon sequence.

   (2) Write all possible resonance structures for **each** of the two specific intermediates which were formed in step **(1)**.

   (3) Identify the carbon that bears the positive charge in each resonance form of each possible original intermediate. Then add the **net** nucleophile to the positive carbon for all possible positively charged carbons found in step **(2)**. The results will show all possible constitutional isomers based on the original molecule. The actual number of **different** isomers is then found by eliminating duplicates formed from chemically equivalent carbons.

3. An example based on a general diene:

   $$R_2C \!=\! CH \!-\! CH \!=\! CHR \ + \ EZ \longrightarrow \qquad (R = \text{specifically carbon})$$

   (1) The diene is **non**symmetrical: the two end carbons are not equivalent (different substituents). The E^+ can add to **two different end carbons** producing **two different** carbon cation **intermediate molecules**.

   Intermediate [A]: $R_2C \!-\! C^+H \!-\! CH \!=\! CHR$ **Intermediate [B]:** $R_2C \!=\! CH \!-\! C^+H \!-\! CHR$

   $\qquad\qquad\qquad\qquad\quad |$ $\qquad\qquad\qquad\qquad\qquad\qquad\qquad\qquad\qquad\qquad\qquad |$

   $\qquad\qquad\qquad\qquad\quad E$ $\qquad\qquad\qquad\qquad\qquad\qquad\qquad\qquad\qquad\qquad\qquad E$

   (2) Each of the **two** intermediates has **two** resonance structures describing the relatively positive carbons:

Intermediate [A]: R$_2$C—C$^+$H—CH=CHR ⟷ R$_2$C—CH=CH—C$^+$HR
                              | |
                              E   E

Intermediate [B]: R$_2$C=CH—C$^+$H—CHR ⟷ R$_2$C$^+$—CH=CH—CHR
                                      | |
                                      E   E

(3) The analysis shows **four** nonidentical carbons in the molecule will bear some positive charge after addition of **E$^+$** to either of the two ends of the conjugated carbon sequence. The possible products are formed by adding **Z:$^-$** to each of these partially positive (electrophilic) carbons in turn to create four products:

Product [A1]: R$_2$C—CH—CH=CHR   **Product [A2]:** R$_2$C—CH=CH—CHR
                          | | | |
                          E   Z   E   Z

Product [B1]: R$_2$C=CH—CH—CHR   **Product [B2]:** R$_2$C—CH=CH—CHR
                              | | | |
                              Z   E   Z   E

4. The products of electrophilic addition are classified as belonging to two types based on the final addition pattern and the final position of the double bond:

   Initial diene with arbitrary numbers: $\overset{\text{\#1}}{R_2C}$=$\overset{\text{\#2}}{CH}$—$\overset{\text{\#3}}{CH}$=$\overset{\text{\#4}}{CHR}$

   a. **Direct addition** (or **1,2-addition**); ***Products [A1] and [B1]***: The remaining double bond is at **one end** of the original four-carbon conjugated diene sequence; the **E** and **Z** are on adjacent carbons counting from the other end of the original diene. The term **direct addition** refers to the fact that the reagent appears to add directly to **only one** of the double bonds. The term **1,2-addition** indicates that **E** and **Z** add to carbons **#1** and **#2** when arbitrarily counting from one end of the diene.

   b. **Conjugate addition** (or **1,4-addition**); ***Products [A2] and [B2]***: The remaining double bond has **rearranged** into the **middle** of the original diene sequence, between carbons **#2** and **#3**. This rearrangement requires a conjugated system, hence the term **conjugate addition**. The term **1,4-addition** refers to the fact that **E** and **Z** are on carbons **#1** and **#4** when counting from one end of the original diene.

16.2.5 REGIOCHEMISTRY FOR ELECTROPHILIC ADDITION TO CONJUGATED SYSTEMS: DETERMINATION OF MAJOR AND MINOR ISOMER PRODUCTS

1. Product distribution for electrophilic addition to conjugated systems can depend on either **thermodynamics** or **kinetics**.

The major constitutional isomer product can be determined if the controlling factor for the reaction is known.

a. If the reaction is under **thermodynamic control**, the major product is determined by the formation of the **most stable product**: usually the product with the **most highly alkyl-**substituted remaining double bond.

b. If the reaction is under **kinetic control**, the major product is determined by the product with the fastest rate of formation: usually the isomer produced through addition of the nucleophile (Z) to the **most stable carbon cation** (or **radical**) formed in the electrophilic addition step of the reaction.

16.2.6 SUMMARY: DETERMINATION OF PRODUCTS FOR ELECTROPHILIC ADDITION TO CONJUGATED SYSTEMS

Step (1) Identify the conjugated double bond system and the **EZ** reagent combination. *Add E⁺ (or E•) to one of the two end carbons of the conjugated carbon sequence* to form one of the allylic intermediates. Form the other allylic intermediate by *adding E⁺ (or E•) to the other end carbon of the sequence.* A total of two possible intermediates have been formed. *Do not form intermediates by addition of* E⁺ *(or E•) to any inside carbons.*

Step (2) *Write all resonance forms of each intermediate constructed in step (1).* For a diene, each intermediate will produce two resonance forms.

Step (3) Identify the carbon that bears the positive charge in each resonance form of each possible original intermediate. *Then add the net nucleophile to the positive carbon for all possible positively charged carbons found in step (2).* These molecules will represent all **potential** products.

Step (4) *If the controlling factor in the reaction is known select the major constitutional isomer product based on:*

a. **Thermodynamic control: most stable product**; molecule with most highly alkyl-substituted remaining double bond(s).

b. **Kinetic control: molecule formed the fastest**; product formed through reaction at the most stable carbon cation (tertiary allylic > secondary allylic > primary allylic).

Step (5) *Determine if the major product from step (4) can have stereoisomers.*

a. If no stereochemistry is possible, the constitutional isomer as written is correct.

b. *If stereoisomers are possible, determine if the EZ reagent shows stereospecificity. Apply the correct stereospecificity as required* to determine the correct stereoisomer.

16.2.7 REPRODUCTION OF SUMMARY TABLE: E–Z REAGENTS FOR ELECTROPHILIC ADDITION

E—Z Reagent	E$^+$	Z:$^-$	Stereospecificity	Comments
H—Cl	**H$^+$**	**Cl$^-$**	None	Trigonal planar cation intermediate
H—Br	**H$^+$**	**Br$^-$**	None	Trigonal planar cation intermediate
H—I	**H$^+$**	**I$^-$**	None	Trigonal planar cation intermediate
H$^+$/H$_2$O	H$^+$	H$_2$O	None	Trigonal planar cation intermediate;
Net result:	**H**	**OH**		fastest rate to form tertiary C—OH
1. H$_2$SO$_4$	H$^+$	H$_2$O	None	**Sequence** of reactions to increase
2. H$_2$O				rate of alcohol formation
Net result:	**H**	**OH**		
H$^+$/ROH	H$^+$	ROH	None	Trigonal planar cation intermediate;
Net result:	**H**	**OR**		**R = alkyl C**; forms **ethers**
Br$_2$	**Br$^+$**	**Br$^-$**	Anti	Bridged cation intermediate
Cl$_2$	**Cl$^+$**	**Cl$^-$**	Anti	Bridged cation intermediate
Br$_2$/H$_2$O	Br$^+$	H$_2$O	Anti	Bridged cation intermediate;
Net result:	**Br**	**OH**		**bromo-alcohol** is formed
Cl$_2$/H$_2$O	Cl$^+$	H$_2$O	Anti	Bridged cation intermediate;
Net result:	**Cl**	**OH**		**chloro-alcohol** is formed
1. B$_2$H$_6$	BH$_3$	H	Syn	Reaction **sequence** forms C—OH
2. H$_2$O$_2$				from C—BH$_2$; H is transferred
Net result:	**OH**	**H**		from BH$_3$ by a **rearrangement**
HBr	**Br·**	**H·**	None	Br· is **electrophile**; forms **opposite**
with peroxides				regiochemistry to HBr ionic
H$_2$	**H**	**H**	Syn	Syn addition due to **catalyst**
with catalyst				surface
RCOOH$\underset{O}{\overset{\parallel}{}}$	**Oxygen**		Syn	Reagent is a **peroxy acid**; product is an **epoxide**
O$_3$	**Oxygen**		None	Complete cleavage of double bond to form **aldehydes** and **ketones**

16.2.8 EXAMPLE: ELECTROPHILIC ADDITION TO A CONJUGATED DIENE

1. Determine the four products for the following reaction; identify the kinetic and thermodynamic majors.

$$H_2C = \underset{\underset{C(CH_3)_3}{|}}{C} - CH = CH_2 \xrightarrow{\text{HBr}}$$

Step 1: The two alkenes of the diene do not have identical substituent groups attached to the sp^2 carbons; the tert-butyl group breaks the symmetry. Therefore, separate addition of an electrophile to each end carbon of the conjugated diene will form different cationic intermediates. Add the **H$^+$** electrophile to

the two end carbons of the diene to produce cation intermediates [A] and [B]:

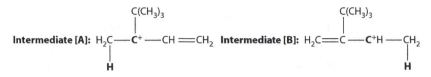

Step 2: Draw the resonance structure(s) for each of the carbon cation
intermediates to determine the position of the other positively
charges carbons.

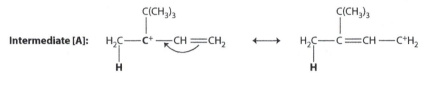

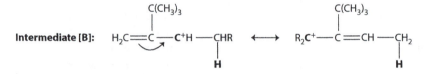

Step 3: Determine all four possible products by adding the **Br:⁻**
nucleophile to each of the partially positive (electrophilic)
carbons.

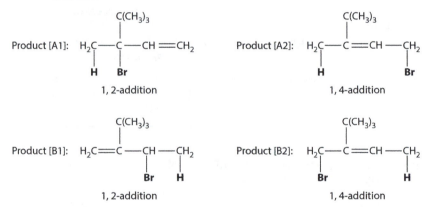

Step 4:

a. The thermodynamic major product is the most stable product: the molecule with most highly alkyl-substituted remaining double bond(s). Products [A2] and [B2] shown above as
the 1,4-addition products both have a tri-substituted remaining double bond; each can be considered the thermodynamic
major.

b. The kinetic major product is the product formed the fastest
through the most stable carbon cation. Product [A1] shown
above is formed through a positively charged tertiary/allylic
carbon; this is the kinetic major.

Step 5: Electrophilic addition of HBr is not stereospecific; no stereochemistry applies.

16.3 ELECTROPHILIC AROMATIC SUBSTITUTION BY AN ADDITION/ELIMINATION MECHANISM

16.3.1 GENERAL REACTION

1. Electrophilic substitution at carbons of an aromatic ring involves the replacement of one C—H bond for a C—E bond: replacement of hydrogen by an electrophile.

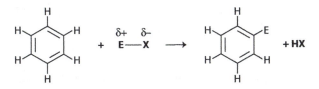

a. In this notation, the **electrophile** is E⁺. The **possible nucleophile** is indicated by **X:⁻** to emphasize the fact that it has no major role in the carbon substitution reaction. The **X:⁻** either acts as a **weak base** to remove the hydrogen as **H⁺** or undergoes no further reaction after the original free electrophile is produced.

b. The general properties of electrophiles for aromatic substitution are similar to those found for electrophilic addition to alkenes. However, aromatic substitution generally requires **stronger** electrophiles and thus the **form** of **EX** is usually different than the forms of **EZ** described in the previous section. A table of electrophile combinations for this reaction type is shown in **Section 16.3.6.**

c. Aromatic carbons do **not** generally react through the standard sp³-carbon substitution mechanisms of **SN1** and **SN2**. This is partially due to the stronger C—H bond formed from sp² carbons and to the extra stabilization of the aromatic ring.

2. Substitution is accomplished by a two reaction process termed **substitution by addition/elimination:**

a. The electrophile **E⁺** first undergoes an **addition** reaction to the pi-electrons of the conjugated double bond system of the ring; one of the aromatic ring carbons forms a new sigma bond to the electrophile.

b. The hydrogen on the carbon that forms the new C—E bond is then removed as **H⁺** in an **elimination** reaction to restore the aromaticity of the ring. The **net** result is **substitution**.

c. The final product of the reaction is never formed by a **net addition** of **E** and **X** to a double bond of the aromatic system. Product formation of this type would destroy the aromaticity of the ring with loss of the stabilization energy.

16.3.2 ELECTROPHILIC AROMATIC SUBSTITUTION: GENERAL MECHANISM

1. The reaction proceeds through two major steps: addition of the electrophile followed by elimination of the hydrogen as H^+.

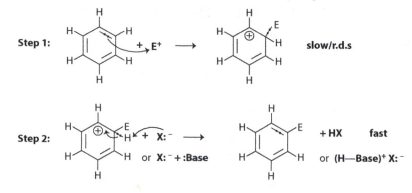

2. **Step 1**, the rate-determining step for the reaction, is the electrophilic addition to the pi-bonding system of the aromatic ring; E^+ is the electrophile, the **pi-electrons** are nucleophilic.

   a. Electron transfer of a pi-electron pair to the electrophile allows formation of a new C—E sigma bond to **one** of the carbons of the aromatic ring; this carbon now has four single bonds and is neutral. As for addition to an alkene, ***the positive charge is not on the carbon bonded to E, but can be placed on one of the neighboring carbons.*** The positive charge is actually placed on the complete ring and is shared by certain other carbons of the conjugated system through resonance structures.

   b. The carbon cation formed in the **r.d.s.** is relatively unstable (high potential energy). This is due not only to the relatively high potential energy of carbon cation intermediates in general, but also to the loss of the considerable aromatic resonance stabilization energy.

   c. The high potential energy carbon cation does, however, have an important stabilization mechanism: ***it has the same general structure as an allylic cation formed from a conjugated triene.*** The neighboring carbon shown bearing the positive charge from the initial electrophilic addition is itself bonded to an sp²-carbon of the conjugated system. Resonance structures can be drawn that show the stabilizing contribution of the ring carbons **ortho (o)** and **para (p)**, but not **meta (m)** to the carbon bearing the electrophile:

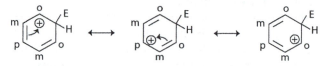

3. **Step 2** is the removal of a hydrogen as H^+ from the carbon bonded to the electrophile. The result is the exchange of **E** for **H** at the carbon forming the original C—E bond: a **substitution**.

a. Proton removal is accomplished through **any weak base** in the reaction; if the **X:⁻** anion from the original **EX** combination acts as a base, the net result is direct formation of the other product, **HX**.

b. Addition of **X:⁻** as a **nucleophile** to one of the positively charged carbons of the ring would result in a net **addition** of **EX** across a double bond. The loss of aromatic resonance stabilization makes this reaction highly unfavorable; ***the possible nucleophile X:⁻ never adds to the aromatic ring.***

16.3.3 DETERMINATION OF THE MAJOR PRODUCT FOR REACTIONS AT SUBSTITUTED AROMATIC RINGS: ANALYSIS OF RESONANCE FORMS

1. The unsubstituted six-carbon aromatic ring, benzene, is symmetrical; a single substitution of one **E** for one **H** will produce no possible constitutional isomers. The benzene ring is planar; ***electrophilic aromatic substitution will not produce stereoisomers.***

2. Multiple substitutions or single substitutions on aromatic rings with attached substituents or functional groups will generally produce more than one possible constitutional isomer product.

    a. An example for substitution on a benzene ring that already has one substituent (functional group) attached:
    S = original substituent/functional group
    E = second substituting electrophile.
    (Arbitrary numbering of carbons is shown.)

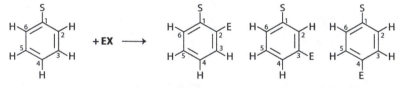

    b. ***Substitution of E for H requires that the reacted carbon have one hydrogen:*** ignoring certain exceptions, ***substitution of E for an existing S will not occur.*** Therefore, the aromatic ring has five remaining positions for possible substitution.

3. Due to the symmetry of the reacting molecule, three possible constitutional isomers can form: **(Numbering of carbons based on the example shown.)**

    a. Exchange of **E** for **H** at either carbon **#2** or carbon **#6** will form a product with the new electrophile on a carbon **ortho** to the carbon bearing the original substituent or functional group.

    b. Exchange of **E** for **H** at either carbon **#3** or carbon **#5** will form a product with the new electrophile on a carbon **meta** to the carbon bearing the original substituent or functional group.

c. Exchange of **E** for **H** at carbon **#4** will form a product with the new electrophile on a carbon **para** to the carbon bearing the original substituent/functional group.

4. The major constitutional isomer(s) are determined by the **rates** of reaction (**kinetic** control). *The favored (major) constitutional isomer product is derived through formation of the most stable carbon cation intermediate in the electrophilic addition step (step 1).* The most stable carbon cation intermediate is determined from resonance structures of the three non-equivalent possibilities:

   a. The carbon cation formed by addition of **E⁺** to an **ortho** carbon (as defined above):

   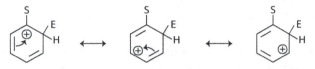

   b. The carbon cation formed by addition of **E⁺** to a **para** carbon (as defined above):

   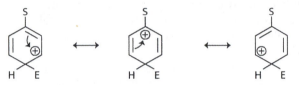

   c. The carbon cation formed by addition of **E⁺** to a **meta** carbon (as defined above):

   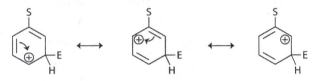

5. The critical feature of the resonance structure analysis for the three possible substitution positions is the role of the carbon that bears the original substituent (**S**).

   a. For each of the three possible carbon cation intermediates, *the carbon directly connected to the original substituent will share the positive charge whenever the new electrophile adds to a carbon that is ortho or para to this substituent-bearing carbon.*

   b. It is also shown that *the positive charge will never be shared by the substituent-bearing carbon whenever the new electrophile adds to a carbon in the meta position.*

   c. It is generally true that *a substituent has the greatest electronic influence on the stability of the carbon cation intermediate whenever it is directly connected to a carbon that can bear a resonance positive charge.*

6. **The original substituent can either stabilize or destabilize a positive charge.**

For a substituent that **stabilizes** positive charge:

a. The greatest electronic influence occurs whenever it is directly connected to a carbon that bears the resonance positive charge. Therefore, the substituent will provide the **greatest stabilization** to a carbon cation formed by addition of an electrophile to a carbon **ortho** or **para** to the substituent-bearing carbon.

b. The result is that electrophilic substitution at a position **ortho** or **para** to the substituent-bearing carbon will have a **faster rate** of reaction. *Substituents that stabilize positive charge favor the formation of constitutional isomers that show an ortho or para substitution pattern.*

7. For a substituent that **destabilizes** positive charge:

a. The greatest electronic influence occurs whenever it is directly connected to a carbon that bears the resonance positive charge. Therefore, the substituent will provide the **greatest destabilization** to a carbon cation formed by addition of an electrophile to a carbon **ortho** or **para** to the substituent-bearing carbon.

b. The result is that electrophilic substitution at a position **ortho** or **para** to the substituent-bearing carbon will have a **slower rate** of reaction; **meta** substitution will be **less disfavored**. *Substituents that destabilize positive charge favor the formation of constitutional isomers that show a meta substitution pattern.*

16.3.4 DETERMINATION OF THE MAJOR PRODUCT FOR REACTIONS AT SUBSTITUTED AROMATIC RINGS: GENERAL RESULTS FOR STABILIZING AND DESTABILIZING SUBSTITUENTS

1. Ring-substituents/functional groups (**S**) that **stabilize** positive charge **donate** electron density through sigma-bonding electrons or through lone-pair (resonance) electron donation; these are termed **activators**.

weakest activator ⟶ strongest activator

Examples: —CR$_3$ (alkyl) < —OH < —OCR$_3$ < —NH$_2$ < —NR$_2$

a. The direct connection of an **oxygen** or **nitrogen** to a carbon bearing a positive charge is especially effective at stabilization due to electron donation through the oxygen or nitrogen **lone pairs**. This effect occurs through an **additional** resonance structure for the carbon cation intermediate: (**example for an amine**)

b. Electron-**donating** substituents **increase the rate** of electrophilic substitution at positions **ortho** and **para** to the carbon to which they are attached; they increase the rate at the meta position to a much lesser degree. **Thus, electron-donating substituents favor formation of constitutional isomers which place the new electrophile in a position ortho or para to the original substituent**.

c. **General notation**: Electron-**donating** substituents are **activators** (increase the rate at all positions) for electrophilic aromatic substitution and are **ortho/para directors**.

2. Ring-substituents/functional groups (**S**) that **destabilize** positive charge **withdraw** (remove) electron density through sigma–bonding electrons or through resonance; these are termed **deactivators**. This usually involves a strongly relatively positive atom directly connected to the ring carbon.

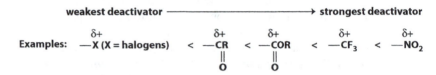

weakest deactivator ⟶ **strongest deactivator**

$$\text{Examples:}\quad \overset{\delta+}{-X}\ (X=\text{halogens}) \;<\; \overset{\delta+}{-\underset{\underset{O}{\|}}{C}R} \;<\; \overset{\delta+}{-\underset{\underset{O}{\|}}{C}OR} \;<\; \overset{\delta+}{-CF_3} \;<\; \overset{\delta+}{-NO_2}$$

a. Electron-**withdrawing** substituents **decrease the rate** of electrophilic substitution at positions **ortho** and **para** to the carbon to which they are attached. Since **the decrease in the rate at the meta position is much less, the net result is that electron-withdrawing substituents show relatively greater rates for electrophile addition to the meta position versus the ortho/para positions**. Thus electron-**withdrawing** substituents favor formation of constitutional isomers that place the new electrophile in a position **meta** to the original substituent.

b. **General notation**: Electron-**withdrawing** substituents are **deactivators** (decrease the rate at all positions) for electrophilic aromatic substitution and are **meta directors** (meta position is less disfavored).

c. Halogens are an exception to this classification: *halogens are (weak) deactivators, but are ortho/para directors*. This change is due to the competition between the electron withdrawing effect of the C—X sigma-bond (based on electronegativity differences) and the electron-donating effect of the lone pairs on the halogen. The final balance shows weak **deactivation**. The **ortho/para** substitution preference is due to the fact that the resonance component has the primary influence on substitution pattern; lone-pair resonance electron donation favors ortho/para positions.

16.3.5 GENERAL SUMMARY: DETERMINATION OF CORRECT MAJOR PRODUCTS FOR ELECTROPHILIC AROMATIC SUBSTITUTION

1. ***Identify the aromatic structure*** in the complete reactant molecule. ***Identify the structure of the E^+*** from the **EX** combination and the ***correct electrophilic*** (potential ring-bonding) ***atom***. Determine if the electrophile will undergo rearrangement. (**See summary table in Section 16.3.6.**)

2. If all positions on the aromatic ring are equivalent, ***exchange E^+ for H^+ at any position***. The isomer product will be correct as written (no stereoisomers are possible).

3. If the aromatic ring has **only one** original substituent/functional group attached, ***identify the substituent classification***: electron-donating groups will be ortho/para directing; halogens will be ortho/para directing; other electron-withdrawing groups will be meta directing. ***Exchange E^+ for H^+ at the position on the aromatic ring, which will produce the ortho/para or meta isomer as necessary***. It is generally not necessary to distinguish between an ortho and a para isomer.

4. If the aromatic ring has **two or more** original substituents, ***analyze the directing tendency for each:***

    a. If the substituents **reinforce** each other (tendencies indicate the same isomer product), write the **one** correct product that agrees with all of the directing tendencies.

    b. If the substituents **oppose** each other (tendencies indicate different possible isomer products), ***follow the directing requirements of the substituent that has the largest effect on carbon cation stability and rate of reaction*** (strongest activator or deactivator). In general, this often means following the tendency of the stronger activator.

16.3.6 SUMMARY TABLE: ELECTROPHILES FOR ELECTROPHILIC AROMATIC SUBSTITUTION

Electrophile Reagent	Resulting E^+	Comments/Formation of E^+ From Reagents
HNO_3/H_2SO_4	$^+NO_2$	$HNO_3 + H_2SO_4 \rightarrow {}^+NO_2 + H_2O + HSO_4^-$
H_2SO_4/SO_3	$^{\delta+}SO_3$	**Aromatic product** is a sulfonic acid: RSO_3H
$Cl_2/AlCl_3$	Cl^+	$Cl_2 + AlCl_3 \rightarrow Cl^+ + AlCl_4^-$
$Br_2/FeBr_3$	Br^+	$Br_2 + FeBr_3 \rightarrow Br^+ + FeBr_4^-$
$R_2C{=}CR_2/H^+$	$R_2C^+{-}CR_2H$	Acid formation of carbon cation: forms most stable carbon cation; **rearrangements** occur
$R_3C{-}OH/H^+$	R_3C^+	Carbon cation from alcohol/acid: forms most stable carbon cation; **rearrangements** occur
$R_3C{-}Cl/AlCl_3$	R_3C^+	$R_3C{-}Cl + AlCl_3 \rightarrow R_3C^+ + AlCl_4^-$ Forms most stable carbon cation; **rearrangements** occur

(Continued)

Electrophile Reagent	Resulting E⁺	Comments/Formation of E⁺ From Reagents
$R_3C-Br/FeBr_3$	$\mathbf{R_3C^+}$	$R_3C-Br + FeBr_3 \rightarrow R_3C^+ + FeBr_4^-$ Forms most stable carbon cation; **rearrangements** occur
$\underset{\overset{\|}{O}}{RC}-Cl/AlCl_3$	$\underset{\overset{\|}{\mathbf{O}}}{\mathbf{RC^+}}$	$\underset{\overset{\|}{O}}{RC}-Cl + AlCl_3 \rightarrow \underset{\overset{\|}{O}}{RC^+} + AlCl_4^-$ **No rearrangements occur**
$\underset{\overset{\|}{O}\ \overset{\|}{O}}{RC}-OCR/AlCl_3$	$\underset{\overset{\|}{\mathbf{O}}}{\mathbf{RC^+}}$	$\underset{\overset{\|}{O}\ \overset{\|}{O}}{RC}-OCR + AlCl_3 \rightarrow \underset{\overset{\|}{O}}{RC^+} + \underset{\overset{\|}{O}}{^-OCR}$ **No rearrangements occur; AlCl₃ = catalyst**

16.4 EXAMPLES: ELECTROPHILIC AROMATIC SUBSTITUTION

Follow steps **1** through **3** to determine the major product for the electrophilic aromatic substitution reactions; assume for the exercises that no complicating reactions occur.

Step 1: Select the correct aromatic ring for substitution if more than one aromatic ring is in the molecule.

Step 2: Select the correct form of the net electrophile; consider any possible rearrangements.

Step 3: Select the correct position on the ring based on the guidelines for the directing effects of all substituents on the aromatic ring. Generally, based on electronic effects, no distinction need be made between ortho versus para. Steric effects, however, may be considered in selecting a preferred position.

1.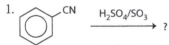

Analysis:

(1) The reactant contains only one aromatic ring.

(2) The reagent combination H_2SO_4/SO_3 produces the electrophile $\overset{\delta+}{SO_3}$ which will form the sulfonic acid as the net product.

(3) The aromatic ring has a cyano substituent: deactivator/meta director. Replace —H with —SO_3H at a position meta to the ring carbon with the cyano group.

Product for Reaction 1:

2. 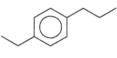 $\xrightarrow{\text{CH}_3\text{CH}_2\text{Cl/AlCl}_3}$?

Analysis:

(1) The reactant contains only one aromatic ring.
(2) The reagent combination $\text{CH}_3\text{CH}_2\text{Cl/AlCl}_3$ produces the net electrophile $^+\text{CH}_2\text{CH}_3$.
(3) The aromatic ring has an alkyl substituent: activator; ortho/para director. Replace —H with —CH_2CH_3 at a position ortho or para to the ring carbon bearing the alkyl group.
Product for Reaction 2:

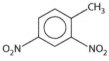

3. 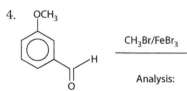 $\xrightarrow{\text{HNO}_3/\text{H}_2\text{SO}_4}$?

Analysis:

(1) The reactant contains only one aromatic ring.
(2) The reagent combination $\text{HNO}_3/\text{H}_2\text{SO}_4$ produces the net electrophile $^+\text{NO}_2$.
(3) The aromatic ring has an alkyl and an additional nitro substituent.
The existing nitro substituent is a deactivator/meta director; the alkyl group is an activator, ortho/para director. In this case, directing influences of the two substituents reinforce. Replace —H with —NO_2 at the position that is meta to the ring carbon with the nitro group and also ortho to the ring carbon bearing the alkyl group.
Product for Reaction 3:

4. $\xrightarrow{\text{CH}_3\text{Br/FeBr}_3}$?

Analysis:

(1) The reactant contains only one aromatic ring.
(2) The reagent combination $\text{CH}_3\text{Br/FeBr}_3$ produces the net electrophile $^+\text{CH}_3$.
(3) The aromatic ring has an alkoxy and an aldehyde substituent.
The aldehyde group is a deactivator/meta director. Due to the directly connected oxygen, the alkoxy group is a strong activator, ortho/para director. The directing influences of the two substituents oppose: in this case follow the directing effect of the strongest activator. Based on

electronic effects —H can be replaced with —CH$_3$ at a position that is either ortho or para to the carbon bearing the methoxy group. Steric effects, however, will not favor the ortho ring position between the existing substituent groups.

Product for Reaction 4:

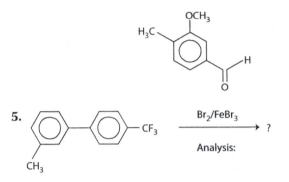

5.

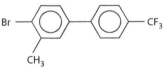

Br$_2$/FeBr$_3$ ⟶ ?

Analysis:

(1) The reactant contains two aromatic rings. One ring has a methyl group, a weak activator; the other ring has a trifluoromethyl group, which is a strong deactivator. Select the ring with the activator; this will have the fastest rate of reaction for substitution. Reaction is favored on the ring with the methyl substituent.

(2) The reagent combination Br$_2$/FeBr$_3$ produces the net electrophile Br$^+$.

(3) The favored aromatic ring has an activator alkyl group which is an ortho/para director. The attached aromatic ring will also have a directing influence, but in this case they reinforce. Replace —H with —Br at the position which is ortho or para to the ring carbon bearing the methyl group.

Product for Reaction 5:

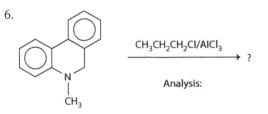

6.

CH$_3$CH$_2$CH$_2$Cl/AlCl$_3$ ⟶ ?

Analysis:

(1) The reactant contains two aromatic rings; note that the ring containing the nitrogen is not aromatic. One aromatic ring has an amine nitrogen directly attached to a ring carbon; the amine group is a strong activator. The other aromatic ring has a weak activator alkyl group directly attached to a ring carbon. For substitution, select the aromatic ring connected to stronger activator amine group.

(2) The reagent combination $CH_3CH_2CH_2Cl/AlCl_3$ produces a net alkyl electrophile. Rearrangement of the initially formed primary carbon cation will form the net secondary electrophile: $CH_3C^+HCH_3$.

(3) The amine activator on the favored aromatic ring is an ortho/para director. Replace —H with —$CH(CH_3)_2$ at a position either ortho or para to the ring carbon bearing the amine group; steric effects will favor the para position.

Product for Reaction 6:

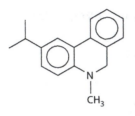

16.5 ADDITIONAL PRACTICE PROBLEMS

Complete the following reactions. For reactions at a conjugated polyene, draw all resonance structures, all possible products, and select the kinetic majors. For electrophilic aromatic substitution, assume monosubstitution, and determine the major product(s).

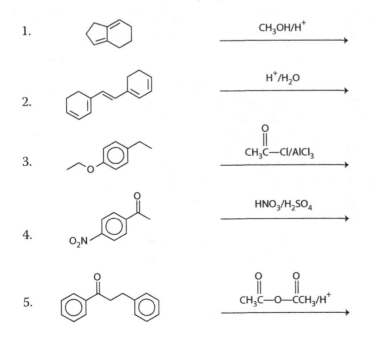

1. CH_3OH/H^+

2. H^+/H_2O

3. $CH_3\overset{O}{\overset{||}{C}}{-}Cl/AlCl_3$

4. HNO_3/H_2SO_4

5. $CH_3\overset{O}{\overset{||}{C}}{-}O{-}\overset{O}{\overset{||}{C}}CH_3/H^+$

6.

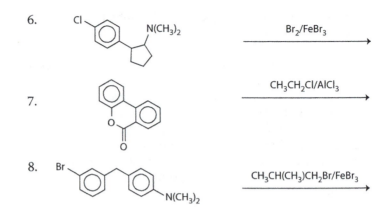

$Br_2/FeBr_3$

7.

$CH_3CH_2Cl/AlCl_3$

8.

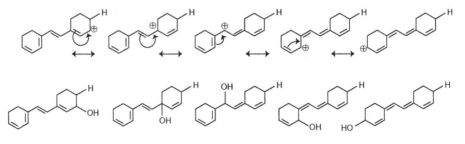

$CH_3CH(CH_3)CH_2Br/FeBr_3$

ADDITIONAL PRACTICE PROBLEMS: ANSWERS

1. The ring system is not aromatic: this reaction is electrophilic addition to a conjugated diane. Add H^+ to each end of the diene, draw the corresponding resonance structures, and draw the four final products using the net nucleophile $-OCH_3$. The kinetic majors are both of the 1,2-addition products that are formed through tertiary/(allylic) carbon cations.

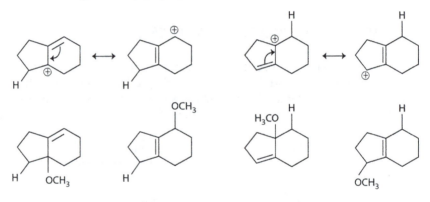

2. The ring system is not aromatic: this reaction is electrophilic addition to a conjugated polyene. This polyene is symmetrical: the same intermediate results by addition of the electrophile to either end of the polyene. Add H^+ to either one end of the polyene; then draw all possible resonance structures. Add the net $-OH$ to each carbon that bears a positive charge to determine all possible products. The one kinetic major is formed through the tertiary/(allylic) carbon cation.

3. The ring system is aromatic; the electrophile is $CH_3\overset{\overset{\displaystyle O}{\|}}{C}{}^+$. For electrophilic aromatic substitution, the ethoxy group is the stronger activator. The electrophile is placed ortho to the ring carbon with the ethoxy group.
   Product for Reaction 3:

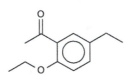

4. The ring system is aromatic, the electrophile is $^+NO_2$; select electrophilic aromatic substution. The nitro group is the stronger deactivator. The electrophile is placed meta to the ring carbon with the nitro group.
   Product for Reaction 4:

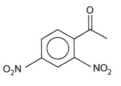

5. Select electrophilic aromatic substution; the electrophile is $CH_3\overset{\overset{\displaystyle O}{\|}}{C}{}^+$. There are two possible aromatic rings. The directly attached ketone carbonyl is a deactivator; the directly attached alkyl group is an activator. The electrophile is placed ortho or para to the ring carbon with the alkyl group; steric effects favor para.
   Product for Reaction 5:

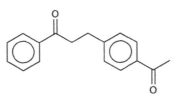

6. Select electrophilic aromatic substution; the electrophile is Br^+. Only one ring is aromatic; the chlorine is a weak deactivator but an ortho/para director. The electrophile is placed ortho to the ring carbon with the chlorine.
   Product for Reaction 6:

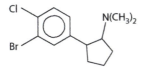

7. Select electrophilic aromatic substution; the electrophile is $CH_3CH_2{}^+$. There are two possible aromatic rings; the ring

composed of the cyclic ester cannot undergo electrophilic aromatic substitution. The directly attached ester carbonyl is a deactivator; the directly attached oxygen is a strong activator. The electrophile is placed ortho or para to the ring carbon bonded to the oxygen; steric effects favor para.

Product for Reaction 7:

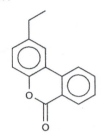

8. Select electrophilic aromatic substution; the major electrophile is produced through a rearrangement of the primary cation to form a tertiary cation: $(CH_3)_3C^+$. There are two possible aromatic rings: the ring with the bromine is weakly deactivated; the ring with the dimethylamino group is strongly activated. The electrophile is placed ortho to the ring carbon bonded to the amino group.

Product for Reaction 8:

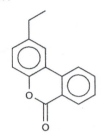

Oxidation/ Reduction Relationships for Carbonyl Carbon

17

17.1 GENERAL CONCEPT OF REDOX REACTIONS

1. A redox reaction describes the transfer of electrons between elements or ions to form new or differently charged elements or ions. For covalent molecules, the reaction describes the alteration in the type of polar covalent bond.
2. The term **redox** is derived from the combination of reactions termed reduction plus oxidation. A complete redox reaction involves the simultaneous process of reduction of one reactant and oxidation of another reactant.
3. **Oxidation** is the <u>removal</u> of electrons or electron share from a specific atom or atomic ion. The specific atom or ion may be independent or may be part of a larger compound or molecule.
   a. For individual elements, or component ions as part of ionic compounds, *oxidation is the direct <u>loss</u> of electrons*; the result is a change in element or ion charge to a more positive or less negative value. Common examples in organic reactions (*these are not balanced equations*):

Zn $\longrightarrow$	Zn^{+2}	element Zn loses 2 electrons
K $\longrightarrow$	K^+	element K loses 1 electrons
Cl^- $\longrightarrow$	Cl	negative ion Cl^- loses 1 electron

   b. Elements in **covalent** molecules are **not** ions and **cannot** change charge. For a **specific atom** in a covalent compound, *oxidation is a <u>decrease</u> in the relative portion of the electron share an atom has in its covalent bonds* with other atoms. An example for the specific atom **H** (*this is not a balanced equation*):

   $$H—H (H_2) \ (+F_2) \longrightarrow H—F (HF)$$
   each **H** has an **equal** share of electrons in the H—H bond   **H** has a **decreased** share of electrons in the H—F bond

4. **Reduction** is the **addition** of electrons or electron share to a specific atom or atomic ion. The specific atom or ion may be independent or part of a larger molecule.

  a. For individual elements, or component ions as part of ionic compounds, **reduction is the direct _gain_ of electrons**; the result is a change in element or ion charge to a more negative or less positive value. Common examples in organic reactions (**these are not balanced equations**):

$Hg^{+2} \longrightarrow Hg$ positive ion Hg^{+2} gains 2 electrons

$Cr^{+6} \longrightarrow Cr^{+3}$ positive ion (actually partly covalent) Cr^{+6} gains 3 electrons

$Br \longrightarrow Br^-$ neutral Br atom gains 1 electron

  b. For a **specific atom** in a covalent compound, **reduction is an _increase_ in the relative portion of the electron share an atom has in its covalent bonds** with other atoms. Example for the specific atom **F** (**this is not a balanced equation**):

  F—F (F_2) (+H_2) $\longrightarrow$ **H—F (HF)**
  each **F** has an **equal** share of   F has an **increased** share of
  electrons in the F—F bond   electrons in the H—F bond

5. The concepts of oxidation and reduction are complementary: **_electron or electron share loss by one atom/ion must result in an electron or electron share gain by another atom/ion_**. A complete redox reaction must involve a combination of oxidation and reduction such that electrons **balance** on each side of the equation.

  a. Ionic compound example: $CuCl_2 + Zn \longrightarrow Cu + ZnCl_2$
  Cu^{+2} (from $CuCl_2$) $\longrightarrow$ Cu positive ion Cu^{+2} gains 2 electrons. $Zn \longrightarrow Zn^{+2}$ (in $ZnCl_2$) element Zn loses 2 electrons. **Electrons balance:** the gain of 2 electrons by Cu^{+2} is balanced by the loss of 2 electrons by Zn.

  b. Covalent compound **conceptual** (partial) example (**does not show how electrons are balanced**):
  $C_{(s)} + O_{2\,(g)} \longrightarrow CO_{2\,(g)}$
  All C atoms in $C_{(s)}$ have equal electron shares in covalent bonding. Each O atom in O_2 has an equal electron share in the covalent bond.

  $\delta\text{-} \quad \delta\text{+} \quad \delta\text{-}$

  In the molecule CO_2 = O=C=O each oxygen gains a share of covalently bonded electrons to carbon; the carbon loses shares of electrons covalently bonded to oxygen.

6. Specific atoms that **lose** electrons/electron share are said to be **oxidized**. The term is specific for the actual atom losing the electrons, but is sometimes applied to the compound or molecule that contains this specific atom.

  a. In the reaction $CuCl_2 + Zn \longrightarrow Cu + ZnCl_2$; **Zn is oxidized**

  b. In the reaction $C_{(s)} + O_{2\,(g)} \longrightarrow CO_{2\,(g)}$; **carbon is oxidized**

7. Specific atoms that **gain** electrons/electron share are said to be **reduced.**
   a. In the reaction $CuCl_2 + Zn \longrightarrow Cu + ZnCl_2$; **Cu is reduced**
   b. In the reaction $C_{(s)} + O_{2\,(g)} \longrightarrow CO_{2\,(g)}$; **oxygen is reduced**

8. The elements or their corresponding molecules can be identified based on the **complementary** result they have on another compound:
   a. A **reducing agent** is a species that causes another atom to be reduced (i.e., to gain electrons). The reducing **agent** supplies the electrons; therefore, the reducing **agent** is the atom (or corresponding molecule) which itself is oxidized.
   b. An **oxidizing agent** is a species that causes another atom to be oxidized (i.e., to lose electrons). The oxidizing **agent** removes the electrons; therefore, the oxidizing **agent** is the atom (or corresponding molecule) which itself is reduced.

17.2 OXIDATION NUMBERS

1. Changes in electron numbers or electron sharing are analyzed through **oxidation numbers**: A complete redox reaction is analyzed by comparing each element in the reactant molecules to the same element in the product molecules. Changes in oxidation number for each element are identified:
   a. If the oxidation number of an element becomes **more positive or less negative** during the reaction, the element is **oxidized**.
   b. If the oxidation number of an element becomes **more negative or less positive** during the reaction, the element is **reduced**.

2. Calculation of oxidation numbers follows the general rules:
   1. All elements that are **not in compounds** (atoms bonded only to other atoms of the same element) are defined as having an oxidation number (ox #) of zero: **ox # = 0.**
   2. Single elemental ions (**not** polyatomic ions) have an oxidation number equal to the charge on the ion (i.e., oxidation number and absolute charge are the same value: **ox # = ion charge).**
   3. Oxidation numbers for nonmetals **in covalent compounds** must be assigned based on relative polarity of the covalent bonds. Certain nonmetals have generally assigned values for **compounds**:
      a. Hydrogen (H) always has **ox #** of (**–1**) when combined with metals and boron, H always has an **ox #** of (**+1**) when combined with any other element.
      b. Oxygen (O) almost always has an **ox #** of (**–2**) when combined with any other element; the only exceptions are when it is combined with itself or with fluorine.

    c. Fluorine (**F**) always has an **ox #** of (**−1**).

    d. Other Group 7A elements (Cl, Br, I) have an **ox #** of (**−1**) except when bonded to oxygen or fluorine or another member of the Group 7A which is above it in the periodic table. In these cases, the ox # is positive and must be calculated from the compound.

4. All atoms not covered by rules (**1**) through (**3**) are considered variable and must be calculated from the compound:

    a. The sum of all ox #'s for all atoms in a **neutral** compound must add up to **zero: sum ox # = 0**.

    b. The sum of all ox #'s for all atoms in a **polyatomic ion** must add up to the **charge** on the ion: **sum ox # = charge on polyatomic ion**.

5. **Calculation process:** add up all the **known ox #'s**, fixed by rules (**1**) through (**3**), for each atom of an element in a neutral compound or polyatomic ion; ___be certain to add the ox #'s using the correct sign___.

    a. To find the **one variable** atom in a **neutral** compound: subtract the total of all known ox #'s from **zero**; this must be the ox # for the **one** variable atom.

    b. To find the **one variable** atom in a **polyatomic ion**: subtract the total of all known ox #'s from the polyatomic ion **charge**; this must be the ox # for the **one** variable atom.

    c. If more than one variable atom occurs in a compound, each variable atom must be isolated and analyzed by further bonding analysis.

6. **Examples:** Determine the ox # of **each** element in the following:

    a. HNO_3 ox # of **H = + 1** and ox # of **O = −2** from rule (3); the molecule is the **neutral**

        sum of known ox #'s = [one H × (+1)] + [three O × (−2)] = [+1] + [−6] = **−5**

        ox # for **N** = (0) − (−5) = **+5**

    b. NO_2^- ox # of **O = −2** (rule (3)); charge on polyatomic ion = (**− 1**)

        sum of known ox #'s = [(two O × (−2))] = **−4**;

        ox # for **N** = (−1) − (−4) = **+3**

    c. $Cr_2O_7^{-2}$ ox # of **O = −2** (rule (3)); charge on polyatomic ion = (**− 2**)

        sum of known ox #'s = [(seven O × (−2))] = **−14**;

        ox # for the **two Cr** = (−2) − (−14) = **+12**

        each Cr = +12 total/2 Cr = **+6 for each Cr**

17.3 OXIDATION/REDUCTION FOR CARBON IN ORGANIC MOLECULES

1. A redox reaction with carbon describes the exchange of reactant carbon bonds for product bonds in which the elements exchanged in the bonding to carbon have differing electronegativities.

a. An **oxidation** of carbon during a reaction involves the exchange of reactant bonds to less electronegative elements (as compared to carbon) for product bonds to more electronegative elements (as compared to carbon). The net result is an *increase in the number of bonds carbon makes to more electronegative elements*.

b. A **reduction** of carbon during a reaction involves the exchange of reactant bonds to more electronegative elements (as compared to carbon) for product bonds to less electronegative elements (as compared to carbon). The net result is a *decrease in the number of bonds carbon makes to more electronegative elements*.

2. A restricted but very useful alternative description for organic reactions involving molecules with only C, H, and O:

a. An *oxidation* of carbon involves an increase *in the number of bonds to oxygen*, most often at the expense of a *decrease in the number of bonds to hydrogen*.

b. A *reduction* of carbon involves a *decrease in the number of bonds to oxygen*, most often at the expense of an *increase in the number of bonds to hydrogen*.

c. *Each exchange of one C—O bond for one C—H bond represents one stage in a redox sequence and results in a two-electron oxidation or reduction*: the oxidation number of carbon changes by $(+2)$ = oxidation or (-2) = reduction.

17.4 OXIDATION/REDUCTION SEQUENCE FOR ALCOHOLS, ALDEHYDES/KETONES, CARBOXYLIC ACIDS, AND DERIVATIVES

1. The redox sequence for primary alcohols is (*R = specifically a carbon group*)
Bond and oxidation number count on alcohol/carbonyl carbon:

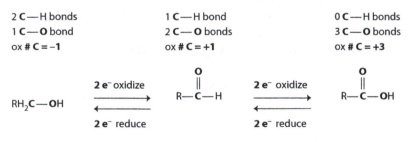

2 **C**—H bonds	1 **C**—H bond	0 **C**—H bonds
1 **C**—**O** bond	2 **C**—**O** bonds	3 **C**—**O** bonds
ox # **C** = –1	ox # **C** = +1	ox # **C** = +3

primary alcohol   **aldehyde**   **carboxylic acid**

2. The redox sequence for secondary alcohols is (*R = specifically a carbon group*)
Bond and oxidation number count on alcohol/carbonyl carbon:

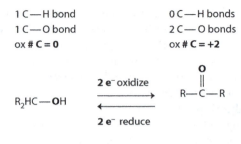

1 C—H bond   0 C—H bonds
1 C—O bond   2 C—O bonds
ox # **C = 0**   ox # **C = +2**

secondary alcohol   ketone

17.5 REAGENTS AND RESULTS FOR <u>OXIDATION</u> REACTIONS OF ALCOHOLS, ALDEHYDES/KETONES, CARBOXYLIC ACIDS, AND DERIVATIVES

17.5.1 COMPLETE OXIDATION

1. Many common **oxidation** reagents follow a general rule for **complete** oxidation of alcohol carbons. Example reagents are $KMnO_4$ and $Cr_2O_7^{-2}$ /H^+/H_2O.
2. Oxidation reactions start with a carbon already bonded to one oxygen (1 **C—O** bond), usually an alcohol. *The reaction always proceeds such that **all** C—H bonds on the original carbon containing the first oxygen are exchanged for C—O bonds; the pattern of bonding to this carbon always follows the functional groups shown.*
3. For complete oxidation, *primary alcohols are converted into carboxylic acids **without stopping at the aldehyde stage**.* The resulting reaction accomplishes a two-stage, 4 e^- oxidation. Oxidation of an aldehyde produces the carboxylic acid in a one-stage, 2 e^- oxidation.
4. Secondary alcohols are converted into ketones. Since a ketone has no more **C—H** bonds available, further oxidation with these reagents is not possible.
5. Tertiary alcohols show no reaction under these conditions. A tertiary alcohol has no **C—H** bonds available and thus cannot oxidize without breaking **C—C** bonds (requiring more severe conditions).
6. **Examples for $KMnO_4$ and $Cr_2O_7^{-2}$/H^+/H_2O:**

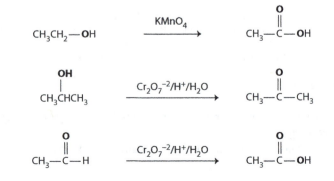

7. The half-reactions for reduction of Mn in $KMnO_4$ and Cr in $Cr_2O_7^{-2}$ are

$$Mn \text{ (ox \# = +7)} \longrightarrow Mn \text{ (ox \# = +3)} + 4\,e^-$$

$$Cr \text{ (ox \# = +6)} \longrightarrow Cr \text{ (ox \# = +3)} + 3\,e^-$$

17.5.2 SELECTIVE OXIDATION OF PRIMARY ALCOHOLS TO ALDEHYDES

1. Partial (selective) single-stage ($2\,e^-$) oxidation of a primary alcohol to an aldehyde can be accomplished by chromium (VI) (ox# +6) reagents that are poisoned to inhibit the second stage of the complete oxidation. The form of these reagents usually involves a complex of chromium (VI) oxide (chromium trioxide) with a weak base such as pyridine.
2. Common examples are pyridinium chlorochromate (PCC): $(C_5H_5NH^+)ClCrO_3^-$; pyridinium dichromate (PDC): $(C_5H_5NH^+)_2Cr_2O_7^{-2}$; and dipyridine chromate (Collin's reagent): $(C_5H_5N)_2CrO_3$.
3. Example for $(C_5H_5N)_2CrO_3$:

$$CH_3CH_2-OH \xrightarrow{(C_5H_5N)_2CrO_3} CH_3-\overset{\overset{\displaystyle O}{\|}}{C}-H$$

17.6 REAGENTS AND RESULTS FOR <u>REDUCTION</u> REACTIONS OF ALCOHOLS; ALDEHYDES/KETONES; CARBOXYLIC ACIDS; AND DERIVATIVES

17.6.1 HYDROGENATION: REDUCTION WITH HYDROGEN

1. Direct reaction with hydrogen (***hydrogenation***) in the presence of a metal catalyst such as Pt, Pd, Ni, Cu, etc., converts ***aldehydes or ketones into the corresponding alcohols; reduction of a carboxylic acid or ester requires much more severe conditions***.
2. Note that hydrogenation is <u>**not** selective</u> for **C=O** if a molecule contains **C=C**; the carbon–carbon double bond will also be reduced under these conditions.
3. Examples:

$$CH_3-\overset{\overset{\displaystyle O}{\|}}{C}-H \xrightarrow[\text{Pt}]{H_2} CH_3-CH_2-OH$$

$$CH_3-\overset{\overset{\displaystyle O}{\|}}{C}-CH_2CH_2CH=CH_2 \xrightarrow{2\,H_2} CH_3-\overset{\overset{\displaystyle OH}{|}}{CH}-CH_2CH_2CH_2-CH_3$$

17.6.2 METAL HYDRIDE REDUCTION OF ALDEHYDES AND KETONES $Li^+ R_3 Al—H$; $Na^+ R_3 B—H$

1. *Metal hydride ($Li^+R_3Al{-}H$; $Na^+R_3B{-}H$) reduction of aldehydes and ketones to the corresponding alcohol is described in Chapter 18 as a Group [1] reaction.*
2. Under these conditions, the reaction **is selective** for **C=O** reduction: *any* **C=C** *will* **remain unreacted.**
3. **Example for sodium borohydride: $Na^+R_3B{-}H$**

$$CH_3 - \overset{\displaystyle O}{\overset{\|}{C}} - CH_2CH_2CH{=}CH_2 \quad \xrightarrow[\text{CH}_3\text{OH}]{\text{Na}^+ \text{R}_3\text{B—H}} \quad CH_3 - \overset{\displaystyle OH}{\overset{|}{CH}} - CH_2CH_2CH{=}CH_2$$

17.6.3 LITHIUM ALUMINUM HYDRIDE REDUCTION OF CARBOXYLIC ACIDS AND ESTERS $Li^+ R_3 Al—H$

1. *Lithium aluminum hydride ($Li^+R_3Al{-}H$) reduction of carboxylic acids or esters to the corresponding primary alcohol is described in Chapter 18 as a Group [6] reaction.*
2. Lithium aluminum hydride reduces **carboxylic acids** or **esters** **completely** to primary **alcohols;** *the reaction does **not** stop at the aldehyde stage*. Note that under these conditions, this reaction also **is selective** for **C=O reduction:** *any* **C=C** *will remain unreacted.*
3. Sodium borohydride **($Na^+R_3B{-}H$)** is **not** sufficiently reactive to reduce carboxylic acids or esters.
4. *Note that carboxylic acids and esters are at the same oxidation state:*

$$CR_3{-}\overset{\displaystyle O}{\overset{\|}{C}}{-}OH \qquad\qquad CR_3{-}\overset{\displaystyle O}{\overset{\|}{C}}{-}OCR_3$$

$$\text{ox \# } \mathbf{C = +3} \qquad\qquad\qquad \text{ox \# } \mathbf{C = +3}$$

5. Examples:

$$CH_3{-}\overset{\displaystyle O}{\overset{\|}{C}}{-}OH \quad \xrightarrow[\text{2. H}^+/\text{H}_2\text{O}]{\text{1. Li}^+ \text{R}_3\text{Al—H}} \quad CH_3CH_2{-}OH \ \ (+H_2O)$$

$$CH_3{-}\overset{\displaystyle O}{\overset{\|}{C}}{-}OCH_3 \quad \xrightarrow[\text{2. H}^+/\text{H}_2\text{O}]{\text{1. Li}^+ \text{R}_3\text{Al—H}} \quad CH_3CH_2{-}OH + CH_3OH$$

A Complete System for Organizing, Identifying, and Solving Carbonyl Reactions

Nucleophilic Addition and Addition/Elimination

18.1 MECHANISMS AND CLASSIFICATION OF CARBONYL REACTIONS

18.1.1 GENERAL CONCEPTS

Nucleophilic addition is a key reaction pathway for the multitude of reactions at the carbonyl carbon ($c=o$). The addition reaction can form the final product directly, or can be the first reaction in a sequence of addition steps followed by various types of elimination steps. The organic compounds (substrates) for this reaction class are all carbonyl containing functional groups: aldehydes, ketones, carboxylic acids, acid chlorides, acid anhydrides, esters, and amides.

A systematic analysis of all nucleophilic addition and addition/elimination reactions at carbonyl is based on an understanding of the nucleophilic addition step in the reaction mechanism that forms the first intermediate. The reactions at carbonyls ($c=o$) initially proceed through a series of similar steps; this can apparently make them difficult to distinguish. In fact, however, the similarity among the reaction types allows a relatively simple classification based on one organizing principle: the structure of the tetrahedral intermediate formed in the initial nucleophilic addition, common to all reactions, directly determines any subsequent reaction steps and therefore the final product.

18.1.2 NUCLEOPHILIC ADDITION TO THE CARBONYL CARBON: $c=o$

The addition of a nucleophile to carbon of $c=o$ occurs for the functional groups aldehydes, ketones, carboxylic acids, and carboxylic acid derivatives shown below.

Notation: R = only carbon (alkyl) **or hydrogen**

Y = various atom groupings describing carbonyl functional groups

$$\underset{\underset{}{R-C-Y}}{\overset{O}{\overset{\|}{}}}$$

Possible Identities of Y	Corresponding Functional Group
Y = carbon (alkyl) or H	ketone or aldehyde
Y = OH	carboxylic acid
Y = OCR₃; SCR₃	ester; thioester
Y = Cl, Br	acid chloride; acid bromide
Y = OOCR	acid anhydride
Y = NR₂	amide

The **nucleophilic addition** reaction to a carbonyl carbon is

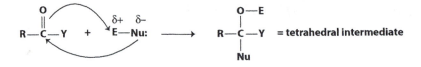

$$Y = -C(alkyl) \text{ or } -H; -OH; -OCR_3; -SCR_3; -Cl; -Br; -OOCR; -NR_2$$

E—Nu = electrophile/nucleophile combination; in most **ENu** combinations, the net **E⁺** (electrophile) is usually **H⁺** or **Metal⁺ˣ**.

The **Nu:** (nucleophile) can be based on **C, H, O, N,** or **S** anions or neutral molecules. In fact, the identities of **−Nu:** are most often the same molecular or atom groupings as those identified for the **−Y** groups above; only the nature of the **E** is significantly different.

Possible Identities of Nu: (δ−)		Corresponding Common E—Nu Combination
—Nu: =	—Carbon (δ−)	Li—CR₃; BrMg—CR₃
—Nu: =	—H (δ−)	Li R₃Al—H; NaR₃B—H
—Nu: =	—OH	H—OH (H₂O)
—Nu: =	—OCR₃; —SCR₃	H—OCR₃ H—SCR₃
—Nu: =	—OOCR	H—OOCR
—Nu: =	—NR₂	H—NR₂

The nucleophilic addition product is very often termed the **tetrahedral intermediate**. **Tetrahedral** refers to the conversion of an sp²-trigonal planar carbonyl carbon to an sp³-tetrahedral carbon. **Intermediate** refers to the fact that the reaction is very often followed by hydrolysis or elimination reactions which lead to other final products. It is true, however, that the addition product formed in this reaction is sometimes the final product. The term **"tetrahedral intermediate"** will generally refer to the addition product of **ENu** to **C=O**, even though in some cases this will be the "tetrahedral product."

The term **nucleophilic** addition for the reaction emphasizes the role of the nucleophile as the major determining factor for final product formation from a variety of available pathways. Further reactions of the nucleophilic addition product (tetrahedral intermediate) ultimately depend on the identity of the nucleophile bonded to the **C=O** carbon.

A neutral resonance form of the carbonyl **C=O** emphasizes the reaction tendency of the two atoms:

The straightforward rule of addition of an **ENu** combination to the double bond of the carbonyl group (**C=O**) is

The **oxygen** of the C=O always gets the electrophile (**E⁺**);
The **carbon** of the C=O always gets the nucleophile (**Nu:**).

The carbon of the **C=O** is δ+; the relatively positive **electrophilic** carbon will always **accept** an electron pair from the nucleophile (**Nu:**); the carbonyl carbon is never nucleophilic. The **oxygen** of the **C=O** is δ–; the relatively negative **nucleophilic** oxygen will always **donate** a lone pair (or a net pi-bonding electron pair) to the electrophile (**E⁺**).

The net result of the addition reaction is a cleavage of the **pi**-bond between the carbon and the oxygen. This opens up a bonding position for the **C—Nu** sigma bond; carbon is converted from sp^2-trigonal planar geometry to sp^3-tetrahedral geometry.

In most carbonyl addition reactions, the net electrophile (**E⁺**) is **H⁺**. This occurs whenever **E—Nu** is **H—Nu**, or when **H⁺/H₂O** is a co-reactant:

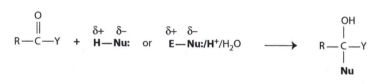

In circumstances where a net **H⁺** is not available (aprotic conditions), the tetrahedral intermediate is usually converted to the final product by a separate reaction with H^+/H_2O:

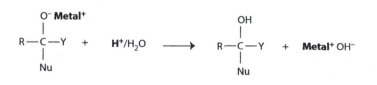

18.1.3 GENERAL MECHANISMS FOR NUCLEOPHILIC ADDITION TO CARBONYL c=o

Mechanisms for **nucleophilic addition** to c=o are general and apply to almost all possible carbonyl functional group types (identity of **Y**) and nucleophile types (identity of **ENu**). The general mechanisms are distinguished by the timing of the addition of (**E⁺**) versus (**Nu:**), and the acid/base conditions or catalyst identity.

General mechanism 1: Reactions where the net electrophile (**E⁺**) is a **metal** ion usually with nucleophiles where the nucleophilic atom is carbon or hydrogen are most often performed under **aprotic** (no protons/no water) conditions. In these cases the reaction is considered to occur in one step and the timing of addition is not a consideration:

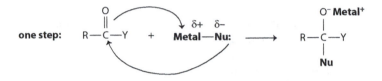

The final product in these reactions is a normal alcohol formed by converting the metal complex through a separate reaction with **H⁺/H₂O**.

General mechanism 2: Reactions where the net electrophile (**E⁺**) is **H⁺** from a strong acid (acidic conditions) are considered to be acid catalyzed. A reaction in strong acid will generally show addition of the **H⁺** electrophile to oxygen first; addition of the nucleophile, usually as a neutral molecule, will occur in a subsequent step. The **H⁺** will be transferred by whichever molecule is the strongest acid under reaction conditions.

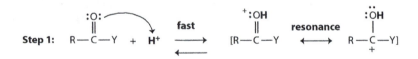

The **H⁺** accepts the lone-pair electrons of the oxygen; an alternative view is net acceptance of the pi-electrons of the c=o. The two resulting resonance structures indicate that the carbon bears a large amount of positive charge.

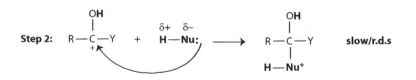

Nucleophilic addition to carbon is the **rate-determining step**. The nucleophile usually exists as a neutral molecule and donates a lone electron pair to the electrophilic carbon formed in step **1**. The conversion of a lone pair to a new sigma bond electron pair places the +**1** formal charge on the nucleophile.

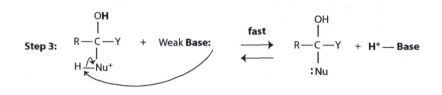

The positively charged product is converted into a neutral molecule by weak base removal of the **H⁺** from the **Nu** portion of the product; the H—Nu bond electron pair remains with the **Nu** to replace the lone pair used in step **2**. The **H⁺** will be removed by whichever molecule is the strongest base in the reaction.

Although the exact forms of the **H⁺** in steps **1** and **3** do not appear identical in this generalized mechanism, the net role of the original proton is as a **catalyst**: it is consumed as **H⁺** in the first step and regenerated as **H⁺** in the last step.

General mechanism 3: For reactions where the net electrophile (E^+) is **H⁺** from a neutral compound or in basic conditions, the reaction may be base catalyzed or uncatalyzed. Under these conditions, the **nucleophile** will generally add to the carbon **first**; addition of the electrophilic **H⁺**, usually from a neutral molecule such as water, will occur in a subsequent step.

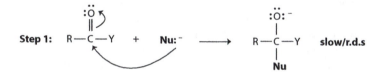

The **nucleophile**, often as an anion especially in basic conditions, transfers an electron pair to the **electrophilic carbon** to form the new sigma bond; the pi-electron pair of the **C═O** is simultaneously transferred to the oxygen to open up a bonding position on carbon.

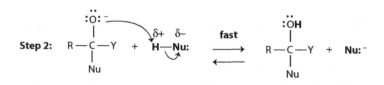

Proton transfer from **any weak** acid in the reaction to the negatively charged carbonyl oxygen produces the neutral tetrahedral intermediate. In the generalization shown, the weak acid is the (**H—Nu:**) molecule; this regenerates the anionic (base) form of the nucleophile.

18.2 CLASSIFICATION OF NUCLEOPHILIC ADDITION/ ELIMINATION REACTIONS THROUGH THE STRUCTURE OF THE TETRAHEDRAL INTERMEDIATE: DETERMINING THE FINAL PRODUCT FOR CARBONYL REACTIONS

Nearly all nucleophilic addition/elimination reaction mechanisms proceed through a **neutral** (anionic is possible) tetrahedral intermediate formed by the nucleophilic addition reaction.

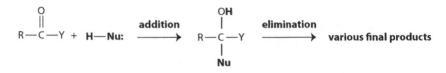

The identity of the final product of the complete reaction depends on the existence and type of subsequent elimination steps, these steps are determined by the structure of the neutral tetrahedral intermediate.

The key carbon in the analysis of the tetrahedral intermediate is the original carbon of the $C=O$, termed the **central carbon**. The determining structure of the tetrahedral intermediate is based on the four atoms directly attached to the central carbon through single bonds. These in turn are identified from the **R**- and **Y**-groups originally attached to the $C=O$ carbon, plus the reacting nucleophile (**Nu:**):

1. One of the four attached atoms must be a carbon or a hydrogen, identified by the definition of the R-group as H or C (alkyl).
2. One of the four atoms **must** be the **oxygen** of the original $C=O$, this oxygen is converted from a double bond to a single bond during the nucleophilic addition reaction.
3. The bonded atom from the **Y**-group is commonly: (Sulfur is also possible.)

—**C** (**carbon**-alkyl); aldehydes and ketones
—**H** (**hydrogen**); aldehydes
—**O** (**oxygen**) from —**OH**; —**OCR₃**; —**OOCR**; carboxylic acids, esters, anhydrides
—**Cl**, —**Br** (**halogens**); carboxylic acid chlorides/bromides
—**N** (**nitrogen**) from —**NR₂**; amides

4. The bonded atom from the nucleophile **Nu:** group is commonly: (Sulfur is possible.)

—**C** (**carbon**) from **ENu:** combinations such as Li—**CR₃**; BrMg—**CR₃**
—**H** (**hydrogen**) from ENu: combinations such as Li R₃Al—**H**; NaR₃B—**H**
—**O** (**oxygen**) from H—**OH** (H₂O); H—**OCR₃**; H—**OOCR**
—**N** (**nitrogen**) from H—**NR₂**

The identification of which type of elimination reaction, if any, a tetrahedral intermediate will undergo is simplified by classification of carbonyl addition/elimination reactions into **six** groups. The organizing principle is the identity of the four single-bonded atoms to the central carbon. The combination of these four atoms affect the stability of the intermediate and direct the subsequent reactions to specific products.

A general process:

1. **Write the correct nucleophilic addition product**; this specifies the structure of the neutral tetrahedral intermediate.
2. **Classify the intermediate as belonging to one of the six groups described**. The description of each group specifies the correct final product based on any subsequent elimination reactions. Since the atoms bonded to the central atom are themselves derived from the identity of the **Nu**- and **Y**-groups, each classification group can also be identified from the structure of the reactants.
3. **Determine the correct final product** by matching the selected group to the corresponding net result described in the following sections.

The six classification groups:

Group [1]: The central carbon of the intermediate is a "normal" (mono-functional) alcohol: the central carbon is bonded to **one oxygen** as hydroxyl (or initially as a metal complex) plus **three carbons** or **hydrogens**. The starting carbonyl compound for reactions in this group **must** be an **aldehyde** or a **ketone**.

$$\begin{array}{c} \text{O(H or metal)} \\ | \\ [\textbf{C or H}] - \text{C} - [\textbf{C or H}] \\ | \\ [\textbf{C or H}] \end{array}$$

Group [2]: The central carbon of the intermediate is an "amino-derivative"/alcohol: it is bonded to **one oxygen** as hydroxyl, **one nitrogen** as an amine (or amine derivative) plus **two carbons** or **hydrogens**. The starting carbonyl compound for reactions in this group **must** be an **aldehyde** or a **ketone**.

$$\begin{array}{c} \text{OH} \\ | \\ [\textbf{C or H}] - \text{C} - [\textbf{C or H}] \\ | \\ \text{NR}_2 \end{array}$$

Group [3]: The central carbon of the intermediate forms an oxaphosphetane ring from a triphenylphosphonium ylide nucleophile specific for the **Wittig** reaction. The starting carbonyl compound for reactions in this group must be an **aldehyde** or a **ketone**. The specific structure of the intermediate is:

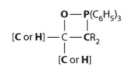

Group [4]: The central carbon of the intermediate is bonded to **two oxygens** as hydroxyl (—**OH**) or ether (—**OCR₃**) functional groups plus **two carbons** or **hydrogens**. The starting carbonyl compound for reactions in this group **must** be an **aldehyde** or a **ketone**.

$$
\begin{array}{c}
\text{OH} \\
| \\
[\textbf{C or H}] - \text{C} - [\textbf{C or H}] \\
| \\
\textbf{O}(\text{H or CR}_3)
\end{array}
$$

Group [5]: The central carbon of the intermediate is bonded to **three** functional groups with any combination of **nitrogen, oxygen** (or sulfur), or **halogen** as bonded atoms; only **one** bond is to **carbon** or **hydrogen**. This group of reactions includes all of the carboxylic acid derivatives: **acids, acid chlorides, anhydrides, esters, and amides**.

$$
\begin{array}{c}
\text{OH} \\
| \\
[\textbf{C or H}] - \text{C} - [\textbf{O, N, S, Cl}] \\
| \\
[\textbf{O, N, S, Cl}]
\end{array}
$$

Group [6]: This group is a sequential combination of a Group [**5**] type reaction followed by a Group [**1**] type reaction; the Group [**5**] reaction, however, involves a carbon or hydrogen nucleophile. This group is specific for the **double** addition of a **carbon** or **hydrogen** based nucleophile to an **ester** (or less commonly) an **acid chloride**.

18.3 DESCRIPTIONS OF GROUP REACTION RESULTS AND MECHANISMS

18.3.1 GROUP [1]

The central carbon of the intermediate is a "normal" (mono-functional) alcohol: the central carbon is bonded to **one oxygen** as hydroxyl (or initially as a metal complex) plus **three carbons** or **hydrogens**.

$$
\begin{array}{c}
\textbf{O}(\text{H or metal}) \\
| \\
[\textbf{C or H}] - \text{C} - [\textbf{C or H}] \\
| \\
[\textbf{C or H}]
\end{array}
$$

18.3.1.1 GENERAL REACTANT REQUIREMENTS

Carbonyl substrate: $\textbf{Y} = \textbf{C}$ (alkyl) or $\textbf{H}$
The starting carbonyl compound must be an aldehyde or a ketone
Nucleophilic atom: must be $-\textbf{C}$ or $-\textbf{H}$
Common nucleophiles as E—Nu combinations (E$^+$ usually = metal):

$$
\begin{array}{c}
\overset{\delta+}{}\;\overset{\delta-}{} \qquad \overset{\delta+}{}\;\overset{\delta-}{} \\
-\textbf{C} \ (\text{carbon}): \ \text{Li} - \textbf{CR}_3; \ \textbf{BrMg} - \textbf{CR}_3; \ \textbf{Na}^{+\,-}\!:\textbf{C}\!\equiv\!\textbf{CR}; \ \textbf{Na}^{+\,-}\!:\textbf{C}\!\equiv\!\textbf{N}
\end{array}
$$

$$
\begin{array}{c}
\overset{\delta-}{} \qquad\qquad\quad \overset{\delta-}{} \\
-\textbf{H} \ (\text{hydrogen}): \ \text{Li}^+\, \text{R}_3\text{Al} - \textbf{H} \,; \ \text{Na}^+\, \text{R}_3\text{B} - \textbf{H}
\end{array}
$$

Reactant requirement explanation: The oxygen of the tetrahedral intermediate is derived from the oxygen of the **C=O**. The three bonded [**C** or **H**] atoms require that both the **R**-group and **Y**-group be carbon or hydrogen (=aldehyde or ketone) and that the nucleophile group must be derived from carbon or hydrogen.

Group [1] Reaction results and mechanism: Overall reaction stops at the tetrahedral product; **no elimination step occurs**. The net product is an **alcohol** sometimes formed directly; most often the alcohol is formed from the metal complex by separate hydrolysis with H^+/H_2O.

Since there is no subsequent elimination, the mechanism is the **one-step** nucleophilic addition described in previous section. (**R** = specifically **C** or **H**)

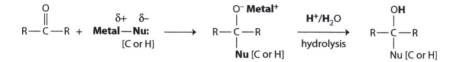

Group [1] Examples: A large variety of organometallic reactions (carbon nucleophile) and metal hydride reductions (hydrogen nucleophile) fall into Group [1]. An important extension of this group is the aldol condensation reaction, for which the nucleophilic carbon is the sp^3 carbon (α-carbon) adjacent to a carbonyl in aldehydes and ketones.

The final product of a Group [1] is a normal alcohol (after reaction with H^+/H_2O if required). To find the net result product, convert the double bond of the **C=O** to a single bond C—O; add a net H to the oxygen to form C—OH, then connect through a single bond the **C** or **H** nucleophile to the central carbon of the original **C=O**.

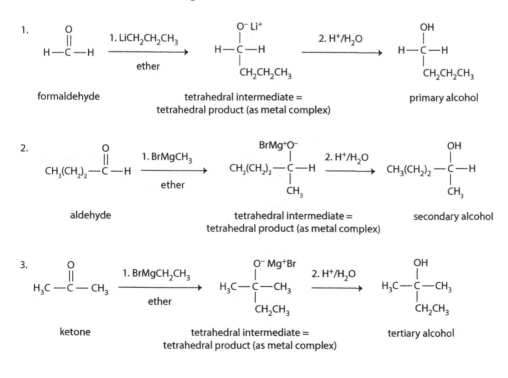

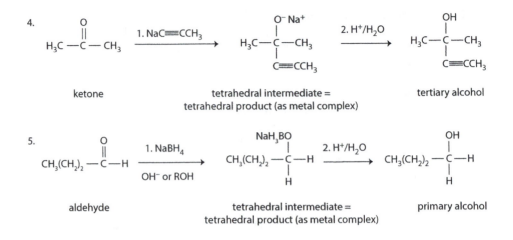

4. ketone → tetrahedral intermediate = tetrahedral product (as metal complex) → tertiary alcohol

5. aldehyde → tetrahedral intermediate = tetrahedral product (as metal complex) → primary alcohol

18.3.2 GROUP [2]

The central carbon of the intermediate is an "amino-derivative"/alcohol. It is bonded to **one oxygen** as hydroxyl, **one nitrogen** as an amine (or amine derivative) plus **two carbons** or **hydrogens**.

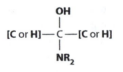

18.3.2.1 GENERAL GROUP REACTANT REQUIREMENTS

Carbonyl substrate: $Y = C$ (alkyl) or H
The carbonyl compound must be an aldehyde or a ketone.
Nucleophilic atom: must be **nitrogen**, divided into two subgroups:
Subgroup [2A]: Nucleophile nitrogen has **two** (or three) hydrogens.
Subgroup [2B]: Nucleophile nitrogen has **only one** hydrogen.
Electrophilic atom $(E^+) = H^+$

Reactant requirements explanation: The oxygen of the tetrahedral intermediate is derived from the oxygen of the $C=O$; two bonded [C or H] atoms, based on the R-group and the Y-group must come from the starting aldehyde or ketone carbonyl; the nitrogen bonded to the central carbon must always come from the nucleophile.

18.3.3 SUBGROUP [2A] REACTION RESULTS AND MECHANISM

The subsequent reaction of the tetrahedral intermediate for Subgroup [2A] is an elimination of both the hydroxyl group from the central carbon plus a proton from the nucleophilic atom to produce $C=Nu$. These products are generally termed **imines**. Elimination occurs because the tetrahedral

amino-alcohol intermediate is relatively unstable as compared to the final product due to the electronegativity of the nitrogen nucleophile.
(R = specifically C or H; nucleophilic atom of Nu = **nitrogen**)

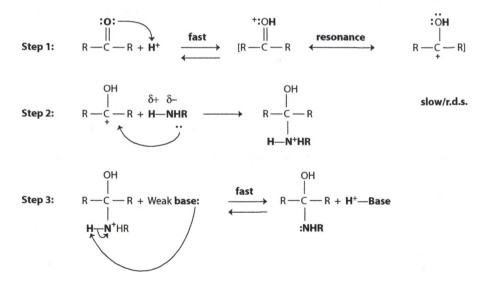

The requirement for Subgroup [2A] is determined by the fact that the nucleophilic atom must have a proton capable of removal for double bond formation in the elimination steps. Since the first proton is removed in the addition step, the original nucleophilic nitrogen must have at least **two** protons.

Common nucleophiles as E—Nu combinations ($E^+ = H^+$)

H—$NHCR_3$ (primary amines); NH_3 (ammonia)
H—NHX (X not C or H):
H—$NHNH_2$ (hydrazine); H—$NHNHCONH_2$ (semicarbazide);
    H—$NHOH$ (hydroxylamine); H—$NHNHC_6H_3(NO_2)_2$ (2,4-dini-trophenylhydrazine)

The reaction usually occurs under acid catalysis. The mechanism for the **addition** reaction is based on that shown in Section 18.1; an example for reaction of a primary amine nucleophile is shown: The complete reaction is readily reversible.

Nucleophilic addition steps: (R = C or H; the weak base = water or amine)

The elimination under acid catalysis conditions shows the following mechanism:

Elimination steps: (R = C or H; the weak base = water or amine)

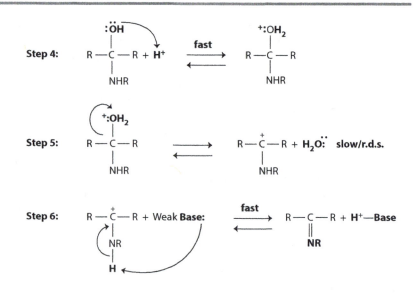

18.3.4 GROUP [2A] EXAMPLES

Primary amines produce imines; other nitrogen reactants form various specific aldehyde and ketone derivatives.

To find the net result product, convert the double bond of the $C=O$ to a double bond $C=NR$; the carbonyl oxygen and the two nitrogen protons are removed as H_2O. The reaction is readily reversible; forward or reverse can be selected by adjusting reaction conditions.

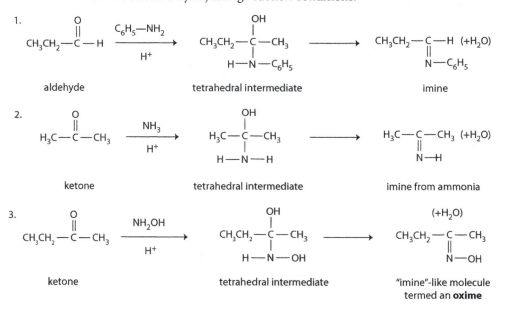

18.3.5 SUBGROUP [2B]: REACTION RESULTS AND MECHANISM

The subsequent reaction of the tetrahedral intermediate for Subgroup [2B] is an elimination of both the hydroxyl group from the central carbon plus a proton from a neighboring carbon to produce **c=c—Nu**. These products are termed **enamines**.

(**R** = specifically **C** or **H**; nucleophilic atom of **Nu** = **nitrogen**)

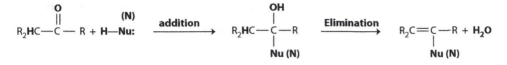

Elimination must occur due to the relative instability of the tetrahedral **amino-alcohol** intermediate. Double bond formation to the nucleophile is not possible; the nitrogen in the tetrahedral intermediate has no proton capable of being removed. (One proton of the reactant was removed in the addition reaction.) The **enamine** product is produced by loss of a proton from a neighboring carbon.

Common nucleophiles as E—Nu combinations (E⁺ = H⁺)

$$H—N—CR_3 \quad \text{(secondary amines)}$$
$$\overset{|}{CR_3}$$

The reaction generally occurs under acid catalysis; the mechanism for the addition steps are identical to Subgroup [2A]. The mechanism for the elimination steps are similar to Subgroup [2A] except that in **step 6**, the weak base removes a proton from a carbon rather than the nitrogen.

Tetrahedral intermediate formed in the addition reaction:

$$\overset{OH}{\underset{\overset{|}{H} \quad \underset{N(CR_3)_2}{}}{R_2C—C—R}} \text{ (no proton on the nitrogen)}$$

Step 6 of elimination reaction

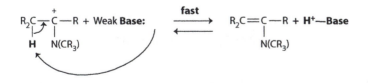

18.3.6 GROUP [2B] EXAMPLES

Nucleophilic nitrogens with one original proton (secondary amines) produce enamines. To find the net result product, convert the double bond of the **c=o** into a single bond **CC—NR₂** with a pi-bond shifted to the neighboring carbon to form a double bond **c =CR₂**. The carbonyl oxygen, the one nitrogen proton, and the proton from the neighboring carbon are removed as H_2O. The reaction is readily reversible; forward or reverse can be selected by adjusting reaction conditions.

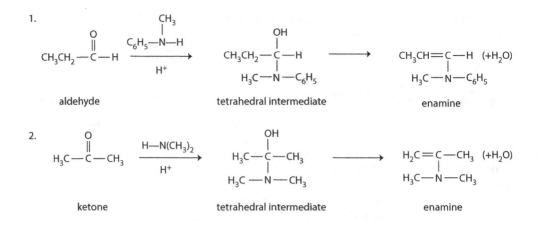

1. aldehyde → tetrahedral intermediate → enamine

2. ketone → tetrahedral intermediate → enamine

18.3.7 GROUP [3]

The central carbon of the intermediate forms an oxaphosphetane ring from a triphenylphosphonium ylide nucleophile, specific for the named **Wittig** reaction. The starting carbonyl compound for reactions in this group must be an aldehyde or a ketone. The specific structure of the intermediate is

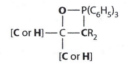

18.3.7.1 GENERAL REACTANT REQUIREMENTS

Carbonyl substrate: **Y** = **C** (alkyl) or **H**; the starting carbonyl compound must be an aldehyde or a ketone.

Nucleophilic atom: this reaction applies **only** to the specific **triphenylphosphonium ylide**, in which **carbon** is the nucleophilic atom in the **ENu** Combination ($E^+ = P$):

$$(C_6H_5)_3 \overset{\delta+}{P} = \overset{\delta-}{CR_2} \quad \longleftrightarrow \quad (C_6H_5)_3 P^+ - \,^-:CR_2$$

R = **C** or **H**; C_6H_5 is the phenyl ring; two resonance structures are shown to indicate the nucleophilic carbon.

Reactant requirement explanation: The final product is formed by an elimination reaction of the intermediate oxaphosphetane ring. In this specific structure, the central carbon is connected to **two** carbons or hydrogens that must come form the carbonyl substrate (aldehyde or ketone). The other two bonds are part of a four-membered ring, one bond specifically to the carbon nucleophile and one to the oxygen of the original carbonyl **c**＝**o**. The electrophilic phosphorous atom is bonded to the oxygen and remains singly bonded to the nucleophilic carbon; this produces the ring structure.

Reaction results: The oxaphosphetane ring intermediate undergoes an elimination of both the oxygen bonded to the central carbon plus the phosphorous bonded to the nucleophilic carbon to generate a $C=Nu$ in the form of a normal $C=C$ bond between the original $C=O$ carbon and the carbon nucleophile. ($R = C$ or H)

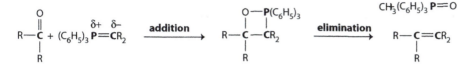

Mechanism

Addition: The mechanism for the addition reaction can be shown as two steps. For simplicity, the two addition versions below are each shown as a combination **one** step.

The nucleophilic electrons of the triphenylphosphonium ylide can be viewed as the **pi**-bonding electrons of the $P=C$ bond; the nucleophilic carbonyl oxygen bonds to the electrophilic phosphorous.

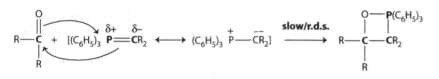

Alternatively, the nucleophilic electrons of the triphenylphosphonium ylide can be viewed as the lone electron pair from a carbon anion from the contributing resonance form.

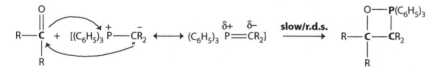

Elimination: Occurs because the oxaphosphetane ring is unstable relative to the final products. The carbonyl-based product that results represents an exchange of the original $C=O$ for a $C=CR_2$. The reaction can be considered **one** step.

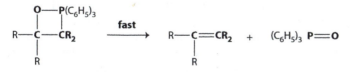

Formation of $(C_6H_5)_3 P=O$. is highly energetically favorable due to the high bond dissociation energy of the $P=O$ bond. The complete reaction is essentially irreversible.

18.3.8 GROUP [3] EXAMPLES

The Wittig reaction through a carbon nucleophile generates an alkene. To find the net result product, convert the double bond of the $C=O$

into a double bond **C═CR₂** based on the carbon nucleophile from the **ENu** triphenylphosphonium ylide combination; the carbonyl oxygen is removed as part of the by product (C₆H₅)₃ P═O.

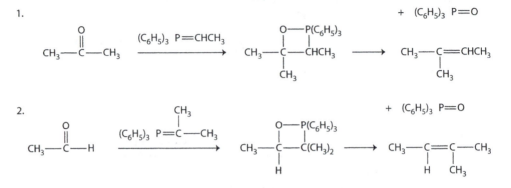

18.3.9 GROUP [4]

The central carbon of the intermediate is bonded to **two oxygens** as hydroxyl (—**OH**) or ether (—**OCR₃**) functional groups plus **two carbons** or **hydrogens**.

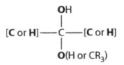

General group reactant requirements
Carbonyl substrate: **Y = C** (alkyl) or **H**. The carbonyl compound must be an aldehyde or a ketone.
Nucleophilic atom: must be oxygen, in the form of water, an alcohol, or a diol.
Common nucleophiles as E–Nu combinations (E⁺ = H⁺):

H — OH (water); H — OCR₃ (alcohol); H — O(CR₂)ₓO — H (diol)

Reactant requirements explanation: One of the two oxygens of the tetrahedral intermediate is derived from the oxygen of the **C═O**; this oxygen will exist as a hydroxyl group. The two bonded [**C** or **H**] atoms, based on the **R**-group and the **Y**-group must both come from the starting aldehyde or ketone carbonyl; the **second** oxygen bonded to the central carbon must always come from the nucleophile **Nu**.
General Group [4] reaction results
The subsequent reaction of the tetrahedral intermediate can be varied; the final results are divided into two subgroups.
Subgroup [4A]: After completion of the addition reaction, the tetrahedral intermediate remains sufficiently stable to become the final product; no subsequent elimination reaction occurs.
Subgroup [4B]: For this group, the final product is derived from a subsequent elimination of the hydroxyl group (as water) from the tetrahedral

intermediate, followed by a second addition reaction of another alcohol oxygen nucleophile. This reaction sequence only applies to tetrahedral intermediates formed from alcohol or diol nucleophiles.

Reactions and mechanism for Subgroup [4A]

General result for Subgroup [4A] ($R = C$ or H; **Nu** based on **O**):

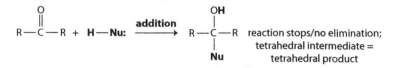

The specific molecules formed for Subgroup [4A] are termed **hydrates** if the nucleophile is **water**; the reaction is generally an equilibrium ($R = C$ or H).

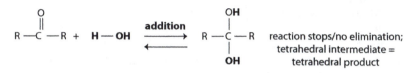

The products of the reaction for **alcohol** or **diol** nucleophiles are termed **hemiacetals** (from aldehydes) or **hemiketals** (from ketones).

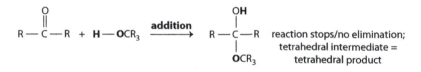

The tetrahedral intermediate formed from the aldehyde/ketone and an alcohol is not generally very stable; most often the reaction will fall into Subgroup [4B]. However, **cyclic** hemiacetals/hemiketals can be stable: the aldehyde/ketone carbonyl $c = o$ plus the alcohol nucleophile are derived from the same molecule to form a **ring**. The resulting product is an internal (intramolecular) ring hemiacetal/hemiketal. An example is the cyclization of glucose.

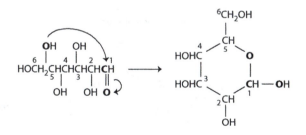

Since there is no subsequent elimination reaction, the mechanisms for the **addition** reaction are based on those shown in Section 18.1. Acid catalysis is shown below; base conditions are also possible. The reaction is readily reversible.

Nucleophilic addition steps ($R = C$ or H; the weak base = water or alcohol):

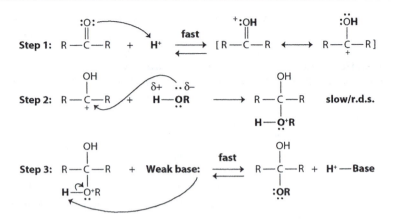

Group [4A] examples

Addition of a water nucleophile will form an equilibrium hydrate. One addition of an alcohol nucleophile will form a hemiacetal or hemiketal. To find the net result product convert the double bond of the **C═O** to a single bond **C–OH** based on the carbonyl oxygen. Add a second single bond **–OH** to the central carbon if the nucleophile is water or a single bond **–OCR₃** to the central carbon if the nucleophile is an alcohol. The reaction is readily reversible; forward or reverse can be selected by adjusting reaction conditions.

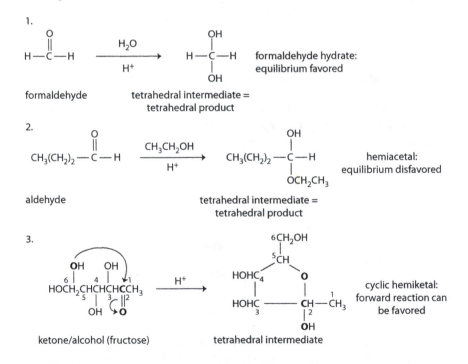

Reaction results for Subgroup [4B]

The final product is derived from a subsequent elimination of the hydroxyl group (as water) from the tetrahedral intermediate, followed by a second addition reaction of another alcohol oxygen nucleophile. This reaction sequence only applies to tetrahedral intermediates formed from alcohol or diol nucleophiles. The complete reaction is readily reversible; the forward or reverse reactions can be favored by adjusting conditions.

$$R—\overset{\overset{\displaystyle O}{\|}}{C}—R + H—Nu: \xrightarrow{\text{addition}} R—\underset{\underset{\displaystyle Nu}{|}}{\overset{\overset{\displaystyle OH}{|}}{C}}—R \xrightarrow[\displaystyle H—Nu:]{\substack{\text{elimination/}\\\text{2nd addition}}} R—\underset{\underset{\displaystyle Nu}{|}}{\overset{\overset{\displaystyle Nu}{|}}{C}}—R + H_2O$$

The products of the reaction are termed **acetals** (from aldehydes) or **ketals** (from ketones) (**R = C or H**).

$$R—\overset{\overset{\displaystyle O}{\|}}{C}—R + H—OCR_3 \xrightarrow{\text{addition}} R—\underset{\underset{\displaystyle OCR_3}{|}}{\overset{\overset{\displaystyle OH}{|}}{C}}—R \xrightarrow[\displaystyle H—OCR_3]{\substack{\text{elimination/}\\\text{2nd addition}}} R—\underset{\underset{\displaystyle OCR_3}{|}}{\overset{\overset{\displaystyle OCR_3}{|}}{C}}—R + H_2O$$

The subsequent elimination occurs due to the higher potential energy of the central carbon bonded to two electronegative oxygens; other than the exception for ring formation in Subgroup [**4A**], this is the more common result. The elimination alone does not produce a stable product (similar to Group 2) because the nucleophilic oxygen is incapable of forming a double bond after the addition. The elimination of water is followed by a **second** addition; the net result of this combination is a substitution of the hemiacetal/hemiketal **−OH** for a second **−OCR₃**. Acetals and ketals are generally more stable than hemiacetals or hemiketals.

Mechanisms for Subgroup [4B]

The mechanism of the **addition** reaction is identical to that shown for Subgroup [**4A**]. The tetrahedral, hemiacetal, or hemiketal thus formed undergoes elimination.

Elimination: An example of the mechanism for elimination starting from the tetrahedral intermediate in **acid** catalysis is shown.

Elimination steps (R = C or H):

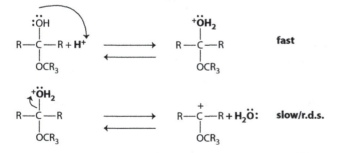

The second addition of an alcohol nucleophilic oxygen produces the final acetal or ketal.

Second addition steps (R = C or H; the weak base = water or alcohol):

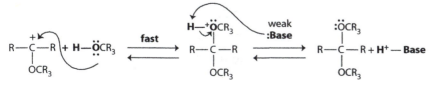

The second alcohol nucleophile need not be the same as the first. A useful group of alcohol reactants are 1,2-diols and 1,3-diols. The addition of both alcohol functional groups to the same carbonyl produces a cyclic acetal or ketal. Note that water is a product of the elimination steps; equilibrium shifting techniques such as removal or addition of water allow the forward or reverse reactions to be favored.

Group [4B] examples

Aldehydes and ketones are converted to the hemiacetal/hemiketal tetrahedral intermediate; elimination followed by a second addition of the alcohol nucleophile then produces an acetal or a ketal. Reaction of two alcohols is usually noted by indicating two molecules of alcohol consumed, or the use of certain diols to form a cyclic compound. To find the net result product convert the double bond of the $C=O$ to **two** single bond $-OCR_3$ to the central carbon based on the alcohol nucleophile. The carbonyl oxygen and the protons on the alcohols are removed as a net H_2O. Forward or reverse reactions can be selected by adjusting reaction conditions.

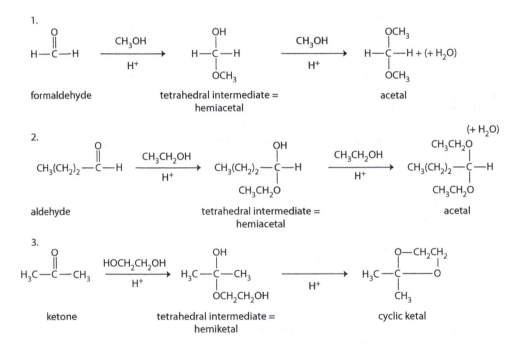

Group [5]: The central carbon of the intermediate is bonded to **three** functional groups with any combination of **nitrogen, oxygen** (or sulfur), or **halogen** as bonded atoms; only **one** bond is to **carbon** or **hydrogen**.

18.3.9.1 GENERAL REACTANT REQUIREMENTS

Carbonyl substrate $R-\overset{\overset{\displaystyle O}{\|}}{C}-Y$:**Y** = **not** C or H; **Y** must be one of the functional groups described as a carboxylic acid or derivative:

Possible Identities of Y	Corresponding Functional Group
Y = —OH	carboxylic acid
Y = —OCR$_3$; —SCR$_3$	ester; thioester
Y = —Cl, —Br	acid chloride; acid bromide
Y = —OOCR	acid anhydride
Y = —NR$_2$	amide

(b) Nucleophilic atom: must be —O (or S); —N ; from **ENu**:

Possible Identities of Nu:	Corresponding Common E—Nu Combination
—Nu: = —OH	H—OH (H$_2$O)
—Nu: = —OCR$_3$: —SCR$_3$	H—OCR$_3$ H—SCR$_3$
—Nu: = —OOCR	H—OOCR
—Nu: = —NR$_2$	H—NR$_2$

Reactant requirement explanation: The **one C** or **H** bonded to the central carbon of the tetrahedral intermediate comes from the **R**-group as part of the original carboxylic acid/derivative. The **three** non-carbon or hydrogen atoms (**O, N, S, Cl,** or **Br**) bonded to the central carbon come from

1. The original **oxygen** of the **C**=**O**.
2. The attached atom of the **Y**-group listed.
3. The nucleophilic atom of the **Nu**-group listed.

General reaction results: The tetrahedral intermediate from Group [5] always reacts through elimination of the atom grouping designated **Y**; the resulting product shows:

1. Reformation of the double bond between the central carbon and the original oxygen to regenerate the carbonyl group **C**=**O**
2. A **net substitution** of **Y** by **Nu** in all cases
3. Formation of **HY** or various acid/base forms of HY as the other product

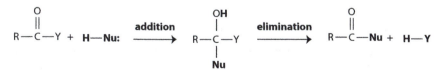

The result is a very general reaction for the formation of amides and lactams (cyclic amides), esters and lactones (cyclic esters), carboxylic acids, and acid anhydrides; acid chlorides cannot be formed by this reaction but can be used as substrates. In the reaction, —**Y** is a good leaving group or is converted to a good leaving group through acid or base catalysis.

The symmetry of the **net** substitution reaction is important: since **Y** is substituted for by **Nu:**, **Y** and **Nu:** belong to the same class of substituents; their role depends on the choice of nucleophile or substrate.

(—Y) Group form in carbonyl compound	**Group form in E—Nu**
—OH	H—OH (H_2O)
—OCR$_3$: —SCR$_3$	H—OCR$_3$ H—SCR$_3$
—OOCR	H—OOCR
—NR$_2$	H—NR$_2$

Reaction direction and free energy: Some exchanges of **Y** for **Nu:** are reversible. Except for certain equilibrium shifts described below, exchange of **Y** for **Nu:** can effectively proceed only in the direction of a more reactive derivative to form a less reactive derivative. Aside from equilibrium shifts, the direction of exchange must always follow a $\Delta G°$ negative spontaneity.

A non-spontaneous ($\Delta G°$ positive) reaction can occur if the equilibrium can be shifted toward the product side by removal of a product molecule and/or deactivation of the reverse reaction through additional proton transfers.

The following sequence lists the potential carbonyl substrates in order of increasing stability and decreasing reactivity moving from **left to right** (arrow direction). The diagram indicates that carbonyl substrates can be converted to products in the direction of the arrows:

1. Each carbonyl compound can be converted through a **Group [5]** reaction to any substituted carbonyl compound listed to its right.
2. Except for equilibrium shifts, a **Group [5]** reaction will not directly convert a carbonyl compound to a substituted carbonyl compound listed to its left.
3. The three common possible equilibrium shift exceptions to rule (2) are shown in the diagram as
   a. An ester/acid equilibrium
   b. Conversion of an amide to an acid by an equilibrium shift
   c. Conversion of an acid to an anhydride by an equilibrium shift.

Most reactive/Least stable ⟶ **Least reactive/most stable**

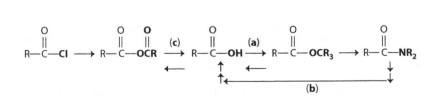

18.3.10 GENERAL ADDITION MECHANISMS FOR GROUP [5]

The mechanisms for the addition reaction for **Group [5]** are shown in Section 18.1. Depending on the specific derivative, strong acid catalysis, strong base, and neutral conditions can all apply.

General elimination mechanisms: Elimination under strong acid catalysis can apply to most carbonyl substrates except acid chlorides. A general mechanism is shown; steps 5 and 6 sometimes occur simultaneously as a single step and are written together (e.g., for acid anhydrides).

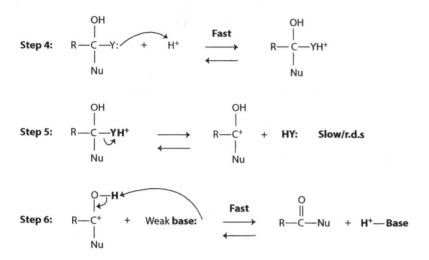

Under neutral conditions (most often applied to acid chlorides), the **elimination** mechanism to form the final product can be considered as one step:

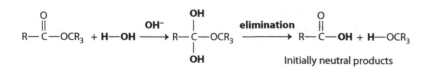

Elimination under strong base conditions usually applies to hydrolysis of esters or amides to the net carboxylic acid.

For the conversion of an ester to a carboxylic acid, $Y = -OCR_3$ and $Nu:^- = OH^-$.

The complete reaction is base catalyzed hydrolysis, or saponification:

$$R\overset{\overset{O}{\|}}{-}C-OCR_3 + H-OH \xrightarrow{OH^-} R\overset{\overset{OH}{|}}{\underset{\underset{OH}{|}}{-}}C-OCR_3 \xrightarrow{elimination} R\overset{\overset{O}{\|}}{-}C-OH + H-OCR_3$$

Initially neutral products

The **elimination** mechanism can be considered to be one step:

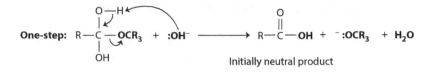

Hydrolysis of an ester proceeds against the thermodynamically favored direction. Strong base conditions produce an equilibrium shift toward formation of the carboxylic acid product due to the additional acid/base reaction of the carboxylic acid with strong base. Deprotonation of the carboxylic acid to form the carboxylate anion deactivates the reverse direction since the carbonyl of the anion is unreactive for nucleophilic addition.

$$\underset{\text{R—C—OH}}{\overset{O}{||}} + OH^- \longrightarrow \underset{\text{R—C—O}^-}{\overset{O}{||}} \text{(unreactive)} + H_2O$$

For the conversion of an amide to a carboxylic acid, $Y = —NR_2$ and $Nu:^- = OH^-$.

The complete reaction is base catalyzed hydrolysis:

$$\underset{\text{R—C—NR}_2}{\overset{O}{||}} + H—OH \xrightarrow{OH^-} \underset{\text{OH}}{\overset{\overset{\text{OH}}{|}}{\underset{|}{\text{R—C—NR}_2}}} \xrightarrow{\text{elimination}} \underset{\text{R—C—OH}}{\overset{O}{||}} + H—NR_2$$
$$\text{Initially neutral product}$$

Elimination requires proton transfer to the leaving nitrogen; R_2N^- is too strongly basic to leave as an anion. Under strong base conditions, proton transfer occurs through a weak acid such as water:

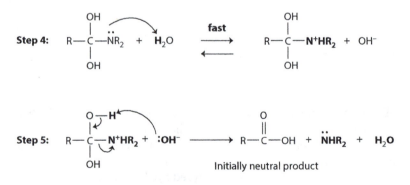

Hydrolysis of an amide also proceeds against the thermodynamically favored direction. Strong base conditions produce the same equilibrium shift toward formation of the carboxylic acid product by deactivating the reverse direction through the carboxylate anion.

$$\underset{\text{R—C—OH}}{\overset{O}{||}} + OH^- \longrightarrow \underset{\text{R—C—O}^-}{\overset{O}{||}} \text{(unreactive)} + H_2O$$

18.3.11 GROUP [5] EXAMPLES

Subject to the thermodynamic limitations, carboxylic acid derivatives are interchanged through a Group [5] reaction. To find the net final product (for acceptable combinations), directly replace the **Y**-group with the **Nu**-group and retain the original $C=O$; the **Y**-group is removed as various acid/base forms of **HY**.

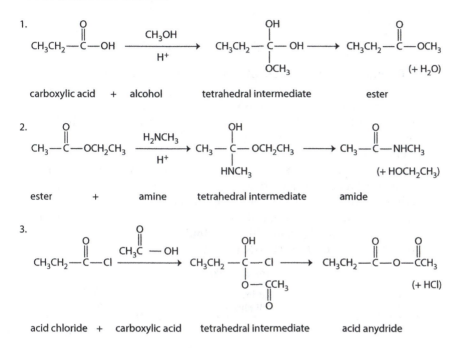

1.

carboxylic acid   +   alcohol   tetrahedral intermediate   ester

2.

ester   +   amine   tetrahedral intermediate   amide

3.

acid chloride   +   carboxylic acid   tetrahedral intermediate   acid anydride

18.3.12 GROUP [6]

This group is a sequential combination of a Group [5] type reaction with however a carbon or hydrogen-based nucleophile, followed a Group [1] type reaction. The initial terahedral intermediate shows the central carbon of the intermediate bonded to an oxygen as hydroxyl (—OH), an oxygen as a (—OCR₃) functional group from the initial **ester** (—Cl if from an acid chloride), and **two** carbons or hydrogens.

$$[\textbf{C or H}]-\underset{\underset{[\textbf{C or H}]}{|}}{\overset{\overset{OH}{|}}{C}}-[\textbf{O, Cl}]$$

This tetrahedral intermediate bonding combination is also found for Group [4]. In this case, the identification of Group [6] must come from the recognition of the starting reactants: the sequential combination of reactions is specific for the **double** addition of a carbon- or hydrogen-based nucleophile to an ester or, less commonly, an acid chloride.

18.3.12.1 GENERAL REACTANT REQUIREMENTS

Carbonyl substrate: **Y = OCR₃** (or **Cl**); the starting carbonyl compound must be an ester or acid chloride.

Nucleophilic atom: must be —**C** (carbon) or —**H** (hydrogen).

Common nucleophiles as E—Nu combinations (E⁺ usually = **metal**):

$$\overset{\delta+ \;\; \delta-}{-C} \text{(carbon): } \overset{\delta+ \;\; \delta-}{Li-CR_3} \; ; \; BrMg-CR_3$$

$$-H \text{(hydrogen): } Li^+ \; R_3Al \overset{\delta-}{-H}$$

Reactant requirement explanation: The central carbon of the tetrahedral intermediate formed in the **first addition** step has bonded **oxygens** from **C=O** and the **OCR₃** of the Y-group. The two bonded [**C** or **H**] atoms are derived from the **R**-group and the **Nu**-group (carbon- or hydrogen-based nucleophile). The resulting intermediate, although appearing similar to Group [**4**], behaves differently due to the high reactivity of the nucleophile; the final product results through elimination and a **second** addition.

General reaction sequence results: A Group [6] reaction involves an addition/elimination/second addition sequence identical to a Group [5] reaction followed by a Group [1] reaction. The final product results from elimination of the Y-group with a net **double** addition of the nucleophile group (—**Nu:**): (**R = C** or **H**; —**Nu** is —**C** or —**H**)

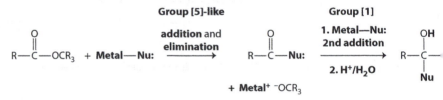

Description of the reaction sequence: The first step is nucleophilic addition to carbonyl; the mechanism is identical to the one-step Group [1] reaction (**R = C** or **H**; —**Nu** is —**C** or —**H**)

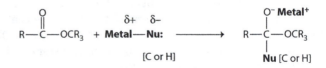

The **elimination** reaction involves loss of the Y-group and reformation of the **C=O** identical to a **Group [5]** reaction. Since the nucleophile is carbon or hydrogen based, the product of this first sequence is an aldehyde or ketone (**R = C** or **H**; —**Nu** is —**C** or —**H**)

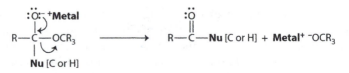

The aldehyde/ketone product formed in this addition/elimination is a good substrate for the **second** addition of another molecule of the nucleophile; the reaction is identical to Group [1]. The final product is a normal alcohol after the metal is replaced by —**H** through a final hydrolysis.

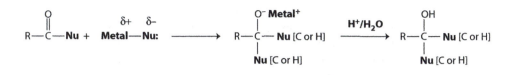

18.3.13 GROUP [6] EXAMPLE

Group [6] is distinguished by noting that the alcohol formed always shows two identical carbon or hydrogen nucleophile groups attached to the same central carbon of the final product. An important extension of the first Group [5]-like reaction with a carbon nucleophile is the Claisen condensation of esters. In this case the nucleophilic carbon is the sp³ carbon (α-carbon) adjacent to a carbonyl in esters.

To find the net reaction product for the complete [Group 6] sequence, directly exchange the **Y**-group with the **Nu**-group. Then convert the resulting ketone or aldehyde **c**=**o** double bond into a single bond C—OH and connect through a single bond the **C** or **H** nucleophile to the central carbon of the original **c**=**o**.

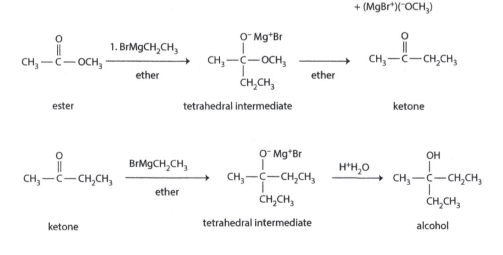

18.4 PROCESS TO SOLVE NUCLEOPHILIC ADDITION/ELIMINATION REACTIONS AT CARBONYL ADDITIONAL PRACTICE EXAMPLES FOR GROUPS [1] THROUGH [6]

The examples use the following steps to determine the correct products for nucleophilic addition and addition/elimination reactions at carbonyl.

1. Identify the carbonyl group in the organic substrate; identify the ENu combination.

2. Match the starting carbonyl substrate and nucleophile type to the descriptions for each Group listed in Section 18.3.
3. Draw the structure of the tetrahedral intermediate formed by the nucleophilic addition reaction.
4. Match the structure of the intermediate to the description of bonded atoms to the central carbon shown in Section 18.3.
5. Use steps (1) through (4) to determine the reacting group classification.
6. Solve for the correct product by following the description of specific reaction results and determination of final net product for the specific classification group selected.

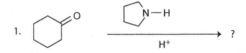

1.

Reactants are a ketone and a secondary amine; the tetrahedral intermediate is shown below: select **Group [2B]**; the product is an enamine.

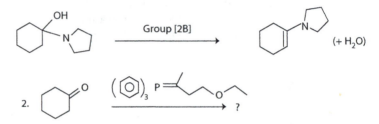

2.

Reactants are a ketone and a triphenylphosphonium ylide; the tetrahedral intermediate is shown below: select **Group [3]**; the product is a derived alkene.

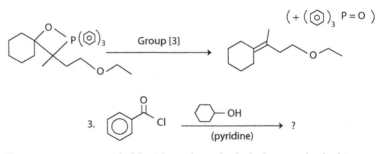

3.

Reactants are an acid chloride and an alcohol; the tetrahedral intermediate is shown below: select **Group [5]**; the product is an ester. (Pyridine is a weak base that neutralizes the HCl by-product.)

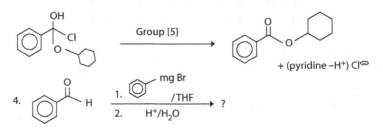

4.

Reactants are an aldehyde and a carbon nucleophile; the tetrahedral intermediate is shown below: select **Group [1]**; the product is an alcohol. Hydrolysis converts the alcohol complex into the neutral alcohol.

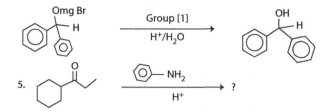

5.
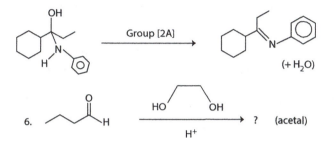

Reactants are a ketone and a primary amine; the tetrahedral intermediate is shown below: select **Group [2A]**; the product is an imine.

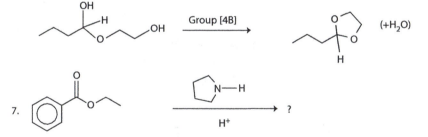

6.

Reactants are an aldehyde and a diol; the tetrahedral intermediate is shown below: select **Group [4B]**; the product is a cyclic acetal. (Cyclization favors the full acetal.)

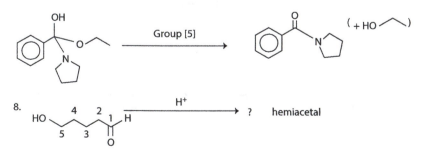

7.

Reactants are an ester and an amine; the tetrahedral intermediate is shown below: select **Group [5]**; the product is an amide.

8.

Reactant is an aldehyde and an alcohol in the same molecule; the tetrahedral intermediate is shown below: select **Group [4A]**. The final product is the tetrahedral intermediate: a cyclic hemiacetal can be stable under certain conditions.

18.5 ADDITIONAL PRACTICE PROBLEMS

Use the techniques described to determine the correct product for each of the following reactions.

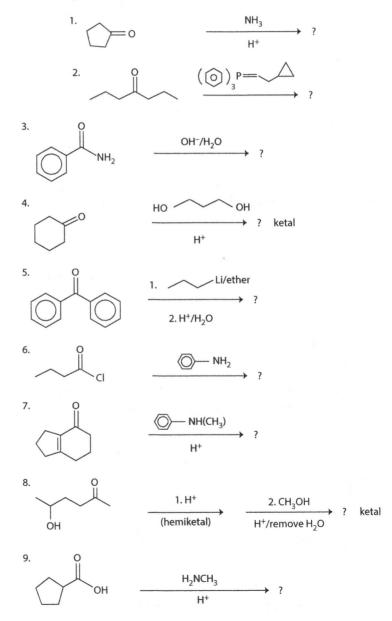

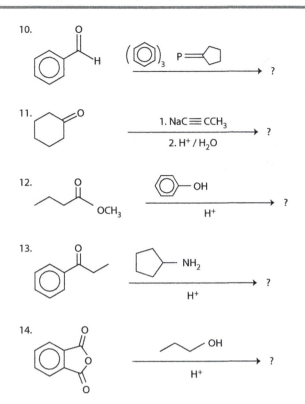

10.

11.

$$\frac{1.\ NaC \equiv CCH_3}{2.\ H^+ / H_2O} \rightarrow\ ?$$

12.

$$\frac{}{H^+} \rightarrow\ ?$$

13.

$$\frac{}{H^+} \rightarrow\ ?$$

14.

$$\frac{}{H^+} \rightarrow\ ?$$

ADDITIONAL PRACTICE PROBLEMS: ANSWERS

The matching practice problem reactants are shown on the previous pages. The answers are shown as first the resulting tetrahedral intermediate to the left of the arrow then the final product to the right of the arrow. The correct group is identified.

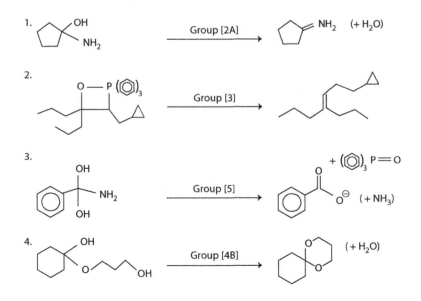

1.

Group [2A]

2.

Group [3]

3.

Group [5]

4.

Group [4B]

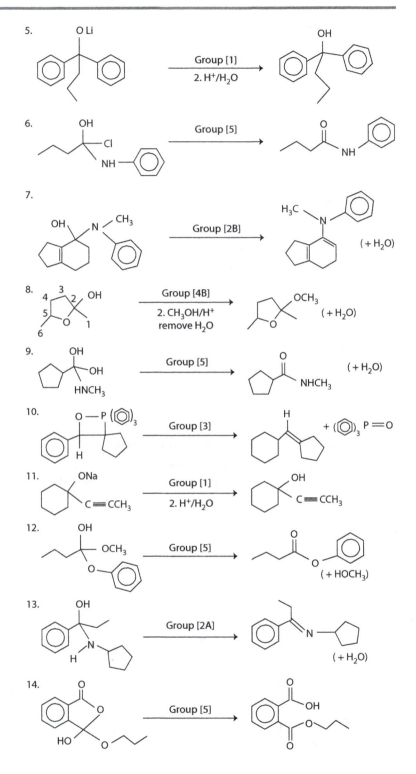

5.

Group [1]
2. H⁺/H₂O

6.

Group [5]

7.

Group [2B]

(+ H₂O)

8.

Group [4B]
2. CH₃OH/H⁺
remove H₂O

(+ H₂O)

9.

Group [5]

(+ H₂O)

10.

Group [3]

11.

Group [1]
2. H⁺/H₂O

12.

Group [5]

(+ HOCH₃)

13.

Group [2A]

(+ H₂O)

14.

Group [5]

A Brief Guideline for Applying Fundamental Concepts in NMR Spectroscopy

19

As a first approach to the topic, the theory and practical techniques of proton NMR (not covered here) can identify four major concepts that are fundamental for generalized spectroscopy problem solving. Practice at applying these four principles can then provide a basis for more comprehensive analysis.

As an initial simplification, application of the four concepts will be considered to operate under "**ideal**" conditions. This refers to conditions that do not produce exceptions, for example, based on instrument limitations, or more complex proton environments and interactions. Also for this introductory approach, application of the principles will not consider circumstances requiring analysis through finer points. It should be noted that analysis exceptions through finer points often provide a great deal of information.

19.1 GENERAL CONCEPTS FOR ¹H NMR INTERPRETATION: NUMBER OF SIGNALS AND PEAK RATIOS

1. An NMR **signal** is a peak or peak combination (split peak).

Principle 1. Under ideal conditions, <u>each</u> <u>set</u> of chemically equivalent hydrogens (protons) will produce <u>one</u> signal in a 1H NMR.

Hydrogen chemical equivalence can be tested, for example, using the same technique used for photohalogenation of an alkane. Hydrogens can be considered chemically equivalent if, when individually reacted through a chemical reaction (e.g., free radical bromination), they produce the same constitutional isomer and stereoisomer. However, the stereoisomer effect of chiral carbons is a separate topic not covered here.

Exceptions and complications will occur due to coincidental overlap of signals (i.e., under nonideal conditions). The electron chemical environment of different sets of chemically equivalent hydrogens may be too similar to allow complete separation in the spectrum.

2. Peak size/signal size is the integrated area under a complete signal (peak or peak combination).

Principle 2. Under ideal conditions, signal _size_ (area under complete signal) is proportional to the _number_ of chemically equivalent hydrogens responsible for the specific signal.

1. **Examples** principles #1 and #2 can be applied to the following molecule to determine the number of signals expected under ideal conditions and the predicted signal integration (peak size) ratio.

(b)   (a)

NC — CH$_2$—O— CH$_3$

A common technique for depiction of the analysis is to label each set of chemically equivalent hydrogens (protons) with a letter (a), (b) as shown. Based on principle #1, all three hydrogens labeled (a) on the methyl group are chemically equivalent. If individually exchanged, for example, a bromine, each exchange would produce the same constitutional and stereoisomer. Repeating this analysis shows that the two methylene hydrogens-(b) are chemically equivalent. However, exchange of a methyl (a)-hydrogen for bromine would not produce the same constitutional isomer as exchange for bromine of a methylene hydrogen-(b). Therefore, set-(a) hydrogens is not chemically equivalent to set-(b) hydrogens. The two sets of chemically equivalent hydrogens would be expected to produce two signals in the proton-NMR.

Principle 2 predicts that the ratio of peak "size" would be proportional to the number of chemically equivalent hydrogens responsible for the specific peak.

The expected integrated area ratio would be: signal (a) = 3 to signal (b) = 2.

Principles #1 and #2 (ideal conditions) are applied to molecules 2 and 3 shown below:

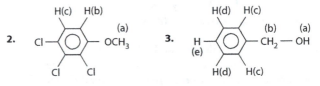

2. Molecule 2 would be expected to produce three signals in the proton-NMR for hydrogens labeled (a), (b), and (c). Note that

H-(b) and H-(c) are not chemically equivalent due to the different substitution position on the aromatic ring (e.g., ortho- vs. meta- as compared to the methoxy group).

The expected ratio would be: signal (a) = 3 to signal (b) = 1 to signal (c) = 1.

3. Molecule 3 would be expected to produce, **under ideal conditions**, five signals in the proton-NMR for the hydrogens labeled (a), (b), (c), (d), and (e).

Equivalency of the two ortho positions and the two meta positions produces the set of **two** H-(c) and **two** H-(d).

The expected ratio would be: signal (a) = 1 to signal (b) = 2 to signal (c) = 2 to signal (d) = 2 to signal (e) = 1.

This molecule, however, represents an exception to principle #1. The actual spectrum shows only three signals; one single peak represents all five aromatic hydrogens (c), (d), and (e). The chemical environments of the five aromatic hydrogens are so similar that the signals do not separate.

19.1.1 EXAMPLES DEMONSTRATING PRINCIPLES #1 AND #2

Use principles #1 and #2 to determine, under ideal conditions, the number of signals expected for the molecules shown, then predict the signal integration (peak size) ratio.

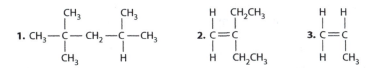

19.1.1.1 EXAMPLE: ANALYSIS FOR MOLECULES 1, 2, AND 3

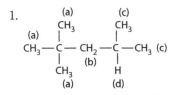

Principle #1: All three methyl groups of the tert-butyl group are equivalent; the nine hydrogens labeled (a) are chemically equivalent. Both methyl groups of the end isopropyl group are equivalent; the six hydrogens labeled (c) are chemically equivalent. The two hydrogens-(b) are chemically equivalent; the hydrogen-(d) is unique. The four sets of chemically equivalent hydrogens, H-(a), H-(b), H-(c), and H-(d) would be expected to produce four signals in the proton-NMR.

Principle #2: The expected integrated area ratio would be:
signal (a) = 9 to signal (b) = 2 to signal (c) = 6 to signal (d) = 1.

2.

Principle #1: The alkene is completely symmetrical: each alkene carbon has identical groups attached. The two alkene hydrogens-(c) are chemically equivalent: individual exchange of each for a bromine would yield the same constitutional and stereoisomer. The six methyl H-(a) hydrogens are equivalent as are the four methylene H-(b) hydrogens. The three sets of chemically equivalent hydrogens, H-(a), H-(b), and H-(c) would be expected to produce three signals in the proton-NMR.

Principle #2: The expected integrated area ratio would be:
signal (a) = 6 to signal (b) = 4 to signal (c) = 2 (or 3/2/1).

3.
```
   (b)    (a)
    H     CH₃
    |     |
    C === C      (group orientations show H-(b) cis- to the methyl group)
    |     |
    H     H
   (c)    (d)
```

Principle #1: This alkene has no stereochemistry since one of the alkene carbons is symmetrical, having identical —**H** atoms attached. However, H-(b) and H-(c) are **not** chemically equivalent: individual exchange of each for a bromine would yield the same constitutional but **not** the same stereoisomer: H-(b) or its exchanged bromine would be cis- to the methyl group; H-(c) or its exchanged bromine would be trans- to the methyl group. The six methyl H-(a) hydrogens are equivalent; the H-(d) hydrogen is unique. The four sets of chemically equivalent hydrogens, H-(a), H-(b), H-(c), and H-(d) would be expected, under ideal conditions, to produce four signals in the proton-NMR. The actual separation of the alkene hydrogens would depend on the instrument.

Principle #2: The expected integrated area ratio would be
signal (a) = 6 to signal (b) = 1 to signal (c) = 1 to signal (d) = 1.

19.2 GENERAL CONCEPTS FOR ¹H NMR INTERPRETATION: SIGNAL POSITION AND PROTON COUPLING

3. The relative position of a signal (peak combination) in a ¹H NMR spectrum is measured on a scale based on parts per million (δ ppm) in a general range of δ ppm **0 → 13**.

Principle #3. Relative position of a signal depends upon the electron chemical environment around the hydrogens generating the ¹H NMR signal. A very general relationship shows that the signal shifts downfield and thus δ ppm value _increases_ with:

1. **Decreasing** electron density around the signal hydrogens which can be caused by electronegative (electron withdrawing) atoms

TABLE 19.1   Table of Approximate Positions for ¹H NMR Hydrogens

Hydrogen Type and Environment		Chemical Shift / δ ppm
R—C—H (with R above and R below)	alkane C—H; R = C or H (small differences for primary, secondary, tertiary)	< 1–2
O—H	alcohol hydroxyl proton (variable)	1–5
(or O) C=C—C—H	sp³ C—H connected to an sp² carbon of an alkene or cabonyl	1.5–2.5
X—C—H	sp³ C—H connected to X with X = N or an sp C (cyano or alkyne carbon) or an sp² aromatic carbon	2–3
Y—C—H	sp³ C—H connected to Y with Y = a halogen or oxygen	3–4
C=C—H	sp² C—H in an alkene	4.5–6.5
Aromatic—H	sp² C—H of an aromatic ring	6.5–8.5
O=C—H	sp² C—H in an aldehyde	9–10
O=C—O—H	O—H in a carboxylic acid	10–13

or groups attached to the carbon bearing the hydrogen producing the signal. A lesser but measurable effect can be detected if the electron withdrawing group is attached to a neighboring carbon.

2. **Proximity** of certain **pi**-electron system configurations in alkenes and aromatics. (Alkyne pi-systems generate an opposite influence.)
3. A very general table of approximately measured positions for selected hydrogens in certain chemical environments is shown in Table 19.1.

A very rough but useful estimating technique for common **C–H** protons can be used for predicting the approximate δ ppm in the absence of a more descriptive table. This is described for the final examples and practice problems in which a complete ¹H NMR is predicted.

19.3 TABLE OF APPROXIMATE POSITIONS FOR ¹H NMR HYDROGENS

The notation indicates the specific —**H** responsible for the signal bonded to its carbon or other atom. Also shown are the other atoms or groups bonded to the connecting carbon (or atom) that influence the electron chemical environment of the −**H** providing the signal.

Neighboring carbons or atoms are defined as carbons or atoms directly bonded to the carbon with the hydrogen responsible for the observed signal. Neighboring protons are then defined as any protons bonded to neighboring atoms.

4. Proton–proton coupling (¹H coupling) viewed as signal (peak) splitting records the interaction of neighboring protons on the signal protons due to aligned or opposed magnetic fields.

Principle #4. For ¹H NMR, nonequivalent protons (hydrogens) on neighboring atoms couple with the protons of the observed signal and split the observed signal into multiple peaks that have a predicted multiplicity and ratio. Coupling is possible for all observed signals in turn.

1. Under ideal conditions, a coupled signal will be split into a specific number of peaks based on the number of neighboring nonequivalent protons.
2. Under ideal conditions, the split peaks of a signal have a specific size ratio pattern dependent on the number of peaks in the split signal.

In general, only protons (hydrogens) on direct neighboring (directly bonded) atoms will couple with each other in a detectable fashion. Coupling with hydrogens farther away can usually not be detected.

Only nonequivalent protons (nonchemically equivalent hydrogens) couple with each other; chemically equivalent protons do not couple and split each other's signals.

The number of peaks in a split signal for a specific signal H(a) = [(# of nonequivalent neighboring protons: H(b), H(c), etc.) + 1)]

Note that the number of peaks in a specific signal does not depend on the number of H(a) protons represented by the signal.

Under ideal conditions, the split peaks of a signal have a specific size ratio pattern dependent on the number of peaks in the split signal (statistically produced).

Multiplicity	Ratio of Peaks
Singlet: one peak/no split	
Doublet: two peaks	Ratio 1:1
Triplet: three peaks	Ratio 1:2:1
Quartet: four peaks	Ratio 1:3:3:1
Quintet: five peaks	Ratio 1:4:6:4:1
Sextet: six peaks	Ratio 1:5:10:10:5:1
Septet: seven peaks	Ratio 1:6:15:20:15:6:1

19.4 EXAMPLES: PREDICTING THE ¹H NMR FOR MOLECULES USING PRINCIPLES #1 THROUGH 4

A very useful exercise is the prediction of the ¹H NMR based on the structure of a specific molecule. Practice at applying principles #1, #2, # 3, and #4 for this type of exercise provides a foundation for more comprehensive problems such as spectrum analysis and molecular structure determination.

Applying the four principles under ideal conditions to a known structure allows the prediction of the following information:

1. The number of signals in the spectrum
2. The peak size (integrated area) signal ratios
3. The approximate position for each signal as a δ ppm value. In the absence of a detailed table of chemical shifts the "Chemical Shift Estimation Method for Common C−H Protons" discussed below can be used; and
4. The multiplicity (singlet, doublet, triplet, etc.) and split peak ratio pattern for each signal

19.4.1 CHEMICAL SHIFT ESTIMATION METHOD FOR COMMON C−H PROTONS

Step 1: Identify the sp³ C−H protons. Start by estimating most simple alkane C−H protons in the range: δ ppm of 1–1.5. This applies to no electronegative atoms or groups directly attached to the carbon bonding the signal hydrogen.

Step 2: For any applicable identified sp³ C−H protons, very roughly add 0.5 δ ppm to the estimated 1–1.5 for an sp² carbon attached to the carbon bonding the signal hydrogen.

Step 3: For any applicable identified sp³ C−H protons, very roughly add 1.0 δ ppm to the estimated 1–1.5 for each nitrogen, aromatic sp² carbon or sp carbon attached to the carbon bonding the signal hydrogen.

Step 4: For any applicable identified sp³ C−H protons, very roughly add 2.0 δ ppm to the estimated 1–1.5 for each oxygen or halogen attached to the carbon bonding the signal hydrogen.

Step 5: For all the identified sp³ C−H protons, electronegative groups bonded to a carbon neighboring the carbon bonding the signal hydrogen will have some downfield effect on the estimation.

Step 6: For any identified sp² C−H protons, estimate 4.5–6.5 δ ppm for alkene protons, 6.5–8.5 δ ppm for aromatic protons, and 9–10 δ ppm for an aldehyde proton.

19.5 EXAMPLES: PREDICTING THE ¹H NMR FROM MOLECULAR STRUCTURE

1. Use ideal conditions to predict the ¹H NMR for the following molecule. The NMR information that can be determined in this

exercise is specified as (1) to (4) based on the four ¹H NMR application principles. Use the "Chemical Shift Estimation Method for Common C−H protons" described previously to determine the approximate δ ppm for the signals:

(b)   (a)

$Cl — CH_2 — O — CH_3$

1. H-(a) and H-(b) represent two sets of chemically equivalent hydrogens. Two signals in the spectrum are predicted.
2. The expected integrated area ratio is: signal (a) = 3 to signal (b) = 2.
3. The sp³C−H(a) hydrogens are attached to a carbon bonded directly to an oxygen. The sp³C−H(b) hydrogens are attached to a carbon bonded directly to an oxygen and a halogen. The chemical shift estimation method assigns to pure alkane hydrogens values of 1–1.5 δ ppm. Step 4 of the estimation method states:

"For any applicable identified sp³C−H protons, very roughly add 2.0 δ ppm to the estimated 1–1.5 for each oxygen or halogen attached to the carbon bonding the signal hydrogen." Based on this method:

The H(a) proton signal is predicted to be (1–1.5) + (2) = 3–3.5 δ ppm.

The H(b) proton signal is predicted to be (1–1.5) + (2) + (2) = 5–5.5 δ ppm.

4. The sp³C−H(a) carbon is connected only to the oxygen. The oxygen does not have a bonded hydrogen; thus there are no neighboring hydrogens to the H(a) signal. The H(a) signal will be a singlet.

The sp³C−H(b) carbon is connected to the chlorine and the oxygen; neither of these atoms has a bonded hydrogen. There are no neighboring hydrogens to the H(b) signal; this signal will be a singlet.

The actual spectrum shows a 3-proton singlet at 3.5 δ ppm matching the predicted H(a) signal and a 2-proton singlet at 5.5 δ ppm matching the predicted H(b) signal.

2. Predict (under ideal conditions) the ¹H NMR information (1) to (4) for the following molecule:

(b)   (a)

$Br — CH_2 — CH_3$

1. H-(a) and H-(b) represent two sets of chemically equivalent hydrogens. Two signals in the spectrum are predicted.
2. The expected integrated area ratio is: signal (a) = 3 to signal (b) = 2.
3. The sp³C−H(a) hydrogens are on a carbon bonded directly to another carbon. The chemical shift estimation method assigns to pure alkane hydrogens values of 1–1.5 δ ppm. The H(a) signal should fall nearly in this 1–1.5 δ ppm alkane range with some downfield shift due to the bromine on the

neighboring carbon as specified by step 5. The H(b) hydrogens are attached to a carbon bonded directly to a halogen. Step 4 of the estimation method predicts the H(b) proton signal will be $(1–1.5) + (2) = 3–3.5$ δ ppm.

4. The $sp^3C–H(a)$ carbon has one neighboring carbon with two protons; these neighboring protons are not chemically equivalent to the H(a) hydrogens. The number of peaks in the split signal for the H(a) hydrogens= [(# of nonequivalent neighboring protons + 1)] = [(2) + 1] = 3 The H(a) signal is predicted to be a triplet. The $sp^3C–H(b)$ carbon has one neighboring carbon with three protons that are nonequivalent to the H(b) hydrogens; the attached bromine has no protons. The number of peaks in the split signal for the H(b) hydrogens = [(# of nonequivalent neighboring protons + 1)] = [(3) + 1] = 4 The H(b) signal is predicted to be a quartet.

Note again that the number of peaks in the H(a) or H(b) signals depends on the number of neighboring nonequivalent protons, not on the number of hydrogens responsible for the signal.

The actual spectrum shows a 3-proton triplet at 1.7 δ ppm matching the predicted H(a) signal and a 2-proton quartet at 3.4 δ ppm matching the predicted H(b) signal. The triplet and quartet show split peak ratios close to those shown in the table.

3. Predict (under ideal conditions) the ^1H NMR information (1) to (4) for the following molecule:

$$\begin{array}{c} \text{(b)} \\ \text{H} \\ \text{(a)} \quad | \quad \text{(a)} \\ CH_3 — C — CH_3 \\ | \\ Cl \end{array}$$

1. H-(a) and H-(b) represent two sets of chemically equivalent hydrogens. Due to the symmetry of the molecule, both methyl groups with their hydrogens are chemically equivalent. Two signals in the spectrum are predicted.

2. The expected integrated area ratio is: signal (a) = 6 to signal (b) = 1.

3. The $sp^3C–H(a)$ hydrogens are on carbons bonded directly to another carbon. The H(a) signal should fall nearly in the 1–1.5 δ ppm alkane range with some downfield shift due to the chlorine on the neighboring carbon (step 5).

    The H(b) hydrogen is bonded to a carbon directly connected to a halogen. Step 4 of the estimation method predicts the H(b) proton signal will be $(1–1.5) + (2) = 3–3.5$ δ ppm.

4. The $sp^3C–H(b)$ carbons have one neighboring carbon with one nonequivalent proton. The number of peaks in the split signal for the H(a) hydrogens = [(# of nonequivalent neighboring protons + 1)] = [(1) + 1] = 2

The H(a) signal is predicted to be a doublet. The $sp^3C-H(b)$ carbon has two neighboring carbons, each with three non-equivalent protons for a total of six neighboring nonequivalent protons. The number of peaks in the split signal for the H(b) hydrogens = [(# of nonequivalent neighboring protons + 1)] = [(6) + 1] = 7

The H(b) signal is predicted to be a septet.

The actual spectrum shows a 6-proton doublet at 1.5 δ ppm matching the predicted H(a) signal and a 1-proton septet at 4.2 δ ppm approximately matching the predicted H(b) signal; the chemical shift of the H(b) proton is farther downfield than predicted. The doublet and septet show split peak ratios close to those shown in the table.

4. Predict (under ideal conditions) the ¹H NMR information (1) to (4) for the following molecule:

$$
\begin{array}{cccc}
 & O & & \\
(c) & \| & (b) & (a) \\
CH_3-&C-O-&CH_2-&CH_3
\end{array}
$$

1. H-(a), H-(b), and H-(c) represent three sets of chemically equivalent hydrogens. Three signals in the spectrum are predicted.
2. The expected integrated area ratio is: signal (a) = 3 to signal (b) = 2 to signal (e) = 3.
3. The chemical shift estimation method assigns to pure alkane hydrogens values of 1–1.5 δ ppm. The $sp^3C-H(a)$ hydrogens are on a carbon bonded directly to another carbon. The H(a) signal should fall in this 1–1.5 δ ppm alkane range.

   The H(b) hydrogens are attached to a carbon bonded to another alkane carbon and to an oxygen.

   Step 4 of the estimation method predicts the H(b) proton signal to be (1–1.5) + (2) = 3–3.5 δ ppm.

   The H(c) hydrogens are attached to a carbon bonded to a carbonyl sp^2 carbon.

   Step 2 of the estimation method states: "add 0.5 δ ppm to the estimated 1–1.5 for an sp^2 carbon attached to the carbon bonding the signal hydrogen."

   The H(c) proton signal is predicted to be (1–1.5) + (0.5) = 1.5 –2.0 δ ppm.

4. The $sp^3C-H(a)$ carbons have one neighboring carbon with two nonequivalent protons. The number of peaks in the split signal for the H(a) hydrogens = [(# of nonequivalent neighboring protons + 1)] = [(2) + 1] = 3

   The H(a) signal is predicted to be a triplet.

The $sp^3C-H(b)$ carbon has one neighboring carbon with three nonequivalent protons; the connected oxygen has no bonded hydrogens. The number of peaks in the split signal for the H(b) hydrogens = [(# of nonequivalent neighboring protons + 1)] = [(3) + 1] = 4

The H(b) signal is predicted to be a quartet.

The sp³C–H(c) carbon has one neighboring carbon with no hydrogens attached; there are no neighboring hydrogens to the H(c) signal.

The H(c) signal is predicted to be a singlet.

The actual spectrum shows:

A 3-proton triplet at 1.2 δ ppm matching the predicted H(a) signal

A 2-proton quartet at 4.0 δ ppm approximately matching the predicted H(b) signal; the chemical shift of the H(b) proton is farther downfield than predicted (see step 5); and

A 3-proton triplet at 2.0 δ ppm matching the predicted H(c) signal. The triplet and quartet show split peak ratios close to those shown in the table.

19.6 ADDITIONAL PRACTICE PROBLEMS

Predict the ¹H NMR for the following molecules; identify predicted information based on the four ¹H NMR application principles described. Use the "Chemical Shift Estimation Method for Common C–H Protons" to determine the approximate δ ppm for the signals. A table of values can be used as confirmation.

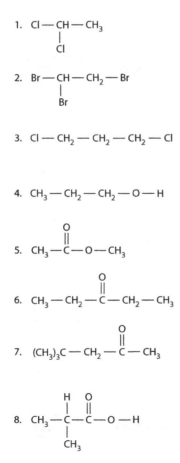

1. $Cl-CH-CH_3$
     $|$
     Cl

2. $Br-CH-CH_2-Br$
     $|$
     Br

3. $Cl-CH_2-CH_2-CH_2-Cl$

4. $CH_3-CH_2-CH_2-O-H$

5. $CH_3-\overset{\overset{\displaystyle O}{\|}}{C}-O-CH_3$

6. $CH_3-CH_2-\overset{\overset{\displaystyle O}{\|}}{C}-CH_2-CH_3$

7. $(CH_3)_3C-CH_2-\overset{\overset{\displaystyle O}{\|}}{C}-CH_3$

8. $CH_3-\overset{\overset{\displaystyle H}{|}}{\underset{\underset{\displaystyle CH_3}{|}}{C}}-\overset{\overset{\displaystyle O}{\|}}{C}-O-H$

19.7 ADDITIONAL PRACTICE PROBLEMS: ANSWERS

$$\text{1.} \quad \underset{(b)}{Cl}-\underset{}{\underset{|}{CH}}-\underset{(a)}{CH_3}$$
$$\underset{Cl}{|}$$

Predicted spectrum:

Signal H(x)	Signal Ratio	δ ppm	Multiplicity
H(a)	3	>1.5	Doublet
H(b)	1	5.0–5.5	Quartet

Actual spectrum:

Signal H(x)	Signal Ratio	δ ppm	Multiplicity
H(a)	3	2.0	Doublet
H(b)	1	5.8	Quartet

$$\text{2.} \quad Br-\underset{(b)}{\underset{|}{CH}}-\underset{(a)}{CH_2}-Br$$
$$\underset{Br}{|}$$

Predicted spectrum:

Signal H(x)	Signal Ratio	δ ppm	Multiplicity
H(a)	2	>3.5	Doublet
H(b)	1	>5.5	Triplet

Actual spectrum:

Signal H(x)	Signal Ratio	δ ppm	Multiplicity
H(a)	2	4.1	Doublet
H(b)	1	5.7	Triplet

$$\text{3.} \quad \underset{(a)}{Cl}-\underset{(a)}{CH_2}-\underset{(b)}{CH_2}-\underset{(a)}{CH_2}-Cl$$

Predicted spectrum:

Signal H(x)	Signal Ratio	δ ppm	Multiplicity
H(a)	4*	3.0–3.5	Triplet
H(b)	2*	> 1.5	Quintet
*ratio = 2/1			

Actual spectrum:

Signal H(x)	Signal Ratio	δ ppm	Multiplicity
H(a)	4	3.7	Triplet
H(b)	2	2.3	Quintet

$$
\begin{array}{cccc}
\text{(a)} & \text{(b)} & \text{(c)} & \text{(d)} \\
\end{array}
$$

4. $CH_3 - CH_2 - CH_2 - O - H$

Predicted spectrum:

Signal H(x)	Signal Ratio	δ ppm	Multiplicity
H(a)	3	1.0–1.5	Triplet
H(b)	2	>1.5	Sextet
H(c)	2	3.0–3.5	Quartet
H(d)	1	1–5*	Triplet
*(δ ppm: refer to table)			

Actual spectrum:

Signal H(x)	Signal Ratio	δ ppm	Multiplicity
H(a)	3	1.0	Triplet
H(b)	2	1.7	Sextet
H(c)	2	3.7	Multiplet: coupling with alcohol protons is variable
H(d)	1	3.3	Triplet

5. $CH_3 - \overset{\text{(b)}}{\underset{}{C}} \overset{O}{\underset{\|}{}} - O - \overset{\text{(a)}}{CH_3}$

Predicted spectrum:

Signal H(x)	Signal Ratio	δ ppm	Multiplicity
H(a)	3*	>3.0–3.5	Singlet
H(b)	3*	>2.0	Singlet
*ratio = 1/1			

Actual spectrum:

Signal H(x)	Signal Ratio	δ ppm	Multiplicity
H(a)	3	3.8	Singlet
H(b)	3	2.0	Singlet

6. $\overset{\text{(a)}}{CH_3} - \overset{\text{(b)}}{CH_2} - \overset{O}{\underset{\|}{C}} - \overset{\text{(b)}}{CH_2} - \overset{\text{(a)}}{CH_3}$

Predicted spectrum:

Signal H(x)	Signal Ratio	δ ppm	Multiplicity
H(a)	6*	1.0–1.5	Triplet
H(b)	4*	2.0	Quartet
*ratio = 3/2			

Actual spectrum:

Signal H(x)	Signal Ratio	δ ppm	Multiplicity
H(a)	6	1.0	Triplet
H(b)	4	2.4	Quartet

7.

$$\text{(a)} \quad \text{(b)} \quad \overset{\displaystyle \overset{O}{\|}}{\text{(c)}}$$
$$(CH_3)_3C \text{—} CH_2 \text{—} C \text{—} CH_3$$

Predicted spectrum:

Signal H(x)	Signal Ratio	δ ppm	Multiplicity
H(a)	9	1.0–1.5	Singlet
H(b)	2	2.0	Singlet
H(c)	3	2.0	Singlet

Actual spectrum:

Signal H(x)	Signal Ratio	δ ppm	Multiplicity
H(a)	9	1.0	Singlet
H(b)	2	2.3	Singlet
H(c)	3	2.1	Singlet

8.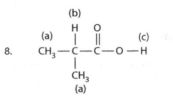

Predicted spectrum:

Signal H(x)	Signal Ratio	δ ppm	Multiplicity
H(a)	6	1.0–1.5	Doublet
H(b)	1	>2.0	Septet
H(c)	1	10–13*	Singlet

*(δ ppm: refer to table)

Actual spectrum:

Signal H(x)	Signal Ratio	δ ppm	Multiplicity
H(a)	6	1.2	Doublet
H(b)	1	2.7	Septet
H(c)	1	12.4	Singlet

Organic Practice Exams

All exams in this book were part of the exams given during organic chemistry semester course-work at Benedictine University. As practice exams, they can be used to test knowledge of the important concepts and the general depth of understanding required. To get the most out of these exercises, do not use them as a topic summary by reading the key first; solve the problems after being prepared for the particular material. After completion, do not go directly to the key. Use course materials as a reference to confirm (or deny!) all answers in order to reinforce the concepts and correct problem areas. At this point, the answers should be identical to the key; areas that still need further work will be identified.

Organic I: Practice Exam 1

1. For each part: (i) Show the complete Lewis structure for the molecule described *includ-ing lone pairs*; **partial credit for a valid structure that fits the bonding rules.** (ii) Draw the electron region (ER) geometry specifically and **only** for the atom stated; use condensed structures or a simple formula for the other atom groupings. (iii) State the name of the ER geometry. (iv) State the name of the atom geometry.

   a. (i) Draw **any** valid structure for C_6H_6BrNO which has an **alkyne** functional group and which contains a **nitrogen** which is a **central** atom with **sp² or sp³** hybridiza-tion. (ii–iv) Then, complete the geometry analysis only for the <u>nitrogen</u> central atom.

Lewis Structure	ER Drawing specifically
(One example shown)	around <u>the Nitrogen Atom</u>

   Name of the ER geometry around Nitrogen:
   Name of the Atom geometry around Nitrogen:

   b. (i) Draw **any** valid structure for $C_4H_2F_4O$ which has an **exactly one** sp-hybridized carbon and which contains an **oxygen** which is a **central** atom with **sp³** hybridiza-tion. (ii– iv) Then, complete the geometry analysis only for the **oxygen** central atom.

Lewis Structure	ER Drawing specifically
(One example shown)	around <u>the Oxygen Atom</u>

   Name of the ER geometry around Oxygen:
   Name of the Atom geometry around Oxygen:

2. Draw structural formulas for molecules which fit the descriptions below. **Any form of structural drawing is acceptable as long as sufficient information is present; *line drawings are preferred*. *Be certain that your answer fits the molecular formula given in the descriptions*. (One example for each is shown.)**

   a. Draw any isomer of $C_7H_9F_2O_2N$ which contains an **amide** functional group.

   b. Draw any isomer of $C_7H_9F_2O_2N$ which contains (i) an **ester** functional groups and (ii) contains at least one **ring**.

c. Draw any isomer of $C_7H_9F_2O_2N$ which contains one **aldehyde** and one **ketone** functional group.

d. Draw any isomer of $C_7H_9F_2O_2N$ which contains a **carboxylic acid** functional group.

3. Folic acid is an enzymatic cofactor required for proper metabolic function; a slightly modified form of its structure is shown below. Complete the questions concerning this molecule.

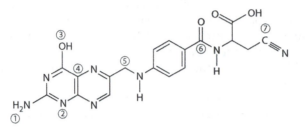

a. How many **total** pi-bonds are in this molecule? _____

b. Complete the following table by describing the hybridization and geometries of each of the specifically identified numbered atoms in this molecule. **Note that the line drawing does not indicate any specific geometries.**
   Hybridization: choose from sp, sp^2, or sp^3
   Geometries: tetrahedral, trigonal planar, linear, pyramidal, and bent.

Specific Atom	Hybridization	Electron Region Geometry	Atom Geometry
Nitrogen # 1			
Nitrogen # 2			
Oxygen # 3			
Carbon # 4			
Carbon # 5			
Carbon # 6			
Carbon # 7			

ORGANIC I: PRACTICE EXAM 0.5A: ANSWERS

1. For each part: (i) Show the complete Lewis structure for the molecule described *including lone pairs*; **partial credit for a valid structure that fits the bonding rules.** (ii) Draw the electron region (ER) geometry specifically and **only** for the atom stated; use condensed structures or a simple formula for the other atom groupings. (iii) State the name of the ER geometry. (iv) State the name of the atom geometry.

   a. (i) Draw **any** valid structure for C_6H_6BrNO which has an **alkyne** functional group and which contains a **nitrogen** which is a **central** atom with sp^2 or sp^3 hybridization. (ii–iv) Then, complete the geometry analysis only for the **nitrogen** central atom.

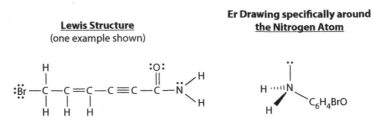

Lewis Structure
(one example shown)

**Er Drawing specifically around
the Nitrogen Atom**

Name of the ER geometry around Nitrogen: tetrahedral
Name of the Atom geometry around Nitrogen: pyramidal

b. (i) Draw **any** valid structure for $C_4H_2F_4O$ which has an **exactly one** sp-hybridized carbon and which contains an **oxygen** which is a **central** atom with sp^3 hybridization. (ii– iv) Then, complete the geometry analysis only for the **oxygen** central atom.

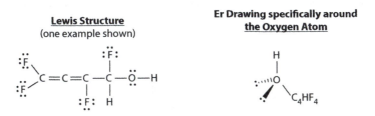

Lewis Structure
(one example shown)

**Er Drawing specifically around
the Oxygen Atom**

Name of the ER geometry around Oxygen: tetrahedral
Name of the Atom geometry around Oxygen: bent (angular)

2. Draw structural formulas for molecules which fit the descriptions below. **Any form of structural drawing is acceptable as long as sufficient information is present; *line drawings are preferred. Be certain that your answer fits the molecular formula given in the descriptions*. (One example for each is shown.)**

3 Rings/π-bonds

a. Draw any isomer of $C_7H_9F_2O_2N$ which contains an **amide** functional group.

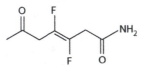

b. Draw any isomer of $C_7H_9F_2O_2N$ which contains (i) an **ester** functional groups and (ii) contains at least one **ring**.

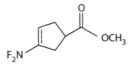

c. Draw any isomer of $C_7H_9F_2O_2N$ which contains one **aldehyde** and one **ketone** functional group.

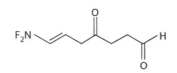

d. Draw any isomer of $C_7H_9F_2O_2N$ which contains a **carboxylic acid** functional group.

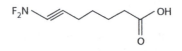

3. Folic acid is an enzymatic cofactor required for proper metabolic function; a slightly modified form of its structure is shown below. Complete the questions concerning this molecule.

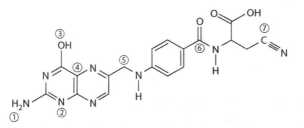

a. How many **total** pi-bonds are in this molecule? ___12___

b. Complete the following table by describing the hybridization and geometries of each of the specifically identified numbered atoms in this molecule. **Note that the line drawing does not indicate any specific geometries.**
   Hybridization: choose from sp, sp^2, or sp^3
   Geometries: tetrahedral, trigonal planar, linear, pyramidal, and bent.

Specific Atom	Hybridization	Electron Region Geometry	Atom Geometry
Nitrogen # 1	sp^3	Tetrahedral	(Trigonal) pyramidal
Nitrogen # 2	sp^2	Trigonal planar	Bent (angular)
Oxygen # 3	sp^3	Tetrahedral	Bent
Carbon # 4	sp^2	Trigonal planar	Trigonal planar
Carbon # 5	sp^3	Tetrahedral	Tetrahedral
Carbon # 6	sp^2	Trigonal planar	Trigonal planar
Carbon # 7	sp	Linear	Linear

Organic I: Practice Exam 2

1. For each part: (i) Show the complete Lewis structure for the molecule described *including lone pairs*; **partial credit for a valid structure that fits the bonding rules**. (ii) Draw the electron region (ER) geometry specifically and **only** for the atom stated; use condensed structures or a simple formula for the other atom groupings. (iii) State the name of the ER geometry. (iv) State the name of the atom geometry.

   a. (i) Draw **any** valid structure for $C_6H_4Cl_3NO$ which has an **alkyne** functional group and which contains a **nitrogen** which is a **central** atom with sp^2 or sp^3 hybridization. (ii–iv) Then, complete the geometry analysis only for the **nitrogen** central atom.

Lewis Structure	**ER Drawing specifically around the Nitrogen Atom**

   Name of the ER geometry around Nitrogen:
   Name of the Atom geometry around Nitrogen:

   b. (i) Draw **any** valid structure for $C_4H_4Cl_2O$ which has an **exactly one sp-hybridized carbon** and which contains an **oxygen** which is a **central** atom with sp^3 hybridization. (ii–iv) Then, complete the geometry analysis only for the **oxygen** central atom.

Lewis Structure	**ER Drawing specifically around the Oxygen Atom**

   Name of the ER geometry around Oxygen:
   Name of the Atom geometry around Oxygen:

2. Draw structural formulas for molecules which fit the descriptions below. **Any form of structural drawing is acceptable as long as sufficient information is present; *line drawings are preferred. Be certain that your answer fits the molecular formula given in the descriptions*.**

   a. Draw any isomer of $C_6H_9O_2N$ which contains an **amide** functional group.

   b. Draw any isomer of $C_6H_9O_2N$ which contains (i) an **ester** functional groups and (ii) contains at least one **ring**.

c. Draw any isomer of $C_6H_9O_2N$ which contains one **aldehyde** and one **ketone** functional group.

d. Draw any isomer of $C_6H_9O_2N$ which contains a **carboxylic acid** functional group.

3. Show the required structures for the two molecules described below; **partial credit for a valid structure that fits the bonding rules.**
    a. Draw **any** valid **complete Lewis** structure for $C_5H_5F_2O_2N$ which fits the following requirement: *only one carbon is sp³ hybridized with tetrahedral atom geometry.* Show covalent bonds with a line and nonbonding pairs of electrons by two dots. All row-2 atoms will be an octet and follow normal neutral bonding rules.

    b. Draw a specific **Lewis** structure **or condensed** structure for the neurotransmitter molecule **acetylcholine**: $(C_7H_{16}O_2N)^{+1}$. Use the following description: acetylcholine contains an **ester** functional group; the nitrogen has a formal charge of +1; and the molecule contains **four** methyl groups: three are bonded to nitrogen and **no** methyl groups are bonded to sp³ carbons.

    c. $C_4H_4F_4N_2O_2$ **All** carbons have **tetrahedral atom** geometry; **both** nitrogens have **bent (angular) atom** geometry.

ORGANIC I: PRACTICE EXAM 0.5B: ANSWERS

1. For each part: (i) Show the complete Lewis structure for the molecule described *including lone pairs;* **partial credit for a valid structure that fits the bonding rules.** (ii) Draw the electron region (ER) geometry specifically and **only** for the atom stated; use condensed structures or a simple formula for the other atom groupings. (iii) State the name of the ER geometry. (iv) State the name of the atom geometry.
    a. (i) Draw **any** valid structure for $C_6H_4Cl_3NO$ which has an **alkyne** functional group and which contains a **nitrogen** which is a **central** atom with sp² or sp³ hybridization. (ii–iv) Then, complete the geometry analysis only for the **nitrogen** central atom.

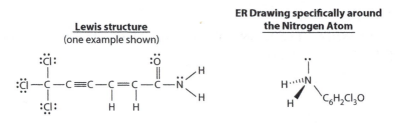

Name of the ER geometry around Nitrogen: tetrahedral
Name of the Atom geometry around Nitrogen: pyramidal

b. (i) Draw **any** valid structure for $C_4H_4Cl_2O$ which has an **exactly one sp-hybridized carbon** and which contains an **oxygen** which is a **central** atom with sp^3 hybridization. (ii–iv) Then, complete the geometry analysis only for the **oxygen** central atom.

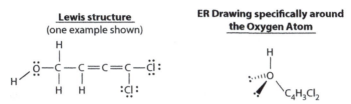

Lewis structure
(one example shown)

ER Drawing specifically around the Oxygen Atom

Name of the ER geometry around Oxygen: tetrahedral
Name of the Atom geometry around Oxygen: bent (angular)

2. Draw structural formulas for molecules which fit the descriptions below. **Any form of structural drawing is acceptable as long as sufficient information is present; *line drawings are preferred. Be certain that your answer fits the molecular formula given in the descriptions*. (One example for each is shown.)**

↗3 Rings/π-bonds

a. Draw any isomer of $C_6H_9O_2N$ which contains an **amide** functional group.

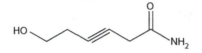

b. Draw any isomer of $C_6H_9O_2N$ which contains (i) an **ester** functional groups and (ii) contains at least one **ring**.

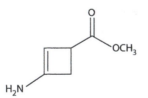

c. Draw any isomer of $C_6H_9O_2N$ which contains one **aldehyde** and one **ketone** functional group.

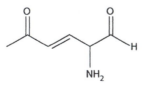

d. Draw any isomer of $C_6H_9O_2N$ which contains a **carboxylic acid** functional group.

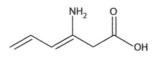

3. Show the required structures for the two molecules described below; **partial credit for a valid structure that fits the bonding rules.**

a. Draw **any** valid **complete Lewis** structure for $C_5H_5F_2O_2N$ which fits the following requirement: *only one carbon is sp³ hybridized with tetrahedral atom geometry.* Show covalent bonds with a line and nonbonding pairs of electrons by two dots. All row-2 atoms will be an octet and follow normal neutral bonding rules.

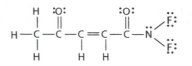

b. Draw a specific **Lewis** structure **or condensed** structure for the neurotransmitter molecule **acetylcholine:** $(C_7H_{16}O_2N)^{+1}$. Use the following description: acetylcholine contains an **ester** functional group; the nitrogen has a formal charge of +1; and the molecule contains **four** methyl groups: three are bonded to nitrogen and **no** methyl groups are bonded to sp³ carbons.

$$CH_3-\overset{\overset{O}{\|}}{C}-O-CH_2-CH_2-\overset{\overset{CH_3}{|}}{\underset{\underset{CH_3}{|}}{\overset{\oplus}{N}}}-CH_3$$

c. $C_4H_4F_4N_2O_2$ **All** carbons have **tetrahedral atom** geometry; **both** nitrogens have **bent (angular) atom** geometry.
Each N must have 3 Electron Regions; contains one double bond.
Each C must have 4 Electron Regions; all single bonds.
One (of many) possible answer:

$$\ddot{O}=\ddot{N}-\overset{\overset{H}{|}}{\underset{\underset{H}{|}}{C}}-\overset{\overset{H}{|}}{\underset{\underset{H}{|}}{C}}-\overset{\overset{:\ddot{F}:}{|}}{\underset{\underset{:\ddot{F}:}{|}}{C}}-\overset{\overset{:\ddot{F}:}{|}}{\underset{\underset{:\ddot{F}:}{|}}{C}}-\ddot{N}=\ddot{O}:$$

Organic I: Practice Exam 3

1. Draw structural formulas for constitutional isomers of $C_6H_{12}O_2$ based on the following descriptions. You need put down only **one** isomer for each part if more than one answer is possible. *Note that the statements do not necessarily give a complete description of all the functional groups in the molecule.*

   a. Draw any **non-branching** (straight chain) **acid**.

   b. Draw any **ester** which also contains a methyl substituent.

   c. Draw any isomer containing both an **aldehyde** and an **alcohol** functional group.

   d. Draw an isomer containing **two alcohol** functional groups, each oxygen being directly connected to a **primary** carbon.

   e. Draw an isomer containing both a **ketone** and an **ether** functional group.

   f. Draw any isomer containing a **cyclopentane** ring.

2. Answer the following questions concerning the four alkanes shown below; select the answers to each question from this molecules list, using the letters to indicate your selection. The alkane answers may be used more than once. *Be careful counting carbons.*

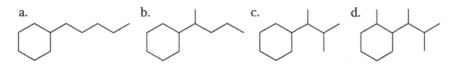

Predict which would have the highest heat of combustion.
Predict which would have the smallest heat of combustion.
Predict which **two** would have the highest boiling points.

Predict which would have the lowest boiling point.

Which compound can exist as a stereoisomeric pair? (diastereomeric pair)

3. Provide a correct **name** for the following compounds:

a.

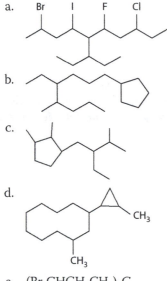

b.

c.

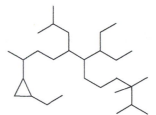

d.

CH₃

CH₃

e. $(Br_2CHCH_2CH_2)_4C$

f.

g.   Name the following compound:

h.   Draw the structure of 1-(2,2-dimethylcyclobutyl)-4-(2,2-diethylbutyl)-2-(2,2,3, 3-tetramethylpentyl)-cyclopentane.

i.   Draw the structure of tricyclo [4.4.1.1.] dodecane.

4. Draw structural formulas for the following compounds:

a.   4,5,7-triethyldecane

b.   3-ethyl-2,4-dimethyl-4-propylheptane

c.   1,4-dicyclohexylcyclooctane

d.   (2,2-dimethylpropyl)-cyclopentane

e.   3-ethyl-5-(1,1-dimethylpropyl)-nonane

f.   6-(1-ethyl-2-methylbutyl)-3,4,5-trimethyl-7-isopropyldecane

5. Consider the following alkane, drawn in perspective:

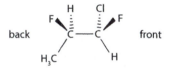

The effective size for groups follows the trend: **methyl > Cl > F > H.**

Draw **Newman diagrams** which describe the following conformations; use a larger circle for the back carbon and a smaller circle for the front carbon. **Hint:** It is easiest to leave the orientation of the back carbon fixed and rotate only the front carbon.
a.   Show the **highest energy** conformation of all possible conformations.
b.   Show the **lowest energy** (most stable) conformation of all possible conformations.
c.   Show the **highest energy staggered** conformation.
d.   Show either of the **two lowest energy eclipsed** conformations.

6. For **each** molecule shown below in a structural formula: draw the two possible stable **chair conformations;** then identify which of the two conformations is **more** stable based on substituent sizes. *Do this as a separate exercise for both molecules.* Note that the structural formulas give a specific cis/trans arrangement for the substituents in each molecule; be careful not to alter these.

a.

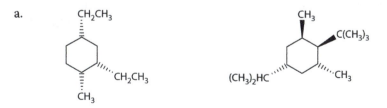

7. Sugar molecules in carbohydrate structures normally exist as a six-membered ring containing one oxygen atom; an example is shown below. Note that the electron pairs on oxygen (bonding and lone) are arranged in a tetrahedral fashion similar to carbon. Answer the following questions using the chair diagram as a guide:

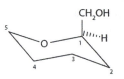

a. If a substituent is placed on carbon **#3** which is **cis** to the −CH$_2$OH group, would it be axial or equatorial?

b. Is an axial substituent on carbon **#4** cis or trans to the −CH$_2$OH group?

c. Identify the number of a ring-carbon which is **anti** to the **hydrogen** on carbon **#1**.

d. Identify the number of a ring-carbon which is **anti** to a lone pair on the **ring**-oxygen.

e. Certain biological sugars are connected through an −OH group attached to carbon **#5**. Show on the molecule drawing given above an −OH group on carbon **#5** which is **trans** to the −CH$_2$OH group, clearly indicating whether it is axial or equatorial.

f. Consider the following sugar molecule. The substituents on carbon #1 have been eliminated for simplicity:

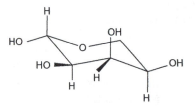

Using the chair template drawn below, show the proper orientation (axial or equatorial) of all the −OH groups **after the above molecule undergoes a ring flip.**

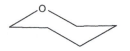

8. Determine the relationship between the two molecules in **each** of the pairs below, using the following descriptions:
identical; conformers; diastereomers; constitutional isomers; and **not isomers** (of each other).

a.

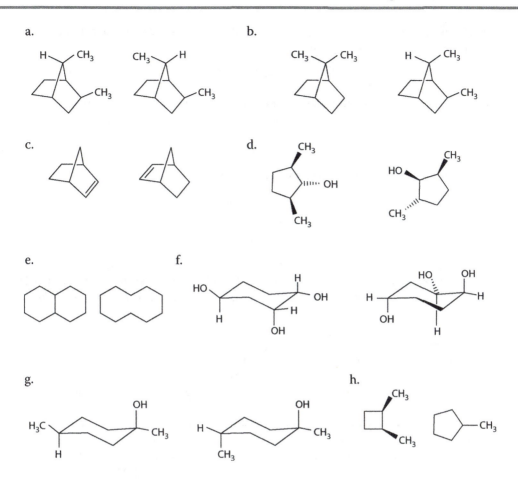

b.

c.

d.

e.

f.

g.

h.

ORGANIC I: PRACTICE EXAM 1A: ANSWERS

1. Draw structural formulas for constitutional isomers of $C_6H_{12}O_2$ based on the following descriptions. You need put down only **one** isomer for each part if more than one answer is possible. ***Note that the statements do not necessarily give a complete description of all the functional groups in the molecule. (One example of each shown.)***

   a. Draw any **non-branching** (straight chain) **acid**: $C_6H_{12}O_2$ has <u>one</u> ring or π-bond.

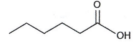

   b. Draw any **ester** which also contains a methyl substituent.

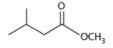

   c. Draw any isomer containing both an **aldehyde** and an **alcohol** functional group.

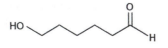

d. Draw an isomer containing **two alcohol** functional groups, each oxygen being directly connected to a **primary** carbon.

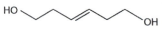

e. Draw an isomer containing both a **ketone** and an **ether** functional group.

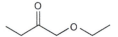

f. Draw any isomer containing a **cyclopentane** ring.

2. Answer the following questions concerning the four alkanes shown below; select the answers to each question from this molecules list, using the letters to indicate your selection. The alkane answers may be used more than once. *Be careful counting carbons.*

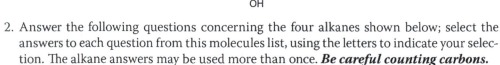

a.   b.   c.   d.

$(C_{11}H_{22})$ $(C_{11}H_{22})$ $(C_{11}H_{22})$ $(C_{12}H_{24})$

Predict which would have the highest heat of combustion.	d. Most carbons
Predict which would have the smallest heat of combustion.	c. Most highly branched = most stable of $C_{11}H_{22}$ molecules
Predict which **two** would have the highest boiling points.	d. Highest MM a. Least branched $C_{11}H_{22}$
Predict which would have the lowest boiling point.	c. Most branched $C_{11}H_{22}$ (diastereomeric pair)
Which compound can exist as a stereoisomeric pair?	d. Two ring substituents

3. Provide a correct **name** for the following compounds:

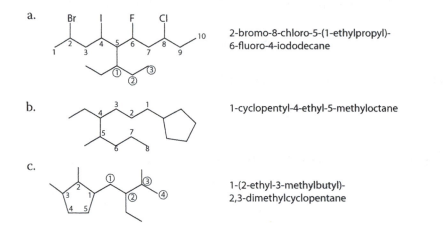

a.

2-bromo-8-chloro-5-(1–ethylpropyl)-6-fluoro-4-iododecane

b.

1-cyclopentyl-4-ethyl-5-methyloctane

c.

1-(2-ethyl-3-methylbutyl)-2,3-dimethylcyclopentane

d.

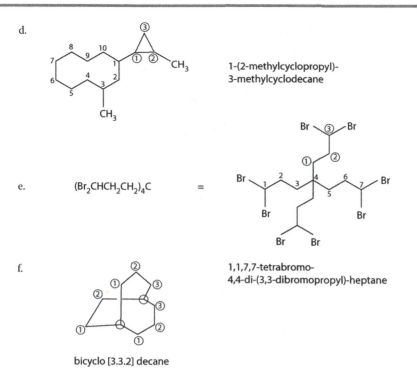

1-(2-methylcyclopropyl)-
3-methylcyclodecane

e.   (Br$_2$CHCH$_2$CH$_2$)$_4$C   =

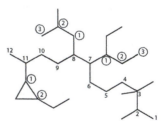

f.

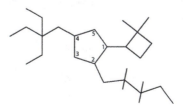

bicyclo [3.3.2] decane

1,1,7,7-tetrabromo-
4,4-di-(3,3-dibromopropyl)-heptane

g. Name the following compound:

11-(2-ethylcyclopropyl)-
7-(1-ethylpropyl)-
2,3,3-trimethyl-8-(2-methylpropyl)-dodecane

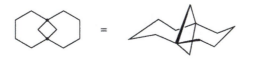

h. Draw the structure of 1-(2,2-dimethylcyclobutyl)-4-(2,2-diethylbutyl)-2-(2,2,3,
3-tetramethylpentyl)-cyclopentane.

i. Draw the structure of tricyclo [4.4.1.1] dodecane.

4. Draw structural formulas for the following compounds:
   a. 4,5,7-triethyldecane

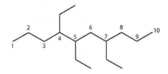

   b. 3-ethyl-2,4-dimethyl-4-propylheptane

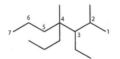

   c. 1,4-dicyclohexylcyclooctane

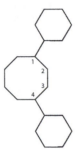

   d. (2,2-dimethylpropyl)-cyclopentane

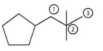

   e. 3-ethyl-5-(1,1-dimethylpropyl)-nonane

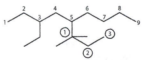

   f. 6-(1-ethyl-2-methylbutyl)-3,4,5-trimethyl-7-isopropyldecane

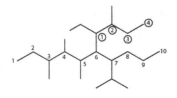

5. Consider the following alkane, drawn in perspective:

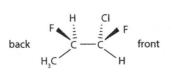

The effective size for groups follows the trend: **methyl > Cl > F > H.**

Draw **Newman diagrams** which describe the following conformations, use a larger circle for the back carbon and a smaller circle for the front carbon. **Hint:** It is easiest to leave the orientation of the back carbon fixed and rotate only the front carbon.

a.   Show the **highest energy** conformation of all possible conformations.
b.   Show the **lowest energy** (most stable) conformation of all possible conformations.
c.   Show the **highest energy staggered** conformation.
d.   Show either of the **two lowest energy eclipsed** conformations.

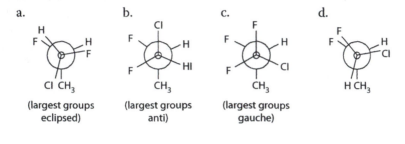

a.   b.   c.   d.

(largest groups   (largest groups   (largest groups
eclipsed)   anti)   gauche)

6. For **each** molecule shown below in a structural formula, draw the two possible stable **chair conformations;** then identify which of the two conformations is **more** stable based on substituent sizes. **Do this as a separate exercise for both molecules.** Note that the structural formulas give a specific cis/trans arrangement for the substituents in each molecule; be careful not to alter these.

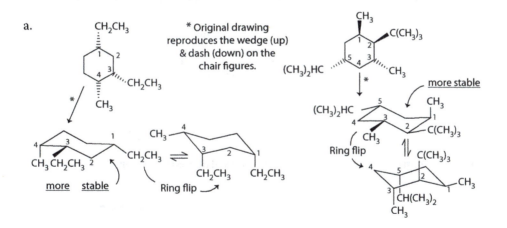

7. Sugar molecules in carbohydrate structures normally exist as a six-membered ring containing one oxygen atom; an example is shown below. Note that the electron pairs on oxygen (bonding and lone) are arranged in a tetrahedral fashion similar to carbon. Answer the following questions using the chair diagram as a guide:

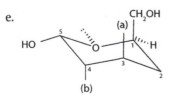

e.

use the diagram as
a visual aid

a. If a substituent is placed on carbon #3 which is **cis** to the —CH₂OH group, would it be axial or equatorial?
   <u>axial</u>
b. Is an axial substituent on carbon #4 cis or trans to the —CH₂OH group?
   <u>trans</u>
c. Identify the number of a ring-carbon which is **anti** to the **hydrogen** on carbon #1.
   <u>carbon #3 or #5</u>
d. Identify the number of a ring-carbon which is **anti** to a lone pair on the **ring**-oxygen:
   <u>C #2 or #4</u> must use equational lone pair (no ring C can be anti to the axial lone pair).
e. Certain biological sugars are connected through an —OH group attached to carbon #5. Show on the molecule drawing given above an —OH group on carbon #5 which is **trans** to the —CH₂OH group, clearly indicating whether it is axial or equatorial.
f. Consider the following sugar molecule. The substituents on carbon #1 have been eliminated for simplicity:

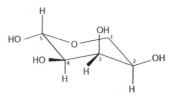

Using the chair template drawn below, show the proper orientation (axial or equatorial) of all the —OH groups **after the above molecule undergoes a ring flip.**

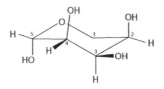

8. Determine the relationship between the two molecules in **each** of the pairs below, using the following descriptions:
   identical; conformers; diastereomers; constitutional isomers; and **not isomers** (of each other).

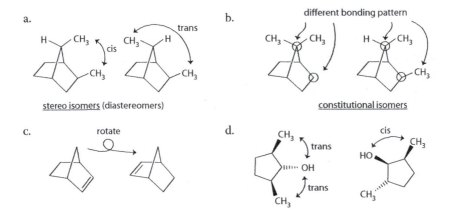

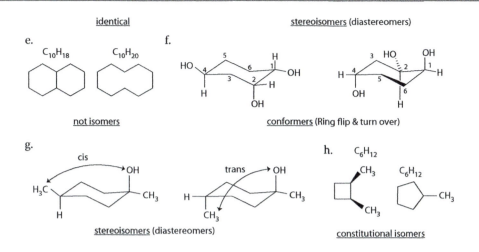

e.   identical

f.   stereoisomers (diastereomers)

not isomers

conformers (Ring flip & turn over)

g.

stereoisomers (diastereomers)

h.   constitutional isomers

Organic I: Practice Exam 4

1. Provide the correct structure or name as required. **Any method of structure drawing is acceptable; _line drawings are preferred_**.
   a. Name the following compound:

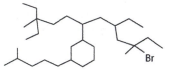

   b. Draw the structure of **cis**-1-(3, 3, 4- trichlorobutyl)-4- (3-propylhexyl)-cyclooctane. **Be sure to indicate stereochemistry using the wedge/dash notation.**

   c. Draw the structure of tricyclo [4.3.2.1.] dodecane.

2. The following molecule represents a protein dipeptide. Complete the questions concerning this molecule.

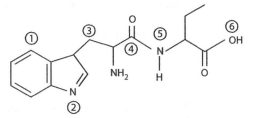

   a. How many **total** pi-bonds are in this molecule? _____
   b. Complete the following table by describing the hybridization and geometries of each of the specifically identified numbered atoms in this molecule. *Note that the structure is a line drawing and does not indicate any specific geometries.*
      Hybridization: choose from sp, sp², or sp³
      Geometries: tetrahedral, trigonal planar, linear, pyramidal, bent.

Specific Atom	Hybridization	Electron Region Geometry	Atom Geometry
Carbon # 1			
Nitrogen # 2			
Carbon # 3			
Carbon # 4			
Nitrogen # 5			
Oxygen # 6			

3. Show the required structures for the molecules described below; **partial credit for a valid structure that fits the bonding rules.**

   a. Draw **any** valid **complete Lewis** structure for $C_4H_6N_2O$ which fits the following requirement: *there are __no__ sp³ hybridized/tetrahedral carbons.*

      Show covalent bonds with a line and nonbonding pairs of electrons by two dots. All row-2 atoms will be an octet and follow normal neutral bonding rules.

   b. Draw a **specific Lewis** structure for the amino acid derivative described: the molecule is $C_6H_{12}N_2O_3$ and contains **one ester** functional group, **one amide** functional group, and **one amine** functional group. The amine functional group and the carbon representing the ester functional group are themselves bonded to the same carbon. The molecule contains only **one methyl** group which is bonded to an sp³ oxygen.

   c. (i) Draw a Lewis structure for $(CNS)^{-1}$. The atoms are listed in alphabetical order and do not necessarily represent the order of bonding; **all atoms should be an octet.** (ii) Then, draw **one** additional resonance structure of this polyatomic ion; **all atoms should be an octet.** (iii) Use the table of electronegativities if necessary; then, indicate which of the two structures will contribute more toward the net structure of the ion (i.e., which is lower in potential energy).

4. Draw structural formulas for molecules which fit the descriptions below. **Any form of structural drawing is acceptable as long as sufficient information is present;** *line drawings are preferred. __Be certain that your answer fits the molecular formula given in the descriptions.__*

   a. Draw any isomer of $C_7H_{12}O_2$ which contains **one aldehyde** and **one ketone** functional group.

   b. Draw any isomer of $C_7H_{12}O_2$ which contains a **carboxylic acid** functional group.

   c. Draw any isomer of $C_7H_{12}O_2$ which contains **__no__ pi-bonds.**

   d. Draw any constitutional isomer of $C_7H_{12}O_2$ which can exist in two **diastereomeric** forms as a diastereomeric pair based on 3-D arrangements specifically due to **substituents on a ring.** ***Show both stereoisomers, indicating spatial orientation using wedges and dashes; credit requires that the structures display a correct stereochemical relationship.***

5. a. Draw a **line drawing** of 1-cyclopropyl-2-methylbutane; number each carbon based on correct nomenclature rules. (b) Next to your line drawing, draw a **Newman**

projection of the **most** stable conformation of this molecule when viewed down the bond axis of carbon #1 in front and carbon #2 at the back; *be certain that the structures for attached groups are clear; use the partial template.*

line drawing   **Newman projection**

b. A **line drawing** of all cis-1, 2, 3-tribromocyclohexane is shown:

i. Use the templates below to draw **both** chair conformations of this molecule.
ii. Then, **indicate which of the two conformations is more stable.**

6. a. A molecule with a chiral carbon is shown in perspective below. **Label** all four groups with a **priority number** (1 = highest to 4 = lowest); then, determine whether the molecule is **R or S.** *No credit for simply taking a guess with no priority numbers indicated.*

b. Draw the structure of the constitutional isomer 1, 4-dibromohexane and identify the chiral carbon.

Then, use the templates to draw the structure of **both enantiomers** of this constitutional isomer: **Label** each group with the correct **priority number**; draw the structures with the **lowest** ranking group pointing **backward** (dashed line); specifically label which one is **R** and which one is **S.**

c. For the Fischer projection shown: Draw a Fischer projection of the **enantiomer** and **one diastereomer** of the molecule shown. Keep most groups in the same location as the given drawing and change only those groups necessary to produce the enantiomer or diastereomer.

Original Structure	The Enantiomer	One Diastereomer

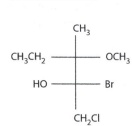

ORGANIC I: PRACTICE EXAM 1B: ANSWERS

1. Provide the correct structure or name as required. **Any method of structure drawing is acceptable; *line drawings are preferred*.**
   a. Name the following compound:

   3-bromo-
   5, 10-diethyl-
   3, 10-dimethyl-
   7 [3-(4-methylpentyl)-cyclohexyl]-dodecane

   b. Draw the structure of <u>cis</u>-1-(3, 3, 4-trichlorobutyl)-4-(3-propylhexyl)-cyclooctane. **Be sure to indicate stereochemistry using the wedge/dash notation.**

   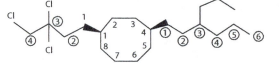

   c. Draw the structure of tricyclo [4.3.2.1.] dodecane.

   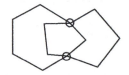

2. The following molecule represents a protein dipeptide. Complete the questions concerning this molecule.

   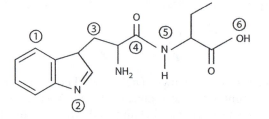

   a. How many **total** pi-bonds are in this molecule? __6__

b. Complete the following table by describing the hybridization and geometries of each of the specifically identified numbered atoms in this molecule. ___Note that the structure is a line drawing and does not indicate any specific geometries___.
Hybridization: choose from sp, sp^2, or sp^3
Geometries: tetrahedral, trigonal planar, linear, pyramidal, bent.

Specific Atom	Hybridization	Electron Region Geometry	Atom Geometry
Carbon #1	sp^2	Trig planar	Trig planar
Nitrogen # 2	sp^2	Trig planar	Bent
Carbon # 3	sp^3	Tetrahedral	Tetrahedral
Carbon # 4	sp^2	Trig planar	Trig planar
Nitrogen # 5	sp^3	Tetrahedral	Pyramidal
Oxygen # 6	sp^3	Tetrahedral	Bent

3. Show the required structures for the molecules described below; **partial credit for a valid structure that fits the bonding rules**.
   a. Draw **any** valid **complete Lewis** structure for $C_4H_6N_2O$ which fits the following requirement: *there are __no__ sp^3 hybridized/tetrahedral carbons*.
   Show covalent bonds with a line and nonbonding pairs of electrons by two dots. All row-2 atoms will be an octet and follow normal neutral bonding rules.

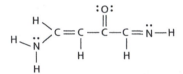

   b. Draw a **specific Lewis** structure for the amino acid derivative described: the molecule is $C_6H_{12}N_2O_3$ and contains **one ester** functional group, **one amide** functional group, and **one amine** functional group. The amine functional group and the carbon representing the ester functional group are themselves bonded to the same carbon. The molecule contains only **one methyl** group which is bonded to an sp^3 oxygen.

   c. (i) Draw a Lewis structure for $(CNS)^{-1}$. The atoms are listed in alphabetical order and do not necessarily represent the order of bonding; **all atoms should be an octet**. (ii) Then, draw **one** additional resonance structure of this polyatomic ion; **all atoms should be an octet**. (iii) Use the table of electronegativities if necessary; then, indicate which of the two structures will contribute more toward the net structure of the ion (i.e., which is lower in potential energy).

   N EN = 3.0   S EN = 2.6

   larger contribution
   ⊖ change on more electronegative element

4. Draw structural formulas for molecules which fit the descriptions below. *Any form of structural drawing is acceptable as long as sufficient information is present; line drawings are preferred. Be certain that your answer fits the molecular formula given in the descriptions.* (One example for each.)

   a. Draw any isomer of $C_7H_{12}O_2$ which contains **one aldehyde** and **one ketone** functional group.

   $C_7H_{12}O_2$ = 2 Rings/π-bonds

   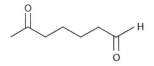

   b. Draw any isomer of $C_7H_{12}O_2$ which contains a **carboxylic acid** functional group.

   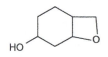

   c. Draw any isomer of $C_7H_{12}O_2$ which contains **no pi-bonds.**

   = 2 Rings

   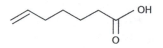

   d. Draw any constitutional isomer of $C_7H_{12}O_2$ which can exist in two **diastereomeric** forms as a diastereomeric pair based on 3-D arrangements specifically due to **substituents on a ring.** *Show both stereoisomers, indicating spatial orientation using wedges and dashes; credit requires that the structures display a correct stereochemical relationship.*

   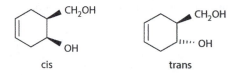

5. a. Draw a **line drawing** of 1-cyclopropyl-2-methylbutane; number each carbon based on correct nomenclature rules. (b) Next to your line drawing, draw a **Newman projection** of the **most** stable conformation of this molecule when viewed down the bond axis of carbon #1 in front and carbon #2 at the back; *be certain that the structures for attached groups are clear; use the partial template.*

   line drawing   **Newmann projection**

   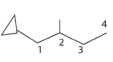   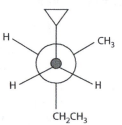

b. A **line drawing** of all cis-1, 2, 3-tribromocyclohexane is shown:

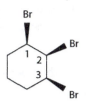

i. Use the templates below to draw **both** chair conformations of this molecule.
ii. Then, **indicate which of the two conformations is more stable.**

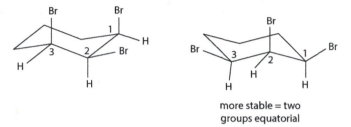

more stable = two
groups equatorial

6. a. A molecule with a chiral carbon is shown in perspective below. **Label** all four groups with a **priority number** (1 = highest to 4 = lowest); then, determine whether the molecule is **R or S**. *No credit for simply taking a guess with no priority numbers indicated.*

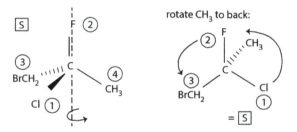

b. Draw the structure of the constitutional isomer 1,4-dibromohexane and identify the chiral carbon.

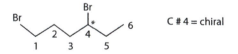

C # 4 = chiral

Then, use the templates to draw the structure of **both enantiomers** of this constitutional isomer: **Label** each group with the correct **priority number; draw** the structures with the **lowest** ranking group pointing **backward** (dashed line); specifically label which one is **R** and which one is **S.**

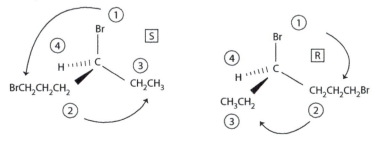

c. For the Fischer projection shown: Draw a Fischer projection of the **enantiomer** and **one diastereomer** of the molecule shown. Keep most groups in the same location as the given drawing and change only those groups necessary to produce the enantiomer or diastereomer.

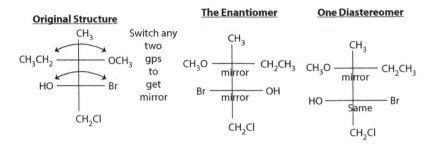

Organic I: Practice Exam 5

1. a. Name the following compound:

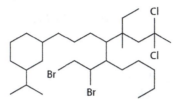

   b. Name the following compound: **name the compound based on the largest ring**. Be sure to include the correct **stereochemistry** indicated in the ring.

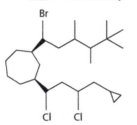

2. a. For the following molecule: Use the line drawing directly and **circle** (or use a clear **asterisk**) to **mark all chiral carbons**; be certain that the identified chiral carbons are clearly marked. (Graded based on number right minus number wrong.)

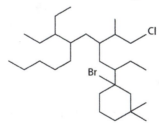

   b. Draw any constitutional isomer of C_9H_{16} which contains at least one **ring** and can exist in two diastereomeric forms as a diastereomeric pair based on 3-D arrangements **specifically due to stereochemistry caused by substituents on the ring**. **Show both stereoisomers**, indicating spatial orientation using wedges and dashes. Credit requires that the structures display a correct stereochemical relationship and that the answer **fits the molecular formula** given. Any form of structural drawing is acceptable as long as sufficient information is present.

c and d. Two molecules, each with **one** chiral carbon are shown in perspective below. For **each** of the two molecules: **Label** all four groups attached to the chiral carbon with a **priority number** (1 = highest to 4 = lowest); then, determine and label whether the molecule is **R or S**. *Credit requires **all** correct priority numbers **and** correct interpretation for R versus S.*

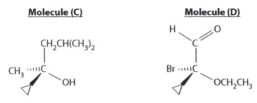

Molecule (C)   Molecule (D)

3. a.   Draw a **line drawing** of *trans*-1-ethyl-3-(2-methylpropyl)-cyclohexane; *use the wedge and dash method to indicate stereochemistry.*

   b.   Use the templates below to draw **both** chair conformations of this molecule; *use condensed structures for attached groups.*
   c.   **Indicate which of the two conformations is more stable.**

4. a.   Draw a **line drawing** of 4-methylheptane; number each carbon based on correct nomenclature rules. (b) Next to your line drawing, draw a Newman projection of the **most** stable conformation of 4-methylheptane when viewed down the bond axis of carbon #3 in front and carbon #4 at the back; *use condensed structures for attached groups and use the partial template.*

line drawing   **Newman projection**

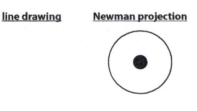

5. Consider the specific molecule shown; this is one specific stereoisomer.
   a.   Use the templates below to draw **both** chair conformations of this **specific** molecule; do **not** consider any other stereoisomer. Draw the **hydrogens** on carbons #1, 2, 4 to help indicate the relative positions of the substituents on these carbons.
   b.   Then, **indicate which of the two conformations is more stable.**
   Limited to no partial credit if the substituents do not show the correct cis/trans arrangement in the chair conformations.

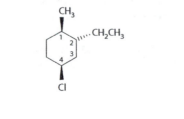

6. For the following molecules: (i) first find the chiral carbon; (ii) then **label** all four groups on the chiral carbon with a **priority number** (1 = highest to 4 = lowest); and (iii) finally determine whether the molecule is **R or S**. *No credit for simply taking a guess with no priority numbers indicated.* Recall that in a Fischer projection, the horizontal lines are coming toward you and the vertical lines are going away from you.

a.   b.

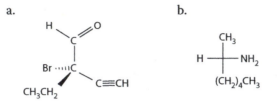

7. a.   Draw a line drawing of the **constitutional** (i.e., do not indicate stereochemistry) isomer: 1-methyl-2-(4-methylpentyl)-cyclohexane.

   b.   This constitutional isomer exists as a diastereomer pair; **each** of the two diastereomers has two possible chair conformations. Of these **four** possible conformations representing either of the diastereomers, show the **most** stable chair conformation. *Use the template and use condensed structures for attached groups.*

   c.   **State whether the molecule you drew is the cis or the trans stereoisomer.**

   For the following parts (d) and (e), draw your answers **directly on the template from part (b); do not remove or replace the methyl and (4-methylpentyl) substituents located for part (b).**

   d.   Place a bromine substituent on the ring **anti** to carbon #3.

   e.   Place a chlorine substituent on the ring **gauche** (60°) to carbon #2.

ORGANIC I: PRACTICE EXAM 1C: ANSWERS

1. a.   Name the following compound:

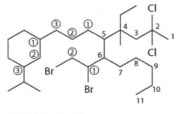

6-(1,2-dibromoethyl)-
2,2-dichloro-4-ethyl-
5-[3-(3-isopropyl cyclohexyl)-propyl]-4-methyl-undecane

   b.   Name the following compound: **name the compound based on the largest ring**. Be sure to include the correct **stereochemistry** indicated in the ring.

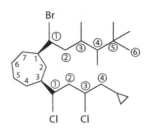

cis-1-(1-bromo-3,4,5,5-tetramethylhexyl)-
3-(1,3-dichloro-4-cyclopropylbutyl)-
cycloheptane

2. a. For the following molecule: Use the line drawing directly and **circle** (or use a clear **asterisk**) to **mark all chiral carbons**; be certain that the identified chiral carbons are clearly marked. (Graded based on number right minus number wrong.)

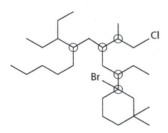

b. Draw any constitutional isomer of C_9H_{16} which contains at least one **ring** and can exist in two diastereomeric forms as a diastereomeric pair based on 3-D arrangements **_specifically due to stereochemistry caused by substituents on the ring_**. **Show both stereoisomers**, indicating spatial orientation using wedges and dashes. Credit requires that the structures display a correct stereochemical relationship and that the answer **fits the molecular formula** given. Any form of structural drawing is acceptable as long as sufficient information is present.

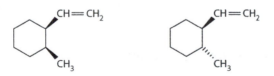

c and d. Two molecules, each with **one** chiral carbon are shown in perspective below. For **each** of the two molecules: **Label** all four groups attached to the chiral carbon with a **priority number** (1 = highest to 4 = lowest); then, determine and label whether the molecule is **R or S**. *Credit requires **all** correct priority numbers **and** correct interpretation for R versus S.*

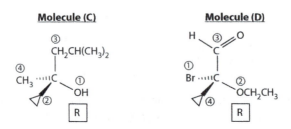

3. a. Draw a **line drawing** of _trans_-1-ethyl-3-(2-methylpropyl)-cyclohexane; **use the wedge and dash method to indicate stereochemistry**.

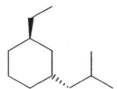

   b. Use the templates below to draw **both** chair conformations of this molecule; **use condensed structures for attached groups**.
   c. **Indicate which of the two conformations is more stable**.

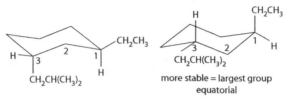

4. (a) Draw a **line drawing** of 4-methylheptane; number each carbon based on correct nomenclature rules. (b) Next to your line drawing, draw a Newman projection of the **most** stable conformation of 4-methylheptane when viewed down the bond axis of carbon #3 in front and carbon #4 at the back; **use condensed structures for attached groups and use the partial template**.

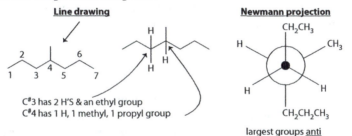

5. Consider the specific molecule shown; this is one specific stereoisomer.
   a. Use the templates below to draw **both** chair conformations of this **specific** molecule; do **not** consider any other stereoisomer. Draw the **hydrogens** on carbons #1, 2, 4 to help indicate the relative positions of the substituents on these carbons.
   b. Then, **indicate which of the two conformations is more stable**.
      Limited to no partial credit if the substituents do not show the correct cis/trans arrangement in the chair conformations.

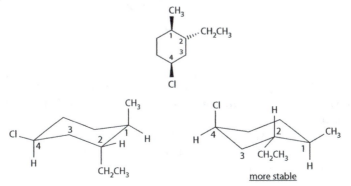

6. For the following molecules: (i) first find the chiral carbon; (ii) then **label** all four groups on the chiral carbon with a **priority number** (1 = highest to 4 = lowest); and (iii) finally determine whether the molecule is **R or S**. *No credit for simply taking a guess with no priority numbers indicated.* Recall that in a Fischer projection, the horizontal lines are coming toward you and the vertical lines are going away from you.

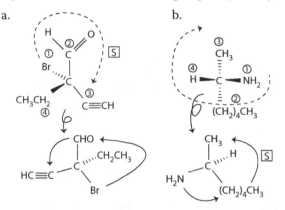

Bottom drawing: Group #4 rotated to point backward.
Top drawing: Keep the original drawing with group #4 pointing forward. Connecting the arrow clockwise with group #4 forward is equivalent to connecting the arrow counterclockwise with group #4 pointing backward.

7. a. Draw a line drawing of the **constitutional** (i.e., do not indicate stereochemistry) isomer: 1-methyl-2-(4-methylpentyl)-cyclohexane.

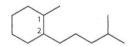

b. This constitutional isomer exists as a diastereomer pair; **each** of the two diastereomers has two possible chair conformations. Of these **four** possible conformations representing either of the diastereomers, show the **most** stable chair conformation. *Use the template and use condensed structures for attached groups.*

c. **State whether the molecule you drew is the cis or the trans stereoisomer.**
most stable = both groups equatorial

answers to d and e*

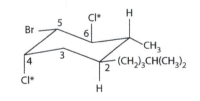

groups are trans

answers to b & c

For the following parts (d) and (e), draw your answers **directly on the template from part (b);** *do not remove or replace the methyl and (4-methylpentyl) substituents located for part (b).*

d. Place a bromine substituent on the ring **anti** to carbon #3.

e. Place a chlorine substituent on the ring **gauche** (60°) to carbon #2.
*two possible answers shown.

Organic I: Practice Exam 6

1. a. Name the compound; be sure to indicate **stereochemistry (E or Z)** of the **one** (and only) **asymmetric** double bond as drawn in the line drawing; *include a designation using an __asterisk__ for the __higher-priority__ group on each double-bonded carbon.*

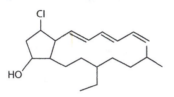

   b. Name the compound: **no stereochemistry** is required.

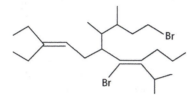

   c. For the molecule drawn above for part (**b**) (**only** the molecule from (b)): Find all chiral carbons in this molecule and *__indicate them directly on the line drawing by circling each chiral carbon (or use an asterisk clearly).__*

2. Show a complete **balanced** acid/base **equation** for the reaction shown. Then, use the table of pKa's and state which **direction** the equilibrium is favored (reactant/left or product/right); indicate your reason by showing the pKa analysis.

$$(CH_3)_3COH \text{ (acid)} + NH_3 \text{ (base)} \longrightarrow$$

3. a. A molecule with **one** chiral carbon is shown in perspective below. **Label** all four groups with a **priority number** (1 = highest to 4 = lowest); then, determine whether the molecule is **R or S**. *No credit for simply taking a guess with no priority numbers indicated.*

   b. Next to the original drawing, use the template to draw the **enantiomer** of this molecule. (**Hint: Keep most groups the same and change only as many necessary to form the enantiomer.**)

   Original Structure   **The Enantiomer**

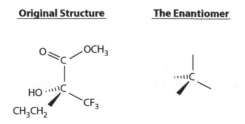

c. A molecule with **two** chiral carbons is shown in perspective below. **Separately,** **label** all four groups on **each** chiral carbon with a **priority number** (1 = highest to 4 = lowest); then, determine **R or S for each chiral carbon**. *No credit for simply taking a guess with no priority numbers indicated.*

d. Next to the original drawing, use the template to draw one **diastereomer** of this molecule. **(Hint: Keep most groups the same and change only as many necessary to form a diastereomer.)**

Original Structure   **One Diastereomer**

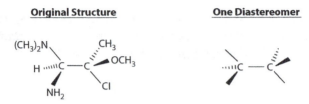

4. a. Use the perspective drawing below to determine R or S for **each** of the two chiral carbons: Assign and **label** all four groups on each chiral carbon with a **priority number** (1 = highest to 4 = lowest); circle one set of numbers for one of the chiral carbons to help keep your analysis clear; then, determine **R or S** for each of the two chiral carbons. *No credit for simply taking a guess with no priority numbers indicated.*

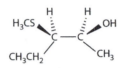

b. Use the molecule shown above as a perspective drawing and convert this specific molecule into a Fischer projection. **Draw the main carbon chain vertical; be certain that the specific stereoisomer shown is correctly reproduced as the Fischer projection.**

c. Use the molecule you have drawn above as a Fischer projection. Draw a Fischer projection of (i) the **enantiomer**; (ii) **one diastereomer**; and (iii) **one constitutional isomer** of the molecule. Keep most groups in the same location as your structure above and change only those groups necessary to produce the isomer requested.

The Enantiomer   **One Diastereomer**   **One Constitutional Isomer**

ORGANIC I: PRACTICE EXAM 1.5A: ANSWERS

1. a. Name the compound; be sure to indicate **stereochemistry (E or Z)** of the **one** (and only) **asymmetric** double bond as drawn in the line drawing; *include a designation using an __asterisk__ for the __higher-priority__ group on each double-bonded carbon.*

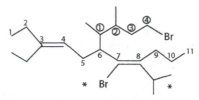

Z-7-bromo-6-(4-bromo-1,2-dimethylbutyl)-
3-ethyl-8-isopropyl 3,7-undecadiene

b. Name the compound: **no stereochemistry** is required.

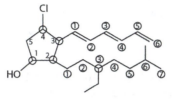

4-chloro-2-(3-ethyl-6-methylheptyl)-
3-(1,3,5-hexatrienyl)-cyclopentanol

c. For the molecule drawn above for **part (b) (only** the molecule from (b)): Find all chiral carbons in this molecule and *indicate them directly on the line drawing by circling each chiral carbon (or use an asterisk clearly).*

2. Show a complete **balanced** acid/base **equation** for the reaction shown. Then, use the table of pKa's and state which **direction** the equilibrium is favored (reactant/left or product/right); indicate your reason by showing the pKa analysis.

$(CH_3)_3COH$ (acid) + NH_3 (base) $\longrightarrow$

$(CH_3)_3COH$ + NH_3 $\rightleftharpoons$ $(CH_3)_3CO^{\ominus}$ + $NH_4^{\oplus}$

pKa = 18 $\xleftarrow[\text{reactant favored}]{}$ pKa = 9.2

3. a. A molecule with **one** chiral carbon is shown in perspective below. **Label** all four groups with a **priority number** (1 = highest to 4 = lowest); then, determine whether the molecule is **R or S**. *No credit for simply taking a guess with no priority numbers indicated.*

b. Next to the original drawing, use the template to draw the **enantiomer** of this molecule. **(Hint: Keep most groups the same and change only as many necessary to form the enantiomer.)**

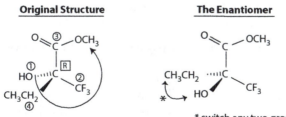

Original Structure

counter clockwise w #4 forward = [R]

The Enantiomer

* switch any two groups

c. A molecule with **two** chiral carbons is shown in perspective below. **Separately, label** all four groups on **each** chiral carbon with a **priority number** (1 = highest to 4 = lowest); then, determine **R or S for each chiral carbon**. *No credit for simply taking a guess with no priority numbers indicated.*

d. Next to the original drawing, use the template to draw one **diastereomer** of this molecule. **(Hint: Keep most groups the same and change only as many necessary to form a diastereomer.)**

Original Structure   **One Diastereomer**

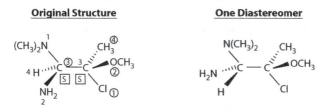

4. a. Use the perspective drawing below to determine R or S for **each** of the two chiral carbons: Assign and **label** all four groups on each chiral carbon with a **priority number** (1 = highest to 4 = lowest); circle one set of numbers for one of the chiral carbons to help keep your analysis clear; then, determine **R or S** for each of the two chiral carbons. *No credit for simply taking a guess with no priority numbers indicated.*

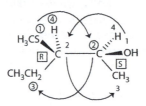

b. Use the molecule shown above as a perspective drawing and convert this specific molecule into a Fischer projection. **Draw the main carbon chain vertical; be certain that the specific stereoisomer shown is correctly reproduced as the Fischer projection.**

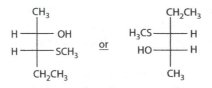

c. Use the molecule you have drawn above as a Fischer projection. Draw a Fischer projection of (i) the **enantiomer**; (ii) **one diastereomer**; and (iii) **one constitutional isomer** of the molecule. Keep most groups in the same location as your structure above and change only those groups necessary to produce the isomer requested.

The Enantiomer   **One Diastereomer**   **One Constitutional isomer**

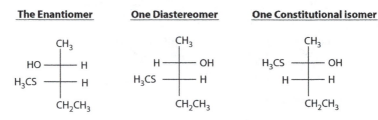

Organic I: Practice Exam 7

1. Provide a correct name for each of the following compounds. No stereochemistry is indicated or necessary for parts (a) through (d); **designate correct stereochemistry** for substituents on parts (e) and (h).

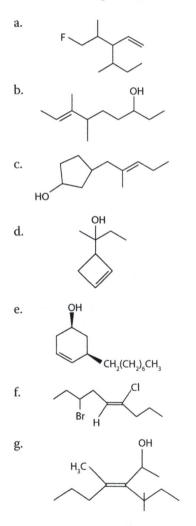

a.

b.

c.

d.

e.

f.

g.

h.

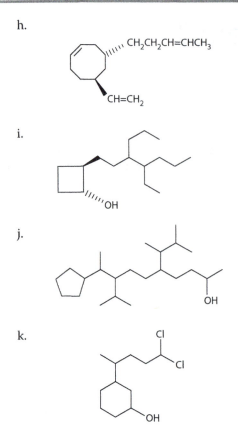

i.

j.

k.

2. For the following two molecules:
   i. Determine the **R** or **S stereochemistry** around the enantiomeric carbon drawn in perspective.
   ii. Next to the molecule drawn, draw a perspective drawing of the **enantiomer** of this molecule. (Draw the enantiomer by making the minimum number of changes.)

   a.

   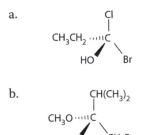

   b.

   CH₃O ····C with CH(CH₃)₂, H, CH₂Br

3. For the following two molecules:
   i. Determine the **R** or **S** stereochemistry around **both** enantiomeric carbons drawn in perspective.
   ii. Next to the molecule drawn, draw a perspective drawing of a **diastereomer** (stereoisomer which is not an enantiomer) of this molecule. (Draw the diastereomer by making the minimum number of changes.)
   iii. Identify all **stereogenic centers** in each molecule and state the **total** number of stereoisomers possible.

a.

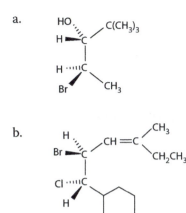

b.

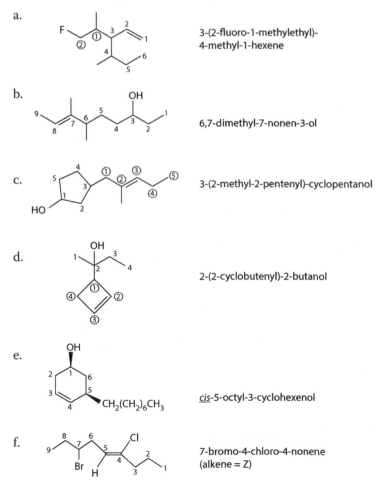

ORGANIC I: PRACTICE EXAM 1.5B: ANSWERS

1. Provide a correct name for each of the following compounds. No stereochemistry is indicated or necessary for parts (a) through (d); **designate correct stereochemistry** for substituents on parts (**e**) and (**h**).

a.

3-(2-fluoro-1-methylethyl)-
4-methyl-1-hexene

b.

6,7-dimethyl-7-nonen-3-ol

c.

3-(2-methyl-2-pentenyl)-cyclopentanol

d.

2-(2-cyclobutenyl)-2-butanol

e.

cis-5-octyl-3-cyclohexenol

f.

7-bromo-4-chloro-4-nonene
(alkene = Z)

g.

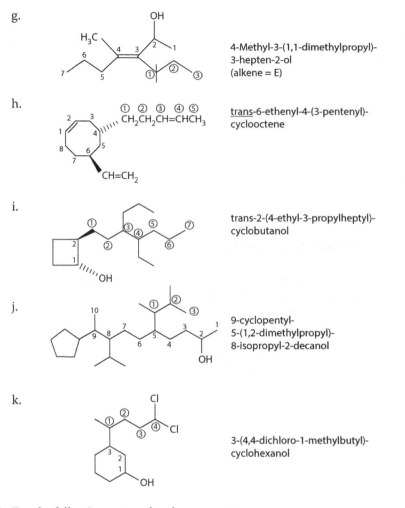

4-Methyl-3-(1,1-dimethylpropyl)-
3-hepten-2-ol
(alkene = E)

h.

trans-6-ethenyl-4-(3-pentenyl)-
cyclooctene

i.

trans-2-(4-ethyl-3-propylheptyl)-
cyclobutanol

j.

9-cyclopentyl-
5-(1,2-dimethylpropyl)-
8-isopropyl-2-decanol

k.

3-(4,4-dichloro-1-methylbutyl)-
cyclohexanol

2. For the following two molecules:
   i.   Determine the **R** or **S stereochemistry** around the enantiomeric carbon drawn in
        perspective.
   ii. Next to the molecule drawn, draw a perspective drawing of the **enantiomer**
        of this molecule. (Draw the enantiomer by making the minimum number of
        changes.)

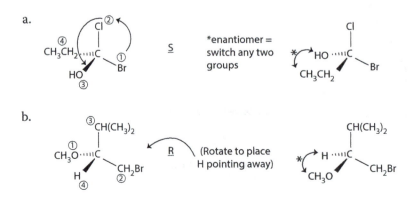

3. For the following two molecules:
   i.   Determine the **R** or **S** stereochemistry around **both** enantiomeric carbons drawn in perspective.
   ii. Next to the molecule drawn, draw a perspective drawing of a **diastereomer** (stereoisomer which is not an enantiomer) of this molecule. (Draw the diastereomer by making the minimum number of changes.)
   iii. Identify all **stereogenic centers** in each molecule and state the **total** number of stereoisomers possible.

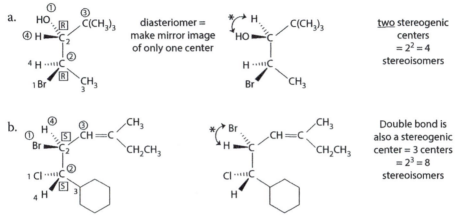

a.   diasteriomer = make mirror image of only one center   <u>two</u> stereogenic centers $= 2^2 = 4$ stereoisomers

b.   Double bond is also a stereogenic center = 3 centers $= 2^3 = 8$ stereoisomers

* mirror image of one chiral carbon: switch any two groups

Organic I: Practice Exam 8

1. a. Name the compound; be sure to indicate **stereochemistry (E or Z)** of the **one** (and only) **asymmetric** double bond as drawn in the line drawing; *include a designation using an __asterisk__ for the __higher-priority__ group on each double-bonded carbon.*

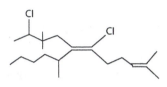

   b. Name the compound: **<u>no</u> stereochemistry** is required.

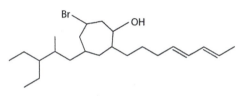

   c. For the molecule drawn above for part (**b**) (**only** the molecule from (b)): Find all chiral carbons in this molecule and **<u>*indicate them directly on the line drawing by circling each chiral carbon (or use an asterisk clearly).*</u>**

2. Show a complete **balanced** acid/base **equation** for the reaction shown. Then, use the table of pKa's and state which **direction** the equilibrium is favored (reactant/left or product/right); indicate your reason by showing the pKa analysis.

$$C_6H_5OH \text{ (acid)} + CH_3NH_2 \text{ (base)} \rightarrow$$

3. a. A molecule with **<u>one</u>** chiral carbon is shown in perspective below. **Label** all four groups with a **priority number** (1 = highest to 4 = lowest); then, determine whether the molecule is **R or S**. *No credit for simply taking a guess with no priority numbers indicated.*

   b. Next to the original drawing, use the template to draw the **enantiomer** of this molecule. **(Hint: Keep most groups the same and change only as many necessary to form the enantiomer.)**

   Original Structure   **The Enantiomer**

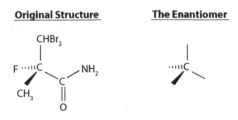

   c. A molecule with **<u>two</u>** chiral carbons is shown in perspective below. **<u>Separately,</u>** **label** all four groups on **<u>each</u>** chiral carbon with a **priority number** (1 = highest to

4 = lowest); then, determine **R or S** for **each** chiral carbon. *No credit for simply taking a guess with no priority numbers indicated.*

d. Next to the original drawing, use the template to draw one **diastereomer** of this molecule. **(Hint: Keep most groups the same and change only as many necessary to form a diastereomer.)**

Original Structure   **One Diastereomer**

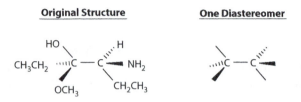

4. a. Use the perspective drawing below to determine R or S for **each** of the two chiral carbons: Assign and **label** all four groups on each chiral carbon with a **priority number** (1 = highest to 4 = lowest); circle one set of numbers for one of the chiral carbons to help keep your analysis clear; then, determine **R or S** for each of the two chiral carbons. *No credit for simply taking a guess with no priority numbers indicated.*

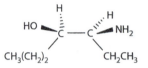

b. Use the molecule shown above as a perspective drawing and convert this specific molecule into a Fischer projection. **Draw the main carbon chain vertical; be certain that the specific stereoisomer shown is correctly reproduced as the Fischer projection.**

c. Use the molecule you have drawn above as a Fischer projection. Draw a Fischer projection of (i) the **enantiomer**; (ii) **one diastereomer**; and (iii) **one constitutional isomer** of the molecule. Keep most groups in the same location as your structure above and change only those groups necessary to produce the isomer requested.

The Enantiomer   **One Diastereomer**   **One Constitutional Isomer**

ORGANIC I: PRACTICE EXAM 1.5C: ANSWERS

1. a. Name the compound; be sure to indicate **stereochemistry (E or Z)** of the **one** (and only) **asymmetric** double bond as drawn in the line drawing; *include a designation using an asterisk for the higher-priority group on each double-bonded carbon.*

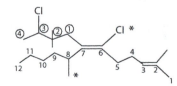

E-6-chloro-7-(3-chloro-2,2-dimethylbutyl)-
2,8-dimethyl-2,6-dodecadiene

b. Name the compound: **<u>no</u> stereochemistry** is required.

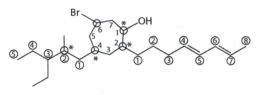

6-bromo-4-(3-ethyl-2-methylpentyl)-
2-(4,6-octadienyl)-cycloheptanol

c. For the molecule drawn above for part (**b**) (**only** the molecule from (b)): Find all chiral carbons in this molecule and ***indicate them directly on the line drawing by circling each chiral carbon (or use an asterisk clearly).***

2. Show a complete **balanced** acid/base **equation** for the reaction shown. Then, use the table of pKa's and state which **direction** the equilibrium is favored (reactant/left or product/right); indicate your reason by showing the pKa analysis.

C_6H_5OH (acid) + CH_3NH_2 (base) $\longrightarrow$

$C_6H_5OH + CH_3NH_2 \rightleftharpoons C_6H_5\overset{\ominus}{O} + CH_3\overset{\oplus}{NH_3}$

pKa = 9.9 $\xrightarrow[\text{product favored}]{}$ pKa = 10.6

3. a. A molecule with **one** chiral carbon is shown in perspective below. **Label** all four groups with a **priority number** (1 = highest to 4 = lowest); then, determine whether the molecule is **R or S**. *No credit for simply taking a guess with no priority numbers indicated.*

   b. Next to the original drawing, use the template to draw the **enantiomer** of this molecule. (**Hint: Keep most groups the same and change only as many necessary to form the enantiomer.**)

Original Structure   **The Enantiomer**

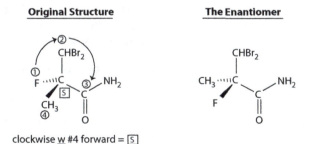

clockwise <u>w</u> #4 forward = S

   c. A molecule with **two** chiral carbons is shown in perspective below. **<u>Separately,</u> label** all four groups on **each** chiral carbon with a **priority number** (1 = highest to 4 = lowest); then, determine **R or S for each** chiral carbon. *No credit for simply taking a guess with no priority numbers indicated.*

   d. Next to the original drawing, use the template to draw one **diastereomer** of this molecule. (**Hint: Keep most groups the same and change only as many necessary to form a diastereomer.**)

Original Structure   One Diastereomer

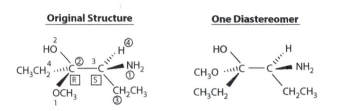

4. a.   Use the perspective drawing below to determine R or S for **each** of the two chiral carbons: Assign and **label** all four groups on each chiral carbon with a **priority number** (1 = highest to 4 = lowest); circle one set of numbers for one of the chiral carbons to help keep your analysis clear; then, determine **R or S** for each of the two chiral carbons. *No credit for simply taking a guess with no priority numbers indicated.*

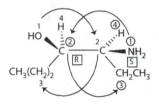

   b.   Use the molecule shown above as a perspective drawing and convert this specific molecule into a Fischer projection. **Draw the main carbon chain vertical; be certain that the specific stereoisomer shown is correctly reproduced as the Fischer projection.**

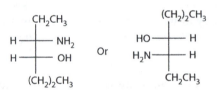

   c.   Use the molecule you have drawn above as a Fischer projection. Draw a Fischer projection of (i) the **enantiomer**; (ii) **one diastereomer**; and (iii) **one constitutional isomer** of the molecule. Keep most groups in the same location as your structure above and change only those groups necessary to produce the isomer requested.

The Enantiomer   The Diastereomer   One Constitutional Isomer

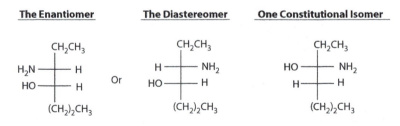

Organic I: Practice Exam 9

1. Consider the following overall reaction:

$$CH_3-\overset{\overset{\displaystyle O}{\|}}{C}-H \ + \ H_2O \ \rightleftharpoons \ CH_3-\overset{\overset{\displaystyle OH}{|}}{\underset{\underset{\displaystyle OH}{|}}{C}}-H$$

a. Which statement describes the probable entropy change in this reaction?
    increases in forward direction
    decreases in forward direction
    stays approximately constant in either direction
    No trend can be predicted without a value for ΔG

b. At certain temperatures, ΔG^0 for the reaction is a **negative** value. What can be determined about the value for enthalpy (ΔH^0) for this reaction? Use the conclusion for entropy determined from part **(a).**
    ΔH^0 must be positive
    ΔH^0 must be exactly zero
    ΔH^0 must be negative
    The sign of ΔH^0 cannot be determined without a value for ΔG^0

c. What can be determined about the value for the equilibrium constant **K under conditions described for part (b)?**
    K must be greater than 1
    K must be less than 1
    K must be exactly equal to zero
    No information about K can be determined without a value for entropy

d. The specific value for **K** at **room temperature** (298 K) is 1.8×10^{-2}. Calculate ΔG^0 for this reaction at 298 K.

e. Based on your answer for the probable direction of entropy change (part (a)): if K were measured at 320 K, how would this value compare to K at 298 K? Consider the effect of the $T\Delta S^0$ term on the values for ΔG^0 and K.
    K would increase with increasing temperature
    K would decrease with increasing temperature
    K would remain exactly the same
    Nothing can be determined without a specific value for ΔG^0
        The forward reaction rate can be increased by addition of acid according to the following mechanism: (Note that H$^+$ (aq) is regenerated in the reaction.)

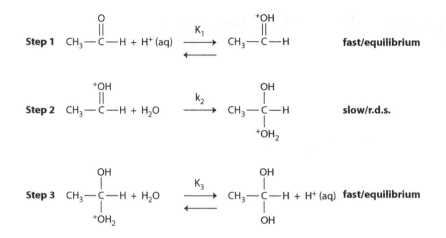

f. Write a rate expression for the **rate-determining step** in the forward direction for this reaction.

g. Write a rate expression for the **overall** rate by correct substitution of intermediate concentrations using the listed starting materials in **Step 1.**

h. The species formed as a **product** in **Step 1** is a very reactive molecule. Which of the following structures represents a valid resonance form for this molecule?

a. CH_3＝$\overset{\displaystyle OH}{C}$—H   b. CH_3—$\overset{\displaystyle OH}{\underset{+}{C}}$—H   c. $^+CH_3$—$\overset{\displaystyle O}{C}$—H   d. CH_3—$\overset{\displaystyle OH}{\underset{+}{C}}$—H

i. How would the rate of the **overall** forward reaction be changed if a different catalyst was added which accelerated **only Step 3** of the complete reaction?

Overall rate would increase

Overall rate would decrease

Overall rate would remain the same

Overall rate could increase or decrease depending on the value for K_3

The rate of the overall reaction can also be accelerated by addition of base instead of acid according to the following mechanism:

Step 1 CH_3—$\overset{\displaystyle O}{\overset{\|}{C}}$—H + OH^- (aq) $\xrightarrow{k_1}$ CH_3—$\overset{\displaystyle O^-}{\underset{\displaystyle OH}{C}}$—H   **slow/r.d.s.**

Step 2 CH_3—$\overset{\displaystyle O^-}{\underset{\displaystyle OH}{C}}$—H + H_2O $\underset{\longleftarrow}{\overset{K_2}{\longrightarrow}}$ CH_3—$\overset{\displaystyle OH}{\underset{\displaystyle OH}{C}}$—H + OH^- (aq)   **fast/equilibrium**

j. Write a rate expression for the **overall** reaction in terms of starting materials listed in **Step 1**.

k. Which species, if any, is acting as a catalyst in this reaction?

l. Draw a representation of the transition state structure for the rate-determining step. Use dashed lines to represent bonds being broken or formed.

m. Noting the value calculated in part (**d**), how would the value of ΔG^0 at 298 K for this mechanism compare to that calculated from part (**d**)?
ΔG^0 will be a larger positive/smaller negative value than the value from part (d)
ΔG^0 will be a larger negative/smaller positive value than the value from part (d)
ΔG^0 will be identical to the value from part (d)
No conclusion about ΔG^0 versus the value from part (d) can be determined

n. Draw an **energy-level diagram** (298 K) for the reaction shown above. Label and show only **relative** energy positions of reactants, products, and intermediates; show in the curves the relative magnitude of activation energies. For the relative positions: the value of K_2 for **Step 2** is greater than 1; also use the sign of the overall K given in part (**d**).

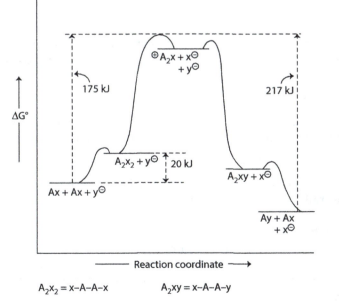

Energy level diagram for problem #2

$A_2x_2 = x-A-A-x$ $A_2xy = x-A-A-y$

2. Use the **potential energy diagram** shown above to answer the following questions. All parts of this problem refer to this diagram.

The **forward** reaction is read on the diagram from **left-to-right; the reverse** reaction is read on the diagram from **right-to-left.**

Addition or subtraction may be required to find some numerical values.

a. What **forward** reaction is depicted by this diagram? that is, write the reaction formula equation for the **forward** reaction as shown in the diagram.

b. What is the value for ΔG^0 for this **forward** reaction? (**Indicate the numerical value plus the correct sign.**)

c. Write the specific formula reaction for the **rate-determining step** in the **forward** direction; exclude (cancel out) unchanged species.

d. Determine the value for activation energy (activation barrier) for this **rate-determining step** in the **forward** direction. (**Indicate numerical value plus correct sign.**)

e. Write the **overall** rate expression for the forward reaction; substitution of starting materials (shown at the left) for intermediates may be necessary; assume any general **k**.

f. Assuming that Y^- and X^- have equal potential energy, choose which of the following species containing **A** is the most stable: (Consider how potential energy and stability are related.)

A_2X_2 A_2X^+ A_2XY

g. Assuming that Y^- and X^- have equal potential energy, and that ΔG^0 is approximately equal to ΔH^0 for this reaction, select which bond is the stronger?

A—Y stronger   A—X stronger   A—Y and A—X are equivalent

h. If the rate of the last step: $A_2XY \rightarrow AY + AX$ is increased, how would the rate of the total **overall** reaction be changed?
Overall rate would **increase**
Overall rate would **decrease**
Overall rate would **remain** the **same**

i. The temperature of the complete reaction is increased. Which statement describes the effect on the rate constant **k** and the activation energy **Ea** for the rate-determining step?
k will increase and Ea will increase
k will increase and Ea will decrease
k will increase and Ea will remain the same
k will remain the same and Ea will decrease

j.   Assume a catalyst is added to the system such that its **only** effect is to increase the rate of the rate-determining step. How will this catalyst affect the value of ΔG^0?
ΔG^0 will be a larger positive/smaller negative value
ΔG^0 will be a larger negative/smaller positive value
ΔG^0 will remain unchanged
No conclusion about ΔG^0 can be determined without a value for ΔH^0

k.   What is the **reverse** reaction depicted by this diagram? that is, write the reaction formula equation for the **reverse** reaction as shown in the diagram.

l.   What is the value for ΔG^0 for this **reverse** reaction? (Indicate **numerical value plus correct sign.**)

m.   Which reaction is spontaneous?

The **forward** reaction   The **reverse** reaction
Both reactions are spontaneous   **Neither** reaction is spontaneous

n.   Write the specific formula reaction for the **rate-determining step** in the **reverse** direction.

o.   Write a rate expression for the **overall reverse** reaction based on starting materials at the right, assuming **Step 4** is an equilibrium.

p.   Calculate the value of K for $AX + AX \longrightleftharpoons A_2X_2$ using the energy value given.

3. Show the three-step mechanism for the reaction of 1-methylcyclohexene with Br_2/H_2O (aqueous solution) to form 2-bromo-1-methylcyclohexanol.

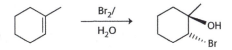

4. Draw structural formulas for the major product for each of the following reactions. **Stereochemistry must be indicated** for those reactions that produce a **stereospecific** product.

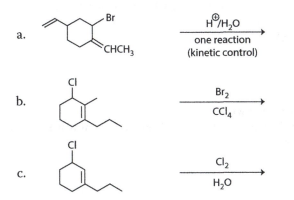

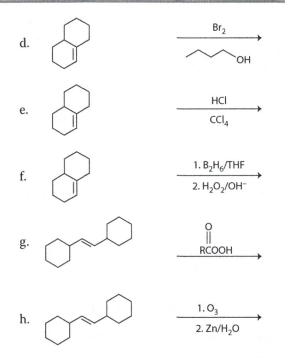

d.

Br_2

CCl_4 (butanol drawn with OH)

e.

HCl

CCl_4

f.

1. B_2H_6/THF

2. H_2O_2/OH⁻

g.

O
‖
RCOOH

h.

1. O_3

2. Zn/H_2O

ORGANIC I: PRACTICE EXAM 2A: ANSWERS

1. Consider the following overall reaction:

$$CH_3-\overset{\overset{\displaystyle O}{\|}}{C}-H + H_2O \rightleftarrows CH_3-\overset{\overset{\displaystyle OH}{|}}{\underset{\underset{\displaystyle OH}{|}}{C}}-H$$

a. Which statement describes the probable entropy change in this reaction?
increases in forward direction
| decreases in forward direction |
stays approximately constant in either direction
No trend can be predicted without a value for ΔG

b. At certain temperatures, ΔG^0 for the reaction is a **negative** value. What can be determined about the value for enthalpy (ΔH^0) for this reaction? Use the conclusion for entropy determined from part **(a)**.
ΔH^0 must be positive
ΔH^0 must be exactly zero
| ΔH^0 must be negative |
The sign of ΔH^0 cannot be determined without a value for ΔG^0

c. What can be determined about the value for the equilibrium constant **K under conditions described for part (b)?**
| K must be greater than 1 |
K must be less than 1

K must be exactly equal to zero

No information about K can be determined without a value for entropy

d. The specific value for **K** at **room temperature** (298 K) is 1.8×10^{-2}. Calculate ΔG^0 for this reaction at 298 K.

$\Delta G^0 = -2.303 \, RT \, \log K$

$\Delta G^0 = -(2.303)(0.008314 \, kJ/mole - K)(298 \, K)(\log (1.8 \times 10^{-2})) = +9.96 \, kJ/mole$

e. Based on your answer for the probable direction of entropy change (part (a)): if K were measured at 320 K, how would this value compare to K at 298 K? Consider the effect of the $T\Delta S^0$ term on the values for ΔG^0 and K.

K would increase with increasing temperature

K would decrease with increasing temperature

K would remain exactly the same

Nothing can be determined without a specific value for ΔG^0

The forward reaction rate can be increased by addition of acid according to the following mechanism: (Note that H^+ (aq) is regenerated in the reaction.)

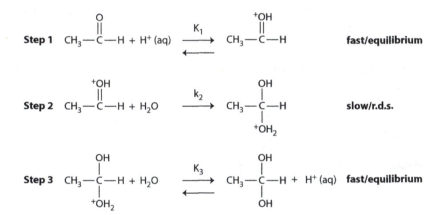

The expression may vary for $[H_2O]$ depending on the format of H^+(aq) or H_3O^+(aq).

f. Write a rate expression for the **rate-determining step** in the forward direction for this reaction.

r.d.s. = step 2   rate = k_2 [CH$_3$C—H] [H$_2$O]
                              ‖
                            $\oplus$OH

g. Write a rate expression for the **overall** rate by correct substitution of intermediate concentrations using the listed starting materials in **Step 1.**

$K_1 = \dfrac{[CH_3C\,(\overset{\oplus}{O}H)H]}{[CH_3C(O)H]\,[H^+(aq)]}$ thus $[CH_3C\,(\overset{\oplus}{O}H)H] = K_1 \, [CH_3C(O)H]\,[H^+(aq)]$

substitute

$r_{overall} = k_2 K_1 \, [CH_3C(O)H]\,[H^+(aq)]\,[H_2O]$

h. The species formed as a **product** in **Step 1** is a very reactive molecule. Which of the following structures represents a valid resonance form for this molecule?

(a) $CH_3{=}\overset{\displaystyle \overset{OH}{|}}{C}{-}H$ (b) $CH_3{-}\overset{\displaystyle \overset{OH}{||}}{\underset{+}{C}}{-}H$ (c) $^{+}CH_3{-}\overset{\displaystyle \overset{O}{||}}{C}{-}H$ (d) $CH_3{-}\overset{\displaystyle \overset{OH}{|}}{\underset{+}{C}}{-}H$

i. How would the rate of the **overall** forward reaction be changed if a different catalyst was added which accelerated **only Step 3** of the complete reaction?
Overall rate would increase
Overall rate would decrease
$\boxed{\text{Overall rate would remain the same}}$ (Step occurs <u>after</u> the r.d.s.)
Overall rate could increase or decrease depending on the value for K_3
The rate of the overall reaction can also be accelerated by addition of base instead of acid according to the following mechanism:

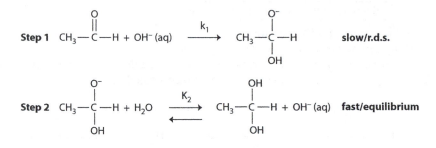

j. Write a rate expression for the **overall** reaction in terms of starting materials listed in **Step** 1.
rate overall = rate r.d.s. = $k_1[CH_3C(O)H][OH^-(aq)]$

k. Which species, if any, is acting as a catalyst in this reaction?
(OH$^-$) consumed in Step 1, regenerated in Step 2

l. Draw a representation of the transition state structure for the rate-determining step. Use dashed lines to represent bonds being broken or formed.

$CH_3{-}\overset{\displaystyle \overset{\delta^-O}{||\!\!:}}{\underset{\displaystyle \underset{\delta^-OH}{:}}{C}}{-}H$ broken (π—bond)
   formed (σ—bond)

m. Noting the value calculated in part (**d**), how would the value of ΔG^0 at 298 K for this mechanism compare to that calculated from part (**d**)?
ΔG^0 will be a larger positive/smaller negative value than the value from part (d)
ΔG^0 will be a larger negative/smaller positive value than the value from part (d)
$\boxed{\Delta G^0 \text{ will be identical to the value from part (d)}}$ $-\Delta G^0$ = state function; independent of mechanism
No conclusion about ΔG^0 versus the value from part (d) can be determined

n. Draw an **energy-level diagram** (298 K) for the reaction shown above. Label and show only **relative** energy positions of reactants, products, and intermediates; show in the curves the relative magnitude of activation energies. For the relative positions: the value of K_2 for **Step 2** is greater than 1; also, use the sign of the overall K given in part (**d**).

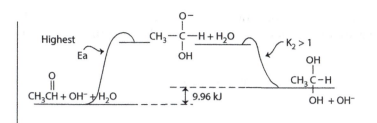

Energy level diagram for problem #2

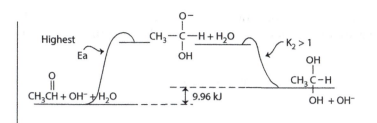

$$A_2x_2 = x{-}A{-}A{-}x \qquad\qquad A_2xy = x{-}A{-}A{-}y$$

2. Use the **potential energy diagram** shown above to answer the following questions. All parts of this problem refer to this diagram.

The **forward** reaction is read on the diagram from **left-to-right;** the **reverse** reaction is read on the diagram from **right-to-left.**

Addition or subtraction may be required to find some numerical values.

a. What **forward** reaction is depicted by this diagram? that is, write the reaction formula equation for the **forward** reaction as shown in the diagram.

$$\text{from diagram}: \ Ax + Ax + y^- \longrightarrow Ay + Ax + x^-$$
$$\text{net reaction}: \ Ax + y^- \longrightarrow Ay + x^-$$

b. What is the value for ΔG^0 for this **forward** reaction? (**Indicate numerical value plus correct sign.**)

$$\Delta G^0 = 175{-}217 = -42 \text{ kJ/mole-reaction}$$

c. Write the specific formula reaction for the **rate-determining step** in the **forward** direction; exclude (cancel out) unchanged species.

$$(\text{Step 2}) \ A_2x_2 + y^{\ominus} \longrightarrow {}^{\oplus}A_2x + x^{\ominus} + y^{\ominus} \ ; \ \underline{\text{net}} = \ A_2x_2 \longrightarrow {}^{\oplus}A_2x + x^{\ominus}$$

d. Determine the value for activation energy (activation barrier) for this **rate-determining step** in the **forward** direction. (**Indicate numerical value plus correct sign.**)

$$Ea = 175 - 20 = 155 \text{ kJ}$$

e. Write the **overall** rate expression for the forward reaction; substitution of starting materials (shown at the left) for intermediates may be necessary; assume any general **k**. Assume Step 1 = Equilibrium (reverse reaction is faster than forward)
rate r.d.s. = $k_2[A_2X_2]$ $K_1 = [A_2X_2]/[AX]^2$ $[A_2X_2] = K_1[AX]^2$
Substitute: rate (overall) = $k_2K_1[AX]^2$ <u>or</u> $r = k[AX]^2$

f. Assuming that **Y⁻** and **X⁻** have equal potential energy, choose which of the following species containing **A** is the most stable: (Consider how potential energy and stability are related.)

A_2X_2 A_2X^+ $\boxed{A_2XY}$ Lowest P.E. (if X⁻ & Y⁻ have equal P.E.)

g. Assuming that **Y⁻** and **X⁻** have equal potential energy, and that ΔG^0 is approximately equal to ΔH^0 for this reaction, select which bond is the stronger?

$\boxed{\text{A—Y stronger}}$ A—X stronger   A—Y and A—X are equivalent

A–Y must be stronger because A_2XY is lower (P.E.) than A_2X_2

h. If the rate of the last step: $A_2XY \rightarrow AY + AX$ is increased, how would the rate of the total **overall** reaction be changed?
Overall rate would **increase**
Overall rate would **decrease**
$\boxed{\text{Overall rate would } \textbf{remain the same}}$ (fast step <u>after</u> r.d.s.)

i. The temperature of the complete reaction is increased. Which statement describes the effect on the rate constant **k** and the activation energy **Ea** for the rate-determining step?
k will increase and Ea will increase
k will increase and Ea will decrease
$\boxed{\text{k will increase and Ea will remain the same}}$ $k \propto T$; T and Ea are not functions of each other
k will remain the same and Ea will decrease

j. Assume a catalyst is added to the system such that its **only** effect is to increase the rate of the rate- determining step. How will this catalyst affect the value of ΔG^0?
ΔG^0 will be a larger positive/smaller negative value
ΔG^0 will be a larger negative/smaller positive value
ΔG^0 will remain unchanged. ΔG^0 = <u>state</u> function; unrelated to rate (path function)
No conclusion about ΔG^0 can be determined without a value for ΔH^0

k. What is the **reverse** reaction depicted by this diagram? that is, write the reaction formula equation for the **reverse** reaction as shown in the diagram.

$$Ay + x^\ominus + Ax \longrightarrow Ax + Ax + y^\ominus$$
$$\underline{\text{NET}}: Ay + x^\ominus \longrightarrow Ax + y^\ominus$$

l.   What is the value for ΔG^0 for this **reverse** reaction? (Indicate **numerical value plus correct sign.**)

Must be the same value with <u>opposite</u> sign as forward reaction $= +42$ kJ/mole-reaction

m.   Which reaction is spontaneous?

$\boxed{\text{The } \textbf{forward} \text{ reaction}}$ (ΔG^0) The **reverse** reaction

Both reactions are spontaneous   **Neither** reaction is spontaneous

n.   Write the specific formula reaction for the **rate-determining step** in the **reverse** direction.

$$A_2xy + x \overset{\cdot\cdot}{\cdot} {}^{\ominus} \longrightarrow \overset{\oplus}{A_2}x + x \overset{\cdot\cdot}{\cdot} {}^{\ominus} + y^{\ominus}$$

$$\underline{NET}: \quad A_2xy \longrightarrow \overset{\oplus}{A_2}x + y^{\ominus}$$

o.   Write a rate expression for the **overall reverse** reaction based on starting materials at the right, assuming **Step 4** is an equilibrium.
(Step 4 is the first step reading in reverse direction)
for <u>reverse</u> direction: $K = [A_2xy]/[Ay][Ax]$; rate r.d.s. $= k[A_2xy]$
Substitute: rate (overall reverse) $= kK[Ay][Ax]$ or $k_{(general)} [Ay][Ax]$

p.   Calculate the value of K for $AX + AX \xrightarrow{\quad} A_2X_2$ using the energy value given.

$$\text{(Step 1 equilibrium) } \Delta G^0 = +20 \text{ kJ} \quad \log K = \frac{\Delta G^0}{-2.303 \, RT}$$

$$\log k = \frac{+20}{-(2.303)(0.008314)(298)} = -3.51, \, K = 10^{-3.51} = 3.1 \times 10^{-4} \text{ kJ}$$

3. Show the three-step mechanism for the reaction of 1-methylcyclohexene with Br_2/H_2O (aqueous solution) to form 2-bromo-1-methylcyclohexanol.

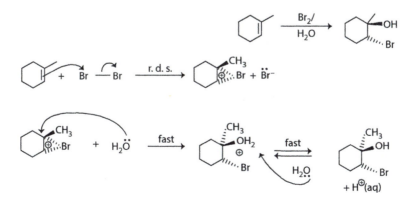

4. Draw structural formulas for the **major** product for each of the following reactions. **Stereochemistry must be indicated** for those reactions that produce a **stereospecific** product.

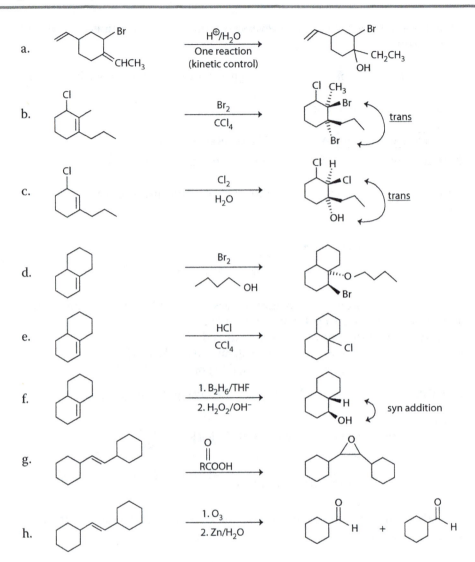

Organic I: Practice Exam 10

1. **Name** the following compounds and **include the extra information requested:**
   a. Name the compound and circle directly on the structure all stereogenic (asymetri-cally substituted) double bonds.

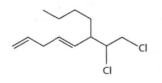

   b. Name the compound and circle directly on the structure all stereogenic (asymetri-cally substituted) double bonds.

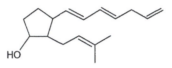

   c. Name the compound and state the stereochemistry E or Z of the stereogenic double bond in the molecule based on the structure as written.

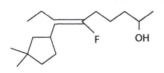

2. a. Draw any **alkene** constitutional isomer of $C_8H_{14}OF_2$ which can exist in two diaste-reomeric forms (a diastereomeric pair) based specifically on a stereogenic double bond. **Show both stereoisomers; credit requires demonstration of the correct stereochemical relationship.**

   b. Draw the structure of molecules [A] and [B] by **first** solving the ozonolysis reaction shown; each of the three products are formed in a **one-to-one-to-one mole ratio.** Select the most probable (most stable) compound.

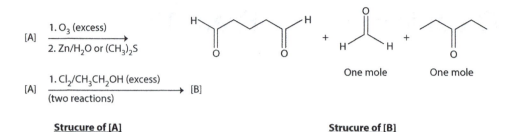

One mole   One mole

[A] $\xrightarrow[\text{2. Zn/H}_2\text{O or (CH}_3)_2\text{S}]{\text{1. O}_3 \text{ (excess)}}$

[A] $\xrightarrow[\text{(two reactions)}]{\text{1. Cl}_2/\text{CH}_3\text{CH}_2\text{OH (excess)}}$ [B]

Strucure of [A]   **Strucure of [B]**

c. Consider a proton transfer reaction for the following two molecules acting as acids. (i) Draw the anion formed for each acid after proton transfer. (ii) For **each** anion structure, draw one **resonance structure** to indicate how the anion might be stabilized (each atom will be an octet); **show how each structure produces the next resonance hybrid by using curved electron arrows** to indicate the changing role of lone pairs and pi-bonding electrons. (iii) From this analysis, select the molecule which will be **more acidic** and **state** the **reason** for your selection.

	Anion	**Resonance Structure**

$CH_3C=C-CH_2-H$

$N=C-CH_2-H$

More acidic and reason:

3. Show complete **balanced** acid/base **equations** for the two reactions shown. Then, use the table of pKa's and state which **direction** the equilibrium is favored (reactant/left or product/right); state your reason by showing the pKa analysis.
   a. CH_3COCH_3 (acid) plus NH_2^- (base)

   b. CF_3CO_2H (acid) plus $C_6H_5O^-$ (base)

4. Complete the following parts for the **electrophilic addition** reaction of Br_2 to 2-hexene:
   a. Show the complete **mechanism** using simple line drawings:
      i.   The mechanism should include **two** steps: r.d.s.; product formation.
      ii. Be certain to show **complete arrow notation** for all electron changes in all steps of the reaction mechanism.
      iii. Be specific when drawing the structure of the **intermediate** which leads to stereospecificity.

   b. Assume that the reaction of Br_2 is **specifically** to **E**-2-hexene. (i) Use the template to draw one of the two possible expected stereoisomeric products; label **each** chiral carbon as **R** or **S**. (ii) Then, state whether the other possible expected stereoisomeric product would be a diastereomer or an enantiomer.

5. Complete the following parts for the **electrophilic addition** reaction of $H^+_{(aq)}/H_2O$ with the molecule shown in the partial equation:

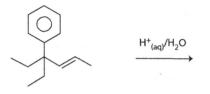

$$H^+_{(aq)}/H_2O$$

a. Show the complete mechanism for the **electrophilic addition** reaction to produce a **rearranged** product: *Follow the directions:* Be certain to show **complete arrow notation** for all electron changes in all steps of the reaction mechanism.

   i. Step 1 is the r.d.s.; *select the direction of this step so that a rearrangement will occur.* Label the specific **electrophilic** and **nucleophilic** atom for this step.

  ii. Show the rearrangement step; in addition, draw the structure of the **transition state** for this step which clearly shows the identity of the atom or group which is migrating.

 iii. Step 3 is the addition of the nucleophile to the rearranged cation; label the specific **electrophilic** and **nucleophilic** atom for this step.

 iv. Step 4 produces the final neutral rearranged product.

b. Draw the structure of the **other** possible **rearranged product** for this reaction; no further mechanism should be shown.

6. Draw structural formulas for the **major** product for the following reactions. *Include stereochemistry where required. However, do **not** indicate a specific stereoisomer if the reaction produces approximately 50%/50% diastereomers; enantiomers are not distinguished.*

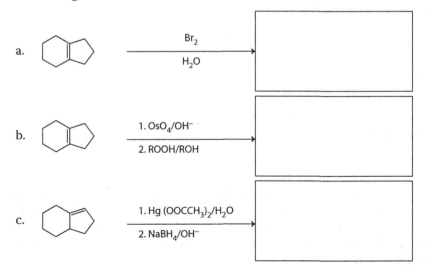

a. $\xrightarrow[\text{H}_2\text{O}]{\text{Br}_2}$

b. $\xrightarrow[\text{2. ROOH/ROH}]{\text{1. OsO}_4/\text{OH}^-}$

c. $\xrightarrow[\text{2. NaBH}_4/\text{OH}^-]{\text{1. Hg (OOCCH}_3)_2/\text{H}_2\text{O}}$

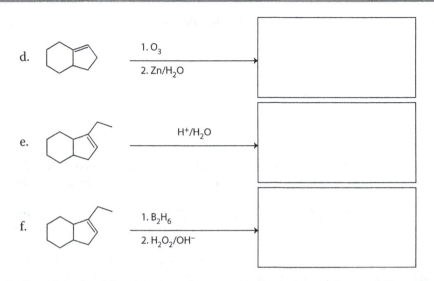

d.   1. O_3
     2. Zn/H_2O

e. H^+/H_2O

f.   1. B_2H_6
     2. H_2O_2/OH^-

7. Complete the following short (one-reaction) syntheses. <u>**General**</u> **Procedure:** Organic carbons must come from the carbon sources listed (use as many molecules of the starting compounds as needed); otherwise, any other reagents can be used. Show the important reagents and products for each of the major steps involved; you do not need to show mechanisms or high-energy intermediates. Balanced equations are not required. **Be certain that your selected reactions produce either the major isomer or (acceptable) one of equivalent or similar major isomers. (Choose the one desired; a synthesis cannot include minor products.)**
   Starting with **2-methyl-1-pentene,** show how you could synthesize
   a.   2-bromo-2-methylpentane

   b.   2-methyl-1-pentanol

   c.   2-methyl-1, 2-pentanediol

   d.   1-bromo-2-methyl-2-pentanol

   e.   2-methylpentane

8. A reaction related to amino acid formation in the living tissue is depicted on the **simplified** potential energy diagram shown below. Use this diagram to answer the following questions about the specific reaction depicted; **use the line drawings exactly as shown in the diagram.** The **forward** reaction is read from **left-to-right**; the **reverse** reaction is read from **right-to-left**.

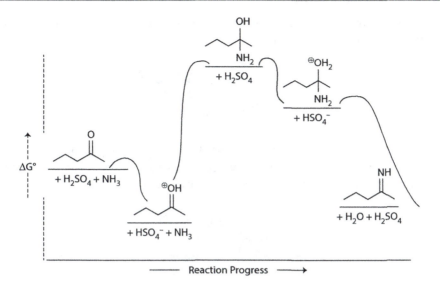

a. Write the net overall **forward** reaction; that is, write the reaction equation for the **forward** reaction as shown in the diagram.

b. Which direction is spontaneous under standard conditions, forward or reverse?

c. Write the reaction for the **specific** step which is the **rate-limiting (slow) step in the forward** direction.

d. Write the reaction for the **specific** step which is the **rate-limiting (slow) step in the reverse** direction.

e. The third step in the forward reaction is an equilibrium; under standard conditions, is the equilibrium constant greater than 1; less than 1; or exactly zero?

ORGANIC I: PRACTICE EXAM 2B: ANSWERS

1. **Name** the following compounds and **include the extra information requested:**
   a. Name the compound and circle directly on the structure all stereogenic (asymetrically substituted) double bonds.

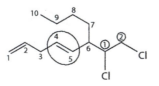

6-(1,2-dichloroethyl)-1,4-decadiene

b. Name the compound and circle directly on the structure all stereogenic (asymetrically substituted) double bonds.

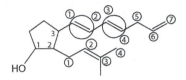

3-(1,3,6-heptatrienyl)-2-(3-methyl-2-butenyl)-cyclopentanol

c. Name the compound and state the stereochemistry E or Z of the stereogenic double bond in the molecule based on the structure as written.

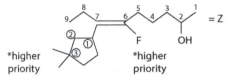

*higher priority   *higher priority

6-fluoro-7-(3,3-dimethylcyclopentyl)-6-nonen-2-ol

2. a. Draw any **alkene** constitutional isomer of $C_8H_{14}OF_2$ which can exist in two diastereomeric forms (a diastereomeric pair) based specifically on a stereogenic double bond. **Show both stereoisomers; credit requires demonstration of the correct stereochemical relationship.**

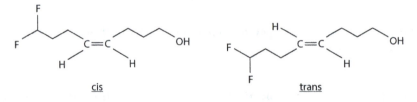

cis   trans

b. Draw the structure of molecules [A] and [B] by **first** solving the ozonolysis reaction shown; each of the three products are formed in a **one-to-one-to-one mole ratio**. Select the most probable (most stable) compound.

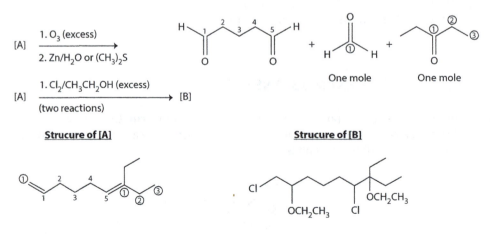

[A] 1. O₃ (excess) / 2. Zn/H₂O or (CH₃)₂S

[A] 1. Cl₂/CH₃CH₂OH (excess) (two reactions) → [B]

One mole   One mole

Strucure of [A]   **Strucure of [B]**

c. Consider a proton transfer reaction for the following two molecules acting as acids with the bold face hydrogen as the proton being transferred. (i) Draw the anion formed for each acid after proton transfer. (ii) For **each** anion structure, draw one **resonance structure** to indicate how the anion might be stabilized (each atom will be an octet); **show how each structure produces the next resonance hybrid by using curved electron arrows** to indicate the changing role of lone pairs and pi-bonding electrons. (iii) From this analysis, select the molecule which will be **more acidic** and **state** the **reason** for your selection.

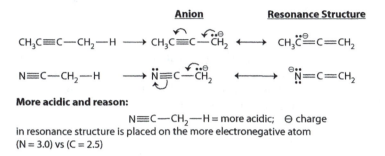

| Anion | Resonance Structure |

More acidic and reason:

$$N \equiv C - CH_2 - H = \text{more acidic;} \quad \ominus \text{charge}$$
in resonance structure is placed on the more electronegative atom (N = 3.0) vs (C = 2.5)

3. Show complete **balanced** acid/base **equations** for the two reactions shown. Then, use the table of pKa's and state which **direction** the equilibrium is favored (reactant/left or product/right); state your reason by showing the pKa analysis.

a. CH_3COCH_3 (acid) plus NH_2^- (base)

$$CH_3COCH_3 + \overset{\ominus}{\underset{..}{N}}H_2 \rightleftharpoons CH_3CO\overset{..}{\underset{..}{C}}H_2 + \overset{..}{N}H_3$$

pKa = 19.2 $\xrightarrow{\text{product favored}}$ pKa = 38

stronger acid $\xrightarrow{\hspace{4cm}}$ weaker acid

b. CF_3CO_2H (acid) plus $C_6H_5O^-$ (base)

$$CF_3\overset{..}{\underset{..}{C}}OOH + C_6H_5\overset{..}{\underset{..}{O}}{:}^{\ominus} \rightleftharpoons CF_3CO\overset{..}{\underset{..}{O}}{:}^{\ominus} + C_6H_5\overset{..}{O}H$$

pKa = 0.18 $\xrightarrow{\text{product favored}}$ pKa = 9.9

stronger acid $\xrightarrow{\hspace{4cm}}$ weaker acid

4. Complete the following parts for the **electrophilic addition** reaction of Br_2 to 2-hexene:
   a. Show the complete **mechanism** using simple line drawings:
      i. The mechanism should include **two** steps: r.d.s.; product formation.
      ii. Be certain to show **complete arrow notation** for all electron changes in all steps of the reaction mechanism.
      iii. Be specific when drawing the structure of the **intermediate** which leads to stereospecificity.

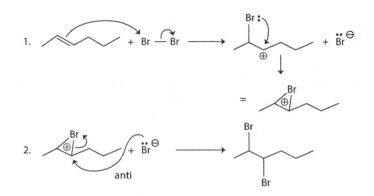

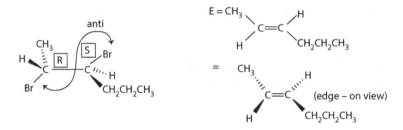

b. Assume that the reaction of Br_2 is **specifically** to **E**-2-hexene. (i) Use the template to draw one of the two possible expected stereoisomeric products; label **each** chiral carbon as **R** or **S**. (ii) Then, state whether the other possible expected stereoisomeric product would be a diastereomer or an enantiomer.

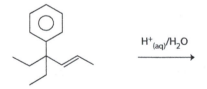

The other product will be the <u>enantiomer</u> ([S, R])

5. Complete the following parts for the **electrophilic addition** reaction of $H^+_{(aq)}/H_2O$ with the molecule shown in the partial equation:

$$H^+_{(aq)}/H_2O$$

a. Show the complete mechanism for the **electrophilic addition** reaction to produce a **rearranged** product: *Follow the directions:* Be certain to show **complete arrow notation** for all electron changes in all steps of the reaction mechanism.
    i.   Step 1 is the r.d.s.; *__select the direction of this step so that a rearrangement will occur.__* Label the specific **electrophilic** and **nucleophilic** atom for this step.
    ii. Show the rearrangement step; in addition, draw the structure of the **transition state** for this step which clearly shows the identity of the atom or group which is migrating.
    iii. Step 3 is the addition of the nucleophile to the rearranged cation; label the specific **electrophilic** and **nucleophilic** atom for this step.
    iv. Step 4 produces the final neutral rearranged product.

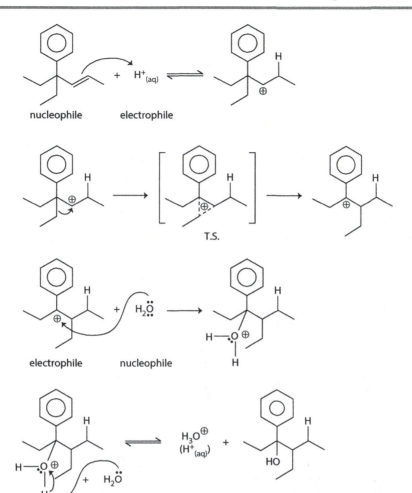

b. Draw the structure of the **other** possible **rearranged product** for this reaction; no further mechanism should be shown.

6. Draw structural formulas for the **major** product for the following reactions. *Include stereochemistry where required. However, do __not__ indicate a specific stereoisomer if the reaction produces approximately 50%/50% diastereomers; enantiomers are not distinguished.*

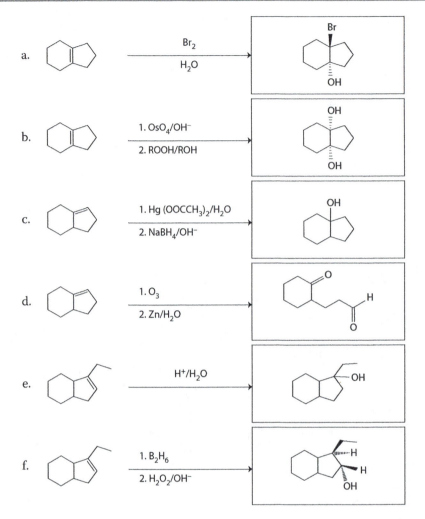

7. Complete the following short (one-reaction) syntheses. **General** **Procedure:** Organic carbons must come from the carbon sources listed (use as many molecules of the starting compounds as needed); otherwise, any other reagents can be used. Show the important reagents and products for each of the major steps involved; you do not need to show mechanisms or high-energy intermediates. Balanced equations are not required. **Be certain that your selected reactions produce either the major isomer or (acceptable) one of equivalent or similar major isomers. (Choose the one desired; a synthesis cannot include minor products.)**

Starting with **2-methyl-1-pentene,** show how you could synthesize
a. 2-bromo-2-methylpentane

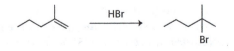

b. 2-methyl-1-pentanol

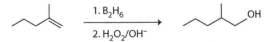

c. 2-methyl-1,2-pentanediol

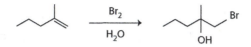

d. 1-bromo-2-methyl-2-pentanol

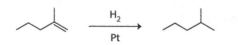

e. 2-methylpentane

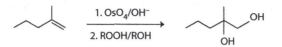

8. A reaction related to amino acid formation in the living tissue is depicted on the **simplified** potential energy diagram shown below. Use this diagram to answer the following questions about the specific reaction depicted; *use the line drawings exactly as shown in the diagram.* The **forward** reaction is read from **left-to-right;** the **reverse** reaction is read from **right-to-left.**

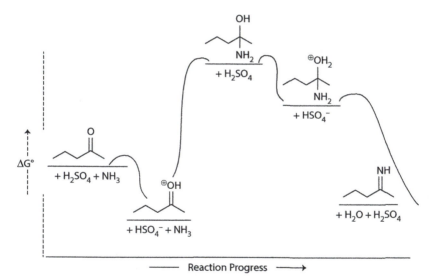

Reaction Progress ⟶

a. Write the net overall **forward** reaction; that is, write the reaction equation for the **forward** reaction as shown in the diagram.

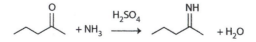

b. Which direction is spontaneous under standard conditions, forward or reverse?
    forward

c. Write the reaction for the **specific** step which is the **rate-limiting (slow) step in the forward** direction.

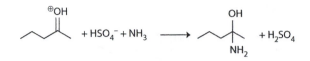

d. Write the reaction for the **specific** step which is the **rate-limiting (slow) step in the reverse** direction.

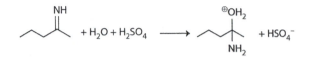

e. The third step in the forward reaction is an equilibrium; under standard conditions, is the equilibrium constant greater than 1; less than 1; or exactly zero?
    greater than 1

Organic I: Practice Exam 11

For mechanisms in this exam, be certain to show <u>complete</u> <u>arrow</u> <u>notation</u> for all electron changes in all reaction steps; <u>label</u> the <u>electrophile</u> and <u>nucleophile</u> <u>where</u> <u>requested</u>.

1. a. Show the complete mechanism for the **electrophilic addition** reaction shown in the partial equation to form the **major** constitutional isomer product; stereochemistry is not required to be shown. The mechanism should show **three** steps: label the specific **electrophilic** atom and **nucleophile** for the first two steps; be specific when drawing the structure of the **intermediate** in the first step which demonstrates the stereochemical requirements.

   b. Determine the structure of molecule **[A]** based on the given products for the ozonolysis reaction:

2. a. Show the complete mechanism for the **electrophilic addition** reaction shown in the partial equation to form the **major** constitutional isomer product. The mechanism should show **three** steps: label the specific **electrophilic** atom and **nucleophile** for the first two steps.

   b. The following reactant molecule will undergo a **<u>rearrangement</u>** reaction. Show the complete mechanism for the **electrophilic addition** reaction shown in the partial equation to form a **rearranged** product: show the formation of the **initial** non-rearranged cation; **identify** the atom/atom grouping which will **migrate** to form a more stable cation and then **draw** the structure of the **<u>transition state</u>** for this step which clearly shows the identity of the atom or group which is migrating; **show** the structure of the **final** rearranged cation; and show the formation of the **final** rearranged product.

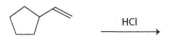

3. Show the structure of the **major** constitutional isomer product for each of the following reactions: ***Draw the correct stereoisomer product if the reaction requires;*** **no** **rearrangements** will occur.

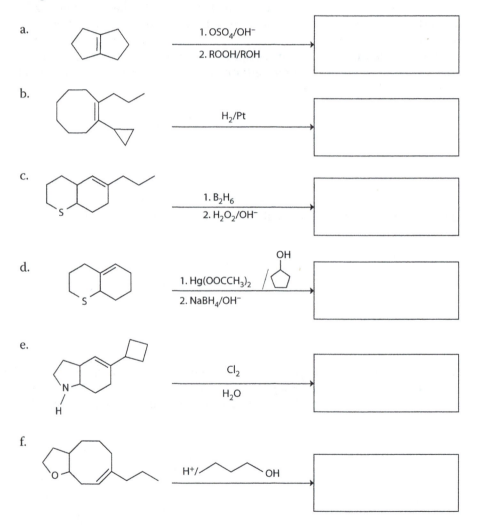

4. Listed below are **three** alkene molecules; also listed are **three** reagents. ***Show the separate reaction of each alkene with each of the three reagents.*** There should be a **total of nine separate** reactions; these reactions are **not** part of a sequence. For reactions of H^+/H_2O and HCl, show **both** of the possible products and indicate the **major** product.

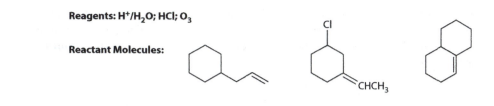

Reagents: H⁺/H₂O; HCl; O₃

Reactant Molecules:

ORGANIC I: PRACTICE EXAM 2C: ANSWERS

For mechanisms in this exam, be certain to show __complete__ __arrow__ __notation__ for all electron changes in all reaction steps; __label__ the __electrophile__ and __nucleophile__ __where__ __requested__.

1. a. Show the complete mechanism for the **electrophilic addition** reaction shown in the partial equation to form the **major** constitutional isomer product; stereochemistry is not required to be shown. The mechanism should show **three** steps: label the specific **electrophilic** atom and **nucleophile** for the first two steps; be specific when drawing the structure of the **intermediate** in the first step which demonstrates the stereochemical requirements.

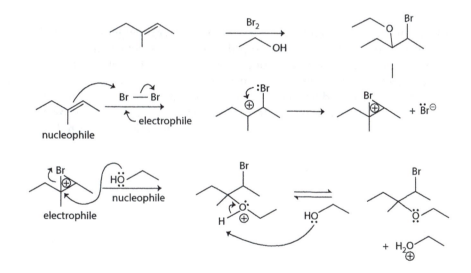

   b. Determine the structure of molecule **[A]** based on the given products for the ozonolysis reaction:

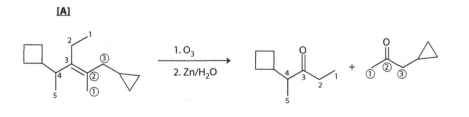

2. a. Show the complete mechanism for the **electrophilic addition** reaction shown in the partial equation to form the **major** constitutional isomer product. The mechanism should show **three** steps: label the specific **electrophilic** atom and **nucleophile** for the first two steps.

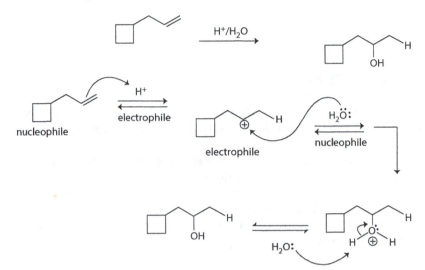

b. The following reactant molecule will undergo a **rearrangement** reaction. Show the complete mechanism for the **electrophilic addition** reaction shown in the partial equation to form a **rearranged** product: show the formation of the **initial** non-rearranged cation; **identify** the atom/atom grouping which will **migrate** to form a more stable cation and then **draw** the structure of the **transition state** for this step which clearly shows the identity of the atom or group which is migrating; **show** the structure of the **final** rearranged cation; and show the formation of the **final** rearranged product.

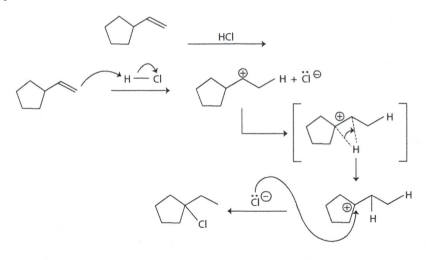

3. Show the structure of the **major** constitutional isomer product for each the following reactions: ***Draw the correct stereoisomer product if the reaction requires***; **no** **rearrangements** will occur.

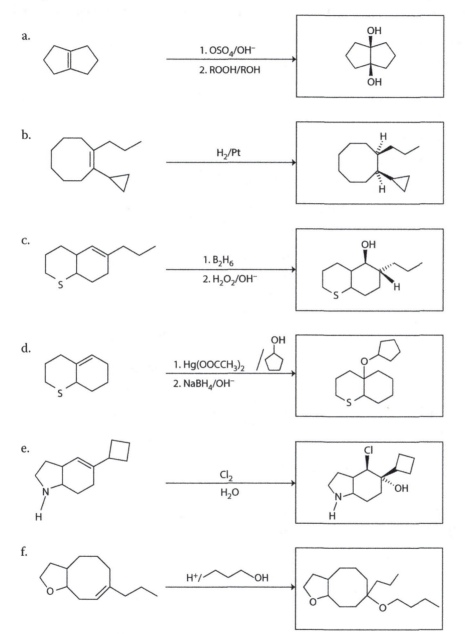

4. Listed below are **three** alkene molecules; also listed are **three** reagents. *Show the sepa-rate reaction of each alkene with each of the three reagents.* There should be a **total** of **nine separate** reactions; these reactions are **not** part of a sequence. For reactions of H⁺/H₂O and HCl, show **both** of the possible products and indicate the **major** product.

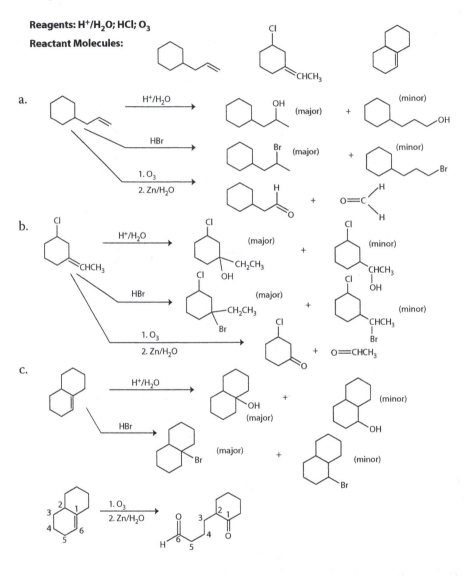

Organic I: Practice Exam 12

For mechanisms in this exam: be certain to show complete arrow notation for all electron changes in all requested steps of the reaction mechanism; if line drawings are used, <u>show hydrogens when they are involved in the reaction step</u>.

1. a. Draw structures for **all constitutional** isomer products for the **<u>free radical halogenation (substitution)</u>** reaction shown below; **do not show duplicates**.

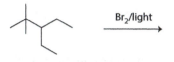

   b. Show **<u>only</u>** the **<u>propagation phase</u>** (2 steps) part of the mechanism for the free radical halogenation reaction shown above in part **(a)** to form the **<u>major</u> constitutional** isomer.

2. Complete the following 2–3 reaction synthesis shown below: start with the organic molecule shown plus **any** other molecules including **<u>any</u> nucleophile** where necessary.

starting with: synthesize

3. a. Show the **<u>complete</u>** mechanism for the **<u>free radical electrophilic addition</u>** reaction shown below to form the **<u>major</u> constitutional** isomer: (i) **initiation** phase; (ii) the **complete propagation** phase (2 steps); **<u>label</u>** the **r.d.s.** in the propagation phase; (iii) **one example** of a termination reaction in the **termination** phase.

HBr/ROOR (peroxides)

   b. Consider any reaction stereospecificity and any formation of any chiral carbons for the reaction from part 3 **(a)**. Then, state
   i.   the total **number** of **<u>chiral</u>** carbons **formed** in this specific reaction _____
   ii. the **<u>total</u> number** of **stereoisomers formed** in this specific reaction _____

4. For the **SN2** reaction shown below in the partial equation: The R/S stereochemical configuration of the one **reacting** chiral carbon is indicated with the structure of the reactant. **(i)** Use the template on the left to draw the **transition state** for this reaction; **be certain that the groups represent the correct stereochemistry**. **(ii)** Then, use the template on the right to draw the correct stereochemical **product;** state the correct **R/S** configuration of the product.

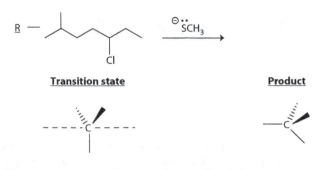

| **Transition state** | | **Product** |

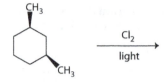

5. Consider the following reaction of cis-1,3-dimethylcyclohexane:

a. Draw structures for the **five** possible **constitutional** isomers, that is, **do not include stereoisomers**.

b. Determine the product percentages for these isomers based on relative hydrogen reaction rates of **tertiary = 5.2 secondary = 3.9 primary = 1**
   Use letter abbreviations to indicate each isomer.

c. How many **total** isomers would be formed in the above reaction if **diastereomers** were **included?**

ORGANIC I: PRACTICE EXAM 2.5A: ANSWERS

For mechanisms in this exam: be certain to show complete arrow notation for all electron changes in all requested steps of the reaction mechanism; if line drawings are used, <u>show hydrogens when they are involved in the reaction step</u>.

1. a. Draw structures for **all** **constitutional** isomer products for the <u>**free radical halogenation (substitution)**</u> reaction shown below; **<u>do not show duplicates</u>**.

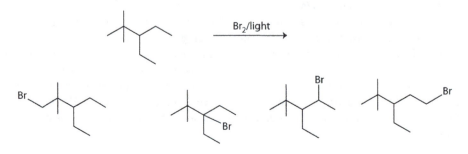

b. Show **only** the **propagation phase** (2 steps) part of the mechanism for the free radical halogenation reaction shown above in part **(a)** to form the **major** constitutional isomer.

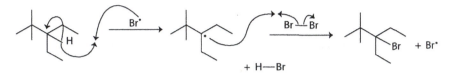

+ H—Br

2. Complete the following 2–3 reaction synthesis shown below: start with the organic molecule shown plus **any** other molecules including **any nucleophile** where necessary.

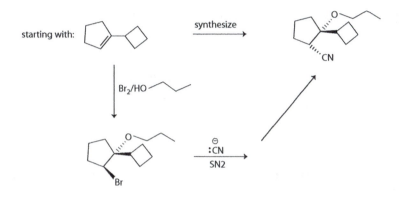

3. a. Show the **complete** mechanism for the **free radical electrophilic addition** reaction shown below to form the **major constitutional** isomer: (i) **initiation** phase; (ii) the **complete propagation** phase (2 steps); **label** the r.d.s. in the propagation phase; (iii) **one example** of a termination reaction in the **termination** phase.

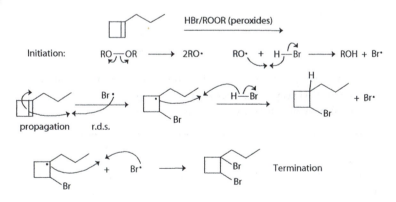

b. Consider any reaction stereospecificity and any formation of any chiral carbons for the reaction from part 3 **(a)**. Then, state
   i. the total **number** of **chiral** carbons **formed** in this specific reaction 2
   ii. **the total** number of **stereoisomers formed** in this specific reaction 4

4. For the **SN2** reaction shown below in the partial equation: The R/S stereochemical configuration of the one **reacting** chiral carbon is indicated with the structure of the reactant. **(i)** Use the template on the left to draw the **transition state** for this reaction; **be certain that the groups represent the correct stereochemistry. (ii)** Then, use the

template on the right to draw the correct stereochemical **product;** state the correct **R/S** configuration of the product.

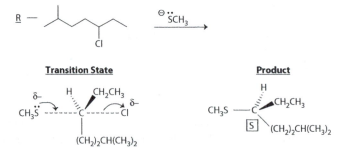

Transition State   **Product**

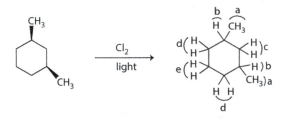

5. Consider the following reaction of cis-1,3-dimethylcyclohexane:

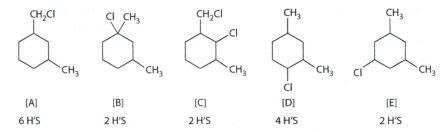

a. Draw structures for the **five** possible **constitutional** isomers, that is, **do not include stereoisomers.**

[A]	[B]	[C]	[D]	[E]
CH$_2$Cl	Cl CH$_3$	CH$_2$Cl	CH$_3$	CH$_3$
6 H'S	2 H'S	2 H'S	4 H'S	2 H'S

b. Determine the product percentages for these isomers based on relative hydrogen reaction rates of **tertiary = 5.2 secondary = 3.9 primary = 1**
    Use letter abbreviations to indicate each isomer.

A: 6H × 1 = 6	% A = 6/47.6 × 100% = 12.6%
B: 2H × 5.2 = 10.4	% B = 10.4/47.6 = 21.8%
C: 2H × 3.9 = 7.8	% C = 7.8/47.6 = 16.4%
D: 4H × 3.9 = 15.6	% D = 15.6/47.6 = 32.8%
E: 2H × 3.9 = 7.8	% E = 7.8/47.6 = 16.4%
47.6	

c. How many **total** isomers would be formed in the above reaction if **diastereomers** were **included?** Radical intermediate is planar.
    A = 1 isomer; B = 2 stereoisomers; C = 2 stereoisomers; D = 2 stereoisomers; and E = 2 stereoisomers. Total = 9

Organic I: Practice Exam 13

1. Owing to symmetry, only **four** total constitutional isomers are possible for the free radical **mono**-halogenation reaction of the molecule shown with Br$_2$/light. Show the structures of **all** constitutional isomers; do not consider stereoisomers. Mechanism is **not** required.

a. Draw the structure of 2,2-dimethyl-3-octene; then, show the reaction equation (reactant and products) of this alkene with 1. O$_3$; 2. Zn/H$_2$O. Show only the product of the complete sequence.

$$\xrightarrow[\text{2. Zn/H}_2\text{O}]{\text{1. O}_3}$$

b. Redraw the structure of 2,2-dimethyl-3-octene; then, show the reaction equation (reactant and product) of this alkene with NBS (*N*-bromosuccinimide); *show the reaction specifically in **allylic** position*.

$$\xrightarrow{\text{NBS}}$$

2. a. Show the complete mechanism for the **electrophilic addition** reaction of HBr in the presence of peroxides (ROOR) to **7-ethyl-4-methyl-4-nonene** to form the **major constitutional** isomer. *The mechanism involves radicals:* Follow the directions for the complete mechanism:

　(i) Initiation phase: **two** reactions for the initiation phase; (ii) propagation phase: **label** the **r.d.s.** in the propagation phase; and (iii) **one** example of a termination reaction for the termination phase. Be certain to show **complete arrow notation** for all electron changes in all steps of the reaction mechanism; **for line drawings: show hydrogens when they are involved in the reaction step.**

　*Note that this reaction is an addition to the double bond, do **not** select substitution.*

b. Assume that reaction of HBr in the presence of peroxides (ROOR) from part (a) is **specifically** to the **specific** stereoisomer **E**-7-ethyl-4-methyl-4-nonene to form the **major** constitutional isomer. Consider the stereochemical requirements of the mechanism and the number of chiral carbons formed; then, determine the total **number** of stereoisomers (all diastereomers and all enantiomers) formed in the reaction with HBr/ROOR.

c. Draw **E**-7-ethyl-4-methyl-4-nonene, clearly showing the correct alkene stereochemistry; then, use the template to draw a correct diastereomer product for the reaction of **E**-7-ethyl-4-methyl-4-nonene in the reaction sequence: 1. OsO$_4$/OH⁻. 2. ROOR/ROH. Show only the product of the complete sequence.

E-7-ethyl-4-methyl-4-nonene   <u>Product of the reaction</u>

3. Complete the following synthesis problem: Starting with 2,5-dimethyl-2-hexene, show how you would synthesize 5-bromo-2,5-dimethyl-3-hexanol. (More than one reaction will be required.)

ORGANIC I: PRACTICE EXAM 2.5B: ANSWERS

1. Owing to symmetry, only **four** total constitutional isomers are possible for the free radical **mono**-halogenation reaction of the molecule shown with Br_2/light. Show the structures of **all** constitutional isomers; do not consider stereoisomers. Mechanism is **not** required.

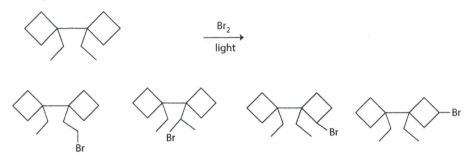

a. Draw the structure of 2,2-dimethyl-3-octene; then, show the reaction equation (reactant and products) of this alkene with 1. O_3; 2. Zn/H_2O. Show only the product of the complete sequence.

b. Redraw the structure of 2,2-dimethyl-3-octene; then, show the reaction equation (reactant and product) of this alkene with NBS (*N*-bromosuccinimide); ***show the reaction specifically in _allylic_ position***.

2. a. Show the complete mechanism for the **electrophilic addition** reaction of HBr in the presence of peroxides (ROOR) to **7-ethyl-4-methyl-4-nonene** to form the **major constitutional** isomer. *__The mechanism involves radicals:__* Follow the directions for the complete mechanism:

(i) Initiation phase: **two** reactions for the initiation phase; (ii) propagation phase: **label** the **r.d.s.** in the propagation phase; and (iii) **one** example of a termination reaction for the termination phase. Be certain to show **complete arrow notation** for all electron changes in all steps of the reaction mechanism; **for line drawings: show hydrogens** when they are involved in the reaction step.

*Note that this reaction is an addition to the double bond, do **not** select substitution.*

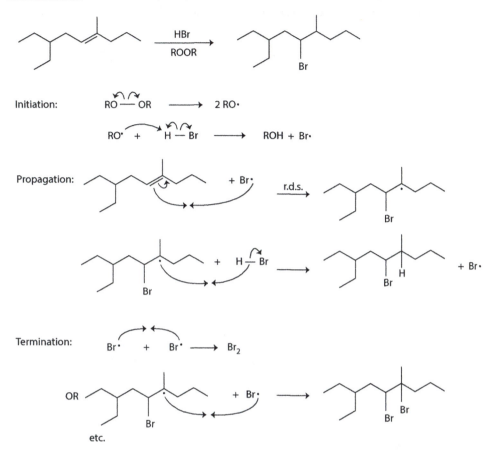

b. Assume that reaction of HBr in the presence of peroxides (ROOR) from part (a) is **specifically** to the **specific** stereoisomer **E**-7-ethyl-4-methyl-4-nonene to form the **major** constitutional isomer. Consider the stereochemical requirements of the mechanism and the number of chiral carbons formed; then, determine the total **number** of stereoisomers (all diastereomers and all enantiomers) formed in the reaction with HBr/ROOR.

2 chiral carbons formed; non-stereospecific reaction,
all diastereomers & enantiomers form = ④

c. Draw **E**-7-ethyl-4-methyl-4-nonene, clearly showing the correct alkene stereochemistry; then, use the template to draw a correct diastereomer product for the reaction of **E**-7-ethyl-4-methyl-4-nonene in the reaction sequence: 1. OsO$_4$/OH⁻. 2. ROOR/ROH. Show only the product of the complete sequence.

E-7-ethyl-4-methyl-4-nonene   **Product of the reaction**

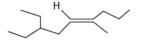
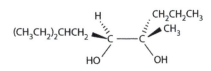

3. Complete the following synthesis problem: Starting with 2,5-dimethyl-2-hexene, show how you would synthesize 5-bromo-2,5-dimethyl-3-hexanol. (More than one reaction will be required.) Synthesis reactions must produce the major isomer as products.

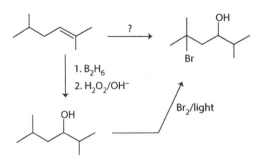

Organic I: Practice Exam 14

1. a. Show the complete mechanism for the **free radical <u>substitution</u> (halogenation)** reaction of **1,4-diethylcyclohexane** with Br_2/light to form the **<u>major</u> <u>mono</u>-bromo constitutional** isomer. Follow the directions for the complete mechanism:
   (i) initiation phase; (ii) propagation phase; **label** the **r.d.s.** in the propagation phase; (iii) **<u>one</u>** example of a termination reaction for the termination phase.
   Be certain to show **complete arrow notation** for all electron changes in all steps of the reaction mechanism; **for line drawings: show <u>hydrogens</u> when they are involved in the reaction step**.

   b. Due to symmetry, only **four** total constitutional isomers are possible for the free radical **<u>mono</u>**-halogenation reaction of 1,4-diethylcyclohexane with Br_2/light. Show the structures of the **<u>other</u> <u>three</u>** constitutional isomers which would form minor products in this reaction; do not consider stereoisomers.

2. a. Show the initiation and propagation phases in the mechanism for the **electrophilic addition** reaction of HBr in the presence of peroxides (ROOR) to **3,5-dimethyl-2-hexene** to form the **major constitutional** isomer. Follow the directions for the complete mechanism: **The mechanism involves radicals:** (i) initiation phase (two separate reactions); (ii) propagation phase; **label** the **r.d.s.** in the propagation phase and **state** the **hybridization** and **geometry** of the carbon radical formed in the r.d.s. of the propagation phase. *<u>The termination phase need not be shown</u>*. Be certain to show **complete arrow notation** for all electron changes in all steps of the reaction mechanism; **if line drawings are used, show hydrogens when they are involved in the reaction step.**

   b. Assume that reaction of HBr in the presence of peroxides (ROOR) from part (a) is **specifically** to the **<u>specific</u>** stereoisomer **<u>Z</u>**-3,5-dimethyl-2-hexene to form the **<u>major</u>** constitutional isomer. Considering any stereochemical requirements of the mechanism, answer the following stereochemical questions:
   i. Draw **<u>Z</u>**-3,5-dimethyl-2-hexene, clearly showing stereochemistry.
   ii. How many chiral carbons are **produced** in this specific reaction? _____
   iii. Does this specific reaction produce diastereomers; if so how many? _____
   iv. Considering all diastereomers and all enantiomers, how many
      total stereoisomers are formed in this specific reaction? _____

3. a. and b. Show the structure of the **major** product for each part of the following two part reaction sequences. Determine the product for the first reaction set; then use this product to complete the second reaction set. Each two part reaction sequence is separate. *<u>Include stereochemistry where required</u>; **no** rearrangements* will occur.

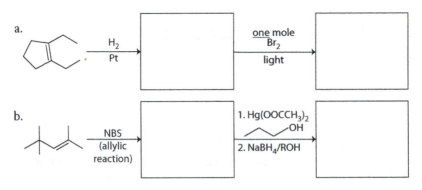

a.

H$_2$ / Pt

one mole Br$_2$ / light

b.

NBS (allylic reaction)

1. Hg(OOCCH$_3$)$_2$ —OH
2. NaBH$_4$/ROH

c. Draw the structure of 4-ethyl-1,4-heptadiene; then show the reaction of this diene with: 1. O$_3$; 2. Zn/H$_2$O. Note that this reaction will form a total of **three** final products.

4-ethyl-1, 4-heptadiene	Products

$$\xrightarrow[\text{2. Zn/H}_2\text{O}]{\text{1. O}_3}$$

d. Complete the following synthesis problem: Starting with 1-cyclobutyl-1-propene, show how you would synthesize 1-(1-bromocyclobutyl)-1,2-propanediol.

ORGANIC I: PRACTICE EXAM 2.5C: ANSWERS

1. a. Show the complete mechanism for the **free radical substitution (halogenation)** reaction of **1,4-diethylcyclohexane** with Br$_2$/light to form the **major mono-bromo constitutional** isomer. Follow the directions for the complete mechanism:

   (i) initiation phase; (ii) propagation phase; **label** the **r.d.s.** in the propagation phase; (iii) **one** example of a termination reaction for the termination phase.

   Be certain to show **complete arrow notation** for all electron changes in all steps of the reaction mechanism; **for line drawings: show hydrogens when they are involved in the reaction step**.

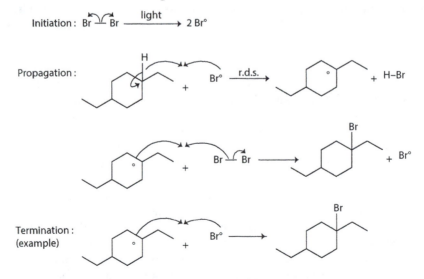

Initiation: Br—Br $\xrightarrow{\text{light}}$ 2 Br°

Propagation: ... Br° $\xrightarrow{\text{r.d.s.}}$ ... + H–Br

... Br—Br ⟶ ... + Br°

Termination: (example) ... Br° ⟶ ...

b. Due to symmetry, only **four** total constitutional isomers are possible for the free radical **mono**-halogenation reaction of 1,4-diethylcyclohexane with Br_2/light. Show the structures of the **other three** constitutional isomers which would form minor products in this reaction; do not consider stereoisomers.

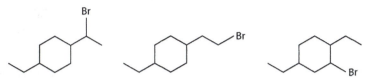

2. a. Show the initiation and propagation phases in the mechanism for the **electrophilic addition** reaction of HBr in the presence of peroxides (ROOR) to **3,5-dimethyl-2-hexene** to form the **major constitutional** isomer. Follow the directions for the complete mechanism: **The mechanism involves radicals:** (i) initiation phase (two separate reactions); (ii) propagation phase; **label** the **r.d.s.** in the propagation phase and **state** the **hybridization** and **geometry** of the carbon radical formed in the r.d.s. of the propagation phase. *__The termination phase need not be shown__.* Be certain to show **complete arrow notation** for all electron changes in all steps of the reaction mechanism; **if line drawings are used, show hydrogens when they are involved in the reaction step.**

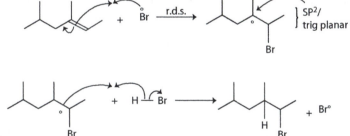

b. Assume that reaction of HBr in the presence of peroxides (ROOR) from part (a) is **specifically** to the **specific** stereoisomer **Z**-3,5-dimethyl-2-hexene to form the **major** constitutional isomer. Considering any stereochemical requirements of the mechanism, answer the following stereochemical questions:

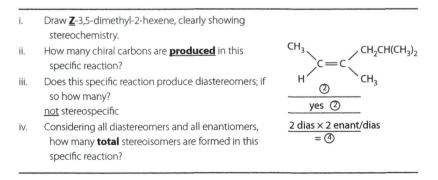

i. Draw **Z**-3,5-dimethyl-2-hexene, clearly showing stereochemistry.

ii. How many chiral carbons are **produced** in this specific reaction?

iii. Does this specific reaction produce diastereomers; if so how many? not stereospecific

iv. Considering all diastereomers and all enantiomers, how many **total** stereoisomers are formed in this specific reaction?

3. a. and b. Show the structure of the **major** product for each part of the following two part reaction sequences. Determine the product for the first reaction set; then use this product to complete the second reaction set. Each two part reaction sequence is separate. ***Include stereochemistry where required***; **no rearrangements** will occur.

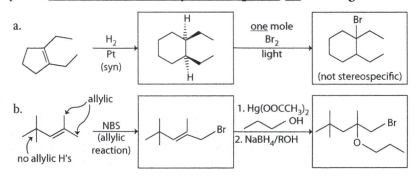

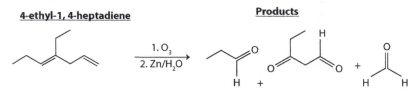

c. Draw the structure of 4-ethyl-1,4-heptadiene; then show the reaction of this diene with: 1. O_3; 2. Zn/H_2O. Note that this reaction will form a total of **three** final products.

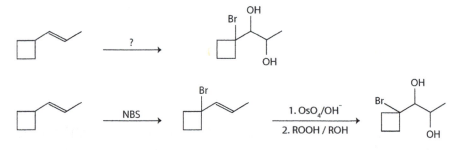

d. Complete the following synthesis problem: Starting with 1-cyclobutyl-1-propene, show how you would synthesize 1-(1-bromocyclobutyl)-1, 2-propanediol.

Organic I: Practice Exam 15

1. a. Name the following compound:

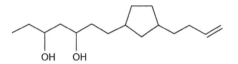

   b. Name the following compound:

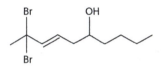

   c. A compound has the molecular formula $C_6H_{14}O_2$. Deduce its structure from the spectra information provided: IR: broad peak at 3300–3400 cm⁻¹; **no** peak at 1720 cm⁻¹
   ¹H-NMR: δ 1.2 (12H) singlet; δ 2.0 (2H) broad singlet. For partial credit, you may want to state some of your conclusions.

2. Each part represents a reaction of one reagent with four possible substrate (reactant) molecules or one substrate molecule with four possible reagent combinations. For **each** reaction shown: (i) Determine the mechanism by which the reaction should occur: **SN1, SN2, free radical halogenation**, etc.; place the answer to the right of the arrow. (ii) Then select the four possible substrates or reagents would react the **fastest** in this reaction; circle the letter answer.

a. **A** or **B** or **C** or **D** $\xrightarrow{\text{NaCN}}$

   A. $CH_3CH_2CH_2OCH_3$ **B.** $CH_3CH_2CH_2OH$ **C.** $CH_3CH_2CH_2Cl$ **D.** $CH_3CH_2CH_2CH_3$

b. CH_3Br $\xrightarrow{\text{A or B or C or D}}$

   A. $CH_3\overset{O}{\overset{||}{C}}OH$/neutral   **B.** $CH_3\overset{O}{\overset{||}{C}}OH/H_2SO_4$ **C.** $CH_3\overset{O}{\overset{||}{C}}OH/H_3PO_4$ **D.** $CH_3\overset{O}{\overset{||}{C}}OH/CH_3\overset{O}{\overset{||}{C}}O^-$ Na⁺

c. $CH_3CH_2CH_2{-}OH$ $\xrightarrow{\text{A or B or C or D}}$

   A. Na Br   **B.** NaSH   **C.** NaCN   **D.** NaCl/H_3PO_4

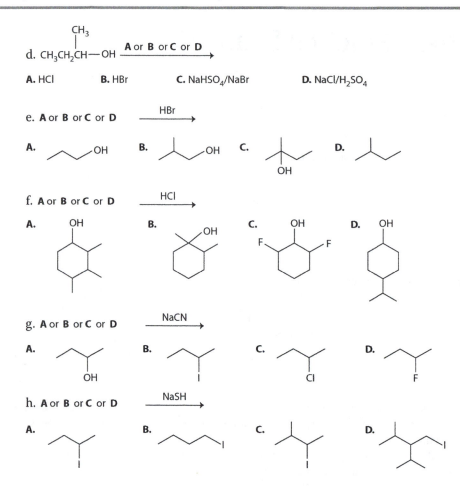

d. $CH_3CH_2CH(CH_3)$—OH $\xrightarrow{\textbf{A or B or C or D}}$

A. HCl **B.** HBr **C.** NaHSO$_4$/NaBr **D.** NaCl/H$_2$SO$_4$

e. **A or B or C or D** $\xrightarrow{\text{HBr}}$

A. ~OH **B.** ~OH **C.** (OH) **D.**

f. **A or B or C or D** $\xrightarrow{\text{HCl}}$

A. OH **B.** OH **C.** OH F, F **D.** OH

g. **A or B or C or D** $\xrightarrow{\text{NaCN}}$

A. OH **B.** I **C.** Cl **D.** F

h. **A or B or C or D** $\xrightarrow{\text{NaSH}}$

A. I **B.** I **C.** I **D.** I

3. Consider the following SN2 reaction interconverting an alkyl chloride and an alcohol:

$$CH_3CH_2CH_2\text{—OH} + HCl \rightleftharpoons CH_3CH_2CH_2\text{—Cl} + H_2O$$

a. Using the given balanced equation and the table of bond dissociation energies (shown below), calculate ΔH^0 for this reaction in the forward direction, that is, conversion of propanol to propyl chloride (1-chloropropane).

* use: prim C−O = 380 kJ
    prim C−Cl = 338 kJ
    H−Cl = 431 kJ
    H−OH = 497 kJ

b. Based on your answer to part **a**, if the reaction is at equilibrium, which organic product predominates?

    alcohol   alkyl chloride   neither (50% of each)

c. Show the **two steps** of the mechanism for the reaction in the forward direction. Include the structure of the transition state of the r.d.s. for this reaction, specifying which bonds are being formed or broken.

e. Write the overall rate expression.
f. Calculate the equilibrium constant at 298 K for this reaction based on the answer to part **a** if ΔS^0 is approximately zero (i.e., if ΔG^0 is approximately $= \Delta H^0$).
Now consider the reaction in a **basic** solution:

$$CH_3CH_2CH_2-OH + Cl^- \rightleftharpoons CH_3CH_2CH_2-Cl + OH^-$$

g. In this case, ΔH^0 can be estimated by using the bond dissociation energies of the C—OH and C—CI only, ignoring the free (nonbonded) anions in the calculation. What would be the calculated estimate for ΔH^0 for the conversion of propanol to propyl chloride in this case?
Use values from part a.

h. Which product will predominate under equilibrium conditions, that is, thermodynamic control?

alcohol   alkyl halide   neither (50%/50% mixture)

f. Which product will predominate if the reaction is not at equilibrium, and rates of reaction determine product formation, that is, kinetic control?

alcohol   alkyl halide   neither (50%/50% mixture)

g. Propyl chloride can also be formed from **free radical chlorination** of propane as one of the possible isomers. Show the **mechanism** of this reaction; include the **initiation** step, the **propagation** steps, and one possible **termination** step.
4. Draw structural formulas for the product for each of the following reactions. If more than one product is possible, draw the structure for the **one major product only**.

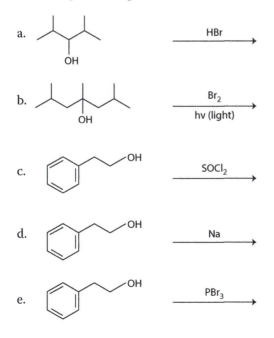

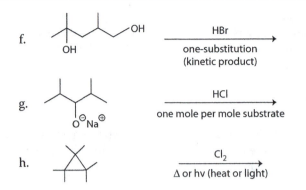

f.

HBr
one-substitution
(kinetic product)

g.

HCl
one mole per mole substrate

h.

Cl₂
Δ or hv (heat or light)

5. Draw structural formulas for the **major product** for each of the following reactions. **Stereochemistry must be indicated** for those reactions that produce a stereospecific product.

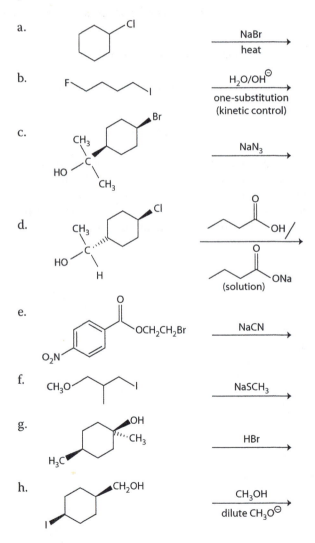

a.

NaBr
heat

b.

H₂O/OH⁻
one-substitution
(kinetic control)

c.

NaN₃

d.

(solution)

e.

NaCN

f.

NaSCH₃

g.

HBr

h.

CH₃OH
dilute CH₃O⁻

6. Use the following molecule with the drawn structure to determine certain features about the ¹H-NMR spectrum of this molecule:

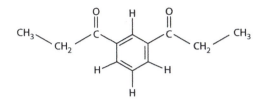

a. Use the above structure directly to label each proton or set of equivalent protons which would give a signal in the proton NMR: do this by labeling each set with a letter: (a), (b), (c), …
b. State which set of protons would show a signal most upfield (lowest δ value): _____
c. State which set of protons would show a signal most downfield (highest δ value): _____
Use the following **example** to help solve the ¹³C-NMR problem:

	(a) (b) (c)	
Example for: **¹³C-NMR**	CH_3 — CH — O — CH_3 $\qquad\ \ $ | $\qquad\ \ CH_3$ $\qquad\ \ $ (a)	(a) = 2 C @ 0–50 ppm (b) = 1 C @ 30–80 ppm (c) = 1 C @ 30–80 ppm

The same molecule used in problem #3 is redrawn below; use this structure to determine certain features about the ¹³C-NMR spectrum for this molecule:

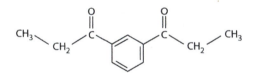

d. Use the above structure directly to label each carbon or set of equivalent carbons which would give a signal in the ¹³C-NMR: do this by labeling each set with a letter: (a), (b), (c), …
e. For **each** signal, state how many carbons would be included in the signal and an **_approximate_** δ value (within 50 ppm)

7. **Synthesis problems; general guidelines:** A valid "paper" synthesis creatively combines known reactions to logically provide a final product. The chosen method need not conform to a preferred **laboratory** synthesis. Use your knowledge of the reactions covered to date in the course to devise a synthesis of the molecule indicated as the final product starting from the stated starting organic compounds; **use as many moles of the starting molecules as desired.** Show the important reagents and products for each of the major steps involved; mechanisms or high-energy intermediates should **not** be included. Common sequences for making often-used products can be condensed by writing several reagents above one reaction arrow. **_Reactions selected must produce either the major isomer or (acceptable) one of near-equivalent major isomers_** (choose the one desired). Minor isomers may not be used as products in a valid synthesis. Once the synthesis of particular intermediate is shown in a problem, you may use this intermediate in a subsequent problem without repeating the synthetic steps. **The Cu/Li reagent will not be needed for these syntheses.**

a.   Starting with **only propane as a carbon source** plus any other reagents, synthesize:

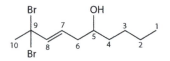

b.   Starting with 1,4,5-trimethylcyclohexene 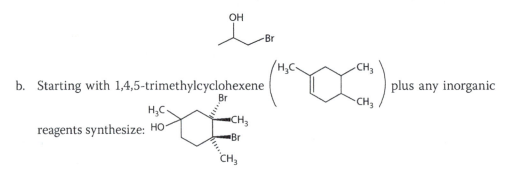 plus any inorganic

reagents synthesize: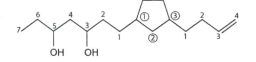

ORGANIC I: PRACTICE EXAM 3A: ANSWERS

1. a.   Name the following compound:

(figure: molecular structure)

1-[3-(3-butenyl) cyclopentyl]-3, 5-heptanediol
   b.   Name the following compound:

(figure: molecular structure)

9,9-dibromo-7-decen-5-ol
   c.   A compound has the molecular formula $C_6H_{14}O_2$. Deduce its structure from the spectra information provided: IR: broad peak at 3300–3400 cm^{-1}; **no** peak at 1720 cm^{-1}
      ^{1}H-NMR: δ 1.2 (12H) singlet; δ 2.0 (2H) broad singlet. For partial credit, you may want to state some of your conclusions.

alcohol; no C = O   12H all equivalent = 4 equivalent methyls

2H equivalent = equivalent –OH

(figure: molecular structure)

2. Each part represents a reaction of one reagent with four possible substrate (reactant) molecules or one substrate molecule with four possible reagent combinations. For **each** reaction shown: **(i)** Determine the mechanism by which the reaction should occur: **SN1, SN2, free radical halogenation**, etc.; place the answer to the right of the arrow. **(ii)** Then select the four possible substrates or reagents would react the **fastest** in this reaction; circle the letter answer.

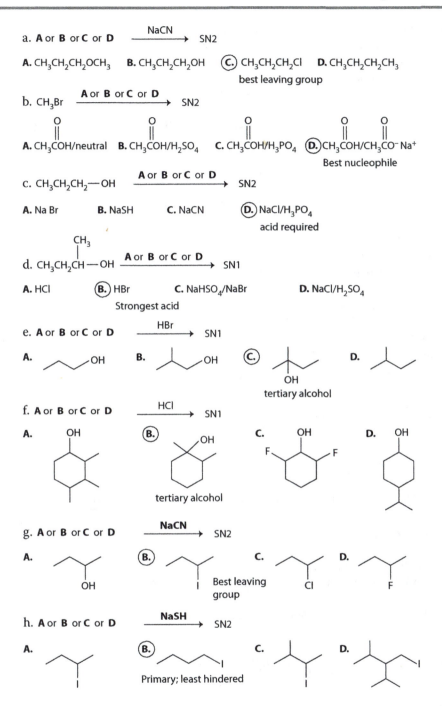

a. **A** or **B** or **C** or **D** $\xrightarrow{\text{NaCN}}$ SN2

A. $CH_3CH_2CH_2OCH_3$ **B.** $CH_3CH_2CH_2OH$ Ⓒ $CH_3CH_2CH_2Cl$ **D.** $CH_3CH_2CH_2CH_3$
best leaving group

b. CH_3Br $\xrightarrow{\text{A or B or C or D}}$ SN2

A. $CH_3\overset{O}{\overset{||}{C}}OH$/neutral   **B.** $CH_3\overset{O}{\overset{||}{C}}OH/H_2SO_4$ **C.** $CH_3\overset{O}{\overset{||}{C}}OH/H_3PO_4$ Ⓓ $CH_3\overset{O}{\overset{||}{C}}OH/CH_3\overset{O}{\overset{||}{C}}O^-$ Na+
Best nucleophile

c. $CH_3CH_2CH_2$—OH $\xrightarrow{\text{A or B or C or D}}$ SN2

A. Na Br   **B.** NaSH   **C.** NaCN   Ⓓ $NaCl/H_3PO_4$
acid required

d. $CH_3CH_2\overset{\overset{\displaystyle CH_3}{|}}{CH}$—OH $\xrightarrow{\text{A or B or C or D}}$ SN1

A. HCl   Ⓑ HBr   **C.** $NaHSO_4/NaBr$ **D.** $NaCl/H_2SO_4$
Strongest acid

e. **A** or **B** or **C** or **D** $\xrightarrow{\text{HBr}}$ SN1

A. [structure with OH]   **B.** [structure with OH]   Ⓒ [structure with OH]   **D.** [structure]
OH
tertiary alcohol

f. **A** or **B** or **C** or **D** $\xrightarrow{\text{HCl}}$ SN1

A. [structure with OH]   Ⓑ [structure with OH]   **C.** [structure with OH, F substituents]   **D.** [structure with OH]
tertiary alcohol

g. **A** or **B** or **C** or **D** $\xrightarrow{\text{NaCN}}$ SN2

A. [structure with OH]   Ⓑ [structure with I]   **C.** [structure with Cl]   **D.** [structure with F]
OH   I Best leaving   Cl   F
                           group

h. **A** or **B** or **C** or **D** $\xrightarrow{\text{NaSH}}$ SN2

A. [structure with I]   Ⓑ [structure with I]   **C.** [structure with I]   **D.** [structure with I]
I   Primary; least hindered   I   I

3. Consider the following SN2 reaction interconverting an alkyl chloride and an alcohol:

$CH_3CH_2CH_2$—OH + HCl $\rightleftharpoons$ $CH_3CH_2CH_2$—Cl + H_2O

a. Using the given balanced equation and the table of bond dissociation energies, calculate ΔH^0 for this reaction in the forward direction, that is, conversion of propanol to propyl chloride (1-chloropropane).

Bonds   Broken   Bonds   formed
Primary   C—O   380 kJ   Primary   C—Cl   338 kJ
           H—Cl   431 kJ   HO—H   497 kJ
                  811 kJ   835 kJ

$$\Delta H = 811 - 835 = -24 \text{ kJ/mole}$$

b. Based on your answer to part **a**, if the reaction is at equilibrium, which organic product predominates?

alcohol | alkyl chloride | neither (50% of each)

$\Delta H^\ominus = \Delta G^\ominus$ if ΔS^0 about zero $\Delta G^\ominus$ = product favored

c. Show the **two steps** of the mechanism for the reaction in the forward direction. Include the structure of the transition state of the r.d.s. for this reaction, specifying which bonds are being formed or broken.

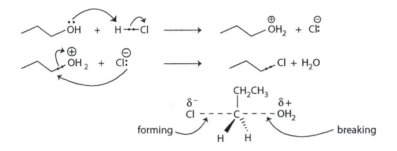

e. Write the overall rate expression.

rate r.d.s. = k [CH$_3$CH$_2$CH$_2$O̊H$_2$] [Cl$^\ominus$]   [CH$_3$CH$_2$CH$_2$O̊H$_2$] = K [CH$_3$CH$_2$CH$_2$OH] [HCl]
                                                                                    ———————————————
                                                                                            [Cl$^\ominus$]

rate (overall) = k$_{(general)}$ [CH$_3$CH$_2$CH$_2$OH] [HCl]

f. Calculate the equilibrium constant at 298 K for this reaction based on the answer to part **a** if ΔS^0 is approximately zero (i.e., if ΔG^0 is approximately = ΔH^0).

$\Delta G^0 \approx 24$ kJ/mole $\Delta G^0 = -RT\ 2.303 \log K$

$$\log K = \frac{-24\,\text{kJ}}{-(2.303)(0.008314)(298)} = +4.2 \quad K = 10^{4.2} = 1.6 \times 10^4$$

Now consider the reaction in a **basic** solution:

CH$_3$CH$_2$CH$_2$—OH + Cl$^-$ $\rightleftharpoons$ CH$_3$CH$_2$CH$_2$—Cl + OH$^-$

g. In this case ΔH^0 can be estimated by using the bond dissociation energies of the C—OH and C—Cl only, ignoring the free (nonbonded) anions in the calculation.

What would be the calculated estimate for ΔH^0 for the conversion of propanol to propyl chloride in this case?

Bond broken	Bond formed
Primary C $-$ O 380 kJ	primary C $-$ Cl 338 kJ

$\Delta H = 380$ kJ $- 338$ kJ $= +42$ kJ/mole

h. Which product will predominate under equilibrium conditions, that is, thermodynamic control?

   boxed:alcohol   alkyl halide   neither (50%/50% mixture)

f. Which product will predominate if the reaction is not at equilibrium, and rates of reaction determine product formation, that is, kinetic control?

   boxed:Alcohol   alkyl halide   neither (50%/50% mixture)
   (OH⁻) better nucleophile and much
   worse leaving group)

g. Propyl chloride can also be formed from **free radical chlorination** of propane as one of the possible isomers. Show the **mechanism** of this reaction; include the **initiation** step, the **propagation** steps, and one possible **termination** step.

Initiation: $Cl_2 \xrightarrow{\text{light}} 2Cl\cdot$

Propagation $CH_3CH_2CH_3 + Cl\cdot \longrightarrow CH_3CH_2\overset{\cdot}{C}H_2 + HCl$

$CH_3CH_2\overset{\cdot}{C}H_2 + Cl_2 \longrightarrow CH_3CH_2CH_2Cl + Cl\cdot$

Termination (example) $CH_3CH_2\overset{\cdot}{C}H_2 + Cl\cdot \longrightarrow CH_3CH_2CH_2Cl$

4. Draw structural formulas for the product for each of the following reactions. If more than one product is possible, draw the structure for the **one major product only.**

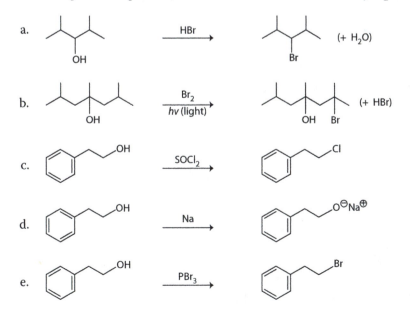

a.   [structure] OH $\xrightarrow{\text{HBr}}$ [structure] Br   (+ H₂O)

b.   [structure] OH $\xrightarrow[hv\,(light)]{Br_2}$ [structure] OH Br   (+ HBr)

c.   [structure] OH $\xrightarrow{SOCl_2}$ [structure] Cl

d.   [structure] OH $\xrightarrow{Na}$ [structure] O⁻Na⁺

e.   [structure] OH $\xrightarrow{PBr_3}$ [structure] Br

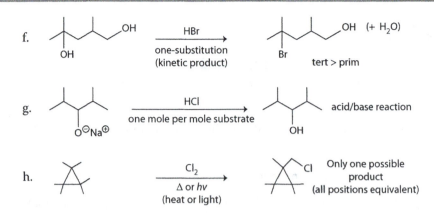

f.
HBr
one-substitution
(kinetic product)
(+ H₂O)
tert > prim

g.
HCl
one mole per mole substrate
acid/base reaction

h.
Cl₂
Δ or hv
(heat or light)
Only one possible product
(all positions equivalent)

5. Draw structural formulas for the **major product** for each of the following reactions. **Stereochemistry must be indicated** for those reactions that produce a stereospecific product.

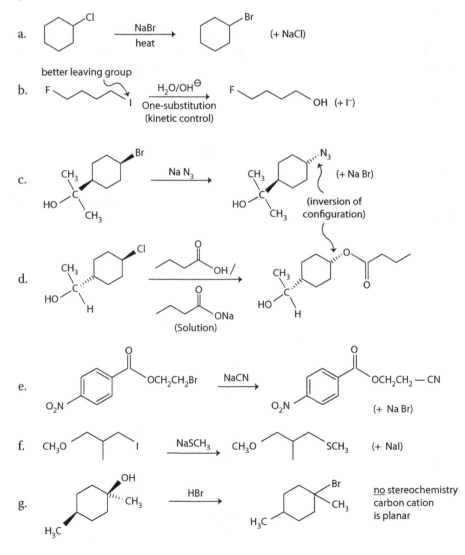

a.
NaBr
heat
(+ NaCl)

b. better leaving group
H₂O/OH⊖
One-substitution
(kinetic control)
OH (+ I⁻)

c.
Na N₃
(+ Na Br)
(inversion of configuration)

d.
(Solution)

e.
NaCN
(+ Na Br)

f. CH₃O
NaSCH₃
CH₃O
SCH₃
(+ NaI)

g.
HBr
no stereochemistry
carbon cation
is planar

h.

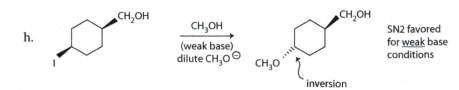

CH₃OH
(weak base)
dilute CH₃O⊖

inversion

SN2 favored
for weak base
conditions

6. Use the following molecule with the drawn structure to determine certain features about the ¹H-NMR spectrum of this molecule:

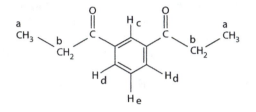

a. Use the above structure directly to label each proton or set of equivalent protons which would give a signal in the proton NMR: do this by labeling each set with a letter: (a), (b), (c), …

b. State which set of protons would show a signal most upfield (lowest δ value): a_____

c. State which set of protons would show a signal most downfield (highest δ value): c or d or e
Use the following **example** to help solve the ¹³C-NMR problem:

Example for:
¹³C-NMR

(a)   (b)   (c)
CH₃ —CH —O —CH₃
        |
        CH₃
        (a)

(a) = 2 C @ 0–50 ppm
(b) = 1 C @ 30–80 ppm
(c) = 1 C @ 30–80 ppm

The same molecule used in problem #3 is redrawn below; use this structure to determine certain features about the ¹³C-NMR spectrum for this molecule:

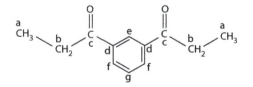

d. Use the above structure directly to label each carbon or set of equivalent carbons which would give a signal in the ¹³C-NMR: do this by labeling each set with a letter: (a), (b), (c), …

e. For **each** signal, state how many carbons would be included in the signal and an **_approximate_** δ value (within 50 ppm):

a. 2C at 0–50 ppm

b. 2C at 10–60 ppm

c. 2C at 170–220 ppm

d. 2C at 120–170 ppm

e. 1C at 120–170 ppm

f. 2C at 120–170 ppm

g. 1C at 120–170 ppm

7. **Synthesis problems; general guidelines:** A valid "paper" synthesis creatively combines known reactions to logically provide a final product. The chosen method need not conform to a preferred **laboratory** synthesis. Use your knowledge of the reactions covered to date in the course to devise a synthesis of the molecule indicated as the final product starting from the stated starting organic compounds; **use as many moles of the starting molecules as desired.** Show the important reagents and products for each of the major steps involved; mechanisms or high-energy intermediates should **not** be included. Common sequences for making often-used products can be condensed by writing several reagents above one reaction arrow. **Reactions selected must produce either the major isomer or (acceptable) one of near-equivalent major isomers** (choose the one desired). Minor isomers may not be used as products in a valid synthesis. Once the synthesis of particular intermediate is shown in a problem, you may use this intermediate in a subsequent problem without repeating the synthetic steps. **The Cu/Li reagent will not be needed for these syntheses.**

a.   Starting with **only propane as a carbon source** plus any other reagents, synthesize:

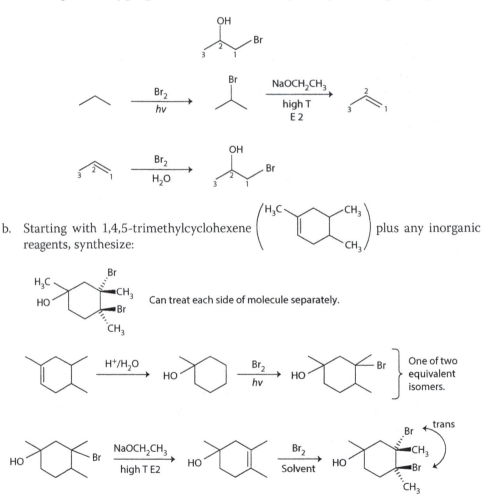

b.   Starting with 1,4,5-trimethylcyclohexene plus any inorganic reagents, synthesize:

Can treat each side of molecule separately.

Organic I: Practice Exam 16

1. Show the **mechanism** for the **strong** base elimination reaction of 1-bromo-2-methylcy-clohexane with $NaOCH_2CH_3$; include the structure of the transition state for the rate-determining step.

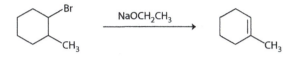

2. Draw structural formulas for the following **constitutional** isomers of $C_6H_{10}O$. Be certain that each answer has the proper atom count. In most cases, more than one answer may be possible; only **one** correct isomer is required in these cases.
   a. Determine the number of **rings plus pi-bonds** (unsaturations) in this molecule; use any method desired.

   b. Draw the structure of any isomer which has **only one sp²** carbon and exactly **two** pi-electrons.

   c. Draw an isomer which has **no pi**-electrons.

   d. Recall that free radical halogenation does **not** occur at sp² carbons. With this in mind, choose an isomer which will give **only one possible product** on free radical **mono**-chlorination.

   e. An isomer which will form a product with **exactly four pi-electrons after dehydration** with concentrated H_2SO_4. (Show the actual isomer **not** the resultant product.)

   f. Draw an isomer with **exactly one** double bond and contains a ring with the **highest ring strain** possible.

g. Draw any isomer which will undergo acid catalyzed dehydration through a bimolecular **E2** mechanism.

3. Each part represents a reaction of one reagent with four possible substrate (reactant) molecules. For **each** reaction shown: **(i)** Determine by which mechanism the reaction should occur: **SN1, SN2, E1, E2, free radical halogenation, electrophilic addition**; place the answer to the right of the arrow. **(ii)** Then select the four possible substrates would react the **fastest** in this reaction; circle the letter answer.

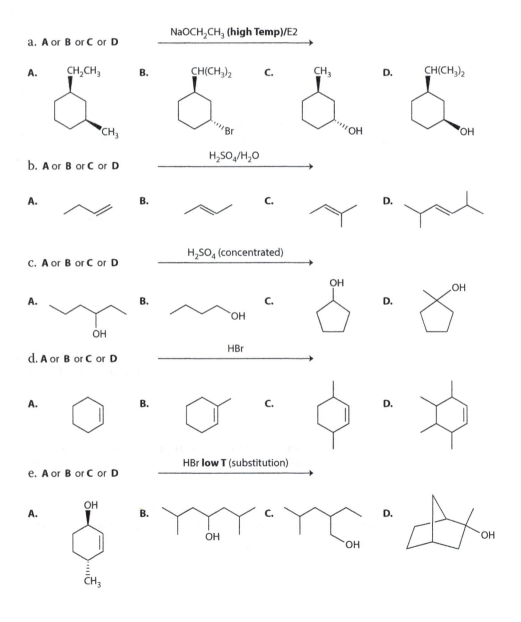

a. **A** or **B** or **C** or **D**   NaOCH$_2$CH$_3$ **(high Temp)/E2** →

A. CH$_2$CH$_3$ ... CH$_3$ B. CH(CH$_3$)$_2$ ... Br   C. CH$_3$ ... OH   D. CH(CH$_3$)$_2$ ... OH

b. **A** or **B** or **C** or **D**   H$_2$SO$_4$/H$_2$O →

A.   B.   C.   D.

c. **A** or **B** or **C** or **D**   H$_2$SO$_4$ (concentrated) →

A. OH   B. OH   C. OH   D. OH

d. **A** or **B** or **C** or **D**   HBr →

A.   B.   C.   D.

e. **A** or **B** or **C** or **D**   HBr **low T** (substitution) →

A. OH ... CH$_3$ B. OH   C. OH   D. OH

4. Draw structural formulas for the **one major** product for each of the reactions below. *Assume that no rearrangements occur,* no stereochemistry is required.

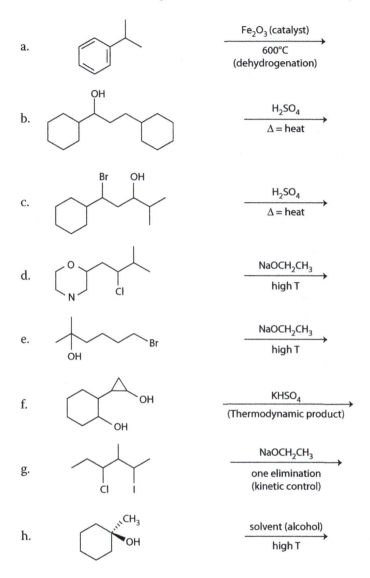

a.   Fe₂O₃ (catalyst)
                                      $\xrightarrow{\hspace{2cm}}$
                                      600°C
                                      (dehydrogenation)

b.   H₂SO₄
                                      $\xrightarrow{\hspace{2cm}}$
                                      Δ = heat

c.   H₂SO₄
                                      $\xrightarrow{\hspace{2cm}}$
                                      Δ = heat

d.   NaOCH₂CH₃
                                      $\xrightarrow{\hspace{2cm}}$
                                      high T

e.   NaOCH₂CH₃
                                      $\xrightarrow{\hspace{2cm}}$
                                      high T

f.   KHSO₄
                                      $\xrightarrow{\hspace{2cm}}$
                                      (Thermodynamic product)

g.   NaOCH₂CH₃
                                      $\xrightarrow{\hspace{2cm}}$
                                      one elimination
                                      (kinetic control)

h.   solvent (alcohol)
                                      $\xrightarrow{\hspace{2cm}}$
                                      high T

5. The following reaction proceeds by a rearrangement:

H₂SO₄ (concentrated)
$\xrightarrow{\hspace{2cm}}$

a. Show the first intermediate directly formed in the first step by proton transfer from H₂SO₄.

b. Show the structure of the carbon cation intermediate formed by elimination of water, being sure to indicate where the charge is located.

c. Show the structure of the most stable **rearranged** carbon cation intermediate, clearly indicating where the charge is located and which species shifted in the rearrangement.

d. Show the final **major rearranged** product of this reaction.

6. Complete the following reaction sequence by identifying the products [A] through [F] using the schematic shown and the additional descriptions below.
Determine the structure of molecules **[A]** through **[E]**.
Complete the sequence by determining the structure of **[F]**.
   [A] is the major product in this reaction.
   [B] is the only isomer possible.
   [C] represents the largest molecular weight compound produced by this cleavage reaction.
   [D] is the major product in this reaction.
   [E] is the only isomer possible.
   [F] has a specific stereochemistry.

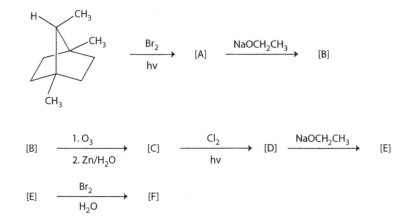

7. Use your knowledge of reactions covered to date to solve the following problem in structural analysis based on reactions or descriptions **(1)** through **(7)**. Identify the structures **[A]** through **[G]** *by first determining the structures of [C] and [D] from the ozonolysis data provided*. The formula of each compound is given as a guide.
Solve Clues (1) through (5) for   ([A], [B], [C], [D], [E])
   (1)   Compound [A] has a formula C_9H_{18}.
   (2)   Reaction of [A] with Br_2/light produces the major isomer **[B]**, with a formula $C_9H_{17}Br$.

(3) Reaction of [B] with a strong base, sodium ethoxide, yields two isomers, **[C]** and **[D]**, which are approximately equal in concentration; each has the formula C_9H_{16}.

(4) *To determine the carbon skeleton, ozonolysis was performed on [C] and [D]. The results are shown below.*

(5) **Both** [C] and [D] form the same product **[E]** when reacted with dilute aqueous acid, H^+/H_2O; [E] has the formula $C_9H_{18}O$.
Complete clues (6) and (7) for   ([F], [G])

(6) If [C] is reacted with $B_2H_6/H_2O_2/OH^-$, **[F]** is formed as a product; [F] is a constitutional isomer of [E].

(7) If [D] is reacted with $B_2H_6/H_2O_2/OH^-$ **[G]** is formed as one specific stereoisomer; [G] is also a constitutional isomer of [E] and [F].

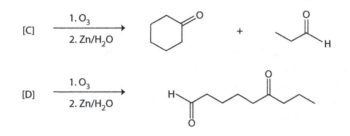

8. **Synthesis problems; general guidelines:** A valid "paper" synthesis creatively combines known reactions to logically provide a final product. The chosen method need not conform to a preferred **laboratory** synthesis. Use your knowledge of the reactions covered to date in the course to devise a synthesis of **each** molecule indicated as the final product starting from the stated starting organic compounds; *use as many moles of the starting molecules as desired.* Show the important reagents and products for each of the major steps involved; mechanisms or high-energy intermediates should **not** be included. Common sequences for making often-used products can be condensed by writing several reagents above one reaction arrow. *Reactions selected must produce either the major isomer or (acceptable) one of near-equivalent major isomers* (choose the one desired). Minor isomers may not be used as products in a valid synthesis. Once the synthesis of particular intermediate is shown in a problem, you may use this intermediate in a subsequent problem without repeating the synthetic steps.

Starting with **cyclohexene as the only carbon source** plus any other reagents desired, synthesize:

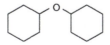

ORGANIC I: PRACTICE EXAM 3B: ANSWERS

1. Show the **mechanism** for the **strong** base elimination reaction of 1-bromo-2-methylcyclohexane with $NaOCH_2CH_3$; include the structure of the transition state for the rate-determining step.

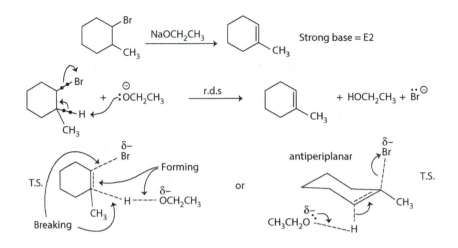

2. Draw structural formulas for the following **constitutional** isomers of $C_6H_{10}O$. Be certain that each answer has the proper atom count. In most cases, more than one answer may be possible; only **one** correct isomer is required in these cases.

   a.   Determine the number of **rings plus pi-bonds** (unsaturations) in this molecule; use any method desired.

OR $[(6 \times 2) + 2] - [10]/2 = 14-10/2 = 4/2 = 2$ (rings + π-bonds)

   b.   Draw the structure of any isomer which has **only one sp^2** carbon and exactly **two** pi-electrons.
        <u>Double bond</u> must be to the oxygen; other "unsaturation" must be a ring, for example,

   c.   Draw an isomer which has **no pi**-electrons: must have <u>two</u> rings.

   d.   Recall that free radical halogenation does **not** occur at sp^2 carbons. With this in mind, choose an isomer which will give **only one possible product** on free radical **mono**-chlorination. All sp^3 C−H must be equivalent.

or any appropriately symmetrical molecule.

e. An isomer which will form a product with **exactly four pi-electrons after dehy-dration** with concentrated H_2SO_4. (Show the actual isomer **not** the resultant product.)

Molecules must have an $-OH$ for dehydration; must have started with only 2π-electrons (dehydration provides the second double bond).

e.g.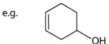

f. Draw an isomer with **exactly one** double bond and contains a ring with the **highest ring strain** possible.

Highest ring strain possible = cyclopropene

e.g.

g. Draw any isomer which will undergo acid catalyzed dehydration through a bimo-lecular **E2** mechanism. E2 dehydration applies to <u>primary</u> <u>alcohols</u>.

e.g.

3. Each part represents a reaction of one reagent with four possible substrate (reactant) molecules. For **each** reaction shown: **(i)** Determine by which mechanism the reaction should occur: **SN1, SN2, E1, E2, free radical halogenation, electrophilic addition**; place the answer to the right of the arrow. **(ii)** Then select the four possible substrates would react the **fastest** in this reaction; circle the letter answer.

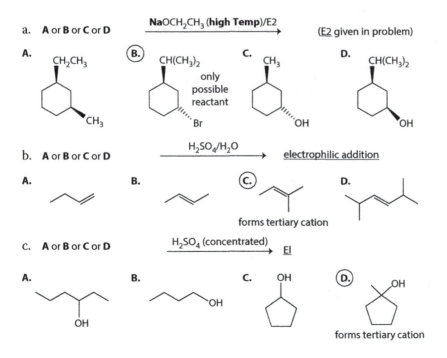

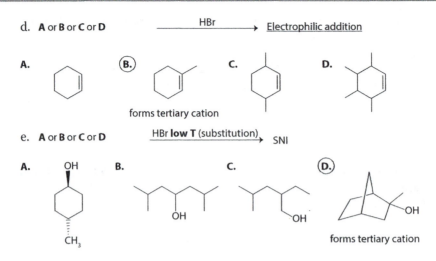

d. **A or B or C or D** $\xrightarrow{\text{HBr}}$ <u>Electrophilic addition</u>

A.   B.   C.   D.

forms tertiary cation

e. **A or B or C or D** $\xrightarrow{\text{HBr low T (substitution)}}$ SNI

A.   OH / CH₃   B.   C.   D.

forms tertiary cation

4. Draw structural formulas for the **one major** product for each of the reactions below. *Assume that no rearrangements occur*; no stereochemistry is required.

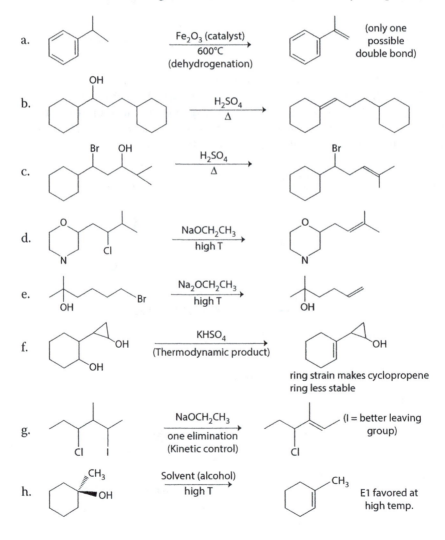

a. $\xrightarrow[\text{600°C}]{\text{Fe}_2\text{O}_3 \text{ (catalyst)}}$ (only one possible double bond)
(dehydrogenation)

b. $\xrightarrow[\Delta]{\text{H}_2\text{SO}_4}$

c. $\xrightarrow[\Delta]{\text{H}_2\text{SO}_4}$

d. $\xrightarrow[\text{high T}]{\text{NaOCH}_2\text{CH}_3}$

e. $\xrightarrow[\text{high T}]{\text{Na}_2\text{OCH}_2\text{CH}_3}$

f. $\xrightarrow[\text{(Thermodynamic product)}]{\text{KHSO}_4}$

ring strain makes cyclopropene ring less stable

g. $\xrightarrow[\substack{\text{one elimination} \\ \text{(Kinetic control)}}]{\text{NaOCH}_2\text{CH}_3}$ (I = better leaving group)

h. $\xrightarrow[\text{high T}]{\text{Solvent (alcohol)}}$ E1 favored at high temp.

5. The following reaction proceeds by a rearrangement:

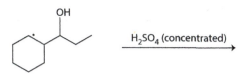

a. Show the first intermediate directly formed in the first step by proton transfer from H_2SO_4.

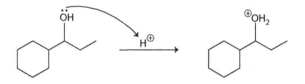

b. Show the structure of the carbon cation intermediate formed by elimination of water, being sure to indicate where the charge is located.

c. Show the structure of the most stable **rearranged** carbon cation intermediate, clearly indicating where the charge is located and which species shifted in the rearrangement.

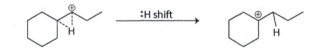

d. Show the final **major rearranged** product of this reaction.

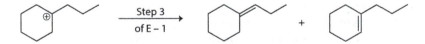

6. Complete the following reaction sequence by identifying the products [A] through [F] using the schematic shown and the additional descriptions below.
Determine the structure of molecules [A] through [E].
Complete the sequence by determining the structure of [F].
   [A] is the major product in this reaction.
   [B] is the only isomer possible.
   [C] represents the largest molecular weight compound produced by this cleavage reaction.
   [D] is the major product in this reaction.
   [E] is the only isomer possible.
   [F] has a specific stereochemistry.

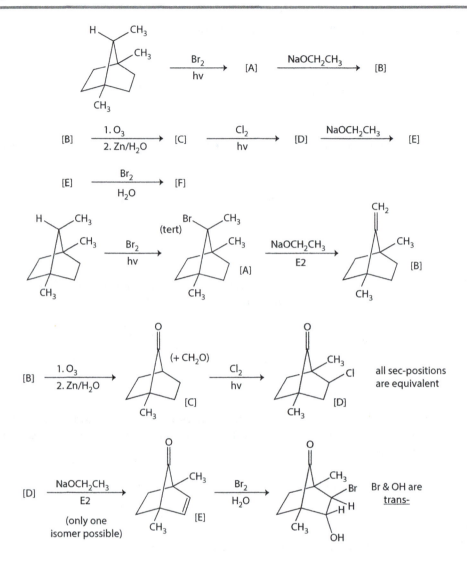

7. Use your knowledge of reactions covered to date to solve the following problem in structural analysis based on reactions or descriptions **(1)** through **(7)**. Identify the structures **[A]** through **[G]** *by first determining the structures of [C] and [D] from the ozonolysis data provided*. The formula of each compound is given as a guide.

Solve clues (1) through (5) for   ([A], [B], [C], [D], [E]).

(1) Compound *[A]* has a formula C_9H_{18}.

(2) Reaction of [A] with Br_2/light produces the major isomer **[B]**, with a formula $C_9H_{17}Br$.

(3) Reaction of [B] with a strong base, sodium ethoxide, yields two isomers, **[C]** and **[D]**, which are approximately equal in concentration; each has the formula C_9H_{16}.

(4) **To determine the carbon skeleton, ozonolysis was performed on [C] and [D]. The results are shown below.**

(5) **Both** [C] and [D] form the same product **[E]** when reacted with dilute aqueous acid, H^+/H_2O; [E] has the formula $C_9H_{18}O$.

Complete clues (6) and (7) for   ([F], [G]).

(6) If [C] is reacted with $B_2H_6/H_2O_2/OH^-$, **[F]** is formed as a product;
[F] is a constitutional isomer of [E].

(7) If [D] is reacted with $B_2H_6/H_2O_2/OH^-$ **[G]** is formed as one specific stereoisomer;
[G] is also a constitutional isomer of [E] and [F].

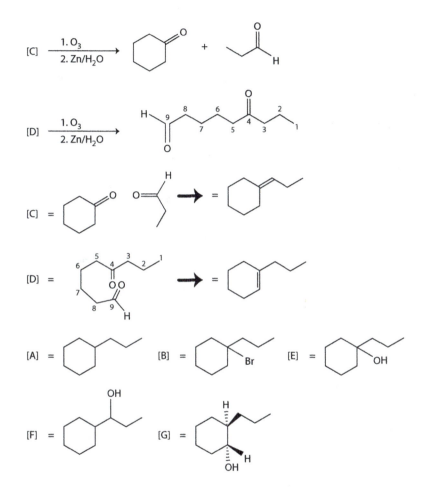

8. **Synthesis problems; general guidelines:** A valid "paper" synthesis creatively combines known reactions to logically provide a final product. The chosen method need not conform to a preferred **laboratory** synthesis. Use your knowledge of the reactions covered to date in the course to devise a synthesis of **each** molecule indicated as the final product starting from the stated starting organic compounds; *use as many moles of the starting molecules as desired*. Show the important reagents and products for each of the major steps involved; mechanisms or high-energy intermediates should **not** be included. Common sequences for making often-used products can be condensed by writing several reagents above one reaction arrow. *Reactions selected must produce either the major isomer or (acceptable) one of near-equivalent major isomers* (choose the one desired). Minor isomers may not be used as products in a valid synthesis. Once the synthesis of particular intermediate is shown in a problem, you may use this intermediate in a subsequent problem without repeating the synthetic steps.

Starting with **cyclohexene as the only carbon source** plus any other reagents desired, synthesize:

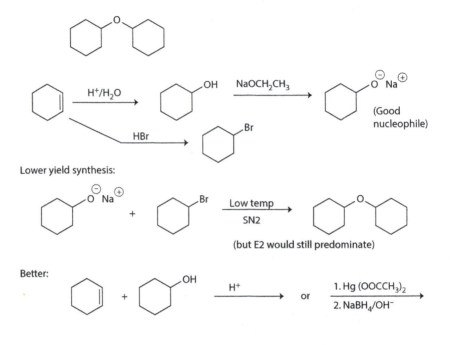

Lower yield synthesis:

Better:

Organic I: Practice Exam 17

1 Show the complete mechanism including r.d.s. transition state for the **bimolecular substitution** reaction (**SN2**) for 2-S-3-R-2-bromo-3-methylpentane using iodide ion as the nucleophile. The name means that using correct numbering, chiral carbon **#2** is **S** and carbon **#3** is **R**. Follow the directions:

a. Use the template to draw chiral carbon **#2**; place the leaving group for the substitution reaction in the position marked. **Label** all four groups on the chiral carbon with a **priority number** (1 = highest to 4 = lowest) to confirm that this carbon is **S**.

b. i. Draw the transition state. Use dashed lines and specific labels to represent bonds breaking and forming. *Be certain to show the geometry (three-dimensional structure) of the reacting electrophilic carbon with its nonreacting attached groups using a perspective drawing (wedges and dashes).*

   ii. Draw the **final product** of the reaction: draw the tetrahedral chiral carbon of the **product** molecule from the reaction in a 3D perspective using the wedge/dash method for the tetrahedron; this should indicate the correct enantiomer product. **Label** all four groups on the chiral carbon with a **priority** number and label the reactive chiral carbon as **R or S**.

   i. **Structure of SN2 transition state**   ii. **Structure of final product**

2. Now show the complete mechanism including r.d.s. transition state for the bimolecular **elimination** reaction (**E2**) for the same molecule from the previous problem: 2-S-3-R-2-bromo-3-methylpentane in the presence of the strong base $NaOCH_2CH_3$. Follow the directions:

a. Use the template to draw the named molecule: place chiral carbon #2 on the left of the drawing and chiral carbon #3 on the right of the drawing. Draw the molecule such that the **two atoms to be eliminated in the transition state are drawn in the plane of the paper**, that is, **not** on the lines with wedges or dashes.

   Be certain that chiral carbon #2 is S and chiral carbon #3 is R.

b. Use the **partial** template to draw the transition state; bond lines breaking or forming are not shown in the template. Use dashed lines **and specific labels** to represent
bonds breaking and forming.

c. Draw the structure (line drawing or wedge/dash structure) of the final product
clearly showing the correct stereochemistry; label the final product as **E** or **Z** with
correct **priority rankings** for each set of groups on the alkene carbons.

3. a. Show the complete mechanism for the **electrophilic addition** reaction of HBr in
the presence of peroxides (ROOR) to 1-methyl**cyclo**pentene to form the **major constitutional** isomer. Follow the directions for the complete mechanism:
   The mechanism involves radicals: (i) initiation phase; (ii) propagation phase;
   label the **r.d.s.** in the propagation phase and **state** the **hybridization** and **geometry**
   of the carbon radical formed in the r.d.s. of the propagation phase; (iii) one example
   of a termination reaction for the termination phase. Be certain to show **complete
   arrow notation** for all electron changes in all steps of the reaction mechanism;
   **if line drawings are used, show hydrogens when they are involved in the
   reaction step.**

   b. Consider the stereochemistry of the reaction to form the **major** constitutional
   product: Use the wedge and dash method on a ring line drawing to **either**:
      i. Draw the **specific** diastereomer formed in this reaction if you consider this reac-
         tion to be stereospecific; **or**
      ii. Draw **both** possible diastereomers formed in this reaction if you consider this
          reaction to be **not** stereospecific. *Do **not** distinguish enantiomers.*

4. a. Show the complete mechanism for the **free radical substitution (halogenation)**
   reaction of 1-pentene (**not** cyclopentene) with Br_2/light to form the **major constitutional** isomer. Do **not consider electrophilic addition across the double bond
   and do not consider allylic rearrangements.** Extra information: **C—H** B.D.E.:
   allylic: 368 kJ; alkyl secondary: 397 kJ; alkyl primary: 410 kJ. Follow the directions for
   the complete mechanism: (i) initiation phase; (ii) propagation phase; **label** the **r.d.s.**
   in the propagation phase; and (iii) one example of a termination reaction for the termination phase.
      Be certain to show **complete arrow notation** for all electron changes in all steps
   of the reaction mechanism; **if line drawings are used, show hydrogens when they
   are involved in the reaction step.**

   b. The reaction to produce the major constitutional isomer will produce a pair of enantiomers. Use the templates below to draw both enatiomers. **Label** all four groups on
   the chiral carbon for each drawing with a **priority number** (1 = highest to 4 = lowest); label the molecule as **R** or **S**. **Draw the structures with the lowest ranking
   group pointing backwards.**

c. Show the structure of all other **minor** constitutional isomer products for free radical substitution. Recall that **hydrogens on sp² carbons are not reacted** for free radical halogenation.

5. a. Show the complete mechanism for the SN1 (**unimolecular**) solvolysis reaction of the molecule shown in the partial equation in the presence of ethanol:

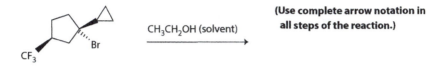

Follow the directions for the complete mechanism: (i) r.d.s.; state the **geometry** and **hybridization** of the reactive carbon **in the intermediate** formed, (ii) reaction of the solvent molecule, and (iii) formation of the final product.

b. Consider the stereochemistry of the reaction: Use the wedge and dash method on a ring line drawing to **either:**
   i. Draw the **specific** diastereomer formed in this reaction if you consider this reaction to be stereospecific; **or**
   ii. Draw **both** possible diastereomers formed in this reaction if you consider this reaction to be **not** stereospecific. **Do** not distinguish enantiomers.

6. Now show the complete mechanism for the E1 (**unimolecular**) solvolysis reaction of the same molecule shown in the previous problem in the presence of ethanol:

Follow the directions for the complete mechanism: (i) r.d.s.; (ii) formation of **any** final product; and (iii) show the **other two** possible **elimination** constitutional isomer products.

7. Show the structure of the **major** product for each part of the following two-part reaction sequences. Determine the product for the first reaction set; then use this product to complete the second reaction set. Each two-part reaction sequence is separate. _**Include stereochemistry where required**_; **no rearrangements** will occur.

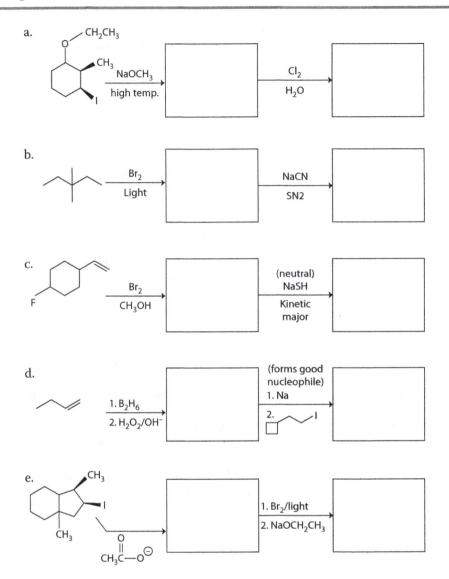

ORGANIC I: PRACTICE EXAM 3C: ANSWERS

1. Show the complete mechanism including r.d.s. transition state for the **bimolecular substitution** reaction (**SN2**) for 2-S-3-R-2-bromo-3-methylpentane using iodide ion as the nucleophile. The name means that using correct numbering, chiral carbon #2 is **S** and carbon #3 is **R**. Follow the directions:

   a. Use the template to draw chiral carbon #2; place the leaving group for the substitution reaction in the position marked. **Label** all four groups on the chiral carbon with a priority number (1 = highest to 4 = lowest) to confirm that this carbon is **S**.

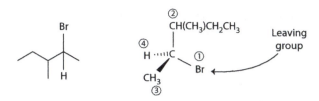

b. i. Draw the transition state. Use dashed lines **and specific labels** to represent bonds breaking and forming. ***Be certain to show the geometry (3D structure) of the reacting electrophilic carbon with its nonreacting attached groups using a perspective drawing (wedges and dashes).***

ii. Draw the **final _product_** of the reaction: draw the tetrahedral chiral carbon of the product molecule from the reaction in a 3D perspective using the wedge/dash method for the tetrahedron; this should indicate the correct enantiomer product. **Label** all four groups on the chiral carbon with a **priority number** and label the reactive chiral carbon as **R or S**.

(i) **Structure of SN2 transition state**   (ii) **Structure of final product**

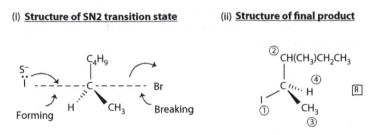

2. Now show the complete mechanism including r.d.s. transition state for the bimolecular **elimination** reaction (**E2**) for the same molecule from the previous problem: 2-S-3-R-2-bromo-3-methylpentane in the presence of the strong base $NaOCH_2CH_3$.
Follow the directions:
a. Use the template to draw the named molecule: place chiral carbon #2 on the left of the drawing and chiral carbon #3 on the right of the drawing. Draw the molecule such that the **two atoms to be eliminated in the transition state are drawn in the plane of the paper,** that is, **not** on the lines with wedges or dashes.
    Be certain that chiral carbon #2 is S and chiral carbon #3 is R.

b. Use the **partial** template to draw the transition state; bonds lines breaking or forming are not shown in the template. Use dashed lines **and specific labels** to represent bonds breaking and forming.

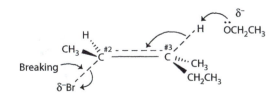

c. Draw the structure (line drawing or wedge/dash structure) of the final product clearly showing the correct stereochemistry; label the final product as **E** or **Z** with correct **priority rankings** for each set of groups on the alkene carbons.

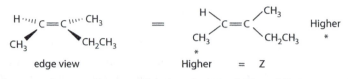

3. a. Show the complete mechanism for the **electrophilic addition** reaction of HBr in the presence of peroxides (ROOR) to 1-methyl**cyclo**pentene to form the **major constitutional** isomer. Follow the directions for the complete mechanism:

   The mechanism involves radicals: (i) initiation phase; (ii) propagation phase; **label** the **r.d.s.** in the propagation phase and **state** the **hybridization** and **geometry** of the carbon radical formed in the r.d.s. of the propagation phase; and (iii) one example of a termination reaction for the termination phase.

   Be certain to show **complete arrow notation** for all electron changes in all steps of the reaction mechanism; **if line drawings are used, show hydrogens when they are involved in the reaction step.**

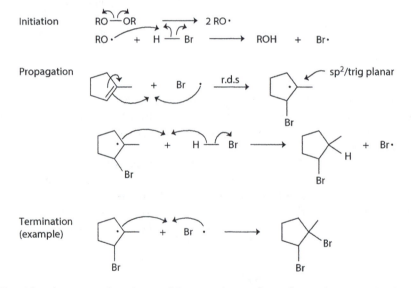

b. Consider the stereochemistry of the reaction to form the **major** constitutional product: Use the wedge and dash method on a ring line drawing to **either:**

   i. Draw the **specific** diastereomer formed in this reaction if you consider this reaction to be stereospecific; **or**

   ii. Draw **both** possible diastereomers formed in this reaction if you consider this reaction to be **not** stereospecific. **Do _not_ distinguish enantiomers.**

   Reaction is _not_ stereospecific:

4. a. Show the complete mechanism for the **free radical substitution (halogenation) reaction** of 1-pentene (**not** cyclopentene) with Br_2/light to form the **major constitutional** isomer. *Do **not** consider electrophilic addition across the double bond and do **not** consider allylic rearrangements.* Extra information: **C–H** B.D.E.: allylic: 368 kJ; alkyl secondary: 397 kJ; alkyl primary: 410 kJ. Follow the directions for the complete mechanism: (i) initiation phase; (ii) propagation phase; **label** the **r.d.s.** in the propagation phase; and (iii) one example of a termination reaction for the termination phase.

   Be certain to show **complete arrow notation** for all electron changes in all steps of the reaction mechanism; **if line drawings are used, show hydrogens when they are involved in the reaction step.**

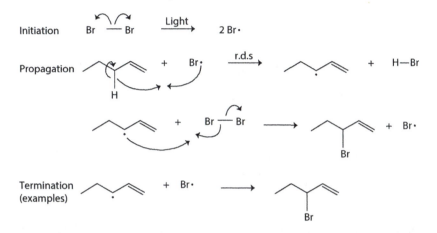

   b. The reaction to produce the major constitutional isomer will produce a pair of enantiomers. Use the templates below to draw both enatiomers. **Label** all four groups on the chiral carbon for each drawing with a priority number (1 = highest to 4 = lowest); label the molecule as **R or S. Draw the structures with the lowest ranking group pointing backwards.**

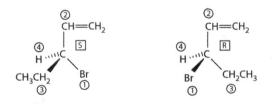

   c. Show the structure of all other **minor** constitutional isomer products for free radical substitution. Recall that **hydrogens on sp² carbons are not reacted** for free radical halogenation.

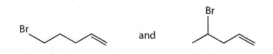

5. a. Show the complete mechanism for the SN1 (**unimolecular**) solvolysis reaction of the molecule shown in the partial equation in the presence of ethanol:

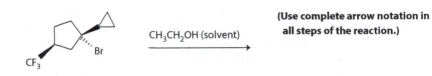

Follow the directions for the complete mechanism: (i) r.d.s.; state the **geometry** and **hybridization** of the reactive carbon **in the intermediate** formed, (ii) reaction of the solvent molecule, and (iii) formation of the final product.

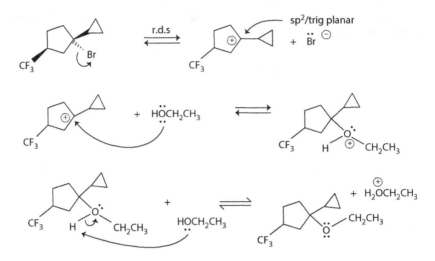

b. Consider the stereochemistry of the reaction: Use the wedge and dash method on a ring line drawing to **either:**

   i. Draw the **specific** diastereomer formed in this reaction if you consider this reaction to be stereospecific; **or**

   ii. Draw **both** possible diastereomers formed in this reaction if you consider this reaction to be **not** stereospecific. **Do *not* distinguish enantiomers.** not stereospecific.

6. Now show the complete mechanism for the E1 (**unimolecular**) solvolysis reaction of the same molecule shown in the previous problem in the presence of ethanol:

Follow the directions for the complete mechanism: (i) r.d.s.; (ii) formation of **any** final product; and (iii) show the **other two** possible **elimination** constitutional isomer products.

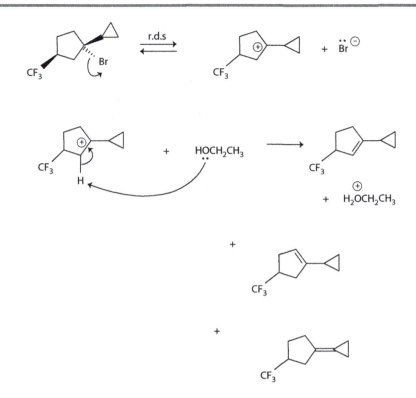

7. Show the structure of the **major** product for each part of the following two-part reaction sequences. Determine the product for the first reaction set; then use this product to complete the second reaction set. Each two-part reaction sequence is separate. _**Include stereochemistry where required; no rearrangements**_ will occur.

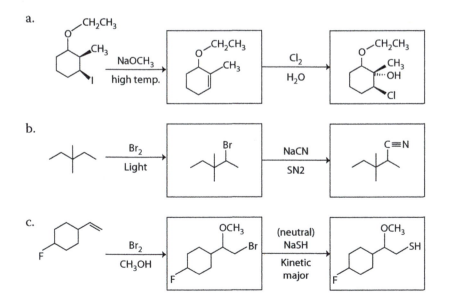

d.

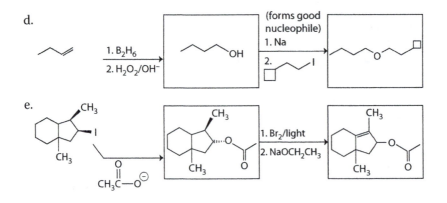

e.

Organic II: Practice Exam 18

1. Provide a correct name for each of the following compounds; **stereochemistry is not required.**

   a.

   b.

   c.

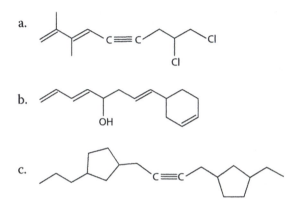

2. Show the complete mechanisms for the following two reactions:
   Complete parts (**a**) and (**b**) for the formation of a ketone from an alkyne: (The reaction is catalyzed by Hg^{+2} but the role of the catalyst should not be shown.)

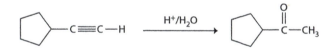

   a. Show the appropriate electrophilic addition reaction including acid/base steps; show the neutral product formed **before** the rearrangement step.
   b. The addition product is then converted to the **final** product showing the reaction by an acid catalyzed rearrangement reaction: show the steps of this rearrangement, including the role of the acid.

   c. Show the mechanism for the 1,4-addition across a conjugated diene:

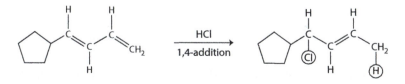

Include all resonance forms which will lead to possible products. Complete the mechanism by using the resonance form which leads to the product shown in the equation.

3. Draw structural formulas for the **major** product in each of the following reactions. Indicate **stereochemistry** where required.

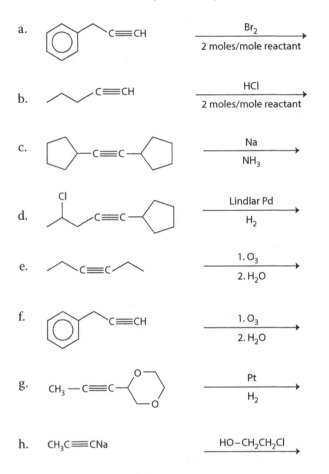

a. C≡CH

$$\xrightarrow[\text{2 moles/mole reactant}]{Br_2}$$

b. C≡CH

$$\xrightarrow[\text{2 moles/mole reactant}]{HCl}$$

c. —C≡C—

$$\xrightarrow[NH_3]{Na}$$

d. (Cl) C≡C—

$$\xrightarrow[H_2]{\text{Lindlar Pd}}$$

e. C≡C

$$\xrightarrow[\text{2. } H_2O]{\text{1. } O_3}$$

f. C≡CH

$$\xrightarrow[\text{2. } H_2O]{\text{1. } O_3}$$

g. $CH_3 - C \equiv C -$

$$\xrightarrow[H_2]{Pt}$$

h. $CH_3C \equiv CNa$

$$\xrightarrow{HO-CH_2CH_2Cl}$$

4. Consider the following electrophilic addition reactions. Indicate the correct product for each reaction:

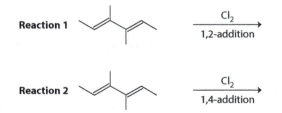

Reaction 1 $\xrightarrow[\text{1,2-addition}]{Cl_2}$

Reaction 2 $\xrightarrow[\text{1,4-addition}]{Cl_2}$

5. Identify the correct products [A], [B], [C], and [D] and [E] in the following reaction sequence. Use the ozone analysis information to solve for the structure of [E] which is used as a second organic reactant in the last Diels–Alder reaction.

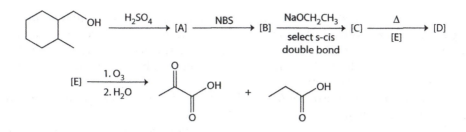

6. Draw structural formulas for the **major** product for each of the following reactions. Indicate correct **stereochemistry** where required. *Molecules can be assumed to adopt correct conformations for reactions.*

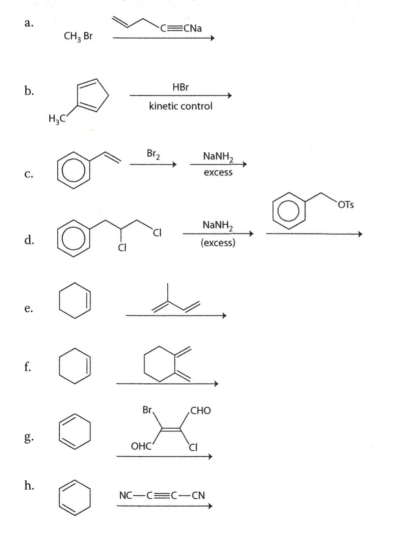

7. **Synthesis problems; general guidelines:** A valid "paper" synthesis creatively combines known reactions to logically provide a final product. The chosen method need not conform to a preferred **laboratory** synthesis. Use your knowledge of the reactions covered to date in the course to devise a synthesis of the molecule indicated as the final product

starting from the stated starting organic compounds; *use as many moles of the starting molecules as desired.* Show the important reagents and products for each of the major steps involved; mechanisms or high-energy intermediates should **not** be included. Common sequences for making often-used products can be condensed by writing several reagents above one reaction arrow.

Reactions selected must produce either the major isomer or (acceptable) one of near-equivalent major isomers (choose the one desired). Minor isomers may not be used as products in a valid synthesis. Once the synthesis of particular intermediate is shown in a problem, you may use this intermediate in a subsequent problem without repeating the synthetic steps.

Using all potential reactions covered through Organic I and II to date, devise a synthesis of each compound in the problem **starting only with 1-propene and ethanol as carbon sources** plus any other reagents. Use the additional information or directions for certain specific products.

a.   and b. Build carbon-skeleton using SN2 reactions of alkyne nucleophiles; form the appropriate alkynes from the starting materials listed.

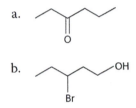

a.

b.

c.   Use a Diels–Alder reaction to form the six-membered ring; consider using two 5-carbon pieces for the Diels–Alder.

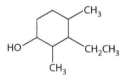

d.   Use two separate Diels–Alder reactions to form the rings; try to use compounds (intermediates or products) synthesized previously in (**a**), (**b**), or (**c**).

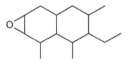

ORGANIC II: PRACTICE EXAM 1A: ANSWERS

1. Provide a correct name for each of the following compounds; **stereochemistry is not required.**

a.

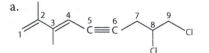

8,9-dichloro-2,3-dimethyl-1,3-nonadien-5-yne

b.

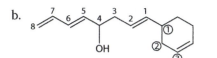

1-(3-cyclohexenyl)-1,5,7-octatrien-4-ol

priority goes to alcohol

c.

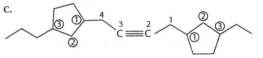

1-(3-ethylcyclopentyl)-4-(3-propylcyclopentyl)-2-butyne

(numbering could be reversed)

2. Show the complete mechanisms for the following two reactions.
   Complete parts (**a**) and (**b**) for the formation of a ketone from an alkyne:

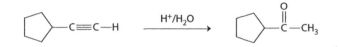

a.   Show the appropriate electrophilic addition reaction including acid/base steps; show the neutral product formed **before** the rearrangement step.
(Role of Hg^{+2} not shown.)

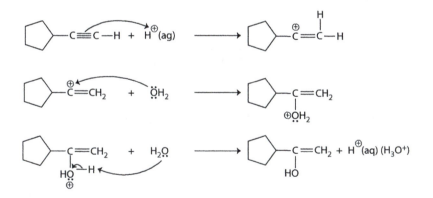

b.   The addition product is then converted to the **final** product showing the reaction by an acid catalyzed rearrangement reaction: show the steps of this rearrangement, including the role of the acid.

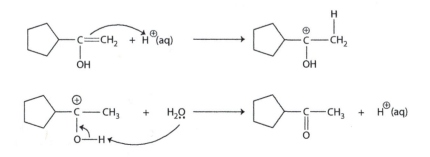

c. Show the mechanism for the 1,4-addition across a conjugated diene:

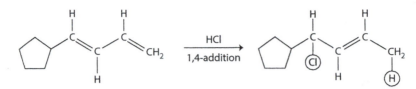

Include all resonance forms which will lead to possible products. Complete the mechanism by using the resonance form which leads to the product shown in the equation.

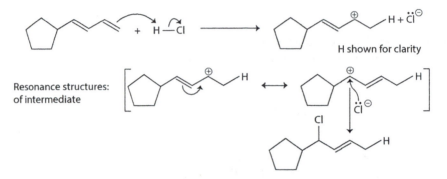

3. Draw structural formulas for the **major** product in each of the following reactions. Indicate **stereochemistry** where required.

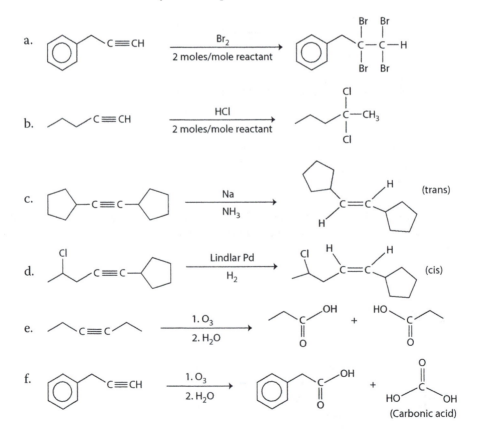

g.

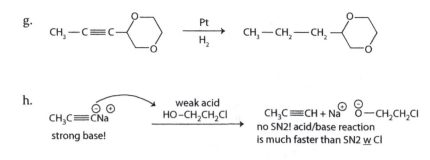

h.

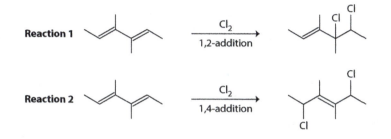

4. Consider the following electrophilic addition reactions. Indicate the correct product for each reaction:

Reaction 1   [structure] $\xrightarrow[\text{1,2-addition}]{Cl_2}$ [product structure with Cl]

Reaction 2   [structure] $\xrightarrow[\text{1,4-addition}]{Cl_2}$ [product structure with Cl]

5. Identify the correct products **[A]**, **[B]**, **[C]**, and **[D]** and **[E]** in the following reaction sequence. Use the ozone analysis information to solve for the structure of **[E]** which is used as a second organic reactant in the last Diels–Alder reaction.

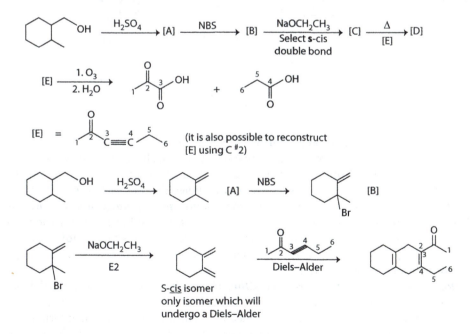

6. Draw structural formulas for the **major** product for each of the following reactions. Indicate correct **stereochemistry** where required. *Molecules can be assumed to adopt correct conformations for reactions.*

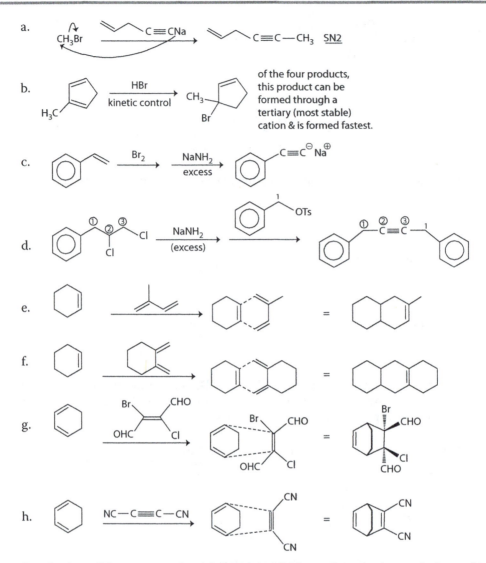

7. **Synthesis problems; general guidelines:** A valid "paper" synthesis creatively combines known reactions to logically provide a final product. The chosen method need not conform to a preferred **laboratory** synthesis. Use your knowledge of the reactions covered to date in the course to devise a synthesis of the molecule indicated as the final product starting from the stated starting organic compounds; *use as many moles of the starting molecules as desired.* Show the important reagents and products for each of the major steps involved; mechanisms or high-energy intermediates should **not** be included. Common sequences for making often-used products can be condensed by writing several reagents above one reaction arrow.

   Reactions selected must produce either the major isomer or (acceptable) one of near-equivalent major isomers (choose the one desired). Minor isomers may not be used as products in a valid synthesis. Once the synthesis of particular intermediate is shown in a problem, you may use this intermediate in a subsequent problem without repeating the synthetic steps.

   Using all potential reactions covered through Organic I and II to date, devise a synthesis of each compound in the problem **starting only with 1-propene and ethanol as**

carbon sources plus any other reagents. Use the additional information or directions for certain specific products.

a.   and b. Build carbon-skeleton using SN2 reactions of alkyne nucleophiles; form the appropriate alkynes from the starting materials listed.

a.

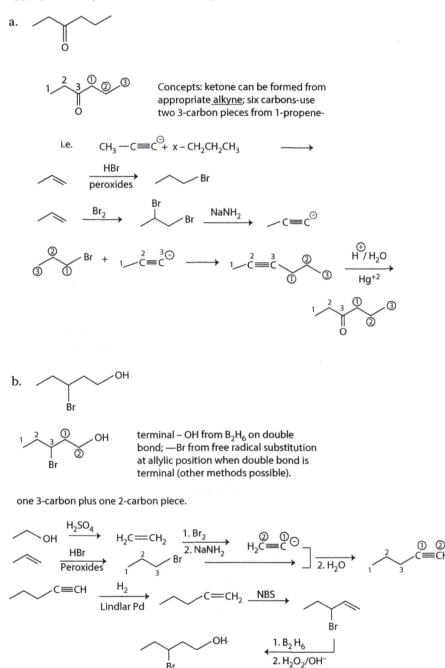

Concepts: ketone can be formed from appropriate alkyne; six carbons-use two 3-carbon pieces from 1-propene-

b.

terminal – OH from B_2H_6 on double bond; —Br from free radical substitution at allylic position when double bond is terminal (other methods possible).

one 3-carbon plus one 2-carbon piece.

c.   Use a Diels–Alder reaction to form the six-membered ring; consider using two 5-carbon pieces for the Diels–Alder.

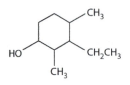

for Diels–Alder:

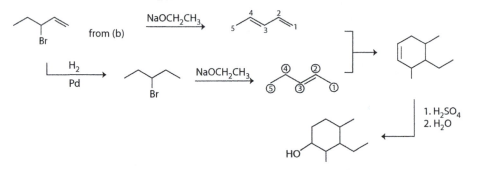

–OH can be added to residual double bond of diene

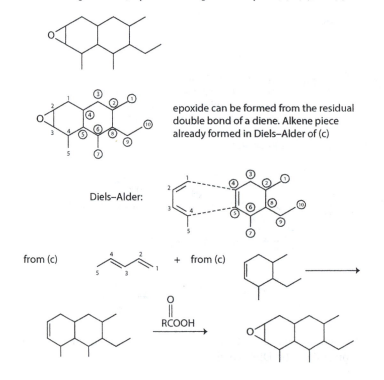

d. Use two separate Diels–Alder reactions to form the rings; try to use compounds (intermediates or products) synthesized previously in (**a**), (**b**), or (**c**).

epoxide can be formed from the residual double bond of a diene. Alkene piece already formed in Diels–Alder of (c)

Diels–Alder:

from (c)   + from (c)

Organic II: Practice Exam 19

1. Provide a correct name for each of the following compounds; **stereochemistry is not required.**

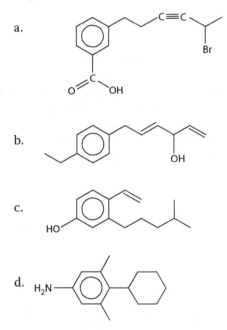

a.

b.

c.

d.

2. Consider the isomers of $C_{10}H_{10}$. Draw structural formulas for isomers which fit the following descriptions. If more than one answer is possible, **only one** isomer needs to be given.

   a. Draw any isomer containing an **aromatic ring** with **exactly six** pi-electrons in the ring. (The isomer may include additional nonaromatic pi-electrons.)

   b. Draw an isomer containing a fully conjugated ring but is **not** aromatic.

   c. Draw an isomer which contains an aromatic ring, but this **ring does not contain exactly six pi-electrons.**

   d. Draw a **non-ring** isomer which contains **no sp²-carbons.**

   e. An isomer which is aromatic and also contains a functional group can be converted into a carbon anionic nucleophile by reaction with $NaNH_2$.

   f. An isomer which is **aromatic** can also undergo a **Diels–Alder** reaction with an **alkene.**

3. Consider the following two electrophilic aromatic substitutions:

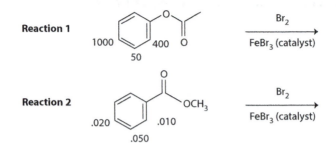

a. Indicate in the above equations the major product for each reaction, being certain to select the proper regioisomer.

    (b)–(c) Show the mechanisms for both reactions. Include the form of the electrophile; the structure of the high-energy intermediate produced (taking into account the correct regiochemistry); the other resonance forms of the intermediate in each case which emphasize the activating and directing properties of the specific substituent; and the final step which regenerates the aromatic ring.

b. Mechanism for **Reaction 1**: include **four** total resonance forms for the intermediate; **one resonance form has the charge located outside the ring.**

c. Mechanism for **Reaction 2**: include **three** total resonance forms for the intermediate.

d. Using the relative rate data shown for the rings above, calculate the rate of **Reaction 1** versus **Reaction 2**; rates at **all** positions must be taken into consideration, not just the rate at the favored position.

4. Answer the following questions **(a)** through **(f)** by picking **one** of the seven molecules shown below which provides the best answer; question **(g)** will require **two** answers. Use the capital letter designations to indicate your answers; letter answers for molecules may be used more than once. *Note that not all molecules are appropriate reactants for every part; first determine the reaction type and then consider only possible substrates for that reaction.*

a.   b.   c.   d.   e.   f.   g.

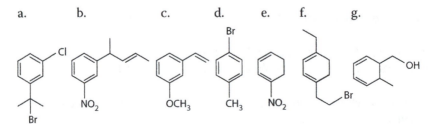

a. Which molecule will have the fastest rate for **free radical substitution** by Br_2/light?

b. Which molecule will have the fastest rate for a **Diels–Alder** reaction with 1,2-dichloroethene? (Two acceptable answers.)

c. Which molecule will show the fastest rate for **SN2** substitution by sodium propyne (i.e., an alkyne anion nucleophile)?

d. Which molecule will produce the most stable carbon cation upon protonation by H_2SO_4?

e. Which molecule will have the fastest rate for **electrophilic aromatic substitution** with Cl_2/$AlCl_3$?

f. Which molecule can be converted into a carbon nucleophile through electrophilic addition of Br_2 followed by reaction with $NaNH_2$?

g. Which **two** molecules can be oxidized by $Cr_2O_7^{-2}$ in H^+/H_2O to produce a substituted benzoic acid derivative?

5. Draw structural formulas for the **one major** product for each of the reactions. **No** stereochemistry is required.

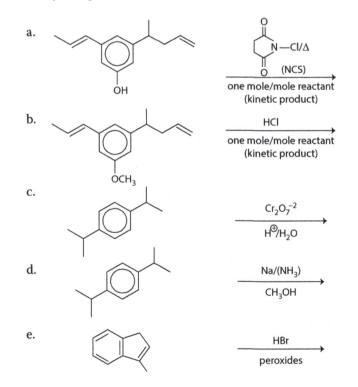

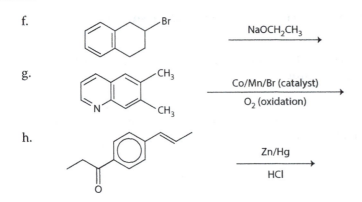

f.

NaOCH$_2$CH$_3$

g.

Co/Mn/Br (catalyst)

O$_2$ (oxidation)

h.

Zn/Hg

HCl

6. Draw structural formulas for the **major** product in each of the following reactions. Indicate stereochemistry where required.

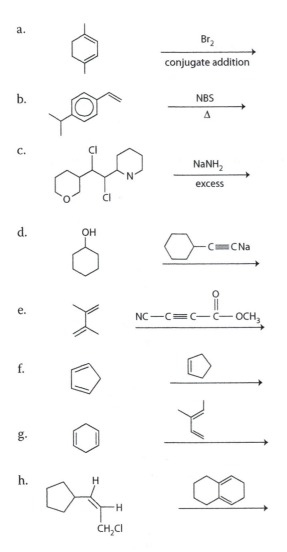

a.

Br$_2$

conjugate addition

b.

NBS

Δ

c.

NaNH$_2$

excess

d.

OH

—C≡CNa

e.

NC—C≡C—C—OCH$_3$

f.

g.

h.

H

—H

CH$_2$Cl

7. Draw structural formulas for the **major** product for each of the following reactions. **All** products should be based on **mono**-substitution. Be certain to consider both the correct form of the electrophile and the correct ring position.

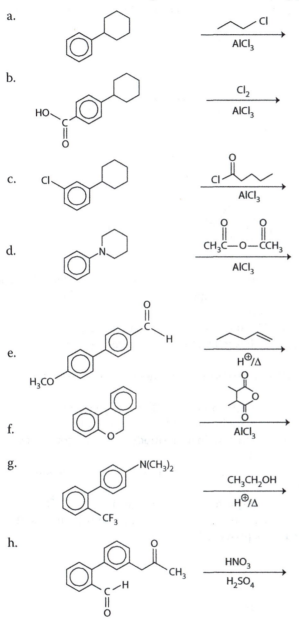

8. Complete the following reaction sequence by identifying the products **[A]** through **[G]** using the schematic shown and the additional descriptions below:
   [A] results from a Friedel–Craft reaction through the acyl chloride at the favored ring position based on electronic and steric effects; the double bond of the acid chloride reagent is not affected.
   [B] is the major isomer of electrophilic addition.

[C] is the product of a reduction reaction.

[D] results from an **internal (intramolecular)** Friedel–Craft reaction.

[E] is the major isomer from this sequence.

[F] is the proper compound which will produce the Diels–Alder product shown in the sequence.

[G] is the major product from this reduction reaction.

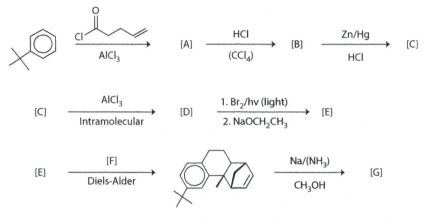

9. Use your knowledge of reactions covered to date to solve the following problem in structural analysis. Identify the structures **[A]** through **[G]**; first determine the structure of **[F]** using the ozonolysis information provided.
   1. Compound **[A]** has a formula $C_{12}H_{14}$ and contains an aromatic ring.
   2. Reaction of **[A]** with **excess** B_2H_6 followed by H_2O_2/OH^- yields the major isomer **[B]**, $C_{12}H_{18}O_2$.
   3. Reaction of **[B]** with **excess NBS** (*N*-bromosuccinimide) produces the major isomer **[C]**, $C_{12}H_{16}Br_2O_2$.
   4. Reaction of **[C]** with **excess** $NaOCH_2CH_3$ produces **[D]**, having a formula $C_{12}H_{14}O_2$.
   5. **[D]** is now converted into a fluoride derivative, **[E]**, by reaction first with toluene-sulfonyl chloride (TsCl), and then with NaF in a thermodynamically controlled SN2 reaction. **[E]** has the formula $C_{12}H_{12}F_2$.
   6. **[E]** is reacted with two moles of Br_2 in an electrophilic addition reaction, followed by double dehydrohalogenation to form **[F]**, $C_{12}H_8F_2$.
   7. If **[F]** is treated with $Hg^{+2}/H^+/H_2O$, **[G]** is formed, with a formula $C_{12}H_{12}O_2F_2$.
   8. To determine the carbon skeleton, ozonolysis was performed on **[F]**. The results are shown below **with no molar ratios provided:**

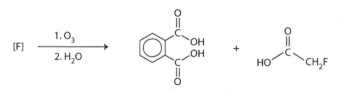

10. **Synthesis problems; general guidelines:** A valid "paper" synthesis creatively combines known reactions to logically provide a final product. The chosen method need not conform to a preferred **laboratory** synthesis. Use your knowledge of the reactions covered to date in the course to devise a synthesis of the molecule indicated as the final product starting from the stated starting organic compounds; *use as many moles of the starting molecules as desired.* Show the important reagents and products for

each of the major steps involved; mechanisms or high-energy intermediates should **not** be included. Common sequences for making often-used products can be condensed by writing several reagents above one reaction arrow. ***Reactions selected must produce either the major isomer or (acceptable) one of near-equivalent major isomers***; (choose the one desired). Minor isomers may not be used as products in a valid synthesis. Once the synthesis of a particular intermediate is shown in a problem, you may use this intermediate in a subsequent problem without repeating the synthetic steps.

Using all potential reactions covered through Organic I and II to date, devise a synthesis of each compound in the problem **starting only with benzene, ethene, plus any two-carbon acid chloride as carbon sources;** use any other reagents as desired. Use electrophilic aromatic substitution in all syntheses for the following compounds:

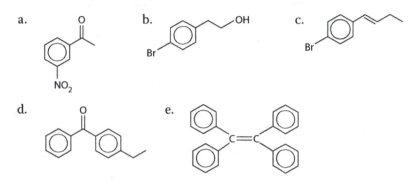

ORGANIC II: PRACTICE EXAM 1B: ANSWERS

1. Provide a correct name for each of the following compounds; **stereochemistry is not required.**

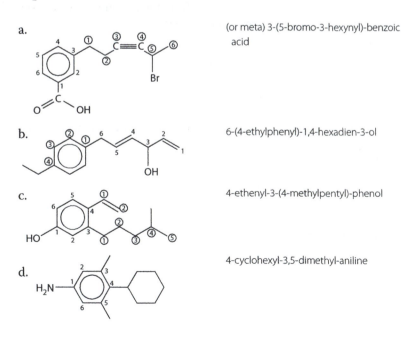

a.

(or meta) 3-(5-bromo-3-hexynyl)-benzoic acid

b.

6-(4-ethylphenyl)-1,4-hexadien-3-ol

c.

4-ethenyl-3-(4-methylpentyl)-phenol

d.

4-cyclohexyl-3,5-dimethyl-aniline

2. Consider the isomers of $C_{10}H_{10}$. Draw structural formulas for isomers which fit the following descriptions. If more than one answer is possible, **only one** isomer needs to be given.

   a. Draw any isomer containing an **aromatic ring** with **exactly six** pi-electrons in the ring. (The isomer may include additional nonaromatic pi-electrons.)

   $C_{10}H_{10} = 6(rings + \pi\text{-bonds})$

   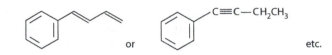

   or   etc.

   b. Draw an isomer containing a fully conjugated ring but is **not** aromatic.

   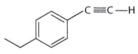

   or   etc.

   (4 π e⁻ in ring)

   c. Draw an isomer which contains an aromatic ring, but this **ring does not contain exactly six pi-electrons.**

   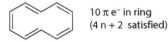

   10 π e⁻ in ring
   (4 n + 2 satisfied)

   d. Draw a **non-ring** isomer which contains **no sp²-carbons.**
      must contain sp-carbons; need three triple bonds

   $HC{\equiv}C-C{\equiv}C-C{\equiv}C-CH_2CH_2CH_2CH_3$ etc.

   e. An isomer which is aromatic and also contains a functional group can be converted into a carbon anionic nucleophile by reaction with $NaNH_2$.
      must have a terminal alkyne.

   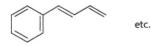

   f. An isomer which is **aromatic** can also undergo a **Diels–Alder** reaction with an **alkene.**
      must have a reactive diene

   etc.

3. Consider the following two electrophilic aromatic substitutions:

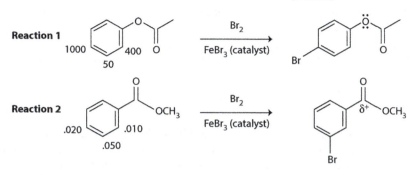

a. Indicate in the above equations the major product for each reaction, being certain to select the proper regioisomer.

(b), (c) Show the mechanisms for both reactions. Include the form of the electrophile; the structure of the high-energy intermediate produced (taking into account the correct regiochemistry); the other resonance forms of the intermediate in each case which emphasize the activating and directing properties of the specific substituent; and the final step which regenerates the aromatic ring.

b. Mechanism for **Reaction 1:** include **four** total resonance forms for the intermediate; **one resonance form has the charge located outside the ring.**

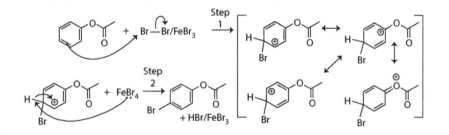

c. Mechanism for **Reaction 2:** include **three** total resonance forms for the intermediate

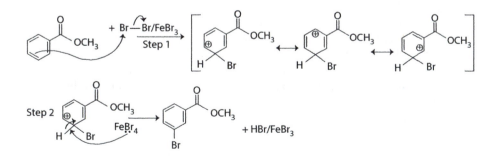

d. Using the relative rate data shown for the rings above, calculate the rate of **Reaction 1** versus **Reaction 2**; rates at **all** positions must be taken into consideration, not just the rate at the favored position.

rate reaction $1 = [(400) \times 2] + [(50) \times 2] + [(1000) \times 1] = 1900$

rate reaction $2 = [(0.010) \times 2] + [(0.050) \times 2] + [(0.020) \times 1] = 0.14$

$$\frac{\text{reaction } 1}{\text{reaction } 2} = \frac{1900}{0.14} = 13{,}570/1$$

4. Answer the following questions **(a)** through **(f)** by picking **one** of the seven molecules shown below which provides the best answer; question **(g)** will require **two** answers. Use the capital letter designations to indicate your answers; letter answers for molecules may be used more than once. *Note that not all molecules are appropriate reactants for every part; first determine the reaction type and then consider only possible substrates for that reaction.*

a. b. c. d. e. f. g.

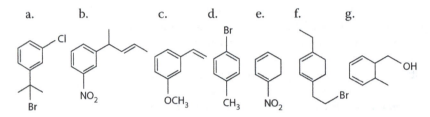

a. Which molecule will have the fastest rate for **free radical substitution** by Br_2/light?
    B (tertiary benzylic hydrogen.)
b. Which molecule will have the fastest rate for a **Diels–Alder** reaction with 1,2-dichloroethene? (Two acceptable answers.)
    F or G (diene (E) has a strong e^--withdrawing group.)
c. Which molecule will show the fastest rate for **SN2** substitution by sodium propyne (i.e., an alkyne anion nucleophile)?
    F (Primary alkyl bromide$-$CH2$-$Br)
d. Which molecule will produce the most stable carbon cation upon protonation by H_2SO_4?
    C (can form a benzylic carbon cation.)
e. Which molecule will have the fastest rate for **electrophilic aromatic substitution** with Cl_2/$AlCl_3$?
    C (has strongest EAS activator $=-OCH_3$)
f. Which molecule can be converted into a carbon nucleophile through electrophilic addition of Br_2 followed by reaction with $NaNH_2$?
    C (has terminal alkene which is converted into terminal alkyne.)
g. Which **two** molecules can be oxidized by $Cr_2O_7^{-2}$ in H^+/H_2O to produce a substituted benzoic acid derivative?
    B and D (have benzylic hydrogens.)

5. Draw structural formulas for the **one major** product for each of the reactions. **No** stereochemistry is required.

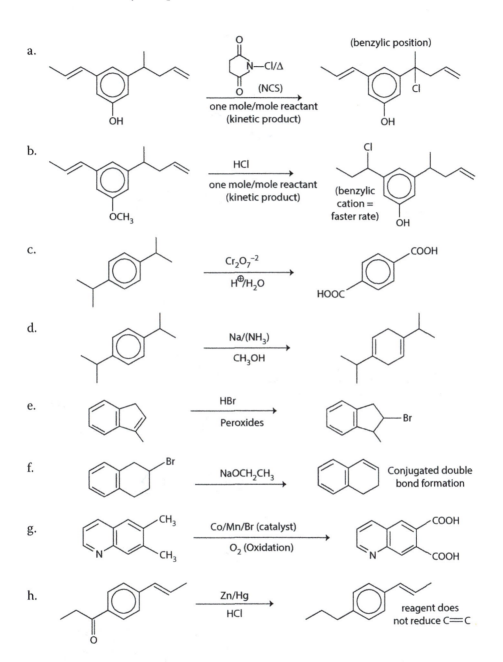

6. Draw structural formulas for the **major** product in each of the following reactions. Indicate stereochemistry where required.

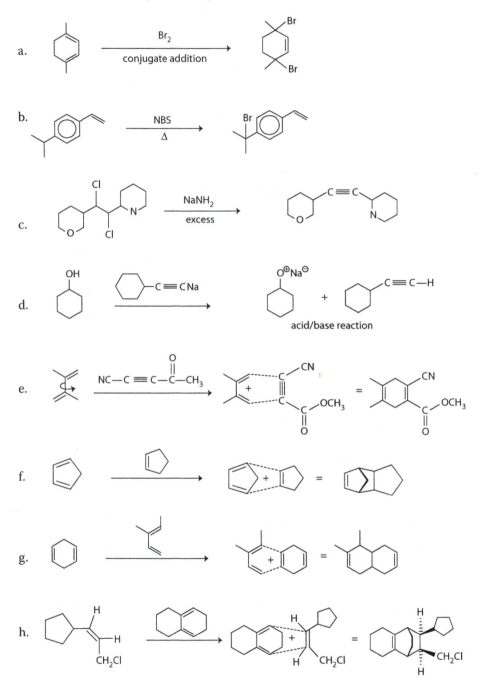

7. Draw structural formulas for the **major** product for each of the following reactions. **All** products should be based on **mono**-substitution. Be certain to consider both the correct form of the electrophile and the correct ring position.

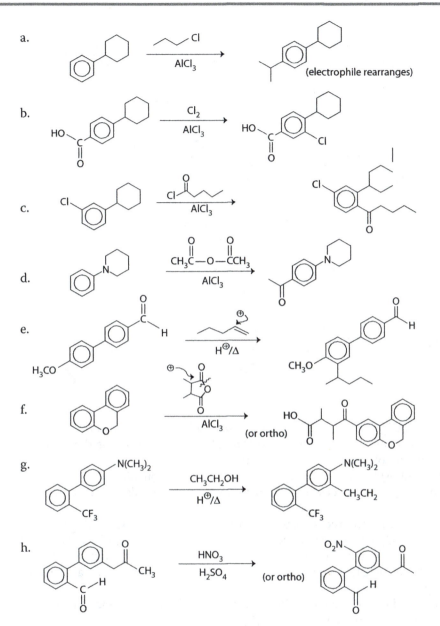

a.   (electrophile rearranges)

b.

c.

d.

e.

f.   (or ortho)

g.

h.   (or ortho)

8. Complete the following reaction sequence by identifying the products [A] through [G] using the schematic shown and the additional descriptions below:

   [A] results from a Friedel–Craft reaction through the acyl chloride at the favored ring position based on electronic and steric effects; the double bond of the acid chloride reagent is not affected.

   [B] is the major isomer of electrophilic addition.

   [C] is the product of a reduction reaction.

   [D] results from an **internal (intramolecular)** Friedel–Craft reaction.

   [E] is the major isomer from this sequence.

   [F] is the proper compound which will produce the Diels–Alder product shown in the sequence.

   [G] is the major product from this reduction reaction.

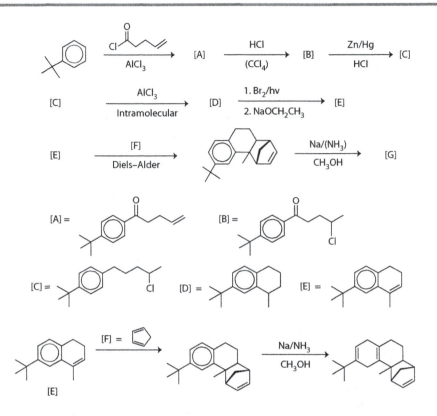

9. Use your knowledge of reactions covered to date to solve the following problem in structural analysis. Identify the structures **[A]** through **[G]**; first determine the structure of **[F]** using the ozonolysis information provided.

1. Compound **[A]** has a formula $C_{12}H_{14}$ and contains an aromatic ring.
2. Reaction of **[A]** with **excess** B_2H_6 followed by H_2O_2/OH yields the major isomer **[B]**, $C_{12}H_{18}O_2$.
3. Reaction of **[B]** with **excess NBS** produces the major isomer **[C]**, $C_{12}H_{16}Br_2O_2$.
4. Reaction of **[C]** with **excess** $NaOCH_2CH_3$ produces **[D]**, having a formula $C_{12}H_{14}O_2$.
5. **[D]** is now converted into a fluoride derivative, **[E]**, by reaction first with TsCl, and then with NaF in a thermodynamically controlled **SN2** reaction. **[E]** has the formula $C_{12}H_{12}F_2$.
6. **[E]** is reacted with two moles of Br_2 in an electrophilic addition reaction, followed by double dehydrohalogenation to form **[F]**, $C_{12}H_8F_2$.
7. If **[F]** is treated with Hg^{+2}/H^+/H_2O, **[G]** is formed, with a formula $C_{12}H_{12}O_2F_2$.
8. To determine the carbon skeleton, ozonolysis was performed on **[F]**. The results are shown below **with no molar ratios provided:**

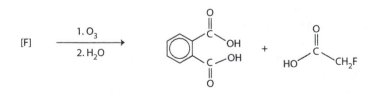

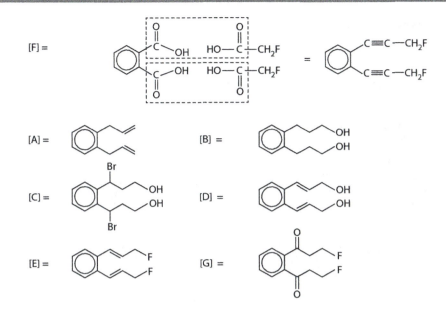

10. **Synthesis problems; general guidelines:** A valid "paper" synthesis creatively combines known reactions to logically provide a final product. The chosen method need not conform to a preferred **laboratory** synthesis. Use your knowledge of the reactions covered to date in the course to devise a synthesis of the molecule indicated as the final product starting from the stated starting organic compounds; *use as many moles of the starting molecules as desired.* Show the important reagents and products for each of the major steps involved; mechanisms or high-energy intermediates should **not** be included. Common sequences for making often-used products can be condensed by writing several reagents above one reaction arrow. *Reactions selected must produce either the major isomer or (acceptable) one of near-equivalent major isomers;* (choose the one desired). Minor isomers may not be used as products in a valid synthesis. Once the synthesis of a particular intermediate is shown in a problem, you may use this intermediate in a subsequent problem without repeating the synthetic steps.

Using all potential reactions covered through Organic I and II to date, devise a synthesis of each compound in the problem **starting only with benzene, ethene, plus any two-carbon acid chloride as carbon sources;** use any other reagents as desired. Use electrophilic aromatic substitution in all syntheses for the following compounds:

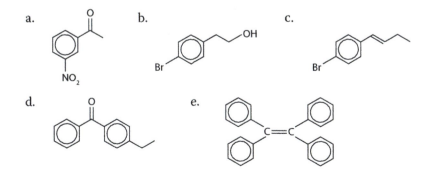

Answers to Synthesis Problem

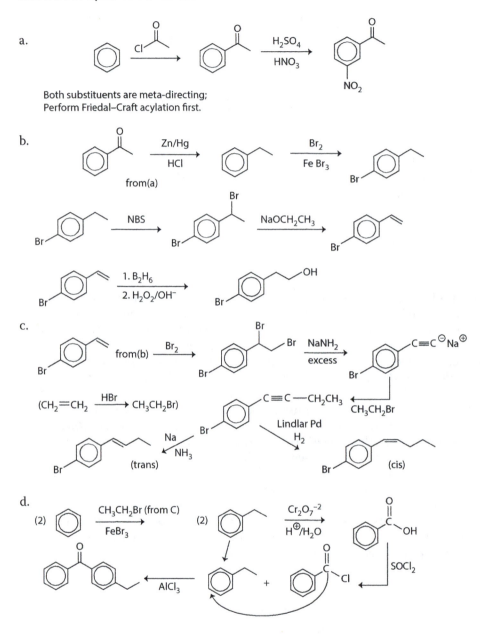

a.

Both substituents are meta-directing;
Perform Friedal–Craft acylation first.

b.

from(a)

c.

(CH$_2$=CH$_2$) $\xrightarrow{\text{HBr}}$ CH$_3$CH$_2$Br

(trans)

(cis)

d.

(2)

e.

(2)

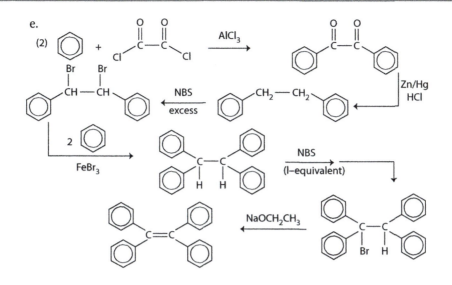

Organic II: Practice Exam 20

For mechanisms in this exam, show complete arrow notation for all steps including arrows to describe resonance structures; show H's whenever they are involved in a mechanistic step.

1. Show the **complete** mechanism for the following **electrophilic aromatic substitution** reaction shown below. **Show** (i) the reaction which produces the correct form of the electrophile; (ii) the electrophile reaction to a favored position producing the **aromatic cation** and then **three additional** resonance structures including **one** which depicts the electronic effect of the substituent group: **be certain to clearly show the location of double bonds and charges;** and (iii) the formation of the final major product; be specific as to the role of the base.

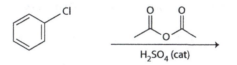

2. a. Name the following compound which is the herbicide "trifluralin"; the base name should be the aromatic portion of the molecule.

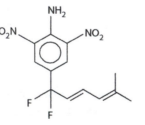

b. Name the following compound which is related to the antibacterial "clofoctol"; the base name should be the aromatic alcohol portion of the molecule.

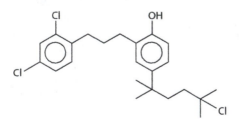

c. Draw structures for compounds [A], [B], and [C] which all have the formula $C_8H_{11}NO$.
Be certain that the molecules fit the required formula.

[A] = **any aromatic** $C_8H_{11}NO$ molecule which contains a **nitrogen** atom as part of the aromatic ring, that is, an electron pair (bonding or lone) from nitrogen must contribute to aromaticity.

[B] = **any aromatic** $C_8H_{11}NO$ molecule which contains an **oxygen** atom as part of the aromatic ring, that is, an electron pair from oxygen must contribute to aromaticity.

[C] = **any aromatic** $C_8H_{11}NO$ molecule which contains **only** carbon atoms in the aromatic ring.

[A]   [B]   [C]

3. Show the complete mechanism for the suggested hypothetical **electrophilic addition** of Ag−OH to the conjugated triene molecule shown. The electrophile is Ag^+ and the nucleophile is OH^-.

   Add the electrophile to the double bond with the ethyl group labeled as 1, 2.

   i.   Show the reaction and structure of the **cation intermediate** formed by electrophilic addition to the 1,2 pi-bond of the conjugated triene to produce a double allylic cation. Then, show the **other two major resonance structures** of this cation (for a total of **three** structures).

   ii. Show the reaction to produce all **three products** which can be formed by addition of the nucleophile.

   iii. ***Identify the major kinetic product***, the favored product when the reaction is under kinetic control.

   Be sure to use complete arrow notation for all steps in the mechanism.

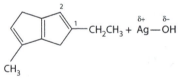

4. Draw structural formulas for the **major organic product** for the following reaction sequences. **Indicate stereochemistry where appropriate**. For electrophilic aromatic substitution, assume mono-substitution and consider proper positioning on the correct ring; for equivalent ortho versus para substitution, either isomer is acceptable.

a.

b.

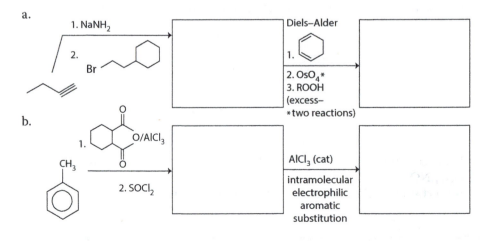

c.

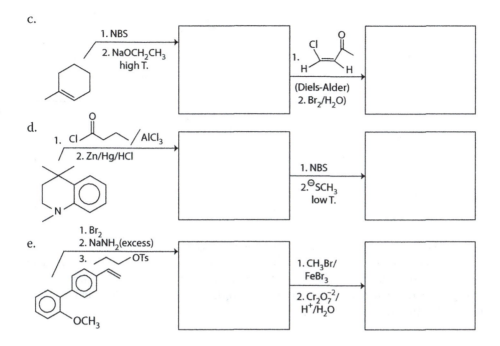

1. NBS

2. NaOCH₂CH₃
high T.

1.

(Diels-Alder)
2. Br₂/H₂O)

d.

1. Cl⟍⟍⟍/AlCl₃

2. Zn/Hg/HCl

1. NBS

2.⊖SCH₃
low T.

e.

1. Br₂
2. NaNH₂(excess)
3. ⟍⟍OTs

1. CH₃Br/
FeBr₃

2. Cr₂O₇⁻²/
H⁺/H₂O

OCH₃

5. Show the structure of the **major** product for each part of the following two-part reaction sequences. Determine the product for the first reaction set; then, use this product to complete the second reaction set. Each two-part reaction sequence is separate. ***Include stereochemistry where required***. **For electrophilic aromatic substitution**, assume **mono**-substitution.

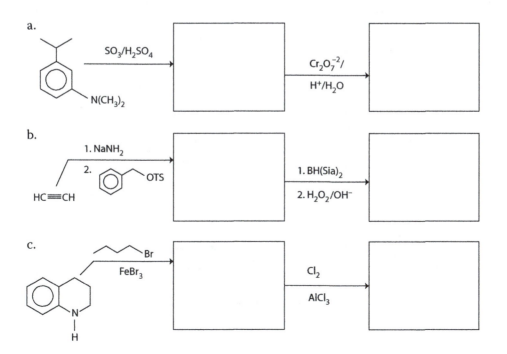

a.

SO₃/H₂SO₄

Cr₂O₇⁻²/
H⁺/H₂O

N(CH₃)₂

b.

1. NaNH₂

2. ⟍OTS

1. BH(Sia)₂

2. H₂O₂/OH⁻

HC≡CH

c.

⟍⟍⟍Br

FeBr₃

Cl₂

AlCl₃

N
|
H

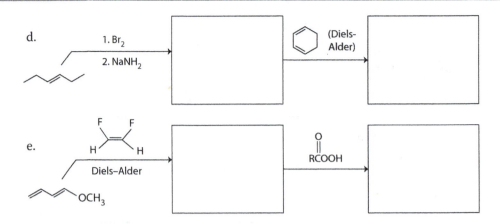

6. Show the structure of the **major** product for each part of the following two-part reaction sequences. Determine the product for the first reaction set; then, use this product to complete the second reaction set. Each two-part reaction sequence is separate. *Include stereochemistry where required*. **For electrophilic aromatic substitution**, assume **mono**-substitution.

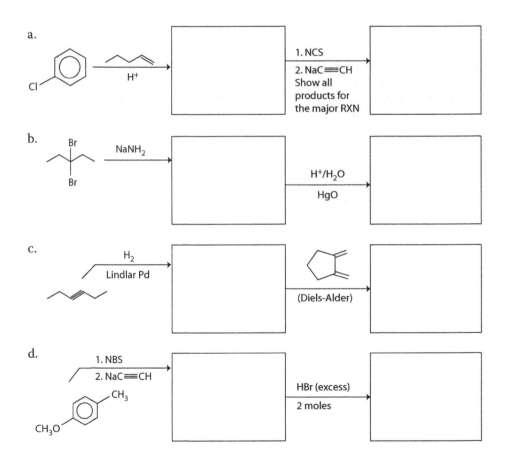

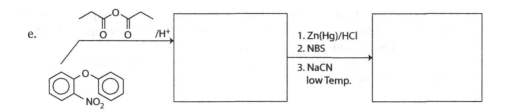

e.   /H$^+$

1. Zn(Hg)/HCl
2. NBS

3. NaCN
    low Temp.

7. Show the structure of the **major** product for each part of the following three-part reaction sequences. Each three-part reaction sequence is separate. ***Include stereochemistry where required***. **For electrophilic aromatic substitution**, assume **mono**-substitution.

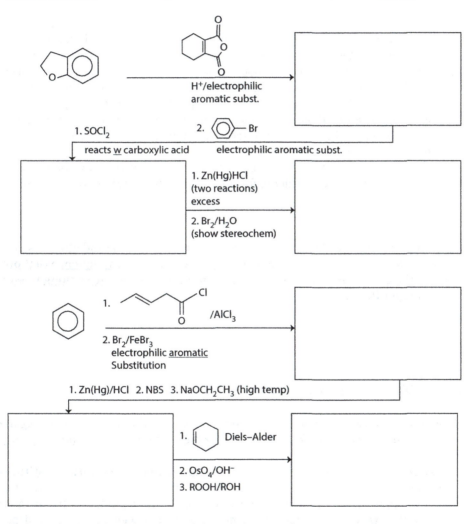

8. Show **only** the following parts of the mechanism for the SN1 substitution reaction of **HBr** with **1-phenyl-2-butanol** which will proceed through a **rearrangement**.
   i. Show the structure of the **initial** (i.e., before a rearrangement) **carbon cation** formed by loss of the leaving group.

 ii. Show the **rearrangement transition state** in the rearrangement step, clearly showing the identity of the group migrating; then, show the **final rearranged cation**.

 iii. Show the **final rearranged substitution product**.

9. This synthesis problem depicts the synthesis of the following molecule starting **only** with **benzene** and excess **ethanol**:

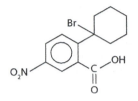

The synthesis is divided into two parts to provide some additional direction; use the **General Procedure:** Organic carbons must come from the carbon sources listed (use as many molecules of the starting compounds as needed); otherwise, any other reagents can be used. Show the important reagents and products for each of the major steps involved. Be certain that your selected reactions produce either the **major** isomer or (acceptable) **one of equivalent or similar major isomers.** (Choose the one desired.) Common reaction combinations may be shortened: indicate a series of reactions by listing all the reagents over the arrow.

 a. Starting **only** with excess ethanol as a carbon source, synthesize **cyclohexene**. Use alkyne anions and a Diels–Alder to form $C-C$ bonds; do **not** use the CuLi reagent.

 b. Starting with the cyclohexene synthesized in part (a), benzene and ethanol synthesize the final molecule drawn. **Be certain to select a reaction sequence that optimizes the correct regiochemistry and eliminates interference among possible reacting positions.**

ORGANIC II: PRACTICE EXAM 1C: ANSWERS

For mechanisms in this exam, show complete arrow notation for all steps including arrows to describe resonance structures; show H's whenever they are involved in a mechanistic step.

1. Show the **complete** mechanism for the following **electrophilic aromatic substitution** reaction shown below. **Show** (i) the reaction which produces the correct form of the electrophile; (ii) the electrophile reaction to a favored position producing the **aromatic cation** and then **three additional resonance** structures including **one** which depicts the electronic effect of the substituent group: **be certain to clearly show the location of double bonds and charges**; and (iii) the formation of the final major product; be specific as to the role of the base.

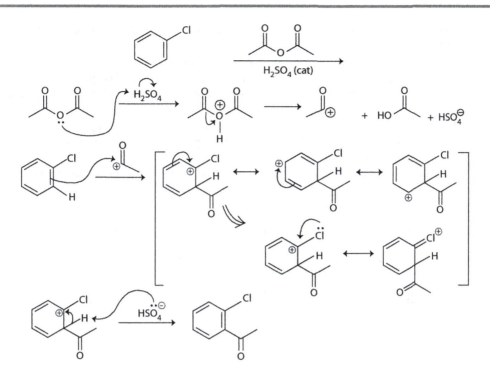

2. a. Name the following compound which is the herbicide "trifluralin"; the base name should be the aromatic portion of the molecule.

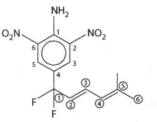

4-(1,1-difluoro-5-methyl-2,4-hexadienyl)-2,6-dinitroaniline

b. Name the following compound which is related to the antibacterial "clofoctol"; the base name should be the aromatic alcohol portion of the molecule.

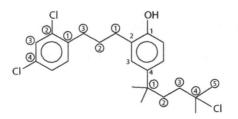

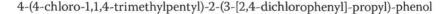

4-(4-chloro-1,1,4-trimethylpentyl)-2-(3-[2,4-dichlorophenyl]-propyl)-phenol

c. Draw structures for compounds [A], [B], and [C] which all have the formula $C_8H_{11}NO$.

Be certain that the molecules fit the required formula.

[A] = **any aromatic** $C_8H_{11}NO$ molecule which contains a **nitrogen** atom as part of the aromatic ring, that is, an electron pair (bonding or lone) from nitrogen must contribute to aromaticity.

[B] = **any aromatic** $C_8H_{11}NO$ molecule which contains an **oxygen** atom as part of the aromatic ring, that is, an electron pair from oxygen must contribute to aromaticity.

[C] = **any aromatic** $C_8H_{11}NO$ molecule which contains **only carbon** atoms in the aromatic ring.

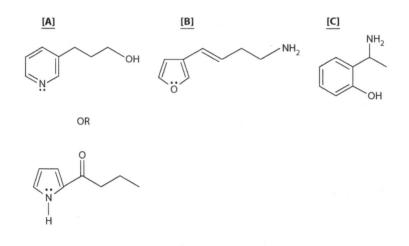

3. Show the complete mechanism for the suggested hypothetical **electrophilic addition** of Ag−OH to the conjugated triene molecule shown. The electrophile is Ag^+ and the nucleophile is OH^-.

   Add the electrophile to the double bond with the ethyl group labeled as 1, 2.

   i. Show the reaction and structure of the **cation intermediate** formed by electrophilic addition to the 1,2 pi-bond of the conjugated triene to produce a double allylic cation. Then, show the **other two major resonance structures** of this cation (for a total of **three** structures).

   ii. Show the reaction to produce all **three products** which can be formed by addition of the nucleophile.

   iii. *Identify the major kinetic product*, the favored product when the reaction is under kinetic control.

   Be sure to use complete arrow notation for all steps in the mechanism.

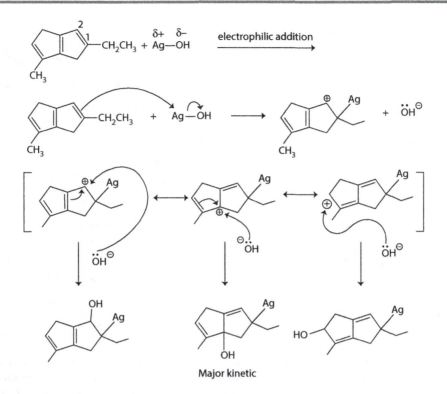

Major kinetic

4. Draw structural formulas for the **major organic product** for the following reaction sequences. **Indicate stereochemistry where appropriate**. For electrophilic aromatic substitution, assume mono-substitution and consider proper positioning on the correct ring; for equivalent ortho versus para substitution, either isomer is acceptable.

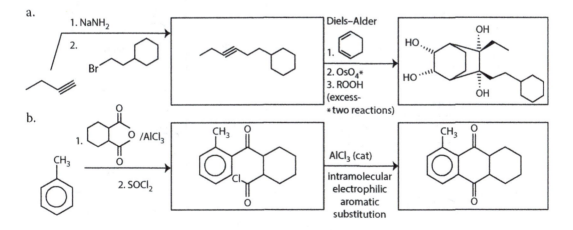

c.

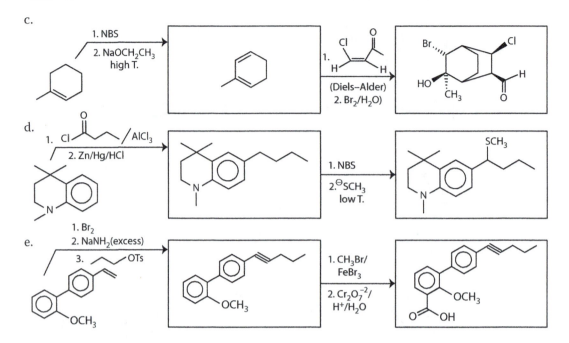

d.

e.

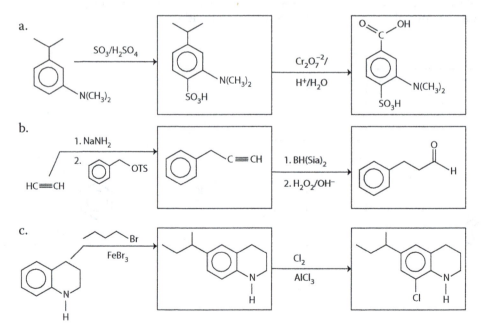

5. Show the structure of the **major** product for each part of the following two-part reaction sequences. Determine the product for the first reaction set; then, use this product to complete the second reaction set. Each two- part reaction sequence is separate. *__Include stereochemistry where required__*. **For electrophilic aromatic substitution**, assume **mono**-substitution.

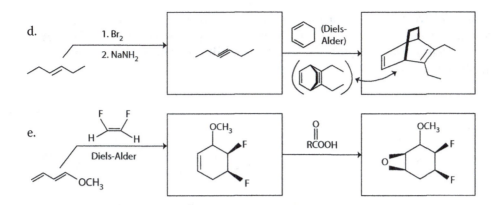

6. Show the structure of the **major** product for each part of the following two-part reaction sequences. Determine the product for the first reaction set; then, use this product to complete the second reaction set. Each two-part reaction sequence is separate. *Include stereochemistry where required*. **For electrophilic aromatic substitution,** assume **mono**-substitution.

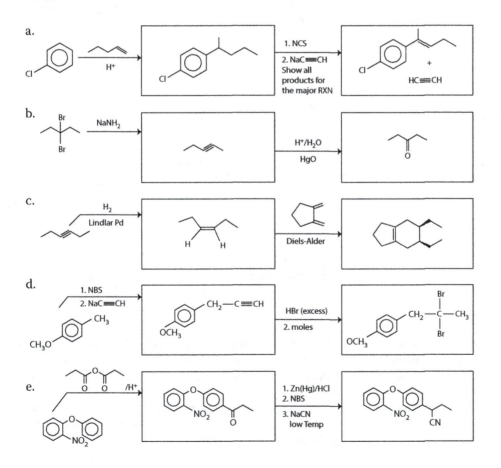

7. Show the structure of the **major** product for each part of the following three-part reaction sequences. Each three-part reaction sequence is separate. *Include stereochemistry where required.* **For electrophilic aromatic substitution,** assume **mono**-substitution.

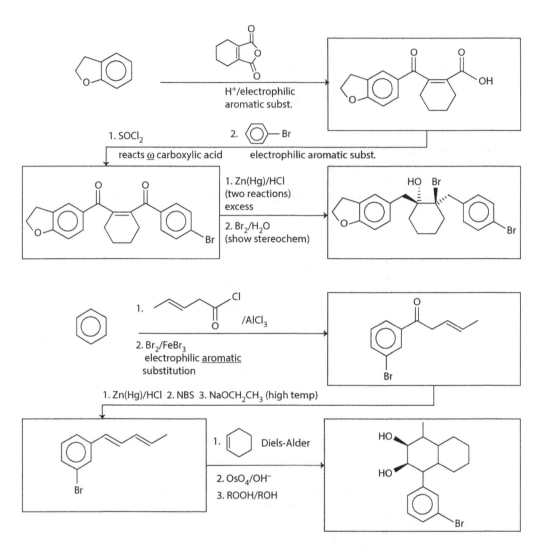

8. Show **only** the following parts of the mechanism for the SN1 substitution reaction of **HBr** with **1-phenyl-2-butanol** which will proceed through a **rearrangement**.

   i.   Show the structure of the **initial** (i.e., before a rearrangement) **carbon cation** formed by loss of the leaving group.

   ii. Show the **rearrangement transition state** in the rearrangement step, clearly showing the identity of the group migrating; then, show the **final rearranged cation.**

   iii. Show the **final rearranged substitution product.**

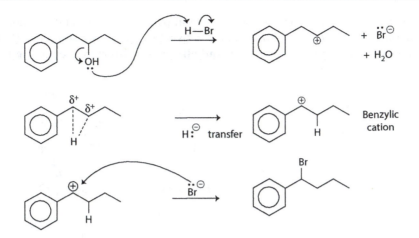

9. This synthesis problem depicts the synthesis of the following molecule starting **only** with **benzene** and excess **ethanol**:

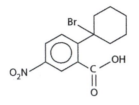

The synthesis is divided into two parts to provide some additional direction; use the **General Procedure:** Organic carbons must come from the carbon sources listed (use as many molecules of the starting compounds as needed); otherwise, any other reagents can be used. Show the important reagents and products for each of the major steps involved. Be certain that your selected reactions produce either the **major** isomer or (acceptable) **one of equivalent or similar major isomers**. (Choose the one desired.) Common reaction combinations may be shortened: indicate a series of reactions by listing all the reagents over the arrow.

a. Starting **only** with excess ethanol as a carbon source, synthesize **cyclohexene**. Use alkyne anions and a Diels–Alder to form C—C bonds; do **not** use the CuLi reagent.

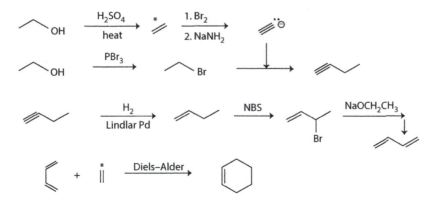

b. Starting with the cyclohexene synthesized in part (a), benzene and ethanol synthe-
    size the final molecule drawn. **Be certain to select a reaction sequence that opti-
    mizes the correct regiochemistry and eliminates interference among possible
    reacting positions**.

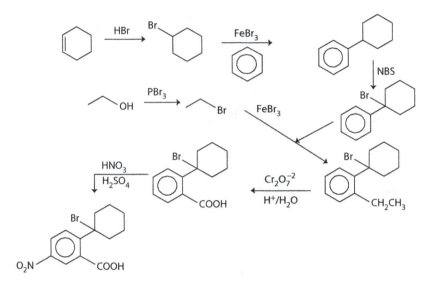

Organic II: Practice Exam 21

1. Provide the correct name for the following compounds:

a.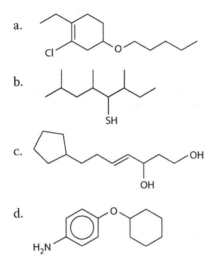

b.

c.

d.

2. Provide the products and mechanisms as indicated for the following reactions. The problem analysis will demonstrate the diverse reactions of substrates and reagents that **superficially** appear similar.

a. Draw the structure of the major product for this Friedel–Craft **electrophilic aromatic substitution (EAS)**:

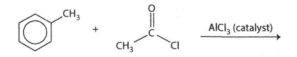

b. Show the **complete** mechanism for the reaction. Include the form of the electrophile and the high-energy carbon cation intermediate produced; show **all three** resonance forms for this intermediate which indicate the activating and directing properties of the substituent.

c. Now, consider the reactions of a **ketone** with a **Grignard reagent**; show the correct product for this **nucleophilic addition to carbonyl**:

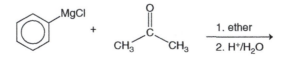

d. Show the mechanism of this reaction; include the form (with charge distribution) of **both** the nucleophilic molecule **and** the electrophilic molecule; the **initial tetrahedral product metal complex**; the formation of the final product by reaction with H^+/H_2O.

e. Another way to form an alcohol is through reaction of a Grignard reagent with an epoxide; show the product of the following reaction:

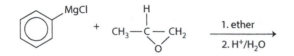

f. Show the mechanism of this reaction; also, indicate the geometry of the bimolecular transition state of the r.d.s. which dictates the regiochemistry of the product.

3. Draw structural formulas for the **one major** product for each of the reactions or reaction sequences. Consider both stereochemistry and regiochemistry where required.

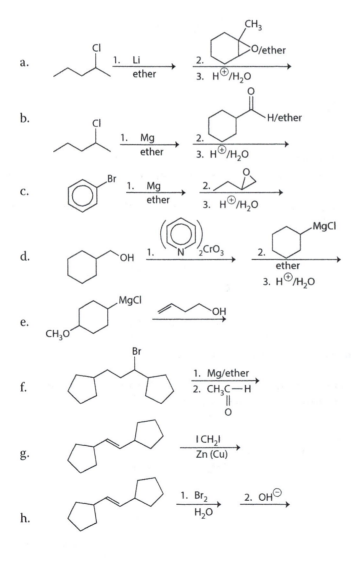

4. Draw structural formulas for the **one major** product for each of the reactions or reaction sequences. Consider both stereochemistry and regiochemistry where required.

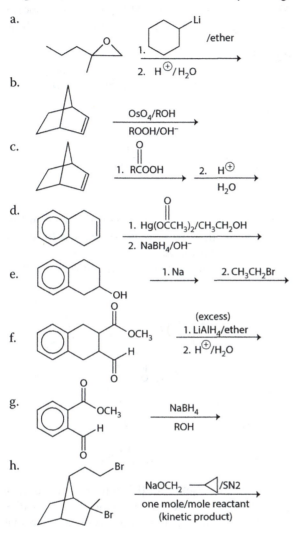

a.

1. (cyclohexyl-Li) /ether
2. $H^{\oplus}/H_2O$

b.

OsO$_4$/ROH
ROOH/OH$^-$

c.

1. RCOOH
2. $H^{\oplus}$
   H_2O

d.

1. Hg(OCCH$_3$)$_2$/CH$_3$CH$_2$OH
2. NaBH$_4$/OH$^-$

e.

1. Na
2. CH$_3$CH$_2$Br

f.

(excess)
1. LiAlH$_4$/ether
2. $H^{\oplus}/H_2O$

g.

NaBH$_4$
ROH

h.

NaOCH$_2$ ◁ /SN2
one mole/mole reactant
(kinetic product)

5. Draw structural formulas for the **one major** product for each of the reactions or reaction sequences. Consider both stereochemistry and regiochemistry where required.

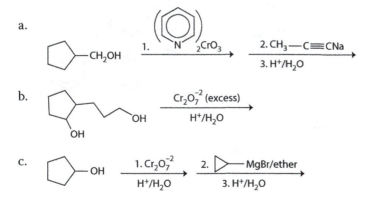

a.

—CH$_2$OH

1. (pyridinium)$_2$CrO$_3$
2. CH$_3$—C≡CNa
3. H$^+$/H$_2$O

b.

OH

Cr$_2$O$_7^{-2}$ (excess)
H$^+$/H$_2$O

c.

—OH

1. Cr$_2$O$_7^{-2}$
   H$^+$/H$_2$O
2. ▷—MgBr/ether
3. H$^+$/H$_2$O

6. Complete the following **reverse** reaction sequences by identifying the compounds **[A]** through **[D]**. First, determine the structure of [D] by the excess HCl analysis reaction shown; then work backward to **[A]**. [A] has the formula $C_8H_6O_4$ and is an important compound used in polymers which form high-strength materials used in boat hulls etc.

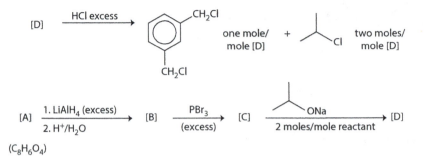

7. Use your knowledge of reactions covered to date to solve the following problems in structural analysis. *Note that all of the reactions in each part must be used to identity an unambiguous structure for the starting compound.*

[A] to [C]: [A] is carvone, the major component in spearmint oil; the molecule contains a six membered ring.

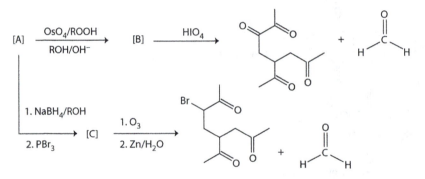

[D] to [F]: [D] is one form of Vitamin B-6. The reaction of [D] to [E] requires only one mole per mole of Collins' reagent, that is, **only one** site in the molecule is oxidized. The reaction of [D] to [F] is an SN2 reaction which will **not** occur at an aromatic position.

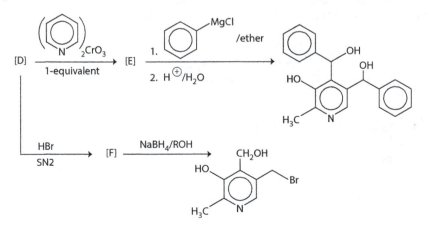

8. Draw structural formulas for the **one major** product for each of the following reactions which review previous reactions. Consider both stereochemistry and regiochemistry where required.

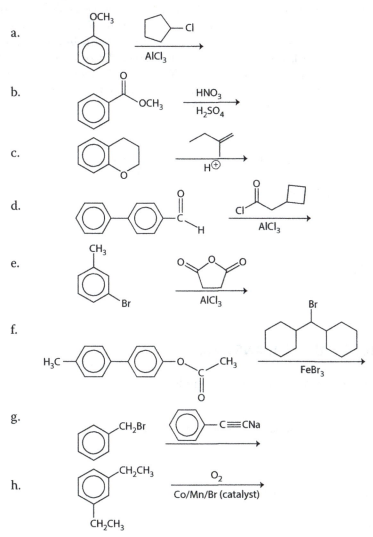

9. **Synthesis problems; general guidelines:** A valid "paper" synthesis creatively combines known reactions to logically provide a final product. The chosen method need not conform to a preferred **laboratory** synthesis. Use your knowledge of the reactions covered to date in the course to devise a synthesis of the molecule indicated as the final product starting from the stated starting organic compounds; *use as many moles of the starting molecules as desired.* Show the important reagents and products for each of the major steps involved; mechanisms or high-energy intermediates should **not** be included. Common sequences for making often-used products can be condensed by writing several reagents above one reaction arrow. *Reactions selected must produce either the major isomer or (acceptable) one of near-equivalent major isomers*; (choose the one desired). Minor isomers may not be used as products in a valid synthesis. Once the synthesis of a particular intermediate is shown in a problem, you may use this intermediate in a subsequent problem without repeating the synthetic steps.

Using all potential reactions covered through Organic I and II to date, devise a synthesis of each compound in the problem **starting only with benzene plus any two-carbon organic molecule (with any desired functional group) as carbon sources** plus any other reagents. Use the additional information or directions for certain specific products.

The following molecules were encountered in a previous exam. For this problem, invent a new synthesis **without using EAS to form carbon–carbon bonds** (i.e., do not use Friedel–Craft reactions); instead, use **only** organometallic reactions such as Grignards. Use **EAS** to form appropriate aromatic c − **X** and C − **N** bonds.

Parts (**a**) through (**d**): Use an epoxide in the synthesis of compound (**b**); consider using a product from (b) in the synthesis of compound (**c**)

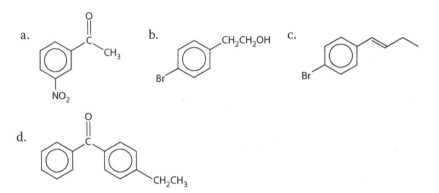

a.   b.   c.

d.

ORGANIC II: PRACTICE EXAM 1.5A: ANSWERS

1. Provide the correct name for the following compounds:

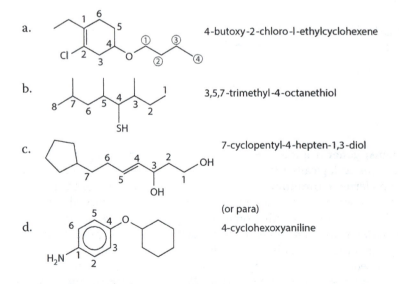

a.   4-butoxy-2-chloro-l-ethylcyclohexene

b.   3,5,7-trimethyl-4-octanethiol

c.   7-cyclopentyl-4-hepten-1,3-diol

(or para)
d.   4-cyclohexoxyaniline

2. Provide the products and mechanisms as indicated for the following reactions. The problem analysis will demonstrate the diverse reactions of substrates and reagents that **superficially** appear similar.

a. Draw the structure of the major product for this Friedel–Craft **EAS**:

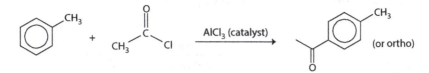

b. Show the **complete** mechanism for the reaction. Include the form of the electrophile and the high-energy carbon cation intermediate produced; show **all three** resonance forms for this intermediate which indicate the activating and directing properties of the substituent.

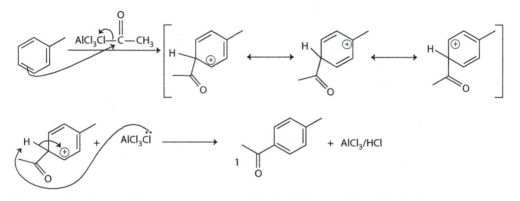

c. Now, consider the reactions of a **ketone** with a **Grignard reagent**; show the correct product for this **nucleophilic addition to carbonyl**:

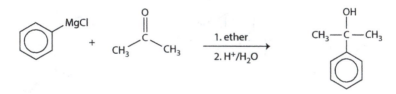

d. Show the mechanism of this reaction; include the form (with charge distribution) of **both** the nucleophilic molecule **and** the electrophilic molecule; the **initial tetrahedral product metal complex**; the formation of the final product by reaction with H^+/H_2O.

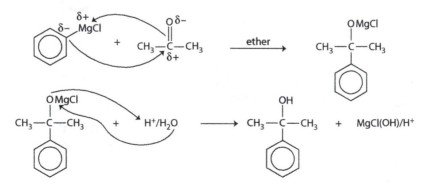

e. Another way to form an alcohol is through reaction of a Grignard reagent with an epoxide; show the product of the following reaction:

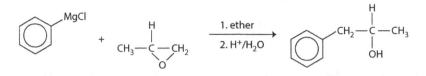

f. Show the mechanism of this reaction; also indicate the geometry of the bimolecular transition state of the r.d.s. which dictates the regiochemistry of the product.

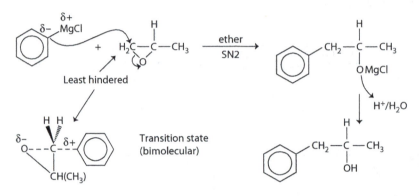

3. Draw structural formulas for the **one major** product for each of the reactions or reaction sequences. Consider both stereochemistry and regiochemistry where required.

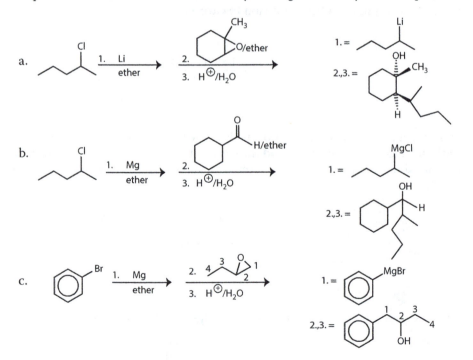

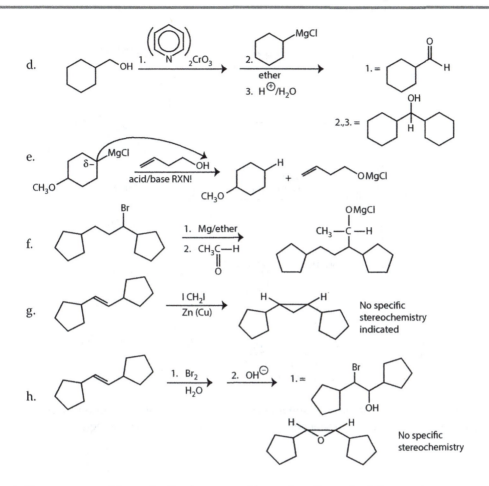

4. Draw structural formulas for the **one major** product for each of the reactions or reaction sequences. Consider both stereochemistry and regiochemistry where required.

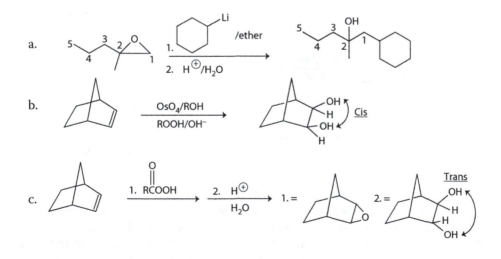

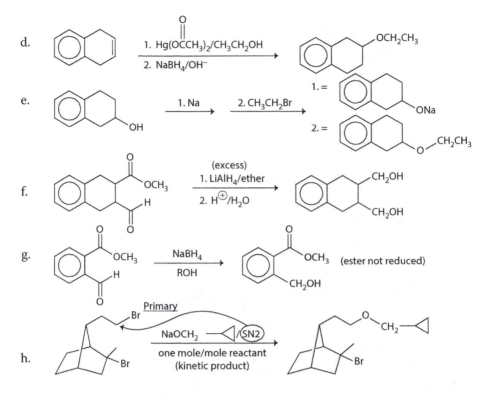

5. Draw structural formulas for the **one major** product for each of the reactions or reaction sequences. Consider both stereochemistry and regiochemistry where required.

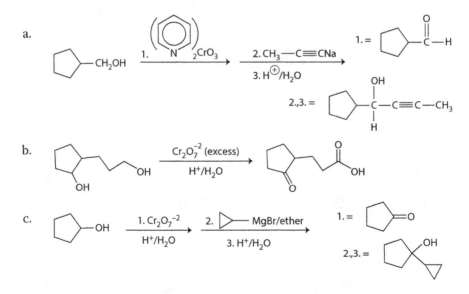

6. Complete the following **reverse** reaction sequences by identifying the compounds [A] through [D]. First, determine the structure of [D] by the excess HCl analysis reaction shown; then work backward to [A]. [A] has the formula $C_8H_6O_4$ and is an important compound used in polymers which form high-strength materials used in boat hulls etc.

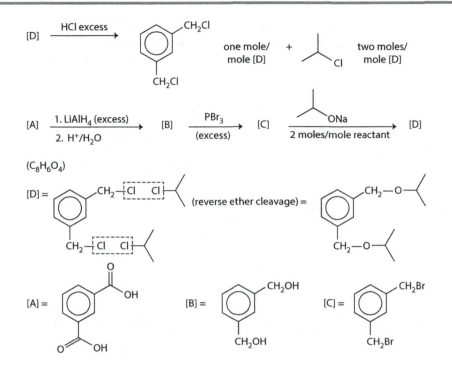

7. Use your knowledge of reactions covered to date to solve the following problems in structural analysis. *Note that all of the reactions in each part must be used to identity an unambiguous structure for the starting compound.*

[A] **to** [C]: [A] is carvone, the major component in spearmint oil; the molecule contains a six membered ring.

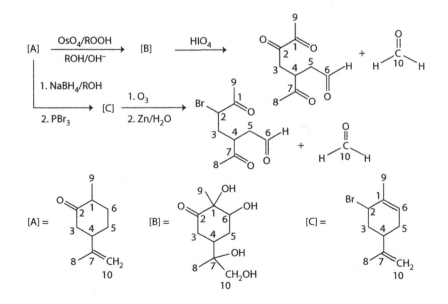

[D] to [F]: [D] is one form of Vitamin B-6. The reaction of [D] to [E] requires only one mole per mole of Collins' reagent, that is, **only one** site in the molecule is oxidized. The reaction of [D] to [F] is an SN2 reaction which will **not** occur at an aromatic position.

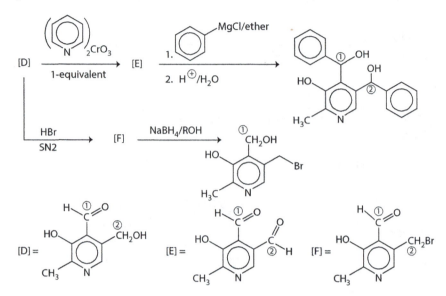

8. Draw structural formulas for the **one major** product for each of the following reactions which review previous reactions. Consider both stereochemistry and regiochemistry where required.

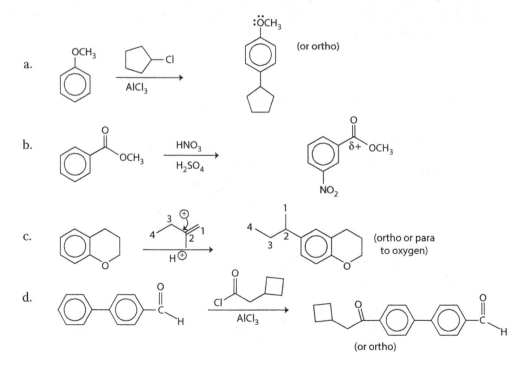

e.

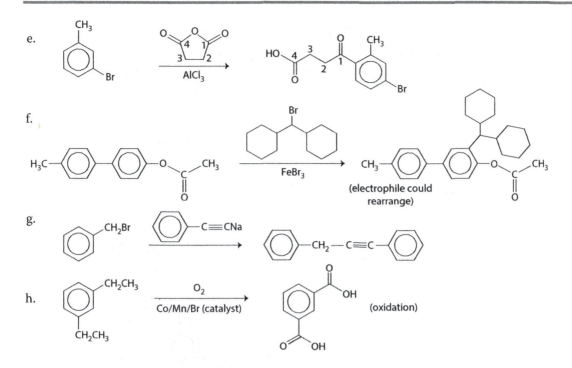

f.

(electrophile could rearrange)

g.

h.

O_2 / Co/Mn/Br (catalyst)

(oxidation)

9. **Answers to Synthesis Problem**

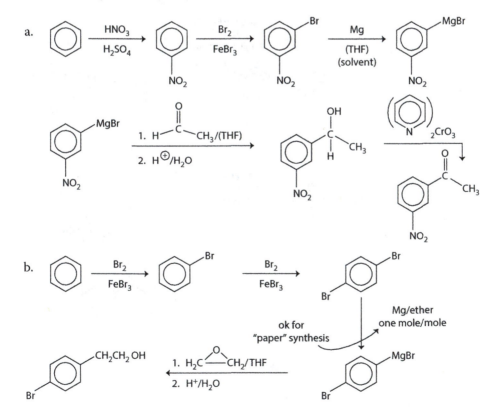

a.

HNO₃ / H₂SO₄

Br₂ / FeBr₃

Mg (THF) (solvent)

1. H—C(=O)—CH₃/(THF)
2. H⊕/H₂O

2 CrO₃

b.

Br₂ / FeBr₃

Br₂ / FeBr₃

ok for "paper" synthesis

Mg/ether one mole/mole

1. H₂C—CH₂/THF (epoxide)
2. H⁺/H₂O

c.

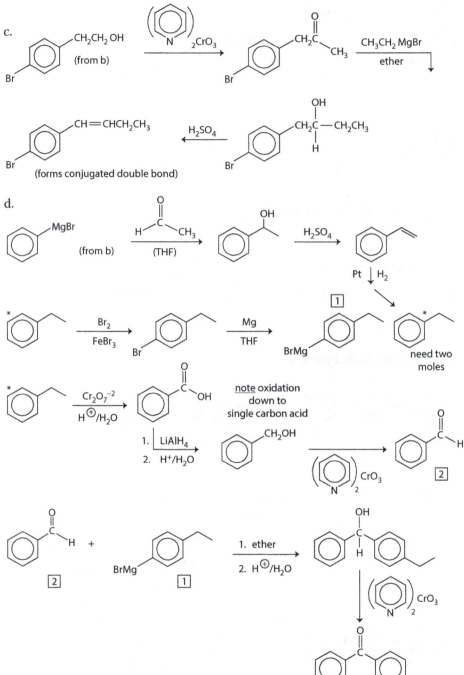

Organic II: Practice Exam 22

1. Provide correct systematic names for the following compounds:

a.

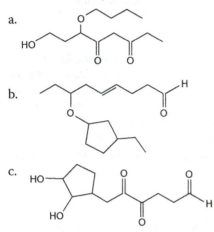

b.

c.

2. a. Show the complete mechanism for the **nucleophilic** aromatic **substitution** reaction of methyl p-chlorobenzoate with ethoxide ion:

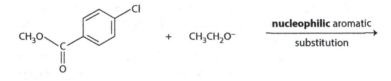

    i. Show the reaction of the nucleophile to the correct position and draw the **initial aromatic anion** formed. Then, draw **three additional resonance** structures for this intermediate, including **one** which depicts the activating effect of the ester substituent. The total will be **four** structures; **be certain to clearly show the location of double bonds and charges.**

    ii. Show the formation of the product for **nucleophilic aromatic substitution.** **Be sure to use complete arrow notation for all steps in the mechanism.**

b. Show the following sequence for the reaction of **Z**-3-methyl-2-pentene:

    i. Draw the structure of **Z**-3-methyl-2-pentene:
       Use a structure which indicates stereochemistry:

    ii. **[A]** is the product of the reaction of **Z**-3-methyl-2-pentene with a **peroxyacid;** indicate stereochemistry using wedges and dashes.

    iii. **[B]** is the **product** of the reaction of molecule [A] with ethoxide ion ($CH_3CH_2O^-$) in an ethanol solvent; use the template to draw the product: label each group for each chiral carbon with a priority number and indicate the R/S designation of each chiral carbon.

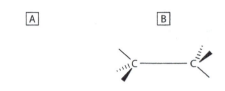

3. a.   Show **only** the **two chain propagation** steps for the **electrophilic addition** reaction of HBr in the presence of peroxides to the **1-pentenyl** side chain of p-(1-pentenyl)-toluene.

   i.   Show the addition of the Br• radical to the correct position (do **NOT** add to the aromatic ring) and draw the **initial radical** formed. Then, draw **three additional resonance** structures for this intermediate which depict the stabilizing effect of the aromatic ring. The total will be **four** structures; **be certain to clearly show the location of double bonds and radicals.**

   ii. Show the formation of the product in the second propagation step. **Be sure to use complete arrow notation for all steps in the mechanism.**

   b.   i.   Show the **mechanism** for the **electrophilic addition** reaction of Hg(OOCCH₃)₂ as the electrophile and cyclohexanol as the nucleophile to **1-butene** to form the initial metal-containing product. **Be sure to use complete arrow notation for all steps in the mechanism.**

   ii. Then, show the reaction of this initial product with NaBH₄/ROH to form the final product of the reaction sequence. Do **NOT** show the mechanism of the NaBH₄/ROH reaction.

4. Show the structure of the **major** product for each part of the following two-part reaction sequences. Each two-part reaction sequence is separate. **Include stereochemistry where required.** For multiple reactions over the same arrow: middle products may be placed outside the box for partial credit; this is not required however.

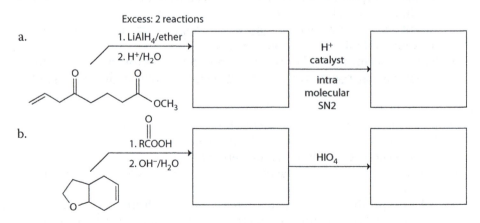

c.

1. NaBH₄/ROH

2. H₂SO₄ (conc.)
high temp

ICH₂ZnI

d.

1. Br₂/H₂O

2. NaOH
(intramolecular)
SN2

—NH₂
/H⁺
catalyst

(acidic
conditions)

e.

1. CH₃Br/FeBr₃

2. NBS

CH₃O

1.

SN2

2. OsO₄/ROH

3. ROOH/OH⁻

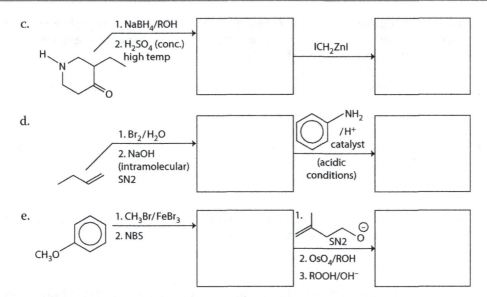

5. Show the structure of the **major** product for each part of the following three-part reaction sequences. Each three-part reaction sequence is separate. **_Include stereochemistry_** **_where required_**. For multiple reactions over the same arrow: middle products may be placed outside the box for partial credit; this is not required however.

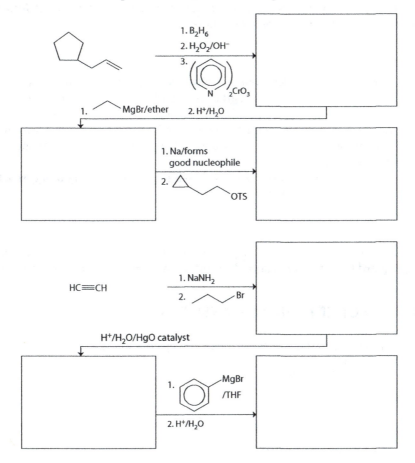

6. Solve the following **short** syntheses: *follow the specific directions for each part*. **Each synthesis must include a carbon–carbon bond formation using the organometallic reaction** specified; each synthesis should take 1–3 reactions.
   a. Start with **any** aldehyde or ketone plus **any** Grignard reagent (organomagnesium bromide) and show **one** of two possible syntheses for 6-methyl-3-heptanone.

   b. Start with **any** aldehyde or ketone plus **any** Grignard reagent (organomagnesium bromide) and show the **other** possible synthesis for 6-methyl-3-heptanone.

   c. Start with the correct epoxide plus the correct Grignard reagent (organomagnesium bromide) and show a synthesis for hexanal.

   d. Start with any **four**-carbon CuLi reagent plus a **four**-carbon correct substrate and show a synthesis for octane.

   e. Start with any **four**-carbon organolithium compound plus a correct **four**-carbon substrate and show a synthesis for octane.

7. Use reactions from the complete course to solve the synthesis problem. Follow the general synthesis procedures: Be certain that **all** organic carbons come from the carbon sources listed, using as many molecules of the starting compounds as needed. Selected reactions must produce either the major isomer or one of equivalent or similar major isomers.
   Synthesis in two parts: Using only excess ethanol, synthesize 2-hexanone
   Follow parts (a) and (b):
   a. Starting **only** with **ethanol**, synthesize **1-butanol** using an **epoxide** reaction to form carbon–carbon bonds.

   b. Starting with the **1-butanol** from part (a) and **ethanol**, synthesize **2-hexanone** using a **Grignard (organomagnesium bromide) reaction** to form carbon–carbon bonds.

ORGANIC II: PRACTICE EXAM 1.5B: ANSWERS

1. Provide correct systematic names for the following compounds:

   a.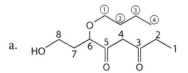

   6-butoxy-8-hydroxy-3,5-octanedione

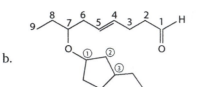

b.

7-(3-ethylcyclopentoxy)-4-nonenal

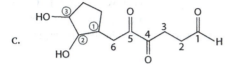

c.

6-(2,3-dihydroxycyclopentyl)-4,5-dioxohexanal

2. a. Show the complete mechanism for the **nucleophilic** aromatic substitution reaction of methyl p-chlorobenzoate with ethoxide ion:

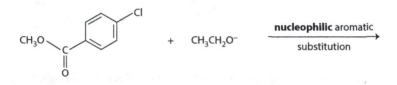

    i. Show the reaction of the nucleophile to the correct position and draw the **initial aromatic _anion_** formed. Then, draw **three additional resonance** structures for this intermediate, including **one** which depicts the activating effect of the ester substituent. The total will be **four** structures; **be certain to clearly show the location of double bonds and charges.**

    ii. Show the formation of the product for **nucleophilic aromatic substitution. Be sure to use complete arrow notation for all steps in the mechanism.**

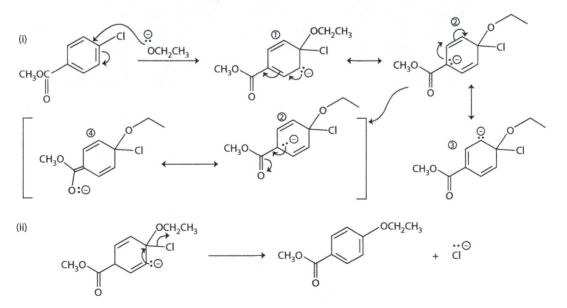

b. Show the following sequence for the reaction of **Z**-3-methyl-2-pentene:

    i. Draw the structure of **Z**-3-methyl-2-pentene:

Use a structure which indicates stereochemistry:

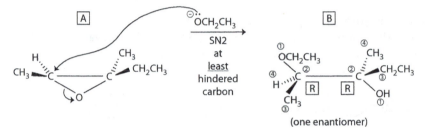

priority * CH₃   CH₂CH₃ * priority

ii. **[A]** is the product of the reaction of Z-3-methyl-2-pentene with a **peroxyacid**; indicate stereochemistry using wedges and dashes.

iii. **[B]** is the **product** of the reaction of molecule [A] with ethoxide ion (CH₃CH₂O⁻) in an ethanol solvent; use the template to draw the product: label each group for each chiral carbon with a priority number and indicate the R/S designation of each chiral carbon.

(one enantiomer)

3. a.   Show **only** the **two chain propagation** steps for the **electrophilic addition** reaction of HBr in the presence of peroxides to the **1-pentenyl** side chain of p-(1-pentenyl)-toluene.

+   HBr/ROOR   **electrophilic addition** ────────────────▶

i. Show the addition of the Br• radical to the correct position (do **NOT** add to the aromatic ring) and draw the **initial radical** formed. Then, draw **three additional resonance** structures for this intermediate which depict the stabilizing effect of the aromatic ring. The total will be **four** structures; **be certain to clearly show the location of double bonds and radicals.**

ii. Show the formation of the product in the second propagation step. **Be sure to use complete arrow notation for all steps in the mechanism.**

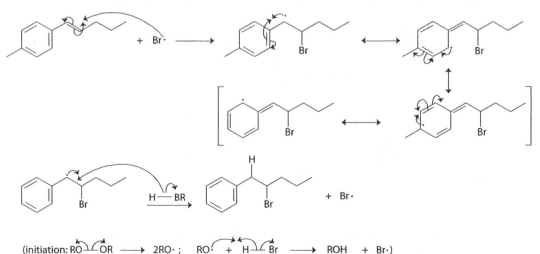

(initiation: RO⌒OR ──▶ 2RO• ;   RO• + H⌒Br ──▶ ROH + Br•)

b. i.   Show the **mechanism** for the electrophilic **addition** reaction of $Hg(OOCCH_3)_2$ as the electrophile and cyclohexanol as the nucleophile to **1-butene** to form the initial metal-containing product. **Be sure to use complete arrow notation for all steps in the mechanism.**

ii. Then, show the reaction of this initial product with $NaBH_4/ROH$ to form the final product of the reaction sequence. Do **NOT** show the mechanism of the $NaBH_4/ROH$ reaction.

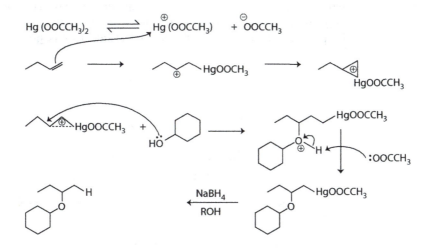

4. Show the structure of the **major** product for each part of the following two-part reaction sequences. Each two-part reaction sequence is separate. _**Include stereochemistry where required**_. For multiple reactions over the same arrow: middle products may be placed outside the box for partial credit; this is **not** required however.

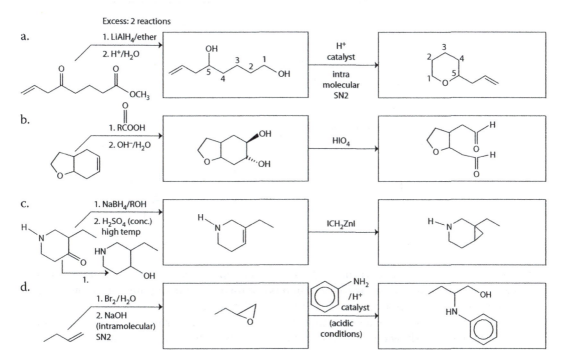

e.

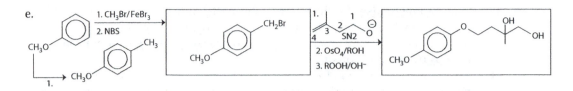

5. Show the structure of the **major** product for each part of the following three-part reaction sequences. Each three-part reaction sequence is separate. **_Include stereochemistry where required_**. For multiple reactions over the same arrow: middle products may be placed outside the box for partial credit; this is <u>not</u> required however.

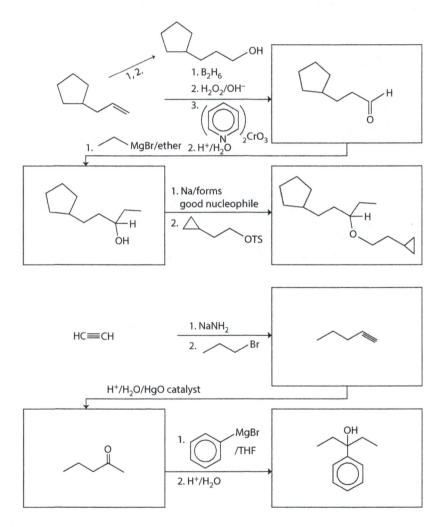

6. Solve the following **short** syntheses: **_follow the specific directions for each part_**. **Each synthesis must include a carbon–carbon bond formation using the organometallic reaction** specified; each synthesis should take 1–3 reactions.

   a. Start with **any** aldehyde or ketone plus **any** Grignard reagent (organomagnesium bromide) and show <u>**one**</u> of two possible syntheses for 6-methyl-3-heptanone.

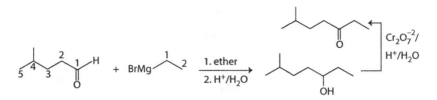

b. Start with **any** aldehyde or ketone plus **any** Grignard reagent (organomagnesium bromide) and show the **other** possible synthesis for 6-methyl-3-heptanone.

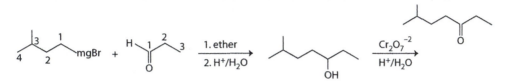

c. Start with the correct epoxide plus the correct Grignard reagent (organomagnesium bromide) and show a synthesis for hexanal.

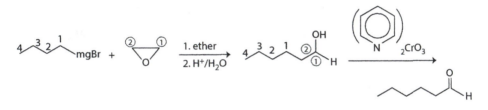

d. Start with any **four**-carbon CuLi reagent plus a **four**-carbon correct substrate and show a synthesis for octane.

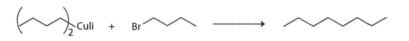

e. Start with any **four**-carbon organolithium compound plus a correct **four**-carbon substrate and show a synthesis for octane.

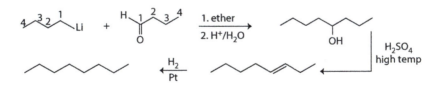

7. Use reactions from the complete course to solve the synthesis problem. Follow the general synthesis procedures: Be certain that **all** organic carbons come from the carbon sources listed, using as many molecules of the starting compounds as needed. Selected reactions must produce either the major isomer or one of equivalent or similar major isomers.

Synthesis in two parts: Using only excess ethanol, synthesize 2-hexanone

Follow parts (a) and (b):

a. Starting **only** with **ethanol**, synthesize **1-butanol** using an **epoxide reaction** to form carbon–carbon bonds.

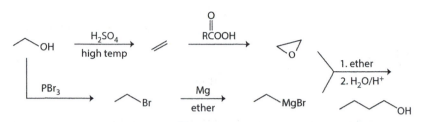

b. Starting with the **1-butanol** from part (a) and **ethanol**, synthesize **2-hexanone** using a **Grignard (organomagnesium bromide) reaction** to form carbon–carbon bonds.

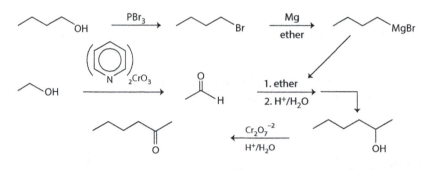

Organic II: Practice Exam 23

1. Provide products and intermediates as required for the following reactions at carbonyl. The problem will demonstrate the mechanistic relationship among nucleophilic additions/substitutions by addition–elimination sequences with similar appearing reagents.
   a. Show the mechanism for the following nucleophilic carbonyl reaction of a ketone:

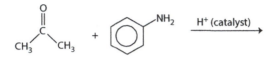

   For the addition phase of the reaction: include the role of the acid catalyst, the initial positively charged addition intermediate, and the neutral tetrahedral intermediate. For the elimination phase of the reaction: include catalyst role, the elimination step to form a positively charged trigonal planar intermediate, and the final product.

   b. Now, show the mechanism of the reaction of the same ketone with a different amine:

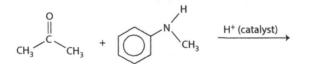

   Abbreviate the mechanism by showing *only the neutral tetrahedral intermediate formed in the addition phase and the final product formed by elimination*; each will occur with appropriate proton transfers which need not be shown as separate steps.

   c. Now, consider the mechanism of the reaction of an amine with an ester:

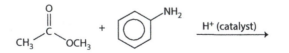

   Show each of the steps in this nucleophilic carbonyl substitution, indicating all intermediates, proton transfers, and the role of the catalyst for the addition phase and the elimination phase.

   d. A Grignard reagent can add **twice** to an **ester;** show the mechanism of the following **nucleophilic addition/elimination/second addition** reaction to the carbonyl of an ester:

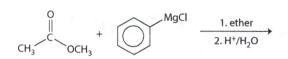

Include **only** the **first addition product** metal complex; the **new carbonyl compound formed by elimination** of an appropriate leaving group; the **second addition product** metal complex; and the **final product** formed by reaction with aqueous acid.

2. Provide a systematic name for the following biomolecules: stereochemistry is not required.

   a. Precursor to glutamic acid, an amino acid, and component of the vitamin folic acid:

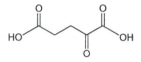

   b. Ester of glucuronic acid, a glucose metabolite:

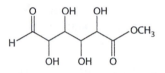

   c. Retinal, acts as a light receptor in the eye:

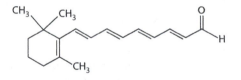

   d. Amide related to prostaglandin E1, associated with pain response.

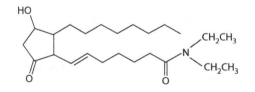

3. Complete the following reverse reaction sequence by identifying compounds **[A]** through **[D]**: The final compound shown is representative of the molecules and bonding found in high-strength polymer materials (aromatic **polyamides**) such as **Kevlar** used in aviation products, body and tank armor, etc. To help in working backward to [C] and [D], the last reaction represents the formation of **amide** bonds through standard nucleophilic carbonyl reactions.

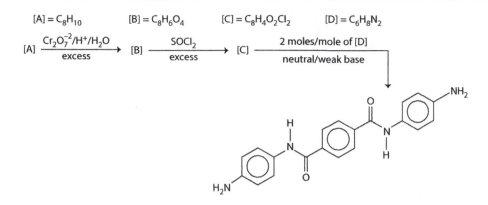

[A] = C_8H_{10} [B] = $C_8H_6O_4$ [C] = $C_8H_4O_2Cl_2$ [D] = $C_6H_8N_2$

[A] $\xrightarrow[\text{excess}]{Cr_2O_7^{-2}/H^+/H_2O}$ [B] $\xrightarrow[\text{excess}]{SOCl_2}$ [C] $\xrightarrow[\text{neutral/weak base}]{\text{2 moles/mole of [D]}}$

4. The following problem demonstrates the interconversion relationships between carboxylic acid derivatives, and the interconversion possibilities to other carbonyl compounds. **Starting with pentanoic acid** and the list of reagents/reactants given, **show the reactions** for all of the transformations indicated. *Do this by following the (a) through (i) sequence,* the starting compound for each reaction will thus have already been synthesized in a previous reaction; almost all transformations represent only one reaction.
Reactants and reagents: pentanoic acid plus; H_2O; CH_3CH_2OH; $SOCl_2$; H_2NCH_3; $LiAlH_4$; Collin's reagent: $(C_5H_5N)_2CrO_3$; phenyl magnesium bromide: C_6H_5-MgBr

Transformations:

a. acid → acid chloride;

b. acid chloride → symmetric acid anhydride

c. acid anhydride → ester

d. ester → amide

e. amide → acid

f. acid → aldehyde

g. aldehyde → secondary alcohol

h. aldehyde → imine

i. ester → tertiary alcohol

5. Draw structural formulas for the **one major** product for each of the reactions or reaction sequences. Consider both stereochemistry and regiochemistry where required.

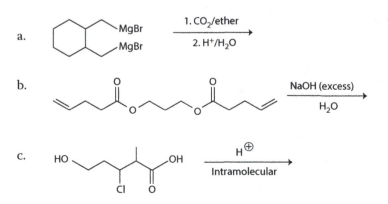

a. $\xrightarrow[\text{2. H}^+/\text{H}_2\text{O}]{\text{1. CO}_2/\text{ether}}$

b. $\xrightarrow[\text{H}_2\text{O}]{\text{NaOH (excess)}}$

c. $\xrightarrow[\text{Intramolecular}]{\text{H}^\oplus}$

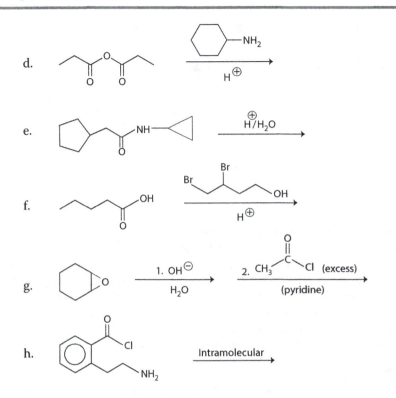

d.

e.

f.

g.

h.

6. Draw structural formulas for the **one major** product for each of the reactions or reaction sequences. Consider both stereochemistry and regiochemistry where required.

a.

b.

c.

d.

e.

f.

g.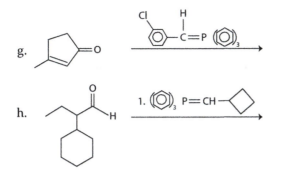

h.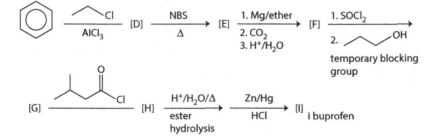

7. Complete the synthesis reaction sequences below. Each part shows a possible synthesis for a prescription or over-the-counter analgesic drug.
   a.   [A] through [C]: **aspirin**

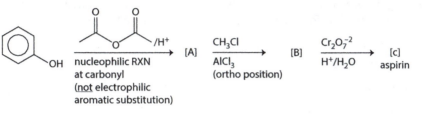

b. **[D] through [I]: ibuprofen;** in Motrin, Nuprin.

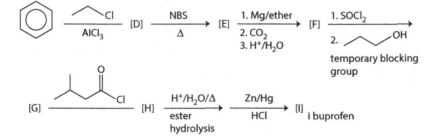

c.   [J] through [L]: **acetaminophen;** nonaspirin analgesics such as Tylenol, Excedrin PM, etc.

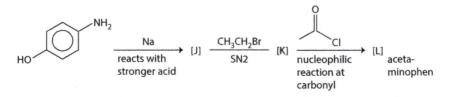

d.   **[M]** through **[Q]: naproxen;** in Naprosyn

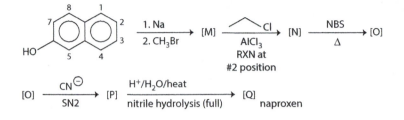

[O] $\xrightarrow[\text{SN2}]{\text{CN}^{\ominus}}$ [P] $\xrightarrow[\text{nitrile hydrolysis (full)}]{\text{H}^+/\text{H}_2\text{O}/\text{heat}}$ [Q] naproxen

8. Show the structure of the **major** product for each part of the following two-part reaction sequences. Each two-part reaction sequence is separate. **_Include stereochemistry where required_**. For multiple reactions over the same arrow: middle products may be placed outside the box for partial credit; this is <u>not</u> required however.

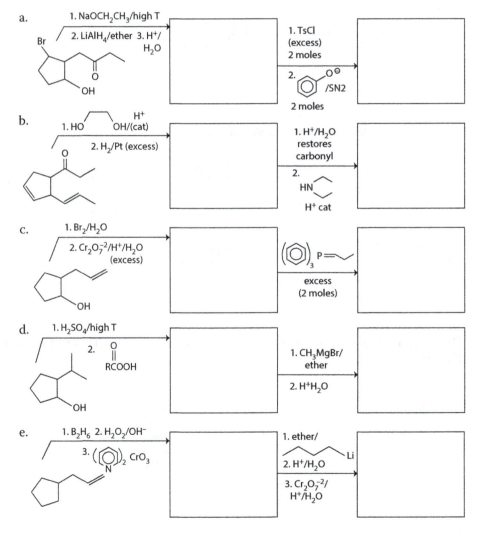

9. **Synthesis**

The following problems will demonstrate concepts for syntheses of derivatives of Alprenolol, an antihypertensive drug:

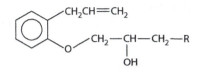

Use the guidelines for synthesis problems stated in the other practice exams. Follow the directions in each problem; the individual parts will serve as a guide to synthesis of certain portions of the molecule. Syntheses of the entire molecules can be thought of as combinations of the pieces demonstrated (with proper attention to reaction sequence and functional group interference).

Provide **four different** syntheses of **2-propenylbenzene** following the guidelines:

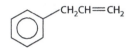

a. **One** method using a Friedel–Craft reaction from **benzene plus propene as carbon sources.**

b. **One** method using a Grignard reaction from **toluene plus any carbonyl compound as carbon sources.**

c., d. **Two** methods in which a Wittig reaction is the last step; use **benzene plus any other appropriate starting compounds.**

e. Starting with phenol and propene as the only carbon sources, provide a synthesis for **1-phenoxy-2-propanol:**

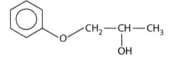

f., g. Starting from any product synthesized in part **(e)** plus any amine, provide a synthesis for the following molecules:

h., i. Starting with any product synthesized in part **(e)** plus any amine, provide **two** syntheses for the following molecule:

h. **One** method using cyanide to add the carbonyl group.

i. **One** method using carbon dioxide to add the carbonyl group.

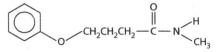

ORGANIC II: PRACTICE EXAM 2A: ANSWERS

1. Provide products and intermediates as required for the following reactions at carbonyl. The problem will demonstrate the mechanistic relationship among nucleophilic additions/substitutions by addition–elimination sequences with similar appearing reagents.
   a. Show the mechanism for the following nucleophilic carbonyl reaction of a ketone:

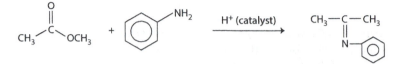

   For the addition phase of the reaction: include the role of the acid catalyst, the initial positively charged addition intermediate, and the neutral tetrahedral intermediate. For the elimination phase of the reaction: include catalyst role, the elimination step to form a positively charged trigonal planar intermediate, and the final product.

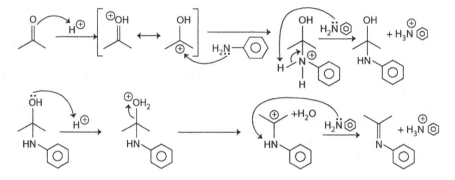

   b. Now, show the mechanism of the reaction of the same ketone with a different amine:

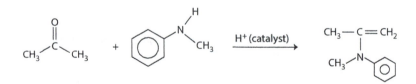

   Abbreviate the mechanism by showing *only the neutral tetrahedral intermediate formed in the addition phase and the final product formed by elimination*; each will occur with appropriate proton transfers which need not be shown as separate steps.

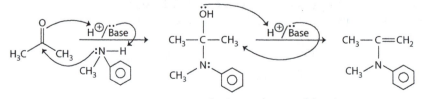

   (amine used as a weak base.)

c. Now, consider the mechanism of the reaction of an amine with an ester:

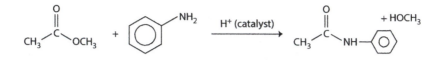

Show each of the steps in this nucleophilic carbonyl substitution, indicating all intermediates, proton transfers, and the role of the catalyst for the addition phase and the
elimination phase.

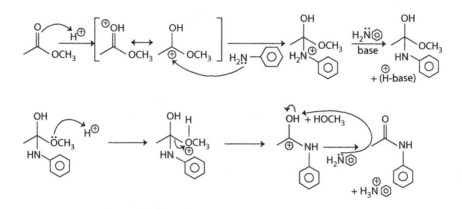

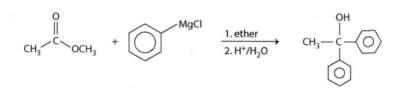

d. A Grignard reagent can add **twice** to an **ester;** show the mechanism of the following
nucleophilic addition/elimination/second addition reaction to the carbonyl of
an ester:

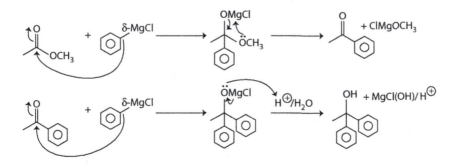

Include **only the first addition product** metal complex; the **new carbonyl compound formed by elimination** of an appropriate leaving group; the **second addition product** metal complex; and the **final product** formed by reaction with aqueous
acid.

2. Provide a systematic name for the following biomolecules: stereochemistry is not required.

a. Precursor to glutamic acid, an amino acid, and component of the vitamin folic acid:

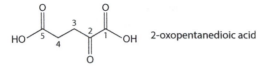

2-oxopentanedioic acid

b. Ester of glucuronic acid, a glucose metabolite:

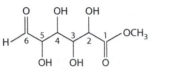

ester = priority
methyl 2,3,4,5-tetrahydroxy-6-oxo-hexanoate

c. Retinal, acts as a light receptor in the eye:

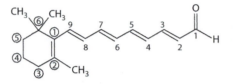

aldehyde = priority
9-(2,6,6-trimethyl-1-cyclohexenyl)-
2,4,6,8-nonatetraenal

d. Amide related to prostaglandin E1, associated with pain response.

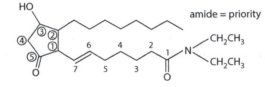

amide = priority   N,N-diethyl-7-(3-hydroxy-2-octyl-5-oxo-
cyclopentyl)-6-heptenamide

3. Complete the following reverse reaction sequence by identifying compounds **[A]** through **[D]**: The final compound shown is representative of the molecules and bonding found in high-strength polymer materials (aromatic **polyamides**) such as **Kevlar** used in aviation products, body and tank armor, etc. To help in working backward to [C] and [D], the last reaction represents the formation of **amide** bonds through standard nucleophilic carbonyl reactions.

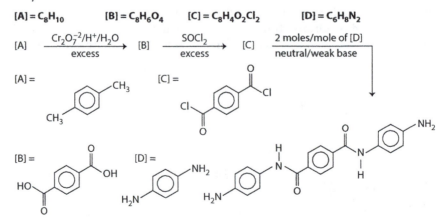

4. The following problem demonstrates the interconversion relationships between carboxylic acid derivatives, and the interconversion possibilities to other carbonyl compounds.

Starting with pentanoic acid and the list of reagents/reactants given, **show the reactions** for all of the transformations indicated. ***Do this by following the (a) through (i) sequence***; the starting compound for each reaction will thus have already been synthesized in a previous reaction; almost all transformations represent only one reaction.
Reactants and reagents: pentanoic acid plus: H_2O; CH_3CH_2OH; $SOCl_2$; H_2NCH_3; $LiAlH_4$; Collin's reagent: $(C_5H_5N)_2CrO_3$; phenyl magnesium bromide: C_6H_5-MgBr

Transformations:

a. acid → acid chloride; **b.** acid chloride → symmetric acid anhydride

c. acid anhydride → ester **d.** ester → amide

e. amide → acid **f.** acid → aldehyde

g. aldehyde → secondary alcohol **h.** aldehyde → imine

i. ester → tertiary alcohol

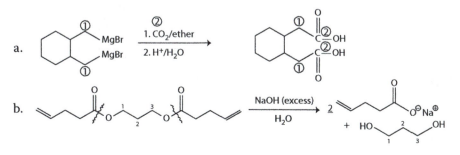

5. Draw structural formulas for the **one major** product for each of the reactions or reaction sequences. Consider both stereochemistry and regiochemistry where required.

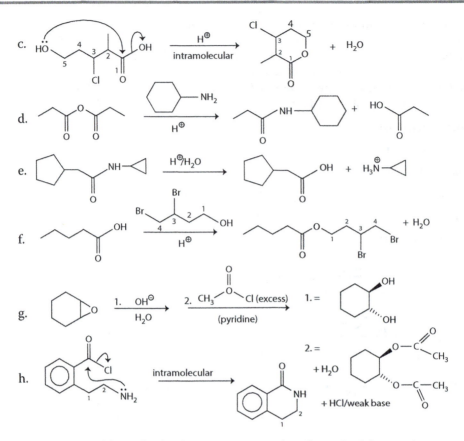

6. Draw structural formulas for the **one major** product for each of the reactions or reaction sequences. Consider both stereochemistry and regiochemistry where required.

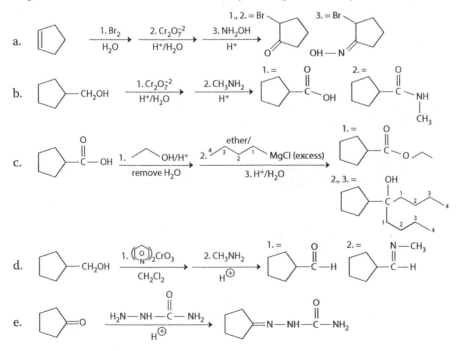

f.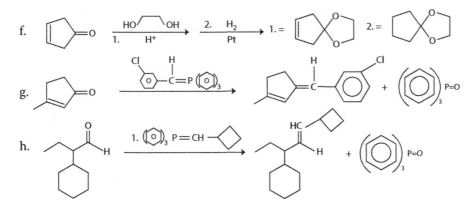

g.

h.

7. Complete the synthesis reaction sequences below. Each part shows a possible synthesis for a prescription or over-the-counter analgesic drug.

a. [A] through [C]: **aspirin**

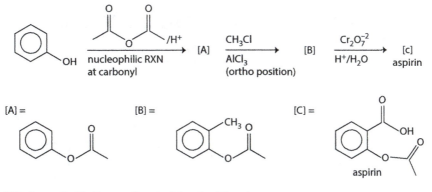

[A] =   [B] =   [C] =

aspirin

b. [D] through [I]: **ibuprofen**; in Motrin, Nuprin.

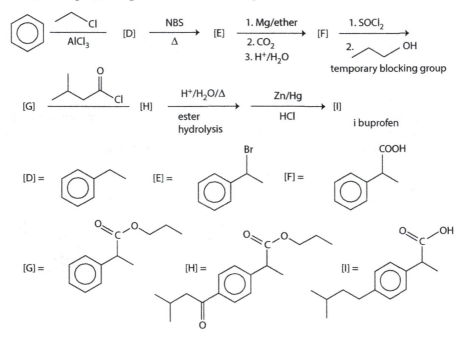

[D] =   [E] =   [F] =

[G] =   [H] =   [I] =

c. [J] through [L]: **acetaminophen**; nonaspirin analgesics such as Tylenol, Excedrin PM, etc.

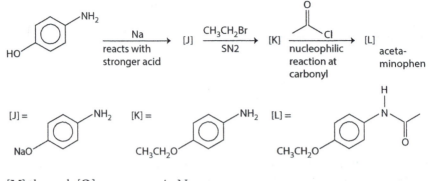

[J] =   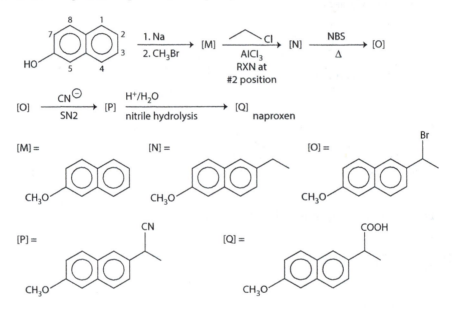   [K] =   ...   [L] =   ...

d. [M] through [Q]: **naproxen**; in Naprosyn

8. Show the structure of the **major** product for each part of the following two-part reaction sequences. Each two-part reaction sequence is separate. ***Include stereochemistry where required.*** For multiple reactions over the same arrow: middle products may be placed outside the box for partial credit; this is <u>not</u> required however.

a.

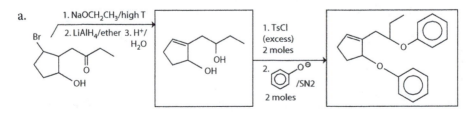

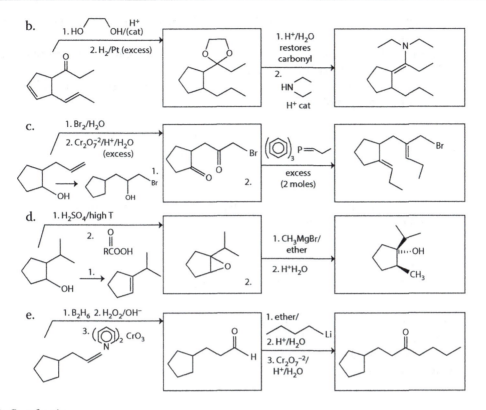

b.
c.
d.
e.

9. Synthesis

The following problems will demonstrate concepts for syntheses of derivatives of Alprenolol, an antihypertensive drug:

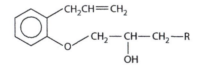

Use the guidelines for synthesis problems stated in the other practice exams. Follow the directions in each problem; the individual parts will serve as a guide to synthesis of certain portions of the molecule. Syntheses of the entire molecules can be thought of as combinations of the pieces demonstrated (with proper attention to reaction sequence and functional group interference).

Provide **four different** syntheses of **2-propenylbenzene** following the guidelines:

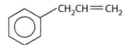

a. **One** method using a Friedel–Craft reaction from **benzene plus propene as carbon sources.**

b. **One** method using a Grignard reaction from **toluene plus any carbonyl compound as carbon sources.**

c., d. **Two** methods in which a Wittig reaction is the last step; use **benzene plus any other appropriate starting compounds.**

e. Starting with phenol and propene as the only carbon sources, provide a synthesis for **1-phenoxy-2-propanol:**

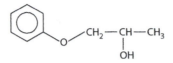

f., g. Starting from any product synthesized in part (e) plus any amine, provide a synthesis for the following molecules:

h., i. Starting with any product synthesized in part **(e)** plus any amine, provide **two** syntheses for the following molecule:

h. **One** method using cyanide to add the carbonyl group.

i. **One** method using carbon dioxide to add the carbonyl group.

10. **Answers to Synthesis Problem**

a.

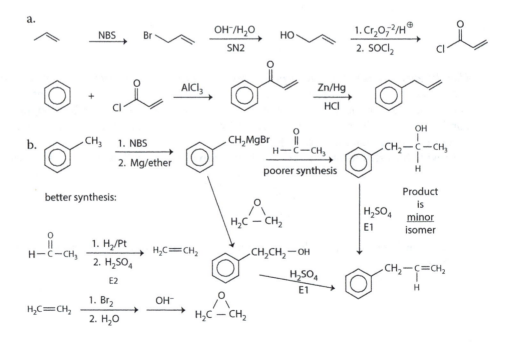

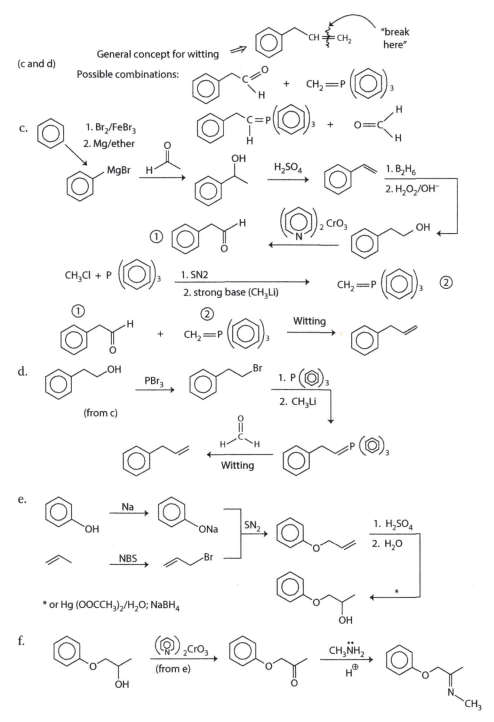

g.

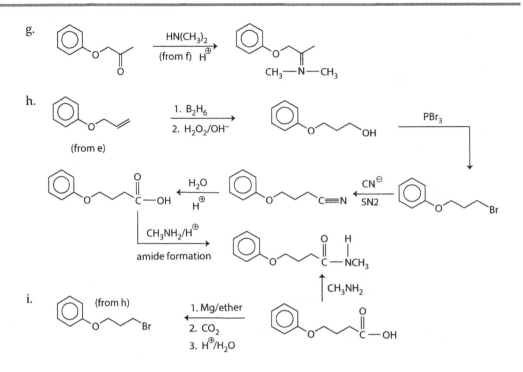

h.

i.

Organic II: Practice Exam 24

1. Provide correct systematic names for the following compounds:

a.

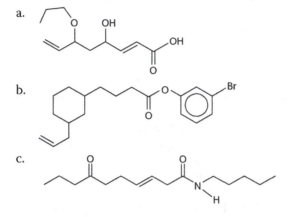

b.

c.

2. ***Be certain to show complete arrow notation for all electron changes in all steps of reaction mechanisms.***
   a. Show the **complete mechanism** for the reaction of **cyclohexanone** with **hydrazine** (H_2NNH_2) to form the **imine** product in the presence of an H^+ catalyst (i.e., an acid mechanism), (i) acid/base reaction involving the catalyst; (ii) addition phase r.d.s.; (iii) formation of the **neutral** tetrahedral intermediate; (iv) acid/base reaction involving the catalyst; (v) elimination of r.d.s.; and (vi) formation of the final **neutral** product.

   b. Draw the structure of the neutral tetrahedral intermediate formed from **cyclohex-anone** and **dimethylamine: $HN(CH_3)_2$**. Then, show **only** the mechanism for the elimination phase (**NOT** the addition phase) for the completion of the reaction to form the **enamine**. (i) Acid/base reaction involving the catalyst; (ii) elimination of r.d.s.; and (iii) formation of the final **neutral** product.

3. Show the **complete mechanism** for the **complete hydrolysis** of the cyclic acetal shown below in the presence of an H^+ catalyst. ***Be certain to show complete arrow notation for all electron changes in all steps of the reaction mechanism.***

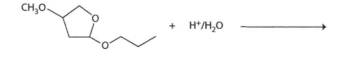

a. Show the mechanism for the hydrolysis of the acetal to form the cyclic hemiacetal; that is, react the **noncyclic** portion of the acetal first. (i) Acid/base reaction involving the catalyst; (ii) elimination reaction; (iii) addition reaction; and (iv) formation of the **neutral** cyclic hemiacetal.

b. Show the mechanism for the hydrolysis of the cyclic hemiacetal to form the final neutral completely hydrolyzed product (containing carbonyl and alcohol functional groups). (i) Acid/base reaction involving the catalyst; (ii) elimination reaction; and (iii) formation of the **neutral** final product.

4. *Be certain to show complete arrow notation for all electron changes in all steps of the reaction mechanisms.*
   a. Show the **complete mechanism** for the **esterification** reaction of butanoyl chloride with ethanol under neutral conditions; **use the list of steps as a help**.
      i. Addition phase r.d.s.; the nucleophile is neutral
      ii. Formation of the **neutral** tetrahedral intermediate (show as two proton transfers together)
      iii. Formation of the final **neutral** final products: elimination/proton transfer together

   b. The product of the reaction in part (a) is **ethyl butanoate.** Show the **complete mechanism** for the strong **base** (OH⁻/H₂O) **hydrolysis** of ethyl butanoate; include the reaction which demonstrates why this reaction is irreversible. (i) Addition of the (strongest) nucleophile; (ii) formation of the **neutral** tetrahedral intermediate in **aqueous** solution; (iii) formation of final **neutral** final products: elimination/proton transfer together; and (iv) all additional acid/base reactions which complete the reaction system and show the reason for non-reversibility.

5. For the following **nucleophilic reactions at carbonyl,** draw the structure of the **neutral tetrahedral intermediate** formed in the addition phase of the reaction. Then, draw the structure of the **final product** of the reaction based on the principles of reactions at carbonyls.

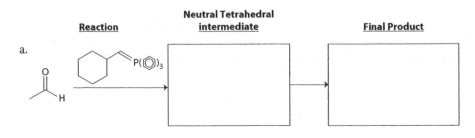

Reaction	Neutral Tetrahedral intermediate	Final Product

a.

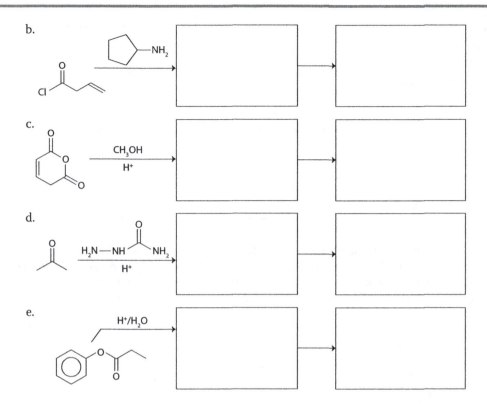

b.

c.

d.

e.

6. For the following **nucleophilic reactions at carbonyl,** draw the structure of the **neutral tet-rahedral intermediate** formed in the addition phase of the reaction. Then, draw the struc-ture of the **final product** of the reaction based on the principles of reactions at carbonyls.

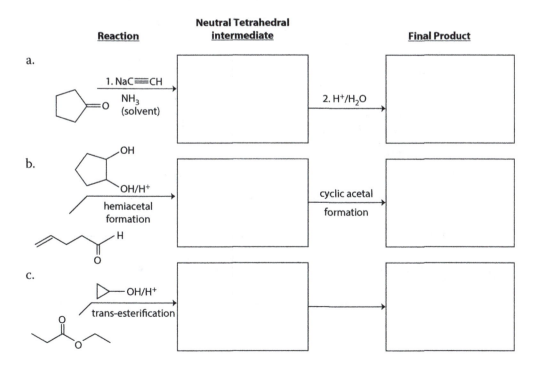

a.

b.

c.

d.

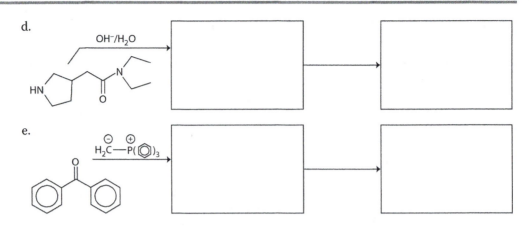

e.

7. Show the structure of the **major** product for each part of the following two-part reaction sequences. Each two-part reaction sequence is separate. _**Include stereochemistry where required**_. For multiple reactions over the same arrow: middle products may be placed outside the box for partial credit; this is not required however.

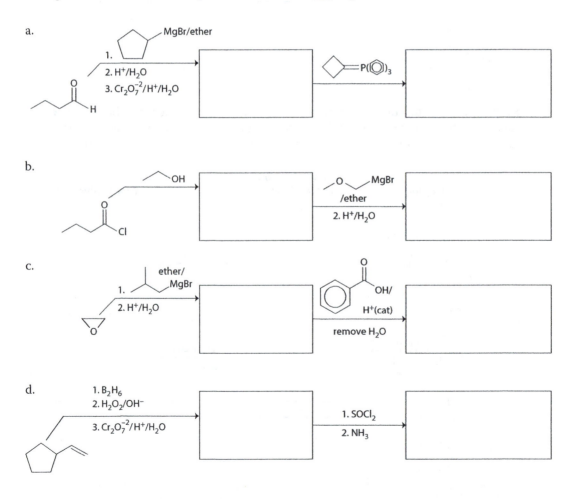

e.

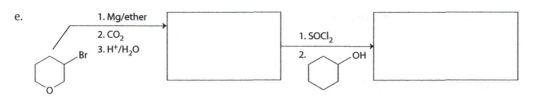

8. Show the structure of the **major** product for the following **six**-part reaction sequence. ***Include stereochemistry where required***. For multiple reactions over the same arrow: middle products may be placed outside the box for partial credit; this is <u>not</u> required however.

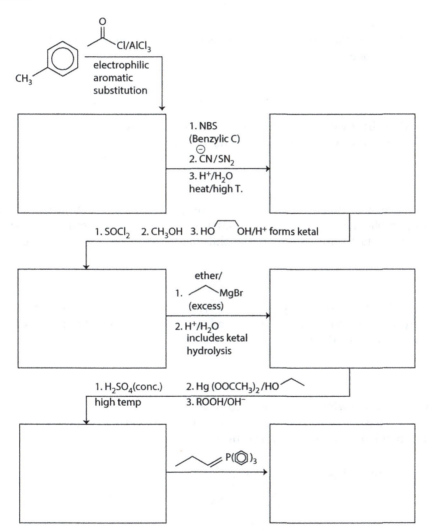

9. For the three parts, draw structures **to the right of the name** for the two compounds which could be reacted to form the named molecule through a nucleophilic carbonyl substitution reaction (i.e., a Group [5] reaction; see Chapter 18). More than one correct combination may be possible.

a. **2-methylbutyl 3-pentenoate**

b. **cyclohexyl p-nitrobenzoate**

c. **N-ethyl-N-propyl-6-oxo-heptanamide**

10. Follow the parts below to show a synthesis for the following compound, related to an insect pheromone:

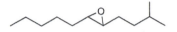

a. Working backward, select the correct alkene and show the **last** reaction in the synthesis which will form the epoxide final product.

b. The alkene selected in part (a) should have 12 carbons. Show a Wittig (using triphenylphosphonium ylide reagent) reaction which can synthesize this alkene in **one** reaction: start with a correct nucleophile Wittig reagent and a correct substrate.

c. Complete the problem by showing a synthesis for the Wittig reagent selected in part (b) from **any** 6-carbon alkene and the substrate selected in part (b) from **any other** 6-carbon alkene.

11. Complete the following parts relating to amino acids:
    a. Draw the structures of the two amino acids which have been bonded through a peptide bond to form the dipeptide shown:

amino acids

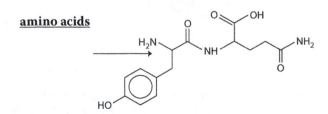

b. Draw the structures of the dipeptide which can be formed from the two amino acids shown:

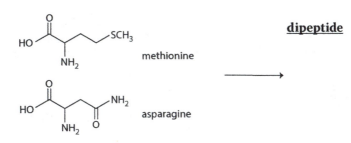

methionine

dipeptide

asparagine

12. Use the general procedures for synthesis to show a synthesis for the lactone shown below starting only with **1,3-butadiene plus carbon dioxide** as carbon sources. **Assume that the double bonds in 1,3-butadiene can be reacted independently.**

ORGANIC II: PRACTICE EXAM 2B: ANSWERS

1. Provide correct systematic names for the following compounds:

a.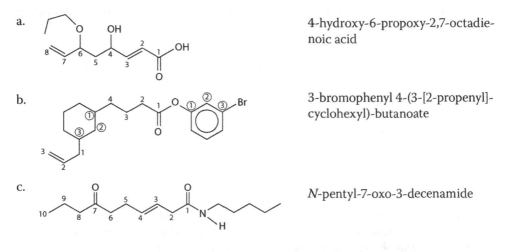

4-hydroxy-6-propoxy-2,7-octadie-noic acid

b.

3-bromophenyl 4-(3-[2-propenyl]-cyclohexyl)-butanoate

c.

N-pentyl-7-oxo-3-decenamide

2. *Be certain to show complete arrow notation for all electron changes in all steps of reaction mechanisms.*

a. Show the **complete mechanism** for the reaction of **cyclohexanone** with **hydrazine (H_2NNH_2)** to form the **imine** product in the presence of an H^+ catalyst (i.e., an acid mechanism), (i) acid/base reaction involving the catalyst; (ii) addition phase r.d.s.; (iii) formation of the **neutral** tetrahedral intermediate; (iv) acid/base reaction involving the catalyst; (v) elimination of r.d.s.; and (vi) formation of the final **neutral** product.

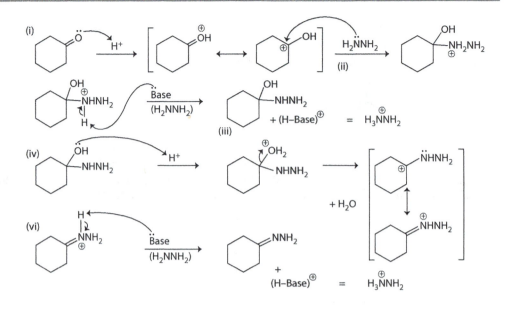

b. Draw the structure of the neutral tetrahedral intermediate formed from **cyclohex-anone** and **dimethylamine: HN(CH₃)₂.** Then, show **only** the mechanism for the elimination phase (**NOT** the addition phase) for the completion of the reaction to form the **enamine.** (i) Acid/base reaction involving the catalyst; (ii) elimination of r.d.s.; and (iii) formation of the final **neutral** product

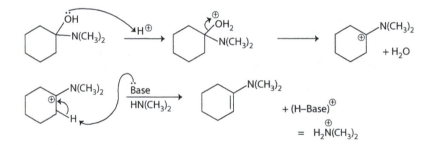

3. Show the **complete mechanism** for the **complete hydrolysis** of the cyclic acetal shown below in the presence of an H⁺ catalyst. **_Be certain to show complete arrow notation for all electron changes in all steps of the reaction mechanism._**

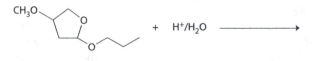

a. Show the mechanism for the hydrolysis of the acetal to form the cyclic hemiacetal; that is, react the **noncyclic** portion of the acetal first. (i) Acid/base reaction involving the catalyst; (ii) elimination reaction; (iii) addition reaction; and (iv) formation of the **neutral** cyclic hemiacetal

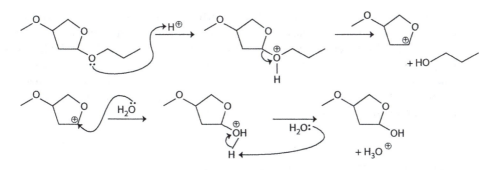

b. Show the mechanism for the hydrolysis of the cyclic hemiacetal to form the final neutral completely hydrolyzed product (containing carbonyl and alcohol functional groups). (i) Acid/base reaction involving the catalyst; (ii) elimination reaction; and (iii) formation of the **neutral** final product.

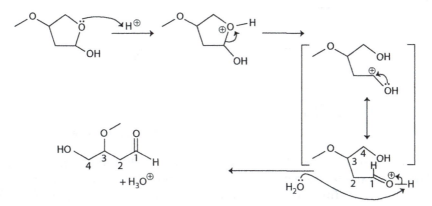

4. *Be certain to show complete arrow notation for all electron changes in all steps of the reaction mechanisms.*
   a. Show the **complete mechanism** for the **esterification** reaction of butanoyl chloride with ethanol under neutral conditions; <u>use the list of steps as a help</u>.
      i. Addition phase r.d.s.; the nucleophile is neutral
      ii. Formation of the **neutral** tetrahedral intermediate (show as two proton transfers together)
      iii. Formation of the final **neutral** final products: elimination/proton transfer together

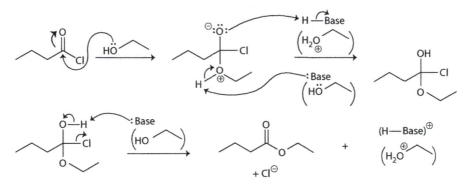

b. The product of the reaction in part (a) is **ethyl butanoate**. Show the **complete mechanism** for the strong **base** (OH⁻/H₂O) **hydrolysis** of ethyl butanoate; include the reaction which demonstrates why this reaction is irreversible. (i) Addition of the (strongest) nucleophile; (ii) formation of the **neutral** tetrahedral intermediate in **aqueous** solution; (iii) formation of the final **neutral** final products: elimination/ proton transfer together; and (iv) all additional acid/base reactions which complete the reaction system and show the reason for non-reversibility.

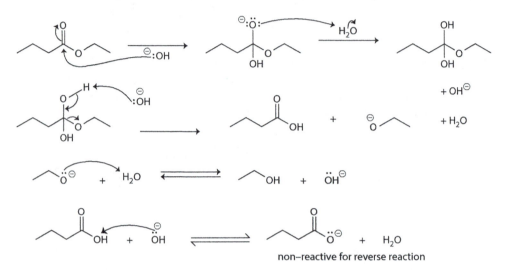

5. For the following **nucleophilic reactions at carbonyl,** draw the structure of the **neutral tetrahedral intermediate** formed in the addition phase of the reaction. Then, draw the structure of the **final product** of the reaction based on the principles of reactions at carbonyls.

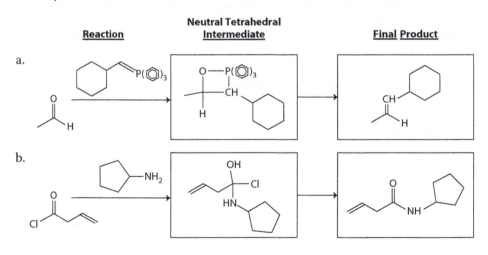

c.

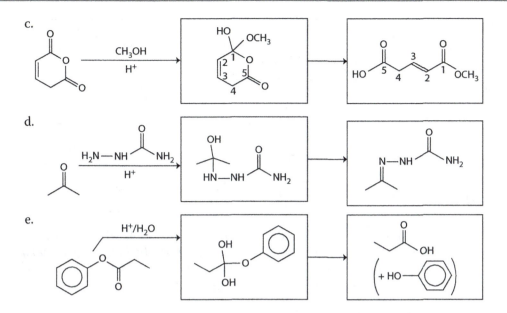

d.

e.

6. For the following **nucleophilic reactions at carbonyl,** draw the structure of the **neutral tetrahedral intermediate** formed in the addition phase of the reaction. Then, draw the structure of the **final product** of the reaction based on the principles of reactions at carbonyls.

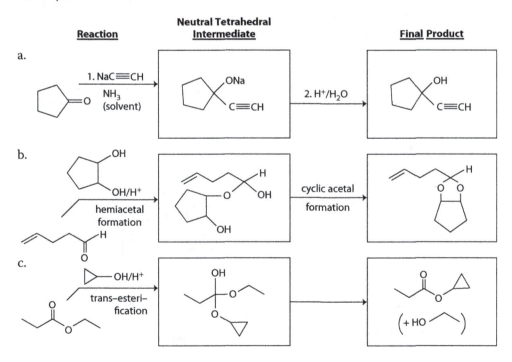

d.

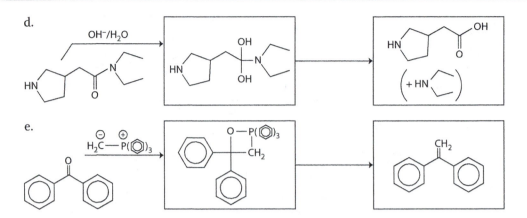

e.

7. Show the structure of the **major** product for each part of the following two-part reaction sequences. Each two-part reaction sequence is separate. ***Include stereochemistry where required.*** For multiple reactions over the same arrow: middle products may be placed outside the box for partial credit; this is <u>not</u> required however.

a.

b.

c.

d.

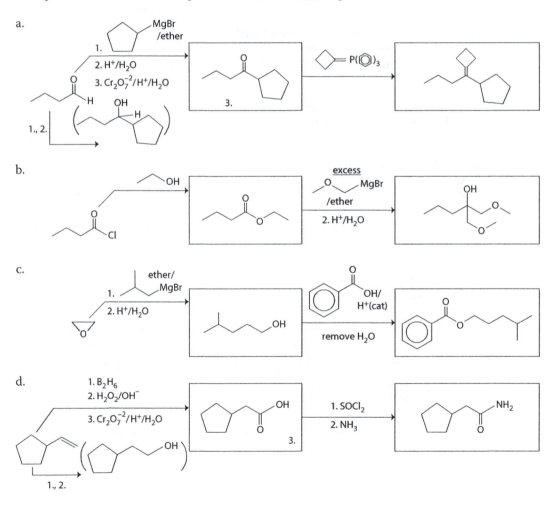

e.

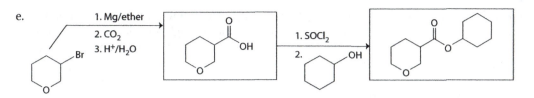

8. Show the structure of the **major** product for the following **six**-part reaction sequence. _**Include stereochemistry where required**_. For multiple reactions over the same arrow: middle products may be placed outside the box for partial credit; this is not required however.

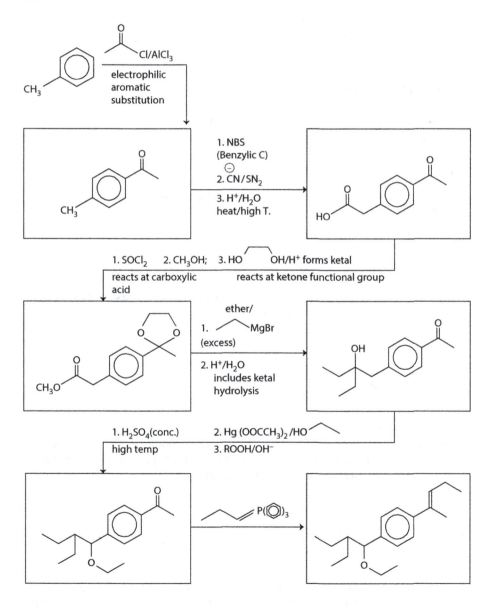

9. For the three parts, draw structures **to the right of the name** for the two compounds which could be reacted to form the named molecule through a nucleophilic carbonyl substitution reaction (i.e., a Group [5] reaction; see Chapter 18). More than one correct combination may be possible.

a.   **2-methylbutyl 3-pentenoate:**

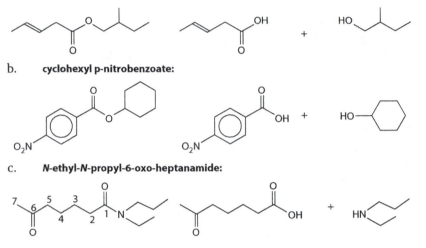

b.   **cyclohexyl p-nitrobenzoate:**

c.   **N-ethyl-N-propyl-6-oxo-heptanamide:**

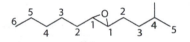

10. Follow the parts below to show a synthesis for the following compound, related to an insect pheromone:

a. Working backward, select the correct alkene and show the **last** reaction in the synthesis which will form the epoxide final product:

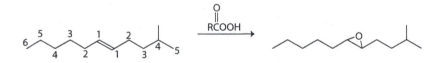

b. The alkene selected in part (a) should have 12 carbons. Show a Wittig (using triphenylphosphonium ylide reagent) reaction which can synthesize this alkene in **one** reaction: start with a correct nucleophile Wittig reagent and a correct substrate:

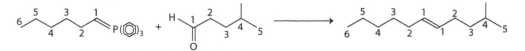

c. Complete the problem by showing a synthesis for the Wittig reagent selected in part (b) from **any** 6-carbon alkene and the substrate selected in part (b) from **any other** 6-carbon alkene.

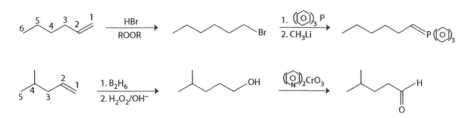

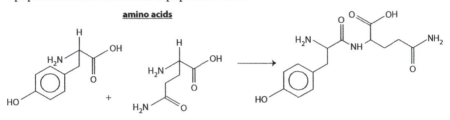

11. Complete the following parts relating to amino acids:
    a. Draw the structures of the two amino acids which have been bonded through a peptide bond to form the dipeptide shown:

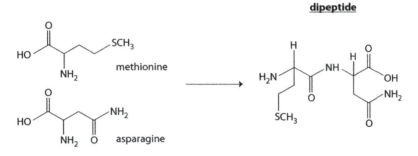

    b. Draw the structures of the dipeptide which can be formed from the two amino acids shown:

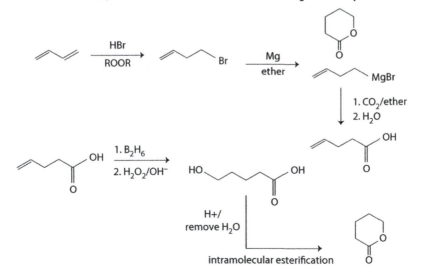

12. Use the general procedures for synthesis to show a synthesis for the lactone shown below starting only with **1,3-butadiene** plus **carbon dioxide** as carbon sources. **Assume that the double bonds in 1,3-butadiene can be reacted independently.**

Organic II: Practice Exam 25

1. Select which of the two molecules shown for each part below will be expected to be the stronger acid:

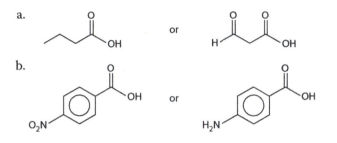

a.

or

b.

or

2. Draw the structure of all **enolate anions** and **neutral enols** which can be formed from the following molecules: (a) has only one enolate anion and one enol; (b) has two enolate anions and two enols.

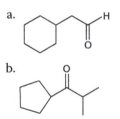

a.

b.

3. The following two reactions are α-**halogenations** of a carbonyl compound through the **enol** form. For each reaction, draw the **enol form** of the reactant, the **first electrophilic addition** product, and the **final** rearranged product.

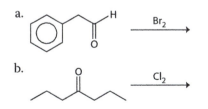

a.

Br$_2$

b.

Cl$_2$

4. Draw structural formulas for the **one major** product for each of the reactions or reaction sequences.

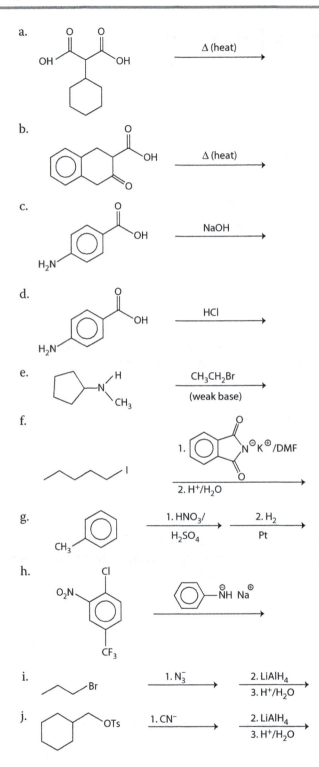

a.   Δ (heat)

b.   Δ (heat)

c.   NaOH

d.   HCl

e. CH_3CH_2Br (weak base)

f.   1. N⁻K⁺/DMF   2. H⁺/H_2O

g.   1. HNO_3/H_2SO_4 2. H_2 Pt

h.

i.   1. N_3^- 2. $LiAlH_4$ 3. H⁺/H_2O

j.   1. CN⁻   2. $LiAlH_4$ 3. H⁺/H_2O

5. Enolate anions from β-diketones or β-ketoesters can be used as nucleophiles in SN2 reactions. Draw the correct products for the following reactions with good SN2 substrates:

a.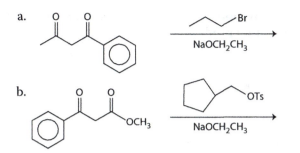

b.

6. Nucleophiles add to α,β-unsaturated ketones or aldehydes; they can react as 1,2- or 1,4-net additions. Draw the correct products for the following reactions:

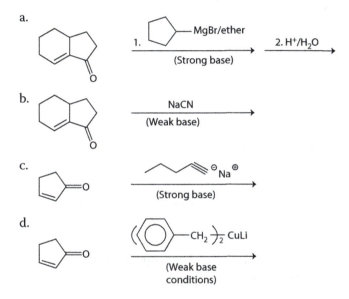

a.

b.

c.

d.

7. For the following **aldol condensation** reactions: show the **neutral addition intermediate** and the α,β-unsaturated ketone or aldehyde **final product**.

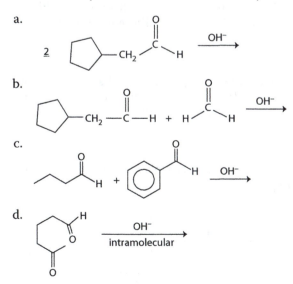

a.

b.

c.

d.

8. For the following **Claisen (ester) condensation** reactions: show the **anionic first addition intermediate** and the β-ketoester **final product**.

a.

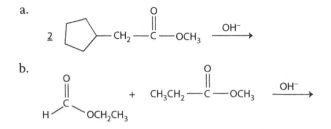

b.

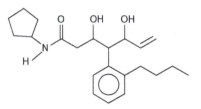

9. Name the following molecule:

10. a.   Draw the structure of 6-methyheptanal and then

b.   Draw the structure of the **neutral enol** which can be formed from this aldehyde.

c.   Show the **product** of the reaction of 6-methyheptanal with Br_2/H^+ in a reaction that proceeds through the enol intermediate.

d.   For the condensation of 2 moles of 6-methyheptanal: show both the **initial aldol** addition product and the **final** product formed through an **elimination** reaction.

e.   The final product from part (**d**) should be an α,β-unsaturated aldehyde. Show the reaction of the α,β-unsaturated aldehyde from part (d) with the weak base nucleophile NaCN/H⁺. Show the product of the reaction, **not** the mechanism.

11. Show the structure of the **major** product for each of the following reactions or reaction sequences. No stereochemistry is required.

a.

heat

b.

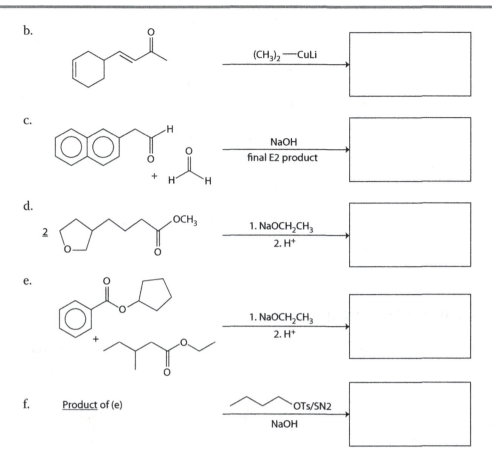

(CH₃)₂ —CuLi

c.

NaOH

final E2 product

d.

2

1. NaOCH₂CH₃

2. H⁺

e.

1. NaOCH₂CH₃

2. H⁺

f.   Product of (e)

OTs/SN2

NaOH

12. Complete the following **two** reaction sequences by indicating the correct **major** product for each part. Place all answers directly in the sequence. No stereochemistry is required. **Note that the sequence [A] → [B] → [C] is _separate_ from the sequence [D] → [E] → [F].**

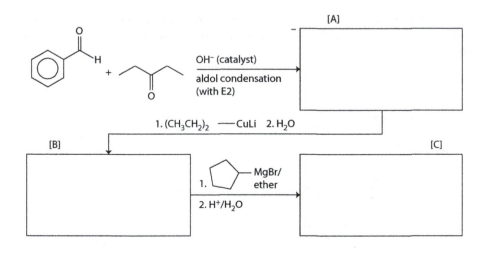

[A]

OH⁻ (catalyst)
aldol condensation
(with E2)

1. (CH₃CH₂)₂ —CuLi   2. H₂O

[B]

[C]

1.   ——— MgBr/ether

2. H⁺/H₂O

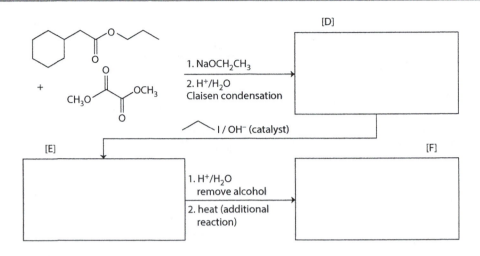

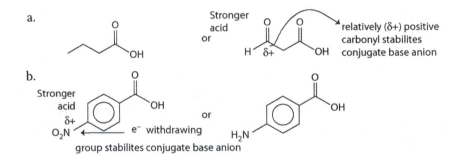

ORGANIC II: PRACTICE EXAM 3A: ANSWERS

1. Select which of the two molecules shown for each part below will be expected to be the stronger acid:

   a.

   Stronger acid or

   relatively (δ+) positive carbonyl stabilites conjugate base anion

   H δ+

   b.

   Stronger acid

   δ+

   O_2N ← e⁻ withdrawing

   or

   H_2N

   group stabilites conjugate base anion

2. Draw the structure of all **enolate anions** and **neutral enols** which can be formed from the following molecules: (**a**) has only one enolate anion and one enol; (**b**) has two enolate anions and two enols.

   a.

   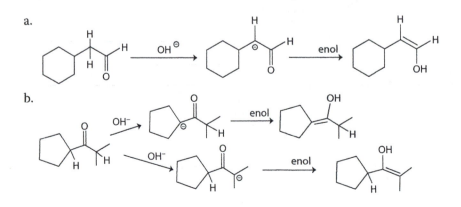

   b.

3. The following two reactions are α-**halogenations** of a carbonyl compound through the **enol** form. For each reaction, draw the **enol form** of the reactant, the **first electrophilic addition** product, and the **final** rearranged product.

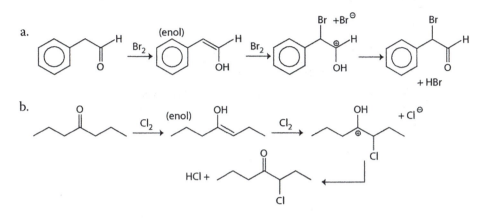

4. Draw structural formulas for the **one major** product for each of the reactions or reaction sequences.

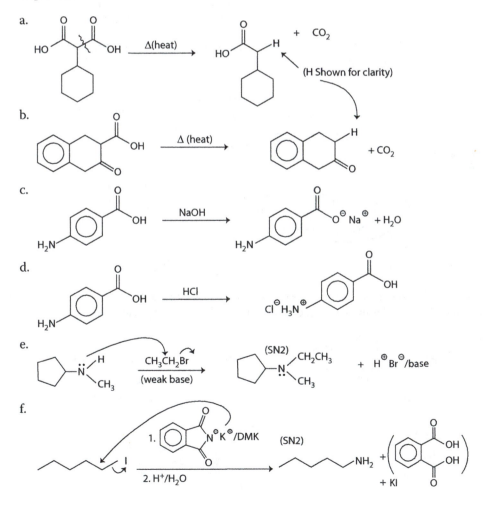

g.

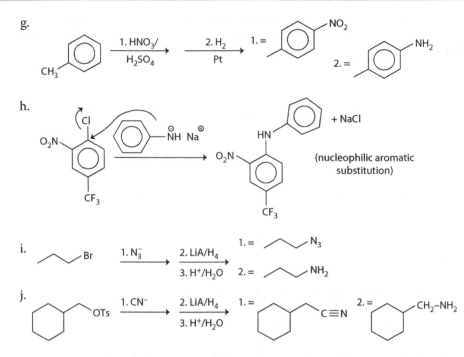

h.

i.

j.

5. Enolate anions from β-diketones or β-ketoesters can be used as nucleophiles in SN2 reactions. Draw the correct products for the following reactions with good SN2 substrates:

a.

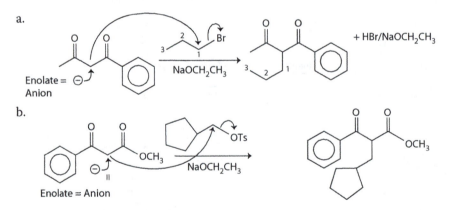

b.

6. Nucleophiles add to α,β-unsaturated ketones or aldehydes; they can react as 1,2- or 1,4-net additions. Draw the correct products for the following reactions:

a.

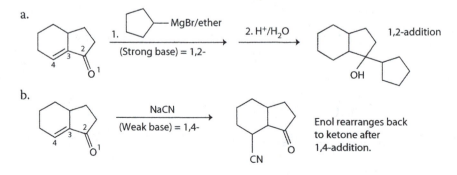

b.

c.

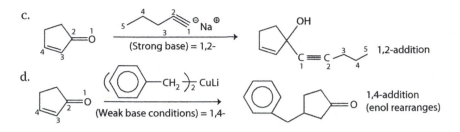

(Strong base) = 1,2-

1,2-addition

d.

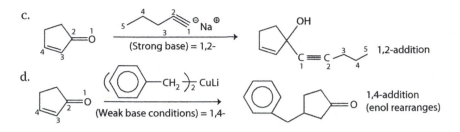

−CH₂ ₂ CuLi

(Weak base conditions) = 1,4-

1,4-addition
(enol rearranges)

7. For the following **aldol condensation** reactions: show the **neutral addition intermediate** and the α,β-unsaturated ketone or aldehyde **final product.**

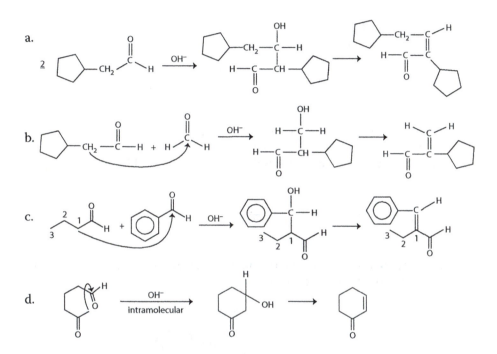

a.

b.

c.

d.   OH⁻   intramolecular

8. For the following **Claisen (ester) condensation** reactions: show the **anionic first addition intermediate** and the β-ketoester **final product**.

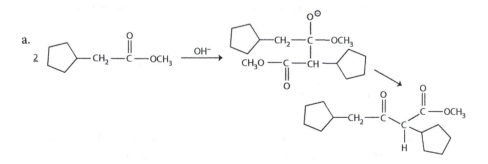

a.

b.

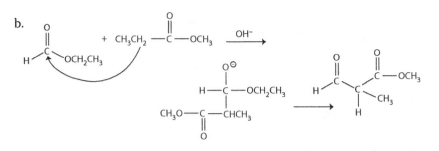

9. Name the following molecule:

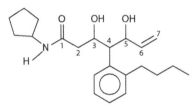

N-cyclopentyl-4-(2-butylphenyl)-3,5-dihydroxy-6-heptenamide

10. a.   Draw the structure of 6-methyheptanal and then
    b.   Draw the structure of the **neutral enol** which can be formed from this aldehyde.

   c.   Show the **product** of the reaction of 6-methyheptanal with Br_2/H^+ in a reaction that
        proceeds through the enol intermediate.

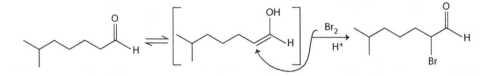

   d.   For the condensation of 2 moles of 6-methyheptanal: show both the **initial aldol**
        addition product and the **final** product formed through an **elimination** reaction.

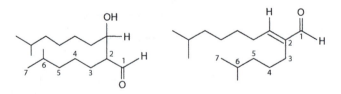

   e.   The final product from part (**d**) should be an α,β-unsaturated aldehyde. Show the
        reaction of the α,β-unsaturated aldehyde from part (d) with the ⎡weak base⎤ nucleo-
        phile $NaCN/H^+$. Show the product of the reaction, **not** the mechanism. 1,4-addition

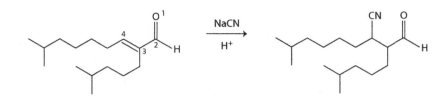

11. Show the structure of the **major** product for each of the following reactions or reaction sequences. No stereochemistry is required.

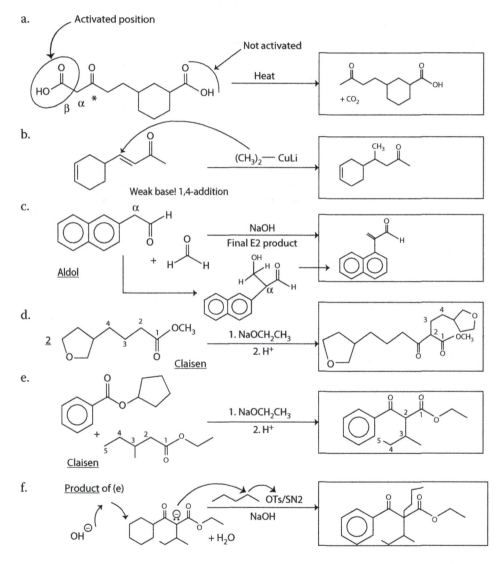

12. Complete the following **two** reaction sequences by indicating the correct **major** product for each part. Place all answers directly in the sequence. No stereochemistry is required. *Note that the sequence [A] → [B] → [C] is separate from the sequence [D] → [E] → [F].*

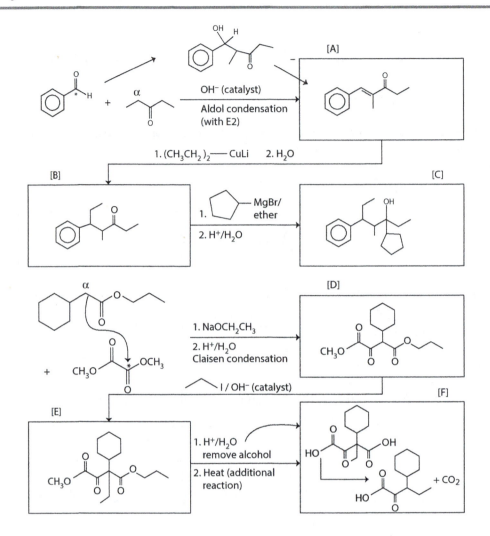

Organic II: Practice Exam 26

1. a. Show the **complete** mechanism for the following strong base-catalyzed **aldol condensation** reaction **followed by an E2 elimination** in base:

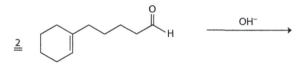

b. (i) The molecules named below are **products** formed by a nucleophilic carbonyl substitution (**Group [5]**); see Chapter 18 reaction. (i) **Draw** the named products to the **right** of the arrow. (ii) Work backward to determine, then draw, the correct **reactants** which will produce the named products; place these to the **left** of the arrow.

    products: 2-chloropentyl   6-phenyl-3-hydroxy-5-oxo-8-nonenoate   and 1-propanol

$$\xrightarrow{\text{H}^+}$$

2. Name the following molecule:

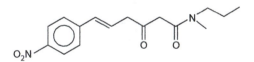

a. Draw the structure of 4,4-dimethyl-3-hexanone and then

b. Draw the structure of the **neutral enol** which can be formed from this ketone.

c. Show the **product** of the reaction of 4,4-dimethyl-3-hexanone with Br_2/H^+ in a reaction that proceeds through the enol intermediate.

d. For the condensation of 2 moles of 4,4-dimethyl-3-hexanone: show both the **initial aldol** addition product and the **final** product formed through an **elimination** reaction.

e. The final product from part (**d**) should be an α, β-unsaturated ketone. Show the
    reaction of the α,β-unsaturated ketone from part (d) with the weak base nucleophile
    NaCN/H⁺. Show the product of the reaction, **not** the mechanism.

3. Show the structure of the **major** product for each of the following reactions or reaction
    sequences. No stereochemistry is required.

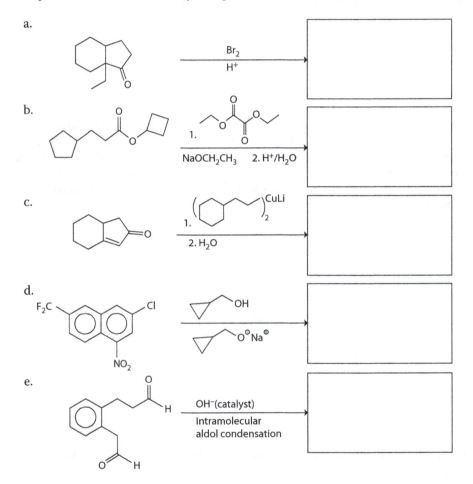

a.

    Br_2
    H^+

b.

    1.
    $NaOCH_2CH_3$ 2. H^+/H_2O

c.

    1.   ⎞CuLi
          ⎠₂

    2. H_2O

d.

    F_2C .Cl

    OH

    $O^⊖ Na^⊕$

    NO_2

e.

    OH^-(catalyst)

    Intramolecular
    aldol condensation

4. Show the structure of the **major** product for each of the following reactions or reaction
    sequences. No stereochemistry is required.

a.

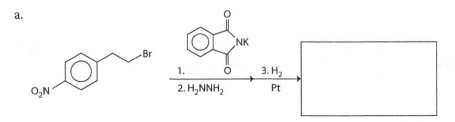

    1.   3. H_2

    2. H_2NNH_2 Pt

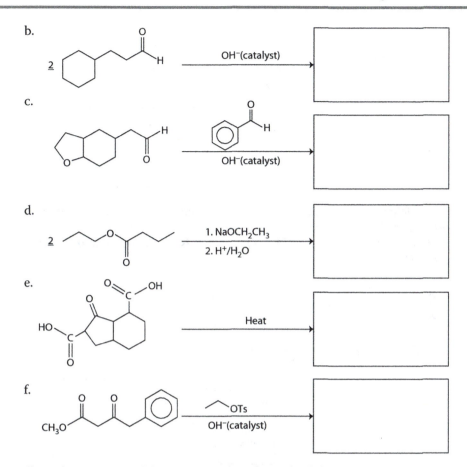

b.

OH⁻(catalyst)

c.

OH⁻(catalyst)

d.

1. NaOCH₂CH₃

2. H⁺/H₂O

e.

Heat

f.

OTs

OH⁻(catalyst)

5. Show the structure of the **major** product for each of the following reactions or reaction sequences. No stereochemistry is required.

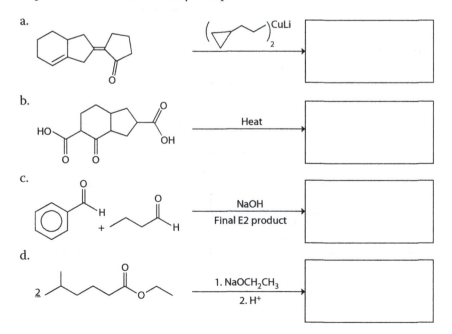

a.

CuLi

b.

Heat

c.

NaOH

Final E2 product

d.

1. NaOCH₂CH₃

2. H⁺

e.

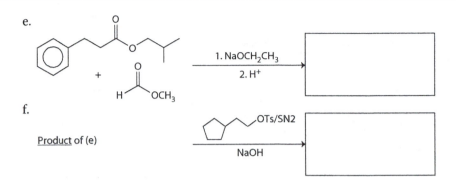

f.

Product of (e)

6. Complete the following **two** reaction sequences by indicating the correct **major** product for each part. Place all answers directly in the sequence. No stereochemistry is required. *Note that the sequence [A] → [B] → [C] is __separate__ from the sequence [D] → [E] → [F].*

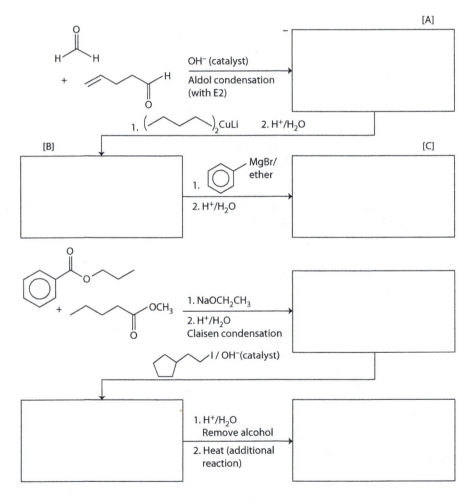

7. For the following **nucleophilic reactions at carbonyl**, draw the structure of the **neutral tetrahedral intermediate** formed in the addition phase of the reaction. Then, draw the structure of the **final product** of the reaction based on the principles of reactions at carbonyls. *Use the additional information shown in some cases over the second arrow.*

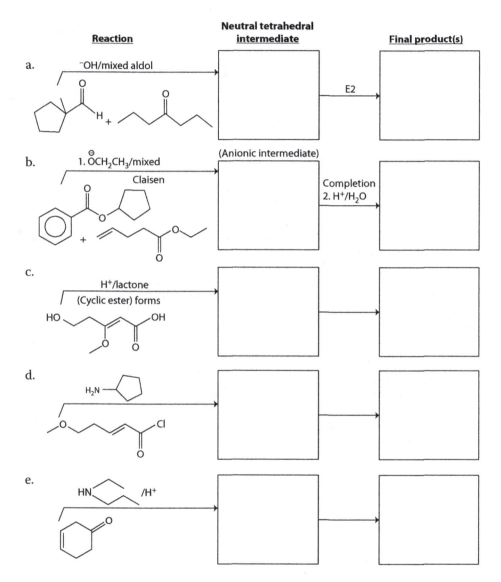

8. Show the structure of the **major** product for each part of the following two-part reaction sequences. Each two-part reaction sequence is separate. ***Include stereochemistry where required.*** For multiple reactions over the same arrow: middle products may be placed outside the box for partial credit; this is <u>not</u> required however.

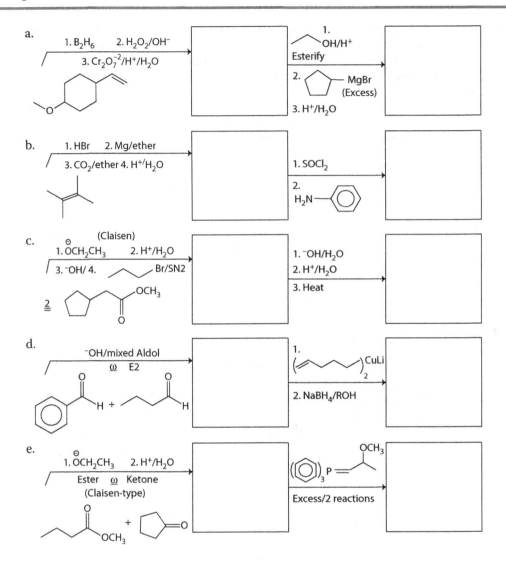

ORGANIC II: PRACTICE EXAM 3B: ANSWERS

1. a. Show the **complete** mechanism for the following strong base-catalyzed **aldol condensation** reaction **followed by an E2 elimination** in base:

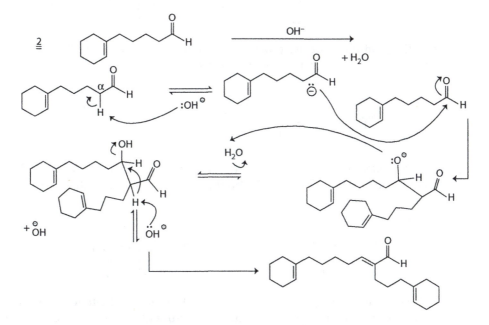

  b. (i) The molecules named below are **products** formed by a nucleophilic carbonyl substitution (**Group [5]**); see Chapter 18 reaction. (i) **Draw** the named products to the **right** of the arrow. (ii) Work backward to determine, then draw, the correct **reactants** which will produce the named products; place these to the **left** of the arrow.
    products: 2-chloropentyl 6-phenyl-3-hydroxy-5-oxo-8-nonenoate   and
    1-propanol

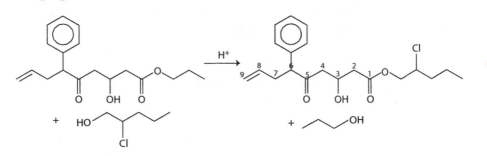

2. Name the following molecule:

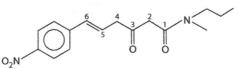

N-methyl-N-propyl-3-oxo-6-(p-nitrophenyl)-5-hexenamide

a. Draw the structure of 4,4-dimethyl-3-hexanone and then
b. Draw the structure of the **neutral enol** which can be formed from this ketone.

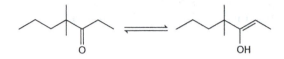

c. Show the **product** of the reaction of 4,4-dimethyl-3-hexanone with Br$_2$/H$^+$ in a reaction that proceeds through the enol intermediate.

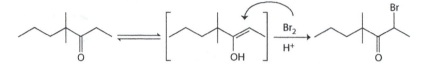

d. For the condensation of 2 moles of 4,4-dimethyl-3-hexanone: show both the **initial aldol** addition product and the **final** product formed through an **elimination** reaction.

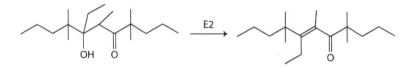

e. The final product from part (d) should be an α,β-unsaturated ketone. Show the reaction of the α,β-unsaturated ketone from part (d) with the weak base nucleophile NaCN/H$^+$. Show the product of the reaction, **not** the mechanism.

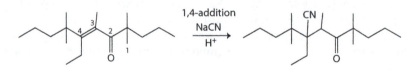

3. Show the structure of the **major** product for each of the following reactions or reaction sequences. No stereochemistry is required.

a.

b.

c.

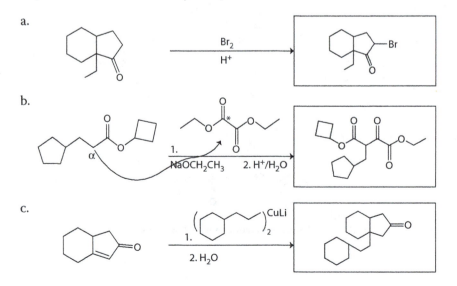

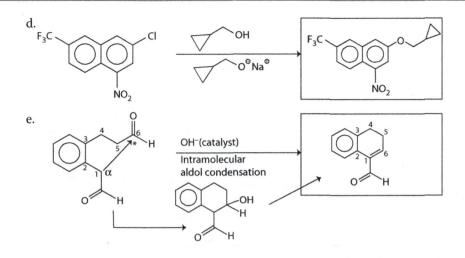

d.

e.

OH⁻(catalyst)

Intramolecular
aldol condensation

4. Show the structure of the **major** product for each of the following reactions or reaction sequences. No stereochemistry is required.

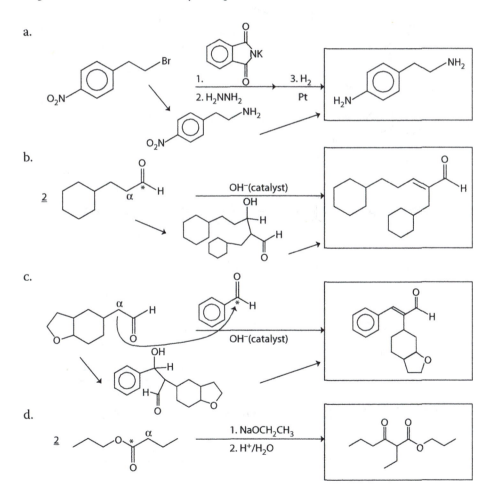

a.

1.

2. H₂NNH₂

3. H₂
Pt

b.

OH⁻(catalyst)
OH

c.

OH⁻(catalyst)

d.

1. NaOCH₂CH₃
2. H⁺/H₂O

e.

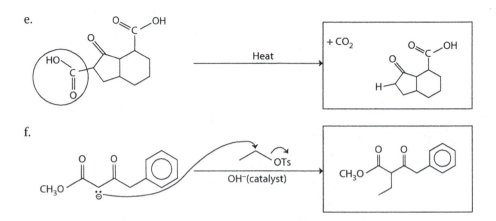

f.

5. Show the structure of the **major** product for each of the following reactions or reaction sequences. No stereochemistry is required.

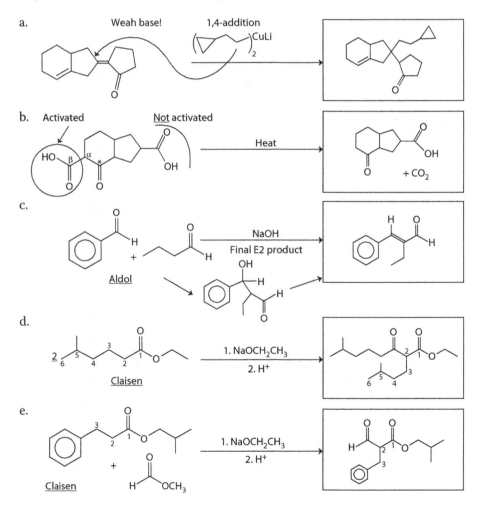

f.

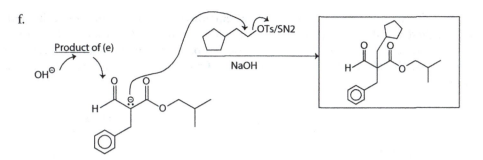

6. Complete the following **two** reaction sequences by indicating the correct **major** product for each part. Place all answers directly in the sequence. No stereochemistry is required. ***Note that the sequence [A] → [B] → [C] is <u>separate</u> from the sequence [D] → [E] → [F].***

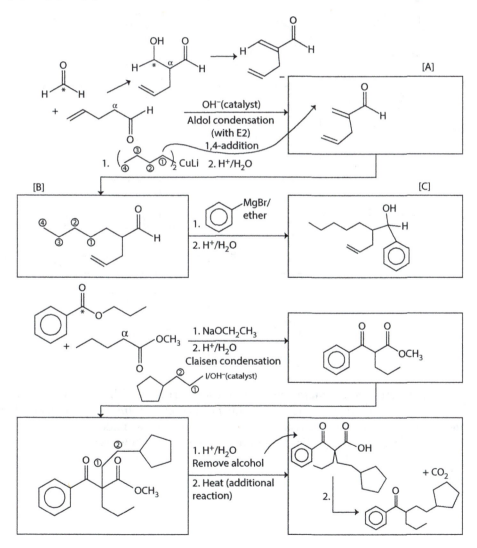

7. For the following **nucleophilic reactions at carbonyl**, draw the structure of the **neutral tetrahedral intermediate** formed in the addition phase of the reaction. Then, draw the structure of the **final product** of the reaction based on the principles of reactions at carbonyls. *Use the additional information shown in some cases over the second arrow.*

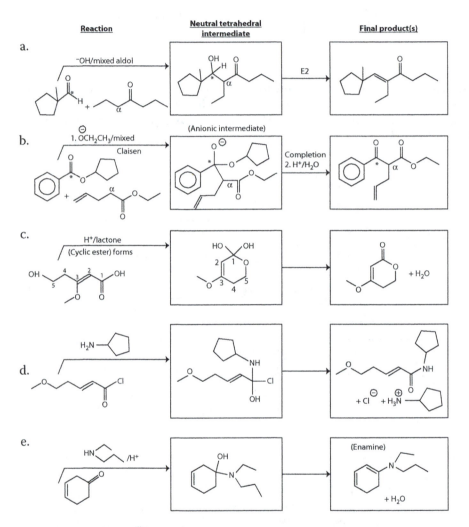

8. Show the structure of the **major** product for each part of the following two-part reaction sequences. Each two-part reaction sequence is separate. *Include stereochemistry where required*. For multiple reactions over the same arrow: middle products may be placed outside the box for partial credit; this is not required however.

a.

b.

c.

d.

e.

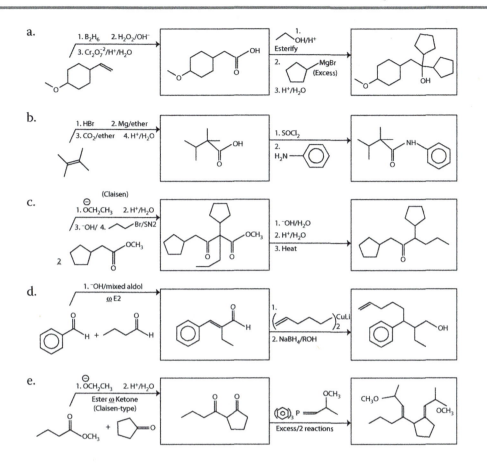

Organic II: Practice Exam 27

1. a. Show the **complete** **mechanism** for the following strong base-catalyzed **Claisen (ester) condensation** reaction; *include the strong base reaction steps responsible for shifting the equilibrium toward the condensation product.*

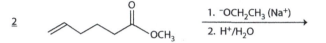

2 ——————→
   1. $^-OCH_2CH_3$ (Na$^+$)
   2. H$^+$/H$_2$O

   b. The molecules named below are **products** formed by a nucleophilic carbonyl substitution **(Group [5])** reaction; see Chapter 18. (i) **Draw** the named products to the **right** of the arrow. (ii) Work backward to determine, then draw, the correct **reactants** which will produce the named products; place these to the **left** of the arrow.
   products: *N*-methyl-7-hydroxy-8-(3-oxocyclohexyl)-3, 5-octadienamide and HCl

   ——————→

2. a. Draw the structure of 4-ethylhexanal. Then, show the **mechanism including electron transfer reaction arrows** for the acid-catalyzed (H$^+$ (aq)) **enolization** to form the enol of 4-ethylhexanal.

   b. Show the **initial aldol addition product** for the reaction of 2 moles of 4-ethylhexanal in the presence of catalytic strong base (OH$^-$). Then, show the **final** stable **product** formed by a further elimination reaction in base.

   Initial addition product   **Final elimination product**

   c. Select the **final** elimination **product** from the aldol condensation described in part (b). Now, show the reaction of **this** molecule with NaN$_3$ (weak base nucleophile)/H$^+$.

3. a.   Determine the structure of benzaldehyde. Now, show the **initial aldol addition product** for the reaction of 1 mole of 4-ethylhexanal with 1 mole of benzaldehyde in the presence of catalytic strong base (OH^-). Then, show the **final** stable **product** formed by a further elimination reaction in base.

Initial addition product   **Final elimination product**

b.   Select the **final** elimination **product** from the aldol condensation described in part (a). Now, show the reaction (not mechanism) of **this** molecule with $(C_6H_5CH_2—)_2$ CuLi (weak base nucleophile).

c.   Show **only** the **product** of the reaction of 4-ethylhexanal with Br_2 in the presence of H^+ catalyst.

4. a.   Draw the structure of the ester isopropyl hexanoate. Now, show the Claisen condensation **reaction product** (**not** mechanism) of 2 moles of isopropyl hexanoate in the presence of (1) a strong base ($NaOCH_2CH_3$); followed by (2) H^+/H_2O.

b.   Select the **product** from the condensation reaction described in part (a). Show both **products** for the following sequential reactions on **this** Claisen condensation molecule: (1) strong acid hydrolysis (H^+/H_2O) and (2) strong heat.

Hydrolysis of Claisen product   **Reaction with strong heat**

c.   Determine the structure of the ester ethyl methanoate (ethyl formate). Now, show the Claisen condensation **reaction product** (**not** mechanism) of 1 mole of isopropyl hexanoate with 1 mole of ethyl methanoate in the presence of (1) a strong base ($NaOCH_2CH_3$); followed by (2) H^+/H_2O.

d.   Select the **product** from the condensation reaction described in part (c). Show both **products** for the following sequential reactions on **this** Claisen condensation molecule: (1) reaction with stoichiometric strong base followed by (2) SN2 with 1-bromopropane.

Anion in strong base   **SN2 with 1-bromopropane**

5. Show the structure of the **major** product for each of the following reactions or reaction sequences. No stereochemistry is required.

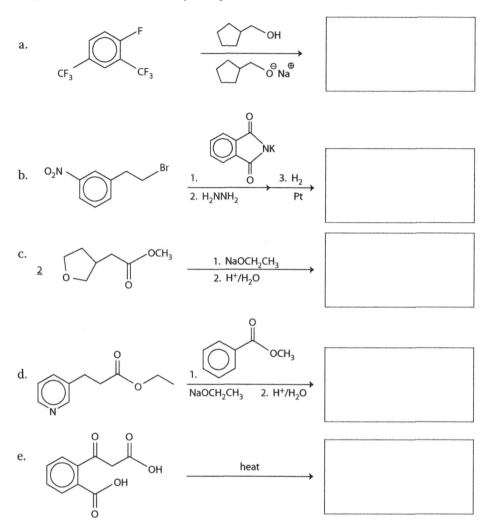

6. Show the structure of the **major** product for each of the following reactions or reaction sequences. No stereochemistry is required.

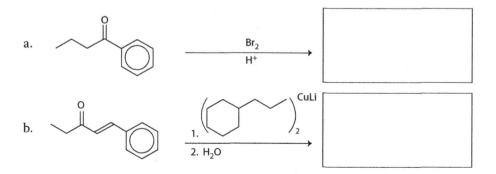

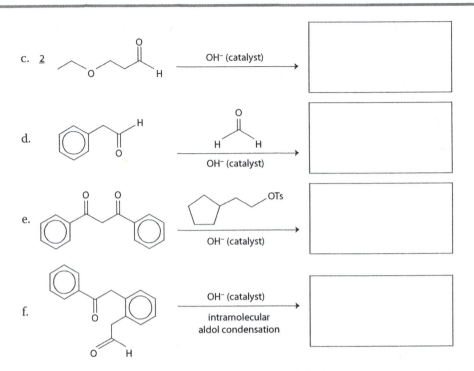

c. 2 [structure] OH⁻ (catalyst) → [ ]

d. [structure] + [structure] OH⁻ (catalyst) → [ ]

e. [structure] + [structure with OTs] OH⁻ (catalyst) → [ ]

f. [structure] OH⁻ (catalyst) / intramolecular aldol condensation → [ ]

7. Show the required structures for the following aldehyde reactions. ***Be careful reading the directions for each part and show only the product or intermediate requested.***

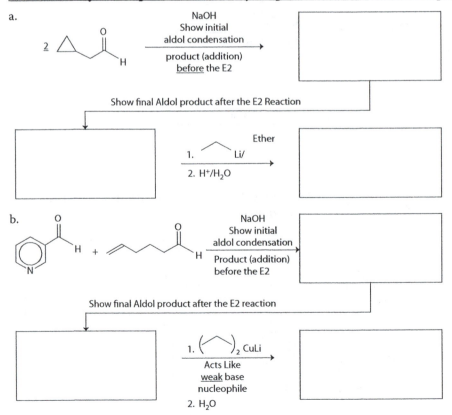

a.

2 [structure]

NaOH
Show initial
aldol condensation
────────────→
product (addition)
<u>before</u> the E2

[ ]

Show final Aldol product after the E2 Reaction

[ ]

1. [structure] Li/ Ether
2. H⁺/H₂O
────────────→

[ ]

b.

[structure] + [structure]

NaOH
Show initial
aldol condensation
────────────→
Product (addition)
before the E2

[ ]

Show final Aldol product after the E2 reaction

[ ]

1. ([structure])₂ CuLi
Acts Like
<u>weak</u> base
nucleophile
2. H₂O
────────────→

[ ]

8. Show the required structures for the following ester or acid reactions. ***Be careful reading the directions for each part and show only the product or intermediate requested.***

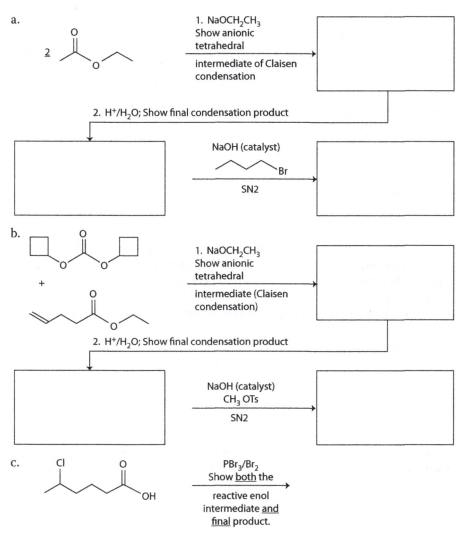

a.

2

1. NaOCH$_2$CH$_3$
Show anionic
tetrahedral

intermediate of Claisen
condensation

2. H$^+$/H$_2$O; Show final condensation product

NaOH (catalyst)

Br

SN2

b.

+

1. NaOCH$_2$CH$_3$
Show anionic
tetrahedral

intermediate (Claisen
condensation)

2. H$^+$/H$_2$O; Show final condensation product

NaOH (catalyst)
CH$_3$ OTs

SN2

c.

Cl

OH

PBr$_3$/Br$_2$
Show <u>both</u> the

reactive enol
intermediate <u>and</u>
<u>final</u> product.

ORGANIC II: PRACTICE EXAM 3C: ANSWERS

1. a. Show the **complete** mechanism for the following strong base-catalyzed **Claisen (ester) condensation** reaction; *include the strong base reaction steps responsible for shifting the equilibrium toward the condensation product.*

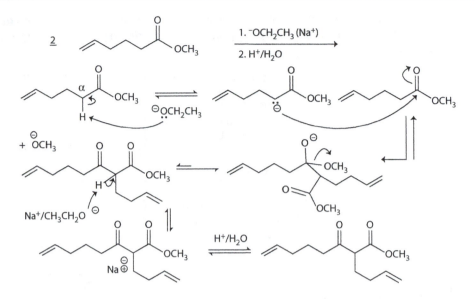

b. The molecules named below are **products** formed by a nucleophilic carbonyl substitution **(Group [5])** reaction; see Chapter 18. **(i) Draw** the named products to the **right** of the arrow. **(ii)** Work backward to determine, then draw, the correct **reactants** which will produce the named products; place these to the **left** of the arrow. **products:** **N-methyl-7-hydroxy-8-(3-oxocyclohexyl)-3, 5-octadienamide and HCl**

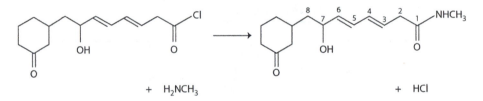

2. a. Draw the structure of 4-ethylhexanal. Then, show the **mechanism including electron transfer reaction arrows** for the acid-catalyzed (H$^+$ (aq)) **enolization** to form the enol of 4-ethylhexanal.

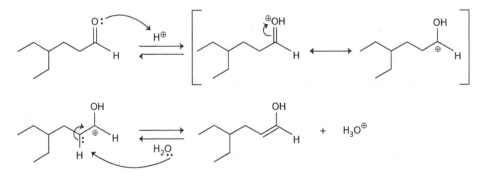

b. Show the **initial aldol addition product** for the reaction of 2 moles of 4-ethylhexanal in the presence of catalytic strong base (OH$^-$). Then, show the **final** stable **product** formed by a further elimination reaction in base.

Initial addition product   **Final elimination product**

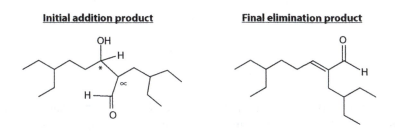

c. Select the **final** elimination **product** from the aldol condensation described in part (b). Now, show the reaction of **this** molecule with NaN$_3$ (weak base nucleophile)/H$^+$.

1,4 - addition
weak base

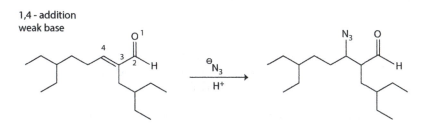

3. a. Determine the structure of benzaldehyde. Now, show the **initial aldol addition product** for the reaction of 1 mole of 4-ethylhexanal with 1 mole of benzaldehyde in the presence of catalytic strong base (OH$^-$). Then, show the **final** stable **product** formed by a further elimination reaction in base.

Initial addition product   **Final elimination product**

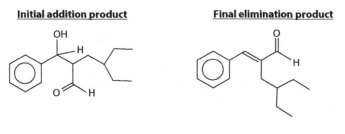

b. Select the **final** elimination **product** from the aldol condensation described in part (a). Now, show the reaction (not mechanism) of **this** molecule with (C$_6$H$_5$CH$_2$—)$_2$ CuLi (weak base nucleophile).

(weak base = 1,4 addition)

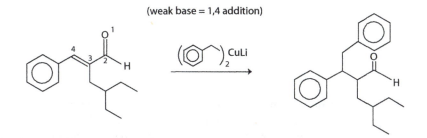

c. Show **only** the **product** of the reaction of 4-ethylhexanal with Br$_2$ in the presence of H$^+$ catalyst.

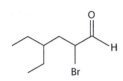

4. a. Draw the structure of the ester isopropyl hexanoate. Now, show the Claisen condensation **reaction product** (**not** mechanism) of 2 moles of isopropyl hexanoate in the presence of (1) a strong base (NaOCH$_2$CH$_3$); followed by (2) H$^+$/H$_2$O.

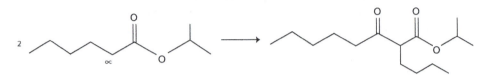

b. Select the **product** from the condensation reaction described in part (a). Show both **products** for the following sequential reactions on **this** Claisen condensation molecule: (1) strong acid hydrolysis (H$^+$/H$_2$O); 2. strong heat

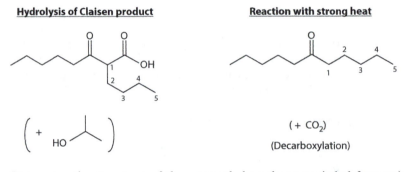

c. Determine the structure of the ester ethyl methanoate (ethyl formate). Now show the Claisen condensation **reaction product** (**not** mechanism) of 1 mole of isopropyl hexanoate with 1 mole of ethyl methanoate in the presence of 1. strong base (NaOCH$_2$CH$_3$); followed by 2. H$^+$/H$_2$O.

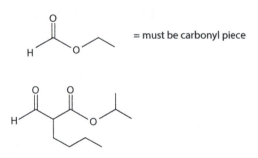

d. Select the **product** from the condensation reaction described in part (c). Show both **products** for the following sequential reactions on **this** Claisen condensation molecule: (1) reaction with stoichiometric strong base followed by (2). SN2 with 1-bromopropane

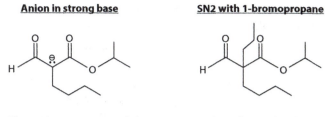

Anion in strong base   **SN2 with 1-bromopropane**

5. Show the structure of the **major** product for each of the following reactions or reaction sequences. No stereochemistry is required.

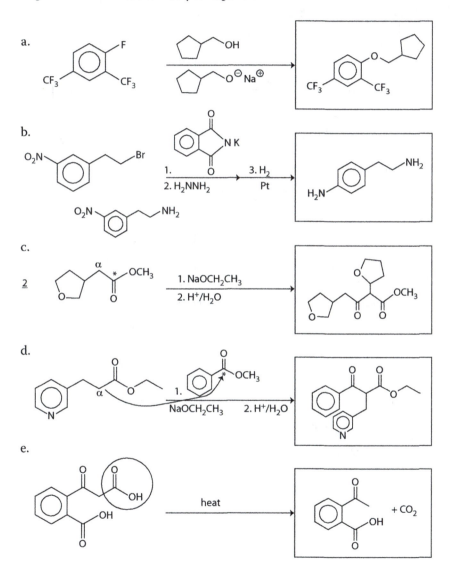

a.

b.

c.

d.

e.

6. Show the structure of the **major** product for each of the following reactions or reaction sequences. No stereochemistry is required.

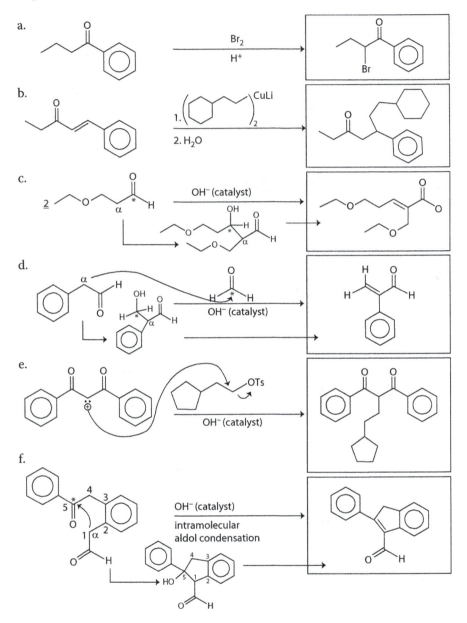

a.

Br₂ / H⁺

b.

1. (CuLi)₂

2. H₂O

c.

2

OH⁻ (catalyst)

d.

OH⁻ (catalyst)

e.

OTs

OH⁻ (catalyst)

f.

OH⁻ (catalyst)

intramolecular aldol condensation

7. Show the required structures for the following aldehyde reactions. ***Be careful reading the directions for each part and show only the product or intermediate requested***.

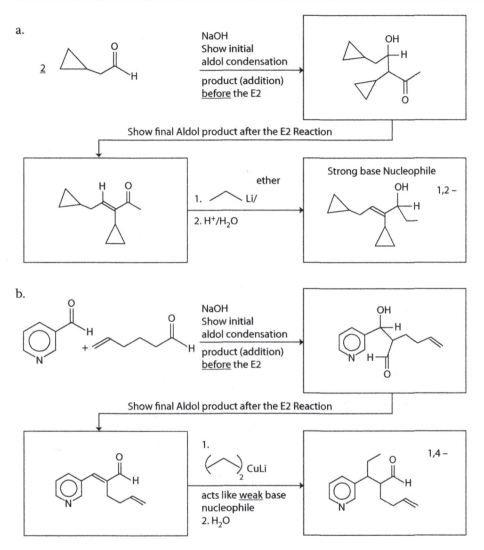

8. Show the required structures for the following ester or acid reactions. ***Be careful reading the directions for each part and show only the product or intermediate requested***.

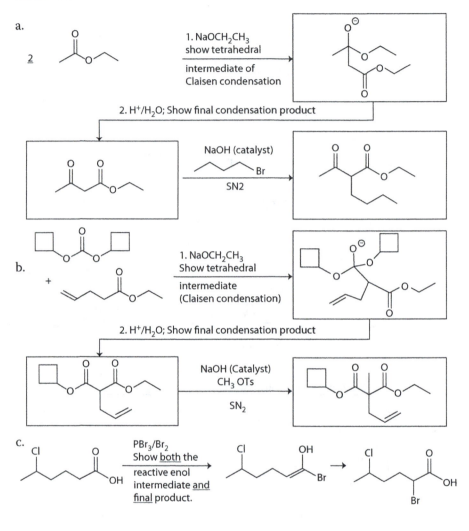

Multiple-Choice
Practice Exams

Organic I: Multiple-Choice Practice Exam 1

1. Which of the following molecules would be expected to have the **highest** boiling point based on the total strength of intermolecular forces between molecules of the same compound?

a.　　　　　b.　　　　　　　　　c.　　　　　　　　d.

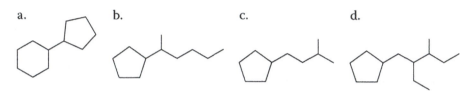

2. Which of the molecules in **question #1** would be expected to have the **lowest** boiling point based on the strength of intermolecular forces? Consider the molecular size and effects of branching. **Use the same lettered answer list as question #1.**
3. Which of the following molecules would be expected to have the highest heat of combustion, that is, the largest negative value for enthalpy of combustion per mole?

a.　　　　　　　b.　　　　　　　c.　　　　　　　d.

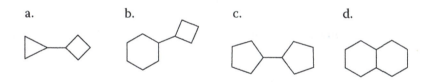

4. Based on the concepts covered in the course **to this point**, which of the following molecules can have a stereoisomer? (**Exclude enantiomers.**)

a.　　　　b.　　　　　c.　　　　　　　　　d.

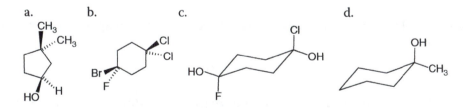

For the following molecules, recall that the **base name as a ring versus a chain is determined by the longest continuous sequence of carbons. That is, compare the longest continuous chain to the number of carbons in the ring.**
List of names for questions #5, #6, and #7
a.　4-(3-sec-butylcyclopentyl)-2-methylbutane
b.　5-cyclopentyl-3-ethyl-2-isopropyloctane
c.　5-cyclopentyl-3-(1,2-dimethylpropyl)-octane
d.　4-cyclopentyl-6-ethyl-7,8-dimethylnonane

    e. 6-cyclopentyl-4-ethyl-2,3-dimethylnonane
    f. 5-cyclopentyl-4-(2-methylbutyl)-nonane
    g. 5-cyclopentyl-8-methyl-6-propyldecane
    h. 6-cyclopentyl-3-methyl-5-propyldecane
    i. 1-(3-methylbutyl)-3-(1-methylpropyl)-cyclopentane
    j. 4-(3-methylbutyl)-1-(1-methylpropyl)-cyclopentane
    k. 1-pentyl-3-butylcylcopentane
    l. 1-(2-methylbutyl)-3-(3-methylpropyl)-cyclopentane
    m. 1-(butyl)-2-(4-methylpentyl)-cyclopentane
    n. 1-(butyl)-2-hexylcyclopentane
    o. 1-(2,3-dimethylbutyl)-3-propylcyclopentane
    p. 1,2-dipentylcyclopentane

5. Select from the list (**directly above**) the correct name for the following molecule:

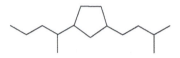

6. Select from the list (**directly above**) the correct name for the following molecule:

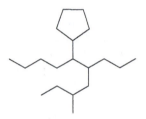

7. Select from the list (**directly above**) the correct name for the following molecule:

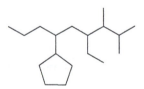

Questions #8–#11 apply to the following **overall** reaction **which may proceed by different mechanisms:**

$$2\ NO_2\ (g) \longrightarrow 2\ NO\ (g) + O_2\ (g)$$

For Questions #8–#10, choose from the following possible rate expressions: (**k** is any combination of appropriate constants; the " **/** " symbol indicates a division):

    a. Rate = k $[NO_2]$
    b. Rate = k $[NO_2]^2$
    c. Rate = k $[NO_2][NO]$
    d. Rate = k $[NO][O_2]$
    e. Rate = k $[NO]^2\ [O_2]$
    f. Rate = k $[NO_2]^2/[O_2]$
    g. Rate = k $[NO_2]^2\ [NO]^2/[O_2]$

8. Select the rate expression for the following mechanism:

$$\text{Step 1: } NO_2 + NO_2 \xrightleftharpoons{K_1} N_2O_4 \quad \textbf{Fast/equilibrium}$$

$$\text{Step 2: } N_2O_4 \xrightarrow{k_2} N_2O_2 + O_2 \quad \textbf{Slow}$$

$$\text{Step 3: } N_2O_2 \xrightarrow{k_3} 2\,NO \quad \textbf{Fast}$$

9. Select the rate expression for the following mechanism:

$$\text{Step 1: } NO_2 \xrightarrow{k_1} NO + O \quad \textbf{Slow}$$

$$\text{Step 2: } O + NO_2 \xrightarrow{k_2} O_2 + NO \quad \textbf{Fast}$$

10. Select the rate expression for the following mechanism:

$$\text{Step 1: } NO_2 \xrightleftharpoons{K_1} N + O_2 \quad \textbf{Fast/equilibrium}$$

$$\text{Step 2: } N + NO_2 \xrightarrow{k_2} 2\,NO \quad \textbf{Slow}$$

11. In a set of experiments, all concentrations were kept approximately constant except for $[NO_2]$. It was found that a plot of $\ln [NO_2]$ versus time gave a straight line with a slope of -0.12. Which mechanism(s) fits these data?
    a. Mechanism shown in **question #8**
    b. Mechanism shown in **question #9**
    c. Mechanism shown in **question #10**
    d. **Both** mechanisms shown in **questions #9 and #10**
    e. **None** of the mechanisms fit the stated data

12. Although vitamin A is critical to human life, its synthesis from the carbon, hydrogen, and oxygen contained in carbohydrates is non-spontaneous: ΔG^0 at body temperature (37°C) is positive. The fact that vitamin A can be synthesized at all means that
    a. The synthesis of vitamin A is not bound by the laws of thermodynamics since it occurs in the living tissue.
    b. Synthesis of vitamin A requires that energy be supplied to the reaction.
    c. The reaction becomes spontaneous because of the presence of enzymes as catalysts.
    d. The entropy contribution to free energy is eliminated in aqueous solutions inside living cells.

13. **Without calculation**, predict the direction of entropy change for the reaction

$$C_5H_{12}O_2 \, (l) \longrightarrow C_5H_8 \, (l) + 2\,H_2O \, (g)$$

    a. ΔS^0 will be positive
    b. ΔS^0 will be negative
    c. ΔS^0 will be exactly zero
    d. No trend in ΔS^0 can be determined without ΔH^0 values

14. A specific reaction is found to be spontaneous at 500°C, but becomes non-spontaneous at the lower temperature of 250°C. Based on the expression for ΔG^0, this probably means that
    a. $\Delta H^0 = 0$
    b. High temperature increases the rate of the reaction

   c. $\Delta S°$ is positive
   d. $\Delta S°$ is negative

Questions #15–#21 refer to the following reaction mechanism:

Step 1: $CH_3OH + H_2Si_4O_7 \rightleftharpoons CH_3OH_2^+ + HSi_4O_7^-$ **fast/equilibrium**

Step 2: $CH_3OH_2^+ + C_7H_8 \longrightarrow C_8H_{11}^+ + H_2O$ **slow**

Step 3: $C_8H_{11}^+ + HSi_4O_7^- \longrightarrow C_8H_{10} + H_2Si_4O_7$ **fast**

15. Which species is/are acting as catalyst(s)?
   a. CH_3OH
   b. $H_2Si_4O_7$
   c. $CH_3OH_2^+$
   d. C_7H_8
   e. $C_8H_{11}^+$
   f. H_2O
   g. C_8H_{10}
16. Identify **two** species which are intermediates in the reaction: ***Place the correct lettered answers in the spaces for questions #16 and #17.*** It may help to add the reactions in the mechanism to determine the overall net reaction.
Use the same lettered answers shown in question #15. Answers may be used more than once.
17. **The other lettered answer for the question described in #16.**
18. Use the mechanism shown above to determine the **overall** rate expression for the reaction and select from the answers below (**k is any combination of appropriate constants; the "/" symbol indicates a division**):
   a. Rate = k $[CH_3OH]$
   b. Rate = k $[CH_3OH]^2$
   c. Rate = k $[CH_3OH][H_2Si_4O_7]$
   d. Rate = k $[CH_3OH][H_2Si_4O_7][C_7H_8]$
   e. Rate = k $[CH_3OH]^2 [H_2Si_4O_7][C_7H_8]$
   f. Rate = k $[CH_3OH][H_2Si_4O_7]^2[C_7H_8]$
   g. Rate = k $[CH_3OH][H_2Si_4O_7][C_7H_8]/[HSi_4O_7^-]$
   h. Rate = k $[HSi_4O_7^-]/[CH_3OH][H_2Si_4O_7][C_7H_8]$
19. Use the mechanism shown above to determine the **overall reaction** by adding the steps. If the mechanism of this overall reaction were composed of a **single step** rather than the three-step mechanism shown, what would be the rate expression in this case? Note that the catalyst can still appear in the rate expression. **Use the same lettered answers shown in question #18. Answers may be used more than once.**
20. If the concentration of the compound $H_2Si_4O_7$ were **increased**, how would the rate of the overall reaction be affected if the mechanism was the three-step mechanism shown on the previous page?
   a. The rate would increase
   b. The rate would decrease
   c. The rate would remain the same
   d. The rate may increase or decrease depending on the rate of **Step 3**
21. An experiment is conducted in which the concentration of CH_3OH can vary while the concentration of all other species remains constant. It is determined that a plot of

1/[CH$_3$OH] (inverse function) versus time is a straight line. Which mechanism, if any, fits this information?
a. The three-step mechanism described by **question #18**
b. The one-step mechanism described by **question #19**
c. Both mechanisms fit this information
d. Neither mechanism fits this information

List of names for questions #22 and #23

a. 2-methyl-3-pentylhexane
b. 2,6-dimethyl-3-propylheptane
c. 2,6-dimethyl-5-propylheptane
d. 2-methyl-5-isopropyloctane
e. 2-methyl-3-propyloctane
f. 7-methyl-4-isopropyloctane
g. 5-isopropyloctane
h. 4-isopropylnonane
i. 6-isopropylnonane

22. Select from the above list the correct name for the following molecule:

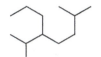

23. Select from the above list the correct name for the following molecule:

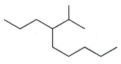

Questions #24 and #25 refer to the following reaction to produce one compound used in the preparation of polyamides for high-strength materials such as Kevlar:

$$C_8H_{10} \text{ (l)} + 3\,O_2 \text{ (g)} \longrightarrow C_8H_6O_4 \text{ (s)} + 2\,H_2O \text{ (l)}$$

24. Kinetic experimental results showed the following: A plot of ln [C$_8$H$_{10}$] versus time was a straight line when [O$_2$] was kept approximately constant; a plot of 1/[O$_2$] was a straight line when [C$_8$H$_{10}$] was kept approximately constant. **Based on the kinetic information**, what is the rate expression for this reaction?
a. Rate = k [C$_8$H$_{10}$]
b. Rate = k [C$_8$H$_{10}$]^2
c. Rate = k [O$_2$]^2
d. Rate = k [C$_8$H$_{10}$][O$_2$]
e. Rate = k [C$_8$H$_{10}$]^2[O$_2$]
f. Rate = k [C$_8$H$_{10}$][O$_2$]^2
g. Rate = k [C$_8$H$_{10}$]^2 [O$_2$]^2
h. Rate = k [C$_8$H$_{10}$]/[O$_2$]
i. Rate = k [C$_8$H$_{10}$]^2/[O$_2$]^2

25. **Without calculation**, predict the direction of entropy change for the reaction shown directly above question #24.
    a. ΔS^0 will be positive
    b. ΔS^0 will be negative
    c. ΔS^0 will be exactly zero
    d. No trend in ΔS^0 can be determined without ΔH^0 values

26. The following compound is an intermediate in organic reactions. Select the valid resonance structure of this compound:

a. H—N—CH₃   b. H—N—CH₃   c. H—N=CH₂   d. H—N—CH₂
        | | |
    CH₃—C—H   CH₂=C+   CH₂=C—H   CH₃—C—H
        +

27. Which statement best applies to the enthalpy (ΔH^0) of the following reaction? Consider the potential energy change associated with bond breaking and formation.

$$SeO_{2\,(g)} + O_{(g)} \longrightarrow SeO_{3\,(g)}$$

    a. Enthalpy is probably positive under standard conditions
    b. Enthalpy is probably negative under standard conditions
    c. Enthalpy is exactly zero at equilibrium concentrations
    d. Enthalpy is equally probable to be positive or negative depending on the sign of the entropy

28. The following reaction has a **positive** value for enthalpy (ΔH^0). What can be determined about the bonding in the molecules? Consider potential energy changes for bond breaking and formation.

$$SF_{3\,(g)} + H_2O_{(g)} \longrightarrow SF_3H_{(g)} + OH_{(g)}$$

    a. The S—H bond is stronger than the O—H bond
    b. The O—H bond is stronger than the S—H bond
    c. Energy is required to form the S—H bond
    d. Energy is released when the O—H bond in water breaks

29. The reaction shown in **question #28** consists of only one elementary reaction step. What is the rate expression for this reaction?
    a. Rate $= k\,[SF_3]$
    b. Rate $= k\,[SF_3]^2$
    c. Rate $= k\,[H_2O]$
    d. Rate $= k\,[SF_3][H_2O]$
    e. Rate $= k\,[SF_3]^2\,[H_2O]$
    f. Rate $= k\,[SF_3][H_2O][SF_3H][OH]$

30. The reaction in **question #28** is found to have a **positive** value for enthalpy change and a small **negative** value for standard entropy change. Under these circumstances, what can be said about the spontaneity of the reaction? Consider the relationship between free energy, entropy, and enthalpy.

a. The reaction will be spontaneous at all temperatures
b. The reaction will be non-spontaneous at all temperatures
c. The reaction will become spontaneous at sufficiently low temperatures
d. The reaction will become spontaneous at sufficiently high temperatures

Questions #31–#34 are conformation problems which concern the following molecule:

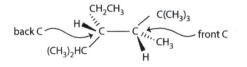

For all answers, consider the orientation of the largest groups on each axis carbon as the **most important criteria.** The Newman projections show different conformations of the same molecule; use this list shown to answer all four questions. **Answers may be used more than once.**

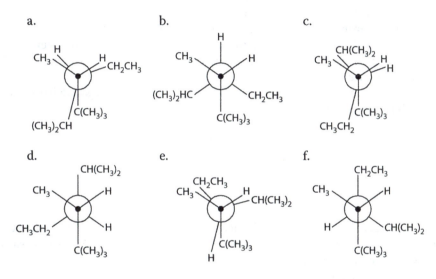

31. Which of the Newman projections shows the **lowest** potential energy conformation (**most stable**) of the molecule shown in the perspective drawing?
32. Which Newman projection shows the **highest** potential energy (**least stable**) conformation of the molecule shown in the perspective drawing?
33. Which Newman projection indicates the **staggered** conformation with the **highest** potential energy (**least stable staggered** conformation) of the molecule shown in the perspective drawing?
34. Which Newman projection shows the **lowest** potential energy **eclipsed** conformation, that is, the **most stable eclipsed** conformation of the molecule shown in the perspective drawing?

Questions #35–#38 refer to molecules shown with specific stereochemistry. Select the one correct answer for each question from the following list of lettered options; answers may be used more than once.
Use the following size relationships: isopropyl > ethyl > bromo
a. Bromo, ethyl, and isopropyl are all equatorial
b. Bromo, ethyl, and isopropyl are all axial
c. Bromo is equatorial; ethyl and isopropyl are axial

    d. Bromo and ethyl are equatorial; isopropyl is axial

    e. Bromo and isopropyl are equatorial; ethyl is axial

    f. Bromo is axial; ethyl and isopropyl are equatorial

    g. Bromo and ethyl are axial; isopropyl is equatorial

    h. Bromo and isopropyl are axial; ethyl is equatorial

35. Draw the **two chair** conformations of the following molecule, shown with specific stereochemistry. Be sure to place all substituents in the correct cis/trans orientation; ***do not consider any other stereoisomer***. What is the specific axial/equatorial orientation of each substituent in the **most stable chair** conformation?

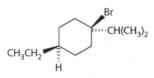

36. What is the specific axial/equatorial orientation of each substituent in the **least stable chair** conformation of the specific molecule shown in **question # 35?**

37. Draw the **two chair** conformations of the following molecule, shown with specific stereochemistry. **Note that this is not the same molecule used in the previous questions**. Be sure to place all substituents in the correct cis/trans orientation; ***do not consider any other stereoisomer***. What is the specific axial/equatorial orientation of each substituent in the **most stable chair** conformation? **Use the same list of answers shown above question #35.**

38. What is the specific axial/equatorial orientation of each substituent in the **least stable chair** conformation of the specific molecule shown in **question #37? Use the same list of answers shown above question #35.**

39. What is the relationship between the following two molecules?

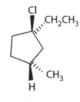

    a. Identical molecules

    b. Conformers of each other

    c. Stereoisomers of each other

    d. Constitutional isomers of each other

    e. Not isomers or conformers of each other: no relationship

40. What is the relationship between the following two molecules?

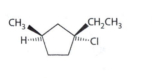

   a. Identical molecules
   b. Conformers of each other
   c. Stereoisomers of each other
   d. Constitutional isomers of each other
   e. Not isomers or conformers of each other: no relationship
41. What is the relationship between the following two molecules?

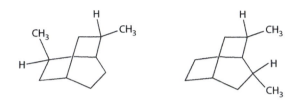

   a. Identical molecules
   b. Conformers of each other
   c. Stereoisomers of each other
   d. Constitutional isomers of each other
   e. Not isomers or conformers of each other: no relationship

ORGANIC I: MULTIPLE-CHOICE PRACTICE EXAM 1: ANSWERS

1.	d	22.	d
2.	c	23.	h
3.	b	24.	f
4.	c	25.	b
5.	i	26.	a
6.	h	27.	b
7.	e	28.	b
8.	b	29.	d
9.	a	30.	b
10.	f	31.	d
11.	b	32.	a
12.	b	33.	b
13.	a	34.	e
14.	c	35.	f
15.	b	36.	c
16.	e or c	37.	e
17.	c or e	38.	h
18.	g	39.	a
19.	d	40.	d
20.	a	41.	c
21.	d		

Organic I: Multiple-Choice Practice Exam 2

Questions #1 and #2 refer to mono-chlorination with Cl2/light by free radical halogenation.
Assume that the relative rate ratios based **only** on substitution pattern for these questions are

Rate for tertiary C−H abstraction: 6
Rate for secondary C−H abstraction: 3
Rate for primary C−H abstraction: 1

Use the following list of lettered answers for **both questions #1 and #2**; the ratios are expressed as the values obtained by direct multiplication of rate and # of equivalent hydrogens, **not** as the simplest ratios.

(a) 6/1	(b) 6/2	(c) 6/4	(d) 12/1	(e) 12/13	(f) 12/6
(g) 12/12	(h) 18/2	(i) 18/6	(j) 18/12	(k) 24/6	(l) 24/18

1. Owing to symmetry of the molecule, the following reaction will produce only **two** constitutional isomers upon **mono**-chlorination with Cl$_2$/light. Use simple arithmetic to calculate the expected **ratio (not** percentage) of the two products (*expressed as major/ minor for ratios other than exactly 1*).

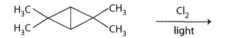

2. Owing to symmetry of the molecule, the following reaction will produce only **two** constitutional isomers upon **mono**-chlorination with Cl$_2$/light. Use simple arithmetic to calculate the expected **ratio (not** percentage) of the two products (*expressed as major/ minor for ratios other than exactly 1*).

3. What is the relationship between the following two molecules?

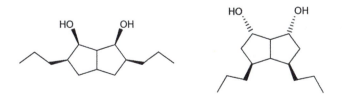

a. Identical molecules
b. Conformers of each other
c. Diastereomers of each other
d. Constitutional isomers of each other
e. Not isomers or conformers of each other: no relationship

4. What is the relationship between the following two molecules?

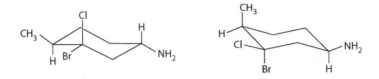

a. Identical molecules
b. Conformers of each other
c. Diastereomers of each other
d. Constitutional isomers of each other
e. Not isomers or conformers of each other: no relationship

5. What is the relationship between the following two molecules?

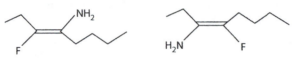

a. Identical molecules
b. Conformers of each other
c. Diastereomers of each other
d. Constitutional isomers of each other
e. Not isomers or conformers of each other: no relationship

6. What is the relationship between the following two molecules?

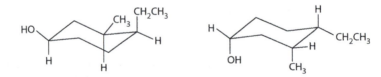

a. Identical molecules
b. Conformers of each other
c. Diastereomers of each other
d. Constitutional isomers of each other
e. Not isomers or conformers of each other: no relationship

7. Determine the relationship for expected boiling points for the following compounds based on **total** strength of intermolecular forces; then, place them in order of the highest boiling point to the lowest boiling point. **Hint:** All three molecules have approximately the same molar mass.

(**1**) $CH_3(CH_2)_4CH_3$ (**2**) $CH_3(CH_2)_4-NH_2$ (**3**) $CH_3-O-(CH_2)_3CH_3$

Highest b.p. ⟷ lowest b.p.
a. (1) > (2) > (3)
b. (1) > (3) > (2)

c. (2) > (1) > (3)
d. (2) > (3) > (1)
e. (3) > (1) > (2)
f. (3) > (2) > (1)

8. Determine the relationship for solubility in **water** for the following compounds based on expected solvent–solute intermolecular forces; then, place them in order of highest water solubility to lowest water solubility.

   (**1**) $CH_3(CH_2)_4CH_3$ (**2**) $CH_3(CH_2)_4-NH_2$ (**3**) $CH_3-O-(CH_2)_3CH_3$

   Highest H$_2$O solubility ⟷ lowest H$_2$O solubility
   a. (1) > (2) > (3)
   b. (1) > (3) > (2)
   c. (2) > (1) > (3)
   d. (2) > (3) > (1)
   e. (3) > (1) > (2)
   f. (3) > (2) > (1)

9. Determine the relationship for expected boiling points for the following compounds based on **total** strength of intermolecular forces; then, place them in order of the highest boiling point to the lowest boiling point. **Hint:** All three molecules have approximately the same molar mass.

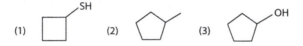

   Highest b.p. ⟷ lowest b.p.
   a. (1) > (2) > (3)
   b. (1) > (3) > (2)
   c. (2) > (1) > (3)
   d. (2) > (3) > (1)
   e. (3) > (1) > (2)
   f. (3) > (2) > (1)

10. Determine the relationship for solubility in **benzene** (C_6H_6), a **nonpolar** hydrocarbon, for the following compounds based on expected solvent–solute intermolecular forces; then, place them in order of highest benzene solubility to lowest benzene solubility.

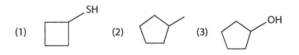

   Highest benzene solubility ⟷ lowest benzene solubility
   a. (1) > (2) > (3)
   b. (1) > (3) > (2)
   c. (2) > (1) > (3)
   d. (2) > (3) > (1)
   e. (3) > (1) > (2)
   f. (3) > (2) > (1)

11. Determine which of the following compounds can **hydrogen bond** between molecules (of the **same** compound)?

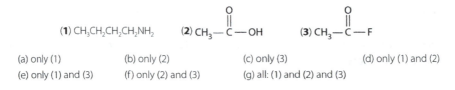

(1) $CH_3CH_2CH_2CH_2NH_2$ (2) $CH_3—C—OH$ (with =O above C)   (3) $CH_3—C—F$ (with =O above C)

(a) only (1)   (b) only (2)   (c) only (3)   (d) only (1) and (2)

(e) only (1) and (3)   (f) only (2) and (3)   (g) all: (1) and (2) and (3)

Questions #12–#21 refer to the following molecules. Use this lettered answer list for **all** questions; answers may be used more than once. Be certain to distinguish between the reactions of the alcohol portion, alkyl halide portion, or alkane portion of the molecule.

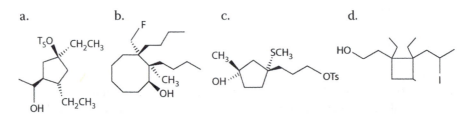

a.   b.   c.   d.

12. Which molecule will react **fastest** in an SN1 (solvolysis) reaction in neutral water solvent?

13. Which molecule will react **fastest** in an SN2 substitution reaction with NaSH as the nucleophile?

14. Which molecule will satisfy **both** of the following criteria: will probably react with NaCN through an **SN2** mechanism and will be converted into the **opposite enantiomer** with 100% inversion of configuration when reacted with NaCN.

15. Which molecule will **most probably** form a four-carbon (butyl) chain when reacted with $(CH_3)_2CuLi$?

16. Which molecule forms the **most** stable cation during an E1 elimination reaction in an appropriate neutral/weak base solvent?

17. Which molecule will have the **most** reactive (reacts the fastest) hydrogen for a free **radical** bromination reaction?

18. Which **alcohol** molecule will react **fastest** with HBr in an SN1 (substitution unimolecular) reaction?

19. Which **alcohol** molecule will react **fastest** with HBr in an SN2 (substitution bimolecular) reaction while producing no stereoisomers?

20. Which **alcohol** molecule will form the **most stable** carbon cation intermediate in an E1 mechanism (elimination unimolecular) when reacted with concentrated H_2SO_4?

21. Which **alcohol** molecule will probably form a tetrasubstituted alkene through an E1 reaction which proceeds through a **rearrangement**?

Questions #22 and #23 refer to the following reaction with Br_2/light:

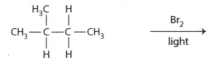

Use the following list of lettered answers to answer **both** questions; the radical symbol (•) is shown with the specific radical atom in each species.

Answers may be used more than once.

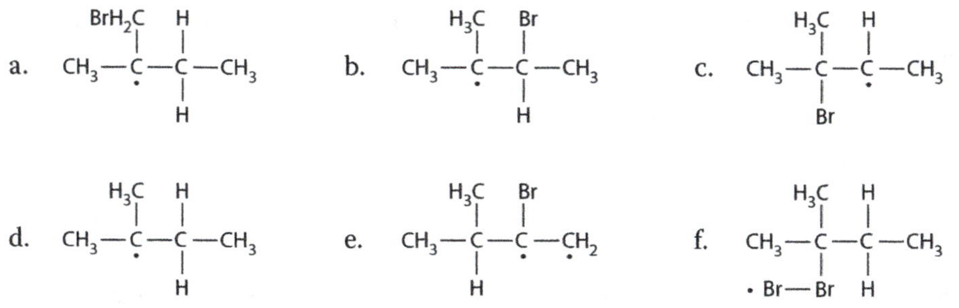

g. Br•   h. H•   i. H_3C•   j. Br—Br•

22. Which radical represents the intermediate product formed in the **first propagation** step of the reaction mechanism that leads to the **major** product?
23. Which radical represents the intermediate product formed in the **second propagation** step; the radical which is regenerated to repeat the chain reaction?

Questions #24 and #25 refer to the reaction of 2-methyl-2-butene with the reagent shown.

The hypothetical reaction can be described as an electrophilic addition to an alkene and proceeds by a mechanism similar to the one described for hydrohalogen (e.g., H–Cl). The electrophilic atom is platinum; the nucleophilic atom is sulfur:

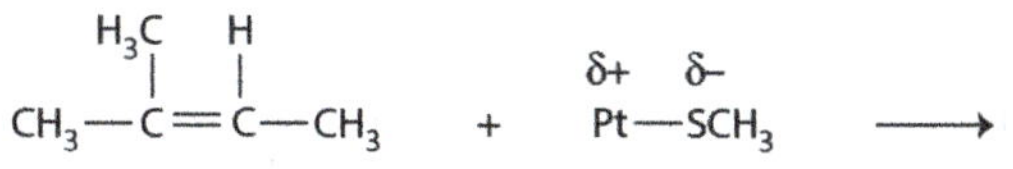

24. Based on the information described, determine which of the following represents a correct possible structure for the **major** high-energy **intermediate** formed during the first (electrophilic addition) step of the reaction?

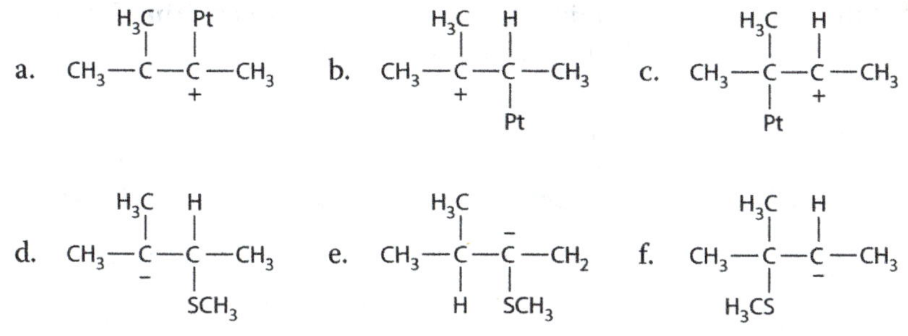

25. Which of the following is the correct **major** constitutional isomer **product** of this reaction?

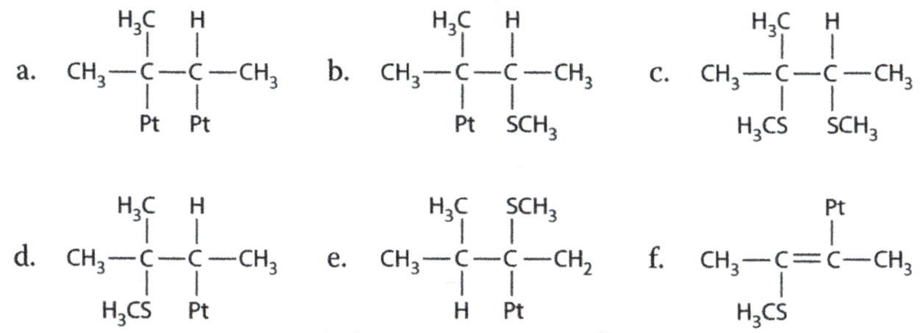

26. Which of the following molecules will form the **most** stable carbon cation upon **addition** of an electrophile (E⁺)?

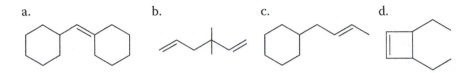

a.   b.   c.   d.

27. The carbon cation shown can rearrange to form a more stable cation intermediate.

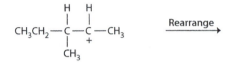

Select from the following list the structure of the **most probable** <u>**rearranged**</u> carbon cation **intermediate:**

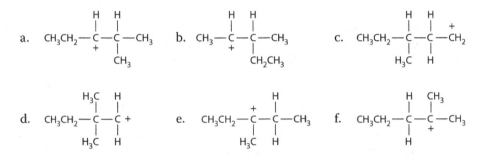

28. Which of the following stereoisomers is most stable, based on the axial/equatorial orientation of the substituent groups in the most stable conformer? **Hint:** Determine which stereoisomer can have all substituent groups equatorial.

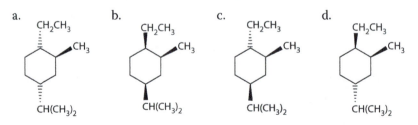

a.   b.   c.   d.

29. How many stereoisomers are possible for the following molecule? Be certain to consider chiral carbons and asymmetric double bonds.

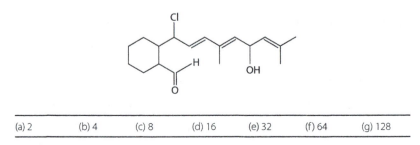

| (a) 2 | (b) 4 | (c) 8 | (d) 16 | (e) 32 | (f) 64 | (g) 128 |

Answer the following three questions for the following molecule shown as a Fischer projection:

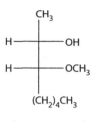

For **each** of the three questions, use the answers shown below directly after question #32. *Answers may be used more than once.*

30. Select from the list below the structure that represents the **enantiomer** of the molecule shown.
31. Select from the list below the structure that represents a **diastereomer** of the molecule shown.
32. Select from the list below the structure that represents a **constitutional isomer** of the molecule shown.

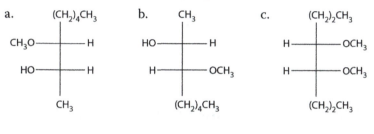

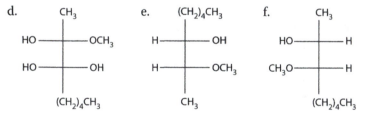

33. Determine the stereochemical designation around the double bond in the following molecule:

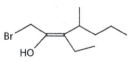

(a) E   (b) Z   (c) Double bond is symmetrical (no stereochemistry)   (d) Depends on conformation

34. Determine the stereochemical designation around the double bond in the following molecule:

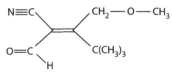

(a) E   (b) Z   (c) Double bond is symmetrical (no stereochemistry)   (d) Depends on conformation

35. Determine the absolute stereochemical designation around the chiral carbon in the following molecule:

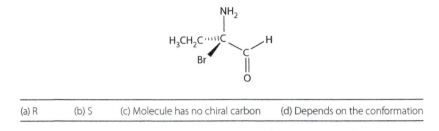

(a) R   (b) S   (c) Molecule has no chiral carbon   (d) Depends on the conformation

36. Determine the absolute stereochemical designation around the chiral carbon in the following molecule:

(a) R   (b) S   (c) Molecule has no chiral carbon   (d) Depends on the conformation

ORGANIC I: MULTIPLE-CHOICE PRACTICE EXAM 2: ANSWERS

1.	g	21.	b or a	41.	
2.	f	22.	d	42.	
3.	d	23.	g	43.	
4.	c	24.	b	44.	
5.	d	25.	d	45.	
6.	b	26.	a	46.	
7.	d	27.	e	47.	
8.	d	28.	c	48.	
9.	e	29.	f	49.	
10.	c	30.	f	50.	
11.	d	31.	b		
12.	a	32.	e		
13.	c	33.	a		
14.	d	34.	a		
15.	c	35.	a		
16.	a	36.	a		
17.	a	37.			
18.	c	38.			
19.	d	39.			
20.	c	40.			

Organic II: Multiple-Choice Practice Exam 1

Select the **one best** answer for the following questions:

1. Which of the following molecules will form the **most stable carbon cation** intermediate after an electrophilic addition reaction with HCl; the addition is to a pi-bond between **carbons #1 and #2** (labeled on each molecule)?

a.   b.   c.   d.   e.

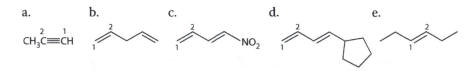

2. Which of the following molecules has the **most stable** pi-bond (i.e., the pi-bond with the largest B.D.E.)?

a.   b.   c.   d.   e.

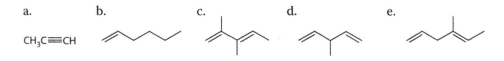

3. Which of the following molecules will have the fastest rate for **SN2 (substitution)** by a propyne anion ($CH_3C \equiv C^-$) as the nucleophile?

a.   b.   c.   d.   e.

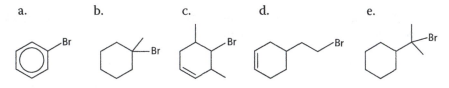

4. Which of the following molecules will have the fastest rate for **SN2 (substitution)** by a propyne anion ($CH_3C \equiv C^-$) as the nucleophile?

a.   b.   c.   d.   e.

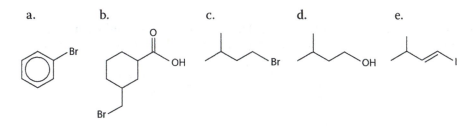

5. Which of the following molecules will have the fastest rate for **electrophilic aromatic substitution** with $Br_2/FeBr_3$?

a.   b.   c.   d.   e.

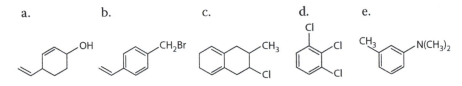

6. Which of the molecules from **question #5** will produce the most stable carbon cation through an **E1 (elimination) solvolysis reaction?** *Use the same lettered answer list shown for question #5; answers may be used more than once.*

7. Which of the following molecules will have the fastest rate for free **radical substitution** with Cl_2/light or NCS?

a.   b.   c.   d.   e.

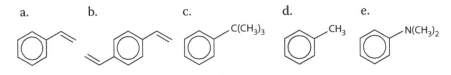

8. Which of the molecules from **question #7** can be **oxidized** by $Cr_2O_7^{-2}/H^+/H_2O$ to produce a substituted benzoic acid derivative? *Use the same lettered answer list shown for question #7; answers may be used more than once.*

9. Which molecule will react fastest with 1-chloro-2-bromoethyne in a **Diels–Alder** (4 + 2) **cycloaddition** reaction? Consider the requirements for the other component of a cycloaddition.

a.   b.   c.   d.   e.

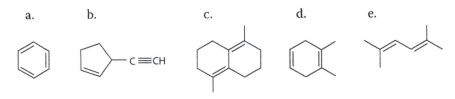

10. Which of the following molecules will form a **bicyclic** compound by a **Diels–Alder** (4 + 2) **cycloaddition** reaction with 2-pentene?

a.   b.   c.   d.   e.

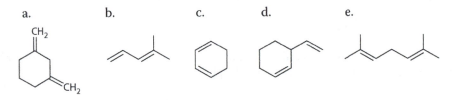

Questions 11–14 refer to the following electrophilic aromatic substitution; assume that the reagent is Br—OH; the electrophile is Br^+; $(OH)^-$ acts **only** as a base in the reaction.

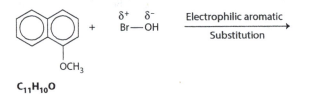

11. Based on your knowledge of this reaction, which of the following **best** represents the **overall rate expression?**
    a. Rate = k [BrOH]
    b. Rate = k [$C_{11}H_{10}O$]
    c. Rate = k [$C_{11}H_{10}O$]/[Br][OH]
    d. Rate = k [$C_{11}H_{10}O$] [BrOH]
    e. Rate = k [$C_{11}H_{10}O$]² [BrOH]²
    f. Rate = k [$C_{11}H_{10}O$]/[BrOH]

12. Which of the following molecules represents the **major** product of the **electrophilic aromatic substitution** reaction?

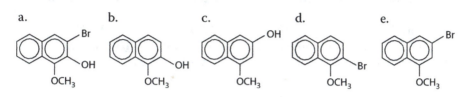

13. Which of the following molecules is the **most probable first aromatic intermediate** in the reaction which produces the **major** product?

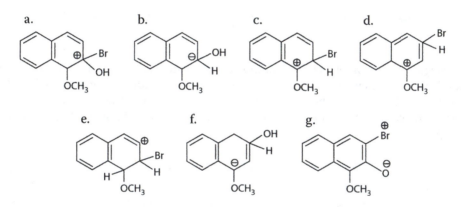

14. Which of the following represents a correct **resonance** form for the aromatic intermediate selected in **question #13?**

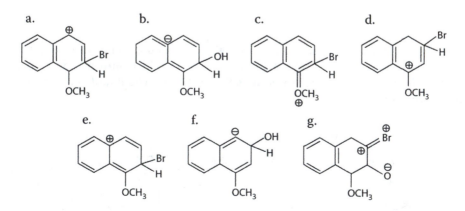

Questions 15–17 refer to the reaction of a carboxylic acid (usually with a catalyst) as an electrophile/nucleophile combination with the electron distribution shown. This reactant will undergo electrophilic addition to 4-ethyl-1,4,8-bicyclo[4.3.0.]nonatriene in a manner identical to normal electrophilic addition to conjugated double-bond systems.

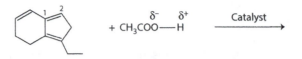

For all questions, the electrophile adds to the double bond between carbons #1 and #2.

All answers will show, for clarity, the specific H from the CH₃COO—H reagent.

15. Which of the following structures represents the **most stable first** intermediate formed by the addition of the electrophile to the **C#1 = C#2** double bond?

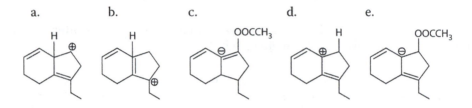

16. **Conjugate addition** refers to the addition pattern which produces a **rearranged** double bond in the final neutral product. After the electrophilic addition step to the **C#1 = C#2** double bond, which of the following molecules represents the final **major** product formed by **conjugate addition across the two double bonds in the five-membered ring?**

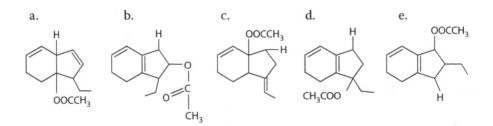

17. After the electrophilic addition step to the **C#1 = C#2** double bond, which of the following molecules represents the final **major** product formed by **conjugate addition across the two double bonds represented by C#1 = C#2 and the double bond in the six-membered ring?**

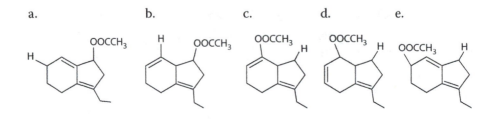

Questions 18 and 19 are nomenclature problems. Use the lettered answer list below for each question; answers may be used more than once.
Nomenclature priority ranking for functional groups:
amides/esters/acids > aldehydes > ketones > alcohols > alkynes/alkenes > halide   and   other substituents

List of names for questions:
   a.   1-(3-methylhexenyl)-3-aminobenzene
   b.   3-(2, 5-heptadienyl)-1-aminobenzene
   c.   m-(4-methyl-1, 4-hexadienyl)-aniline
   d.   m-(3-methyl-2, 5-hexadienyl)-aniline
   e.   o-(3, 6-hexadienyl)-aniline
   f.   o-heptadienylnitrobenzene
   g.   p-(2, 4, 6-heptatrienyl)-benzaldehyde
   h.   p-(3, 5, 7-octatrienyl)-benzene
   i.   3-(2, 4, 6-heptatrienyl)-1-hydroxybenzene
   j.   m-(1, 3, 5-heptatrienyl)-phenol
   k.   p-(2, 4, 6-heptatrienyl)-phenol
   l.   m-(2, 4, 6-heptenyl)-phenol
   m.   m-(3, 5, 7-octenyl)-phenol
   n.   p-(3, 5, 7-octatrienyl)-phenol

18. Select the correct name for the compound shown below:

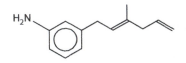

19. Select the correct name for the compound shown below:

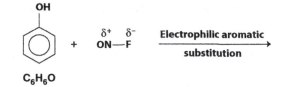

Questions 20–24 refer to the following electrophilic aromatic substitution; the hypothetical reagent has the charge distribution oriented as shown.

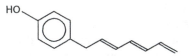

20. Which of the following molecules represents the **major** product of the reaction?

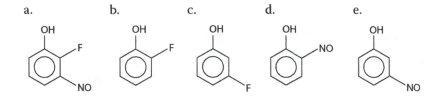

21. Which of the following molecules is the **most probable first aromatic intermediate** in the reaction which produces the **major** product?

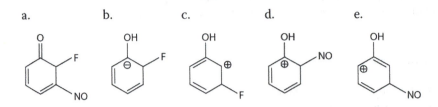

22. Which of the following represents a correct **resonance** form for the aromatic intermediate selected in **question #21?**

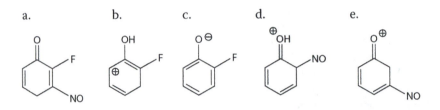

23. Based on your knowledge of this reaction, which of the following **best** represents the **overall rate expression?**
    a. rate = k [C_6H_6O]
    b. rate = k [C_6H_6O] [NOF]
    c. rate = k [NOF]
    d. rate = k [C_6H_6O] [NOF]^2
    e. rate = k [C_6H_6O]/[NO] [F]

24. The rate of this reaction **decreases** if chlorobenzene is used as a reactant in place of phenol. Why does this occur?
    a. The aromatic intermediate is converted into the final product more slowly because chlorine stabilizes negative charge better than hydroxyl.
    b. Chlorine does not stabilize the aromatic intermediate as well as hydroxyl does.
    c. The total kinetic energy of chlorobenzene is less than that of phenol at any specific temperature.
    d. A chlorine substituent can activate the ring in the meta-positions only.
    e. The rate of the reaction is inversely proportional to the bond dissociation energies of C—F versus C—O.

25. Which of the following molecules will react FASTEST with 1,3-cyclopentadiene in a DIELS–ALDER reaction?

a.   b.   c.   d.

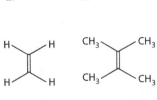
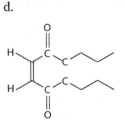

26. Which of the following molecules will react FASTEST with 1,2-dibromoethene in a DIELS–ALDER reaction?

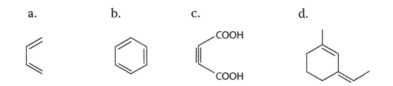

a.   b.   c.   d.

ORGANIC II: MULTIPLE-CHOICE PRACTICE EXAM 1: ANSWERS

1.	d	11.	d	21.	d
2.	c	12.	d	22.	d
3.	d	13.	c	23.	b
4.	c	14.	e	24.	b
5.	e	15.	d	25.	d
6.	b	16.	d	26.	a
7.	d	17.	e		
8.	d	18.	d		
9.	e	19.	k		
10.	c	20.	d		

Organic II: Multiple-Choice Practice Exam 2

ORGANIC CHEMISTRY COMPREHENSIVE EXAM

This short exam is written in the style of a multiple-choice standardized exam. The suggested time limit is 100 minutes. A key is provided.

1. Which species represents an intermediate formed during the reaction of propane with Br_2/light?

   a. CH₃ĊHCH₃   b. CH₃C⁺HCH₃   c. CH₃ĊHBrCH₃   d. CH₃C⁻HCH₃   e. CH₃ĊHCH₂Br

2. Which compound will show a P. E. diagram with three peaks (maxima) of equal energy and three valleys (minima) of equal energy for 360° conformational rotation around the C−C bond?

   a. ClH_2C-CH_2Cl b. ClH_2C-CBr_3 c. $Cl_2HC-CHCl_2$ d. $ClH_2C-CHBrCl$

3. Which statement describes the stereoisomer **product** distribution for the following reaction?

   a.   4 stereoisomers: both diastereomers and enantiomers (2 diastereomers each with its enantiomer)
   b.   2 stereoisomers as an enantiomer pair
   c.   1 enantiomer only
   d.   2 diastereomers only
   e.   1 meso compound

4. Determine the stereochemistry for the two chiral carbons in the molecule shown as a Fischer projection to the right ⎯⎯⎯→ :

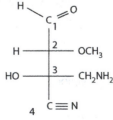

   a. 2R, 3R   b. 2R, 3S   c. 2S, 3R   d. 2S, 3S   e. 2, 3-meso

5. Which of the following is a correct synthesis (i.e., produces the favored **major** product) for trans-(E)-2-hexene starting from propene as one of the reactants plus any other reagents?

a. Propene $\xrightarrow[\text{2. NaNH}_2 \text{ (excess)}]{\text{1. Br}_2}$ $\xrightarrow{\text{3. CH}_3\text{CH}_2\text{CH}_2\text{Br}}$ $\xrightarrow[\text{NH}_3]{\text{4. Na}}$

b. Propene $\xrightarrow[\text{2. TsCl}]{\text{1. H}^+/\text{H}_2\text{O}}$ $\xrightarrow[\text{NaNH}_2]{\text{2. HC}\equiv\text{CCH}_3}$ $\xrightarrow{\text{3. H}_2/\text{Pt}}$

c. Propene $\xrightarrow{\text{1. HCl}}$ $\xrightarrow[\text{NaNH}_2]{\text{2. HC}\equiv\text{CCH}_3}$ $\xrightarrow{\text{3. H}_2/\text{Lindlar Pd}}$

d. Propene $\xrightarrow{\text{1. Br}_2/\text{light}}$ $\xrightarrow[\text{3. H}_2/\text{Lindlar Pd}]{\text{2. HC}\equiv\text{CCH}_3}$ $\xrightarrow{\text{4. NaNH}_2}$

e. All syntheses will form trans-(E)-2-hexene.

6. What is the relationship between the following two molecules?

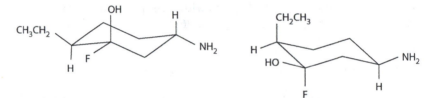

a. Identical molecules
b. Conformers of each other
c. Stereoisomers of each other
d. Constitutional isomers of each other
e. Not isomers or conformers of each other: no relationship

7. Which of the **following** molecules will form the **most** stable cation upon **addition** of an electrophile (E$^+$)?

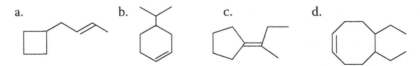

8. Which of the following is a correct synthesis (i.e., produces the favored **major** product) for trans-2-butylcyclopentanol starting with butylcyclopentane?

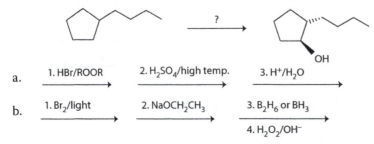

a. $\xrightarrow{\text{1. HBr/ROOR}}$ $\xrightarrow{\text{2. H}_2\text{SO}_4/\text{high temp.}}$ $\xrightarrow{\text{3. H}^+/\text{H}_2\text{O}}$

b. $\xrightarrow{\text{1. Br}_2/\text{light}}$ $\xrightarrow{\text{2. NaOCH}_2\text{CH}_3}$ $\xrightarrow[\text{4. H}_2\text{O}_2/\text{OH}^-]{\text{3. B}_2\text{H}_6 \text{ or BH}_3}$

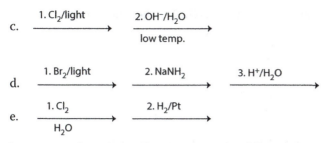

c. $\xrightarrow[\text{low temp.}]{\text{1. Cl}_2\text{/light}}$ $\xrightarrow{\text{2. OH}^-\text{/H}_2\text{O}}$

d. $\xrightarrow{\text{1. Br}_2\text{/light}}$ $\xrightarrow{\text{2. NaNH}_2}$ $\xrightarrow{\text{3. H}^+\text{/H}_2\text{O}}$

e. $\xrightarrow[\text{H}_2\text{O}]{\text{1. Cl}_2}$ $\xrightarrow{\text{2. H}_2\text{/Pt}}$

9. Determine the relationship between the following two molecules shown as Fischer projections.

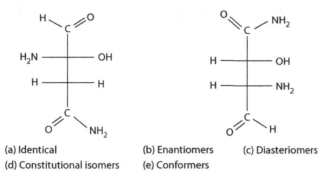

(a) Identical   (b) Enantiomers   (c) Diasteriomers
(d) Constitutional isomers   (e) Conformers

10. Which of the following molecules represents one of the two enantiomers formed as the major constitutional isomer product of electrophilic addition of Cl_2/H_2O specifically to **Z**-3-methyl-2-heptene?
    a. 2S, 3R-3-chloro-3-methyl-2-heptanol
    b. 2R, 3R-2-chloro-3-methyl-3-heptanol
    c. 2S, 3R-2-chloro-3-methyl-3-heptanol
    d. 2R, 3S-3-chloro-3-methyl-2-heptanol
    e. Z-3-chloro-3-methyl-2-hepten-2-ol

11. Which of the following stereoisomers is most stable, based on the axial/equatorial orientation of the substituent groups in the most stable conformer? (**Hint:** determine which stereoisomer can have all groups equatorial.)

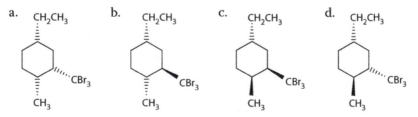

12. How many chiral carbons in the molecule are shown?

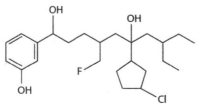

a. 2   b. 3   c. 4   d. 5   e. 6

13. How many signals would be expected for the ^{13}C NMR of the molecule shown?

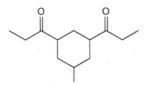

a. 4   b. 6   c. 8   d. 10   e. 12

14. A certain compound contains only carbon, hydrogen, and oxygen. The IR of this compound shows a strong, broad peak at 3200–3400 cm^{-1}; there is no peak between 1680 and 1740 cm^{-1}. The NMR of this compound shows a 12-proton singlet at δ 1.2 and a two-proton singlet at δ 2.0. Which of the following molecules best fits the structure of this compound?

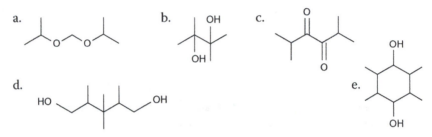

15. Determine the relationship between the following two molecules: (Hint: First determine the *R/S* configuration of each chiral carbon.)

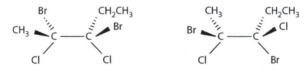

a. Identical   b. Enantiomers   c. Diasteriomers
d. Constitutional isomers   e. Conformers

16. Determine the relationship between the following two molecules: (Hint: First determine the *R/S* configuration of each chiral carbon.)

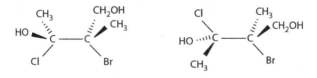

a. Identical   b. Enantiomers   c. Diasteriomers
d. Constitutional isomers   e. Conformers

17. Determine the stereochemical designation of the double bonds in the following molecule:

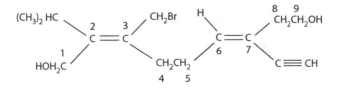

a. 2E, 6E   b. 2E, 6Z   c. 2Z, 6E   d. 2Z, 6Z

Questions #18–#22 refer to the following molecules. *Use this lettered answer list for all questions; answers may be used more than once.* Be certain to distinguish between the reactions of the alcohol portion, alkyl halide portion, or alkane portion of the molecule.

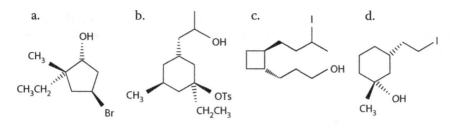

18. Which molecule will react **fastest** in an SN1 (solvolysis) reaction in neutral water solvent?
19. Which molecule will react **fastest** in an SN2 substitution reaction with NaSH as the nucleophile?
20. Which **alcohol** molecule will react **fastest** with HBr in an SN1 (substitution unimolecular) reaction?
21. Which **alcohol** molecule will react **fastest** with HCl in an SN2 (substitution bimolecular) reaction?
22. Which **alcohol** molecule will form a rearranged carbon cation in an E1 reaction with H_2SO_4?
23. What are the hybridizations of the carbons, the nitrogen, and the oxygen in the following molecule?

$$H_2C = CH - NH - OH$$

	$\underline{C}$	$\underline{N}$	$\underline{O}$
(a)	sp^3	sp^2	sp^2
(b)	sp^2	sp^3	sp^3
(c)	sp	sp^2	sp^3
(d)	sp^2	sp^2	sp^3

24. How many **mono**-chlorination constitutional isomer products can be formed from the following molecule upon reaction with Cl_2/light?

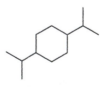

a. 2   b. 4   c. 6   d. 8   e. 10

25. Which of the labeled hydrogens in the molecule shown is most acidic? Select a, b, c, or d directly from the labels shown.

$$HC \equiv C - CH_2CH_2 - CH = CH - CH_3$$
a.   b.   c.   d.

26. Which of the following alcohol compounds would be expected to react through a **rearrangement** reaction during an **SN1** reaction with HBr?

    a. 4-ethyl-5-methyl-2-hexanol
    b. 5,5-dimethyl-3-hexanol
    c. 4,4-dimethyl-3-hexanol
    d. 4,4-dimethylcyclohexanol

27. Which of the following represents a correct possible synthesis for the **specific stereoisomer** _S-2-bromo-3-ethylpentane_ starting from the **specific stereoisomer** _R-3-ethyl-2-pentanol_?

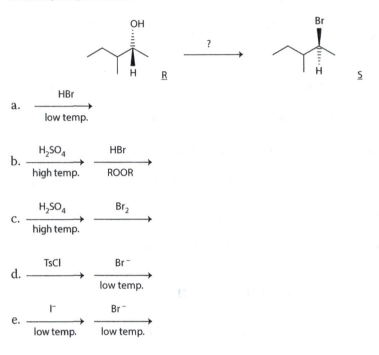

    a. $\xrightarrow[\text{low temp.}]{\text{HBr}}$

    b. $\xrightarrow[\text{high temp.}]{\text{H}_2\text{SO}_4}\xrightarrow[\text{ROOR}]{\text{HBr}}$

    c. $\xrightarrow[\text{high temp.}]{\text{H}_2\text{SO}_4}\xrightarrow{\text{Br}_2}$

    d. $\xrightarrow{\text{TsCl}}\xrightarrow[\text{low temp.}]{\text{Br}^-}$

    e. $\xrightarrow[\text{low temp.}]{\text{I}^-}\xrightarrow[\text{low temp.}]{\text{Br}^-}$

28. What is the product of the following reaction sequence:

$$CH_3CH_2Br \xrightarrow{CH_3C\equiv CNa} \xrightarrow[NH_3]{Na} ?$$

    a. $CH_3CH_2CH_2CH_2CH_3$
    b. $(CH_3CH_2)^-Na^+ + CH_3C\equiv CBr$
    c. $CH_3CH_2CH_2C\equiv CH$
    d. $trans(E)CH_3CH=CHCH_2CH_3$
    e. $cis(Z)CH_3CH=CHCH_2CH_3$

29. What is the **specific major** product of the **E2** (bimolecular elimination) reaction of the **specific stereoisomer** 3R, 4S-3-methyl-4-bromoheptane with the base **OCH_3^-**?
    a. E-3-methyl-3-heptene
    b. Z-3-methyl-3-heptene
    c. E-3-methyl-4-heptene
    d. Z-3-methyl-4-heptene
    e. 50% E/50% Z -3-methyl-3-heptene

30. Which of the Newman projections depicts the lowest potential energy conformation of 2,3-dimethylhexane based on rotation around the carbon #3– carbon #4 bond axis (numbers based on correct nomenclature numbering)?

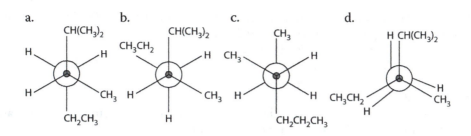

31. Which species represents an intermediate formed in the r.d.s. propagation step for the reaction of 2-methyl-1-butene with HBr in the presence of peroxides (ROOR)?

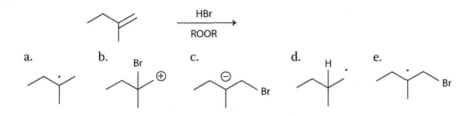

32. Which of the figures represents the r.d.s. transition state for the reaction of $NaOCH_3$ with $CH_3CH_2 - Br$ at low temperature?

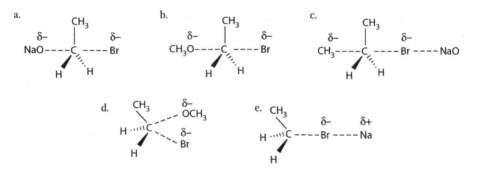

Questions #33–#38 refer to the following hypothetical potential energy diagram:

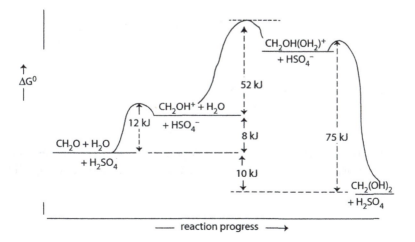

33. Based on the diagram, determine the activation energy (Ea) specifically for the **rate-determining step (r.d.s.) for the <u>forward</u> reaction.**
    a. 8 kJ   b. 52 kJ   c. 60 kJ   d. 70 kJ   e. 75 kJ
34. Based on the diagram, determine the value for AG^0 for the overall complete reaction (balanced equation) in the **<u>forward</u> direction.**
    a. 8 kJ   b. 10 kJ   c. 52 kJ   d. −10 kJ   e. −75 kJ
35. Based on the diagram, determine the activation energy (Ea) specifically for the **rate-determining step (r.d.s.) for the <u>reverse</u> reaction.**
    a. 75 kJ   b. 70 kJ   c. 52 kJ   d. −52 kJ   e. −75 kJ
36. Which of the molecules represents a catalyst in this reaction?
    a. H_2O b. H_2SO_4 c. $CH_2OH(OH_2)^+$ d. CH_2OH^+ e. CH_2O
37. Based on the diagram, which of the following species are organic **intermediates(exclude catalysts)** in the reaction?
    a. $CH_2OH(OH_2)^+$ and $CH_2(OH)_2$
    b. CH_2OH and CH_2O
    c. CH_2OH^+ only
    d. $CH_2OH(OH_2)^+$ and CH_2OH^+
    e. $CH_2OH(OH_2)^+$ only
38. Which reaction below represents the net balanced equation?

    a. $CH_2O + H_2O \longrightarrow CH_2(OH)_2$

    b. $CH_2O + CH_2OH^+ + H_2SO_4 \longrightarrow CH_2(OH)_2 + HSO_4^-$

    c. $CH_2O + CH_2OH^+ + CH_2OH(OH_2)^+ \longrightarrow CH_2(OH)_2 + H_2SO_4$

    d. $CH_2O + H_2O + CH_2OH^+ \longrightarrow CH_2OH(OH_2)^+ + CH_2(OH)_2 + H_2SO_4$

39. What is the expected **final product** for the electrophilic addition of a H^+/H_2O in the presence of an Hg^{+2} catalyst to 1-pentyne?

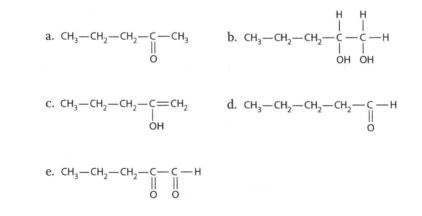

40. Which **<u>one</u>** of the molecules shown below is **not** one of the correct possible products for the reaction of Br_2 with 1-phenyl-4-methyl-1, 3-pentadiene in an electrophilic addition reaction?

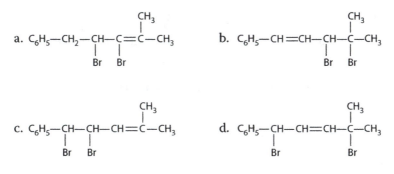

41. Which molecule will react fastest with 1,2-dichloroethene in a **Diels–Alder** $(4+2)$ **cycloaddition** reaction? Consider the requirements for the other component of a cycloaddition.
    a. 4-bromoaniline
    b. 2,5-dimethyl-1,5-hexadiene
    c. 1,4-cyclooctadiene
    d. 2,4-dimethyl-1,3-heptadiene
    e. 4-bromo-2-pentyne

42. An **electrophile** can be formed by the combination of the reagents shown below. What is the **specific structure** of this electrophile when used in **electrophilic aromatic substitution?**

$$(CH_3)_2CH-\overset{\overset{O}{\|}}{C}-O-\overset{\overset{O}{\|}}{C}-CH(CH_3)_2 \xrightarrow{\text{AlCl}_3 \text{ or } \text{H}^+ \text{ catalyst}}$$

a. $(CH_3)_2CH-\overset{\overset{O}{\|}}{C}-O^+$ b. $(CH_3)_2CH-\overset{\overset{O}{\|}}{C}-O^-$ c. $(CH_3)_2CH-\overset{\overset{O}{\|}}{C}-O-C^+$

d. $(CH_3)_2C^+-\overset{\overset{O}{\|}}{C}-OH$ e. $(CH_3)_2CH-\overset{\overset{O}{\|}}{C}^+$ f. $(CH_3)_2CH-\overset{\overset{O}{\|}}{C}-O-Cl^+$

43. Determine the potential effect on reaction rate and electrophile substitution position for the sulfonyl group ($-SO_3H$) substituent as shown below. It may help to determine the bonding/dipole/lone pair configuration around the sulfur.

$$CH_6H_5-SO_3H \xrightarrow[\text{substitution}]{\text{electrophilic aromatic}}$$

The $-SO_3H$ substituent will be
a. Activator; ortho/para-directing
b. Activator; meta-directing
c. Deactivator; meta-directing
d. Deactivator; ortho/para-directing
e. No effect on rate; all positions equivalent

44. The following carbonyl reactions are written in one specific direction. Which of the following general reactions would be thermodynamically/equilibrium favored?

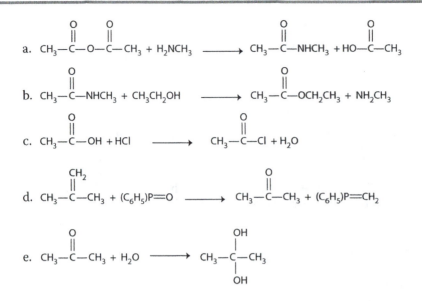

a. $CH_3-\overset{O}{\underset{||}{C}}-O-\overset{O}{\underset{||}{C}}-CH_3 + H_2NCH_3 \longrightarrow CH_3-\overset{O}{\underset{||}{C}}-NHCH_3 + HO-\overset{O}{\underset{||}{C}}-CH_3$

b. $CH_3-\overset{O}{\underset{||}{C}}-NHCH_3 + CH_3CH_2OH \longrightarrow CH_3-\overset{O}{\underset{||}{C}}-OCH_2CH_3 + NH_2CH_3$

c. $CH_3-\overset{O}{\underset{||}{C}}-OH + HCl \longrightarrow CH_3-\overset{O}{\underset{||}{C}}-Cl + H_2O$

d. $CH_3-\overset{CH_2}{\underset{||}{C}}-CH_3 + (C_6H_5)P{=}O \longrightarrow CH_3-\overset{O}{\underset{||}{C}}-CH_3 + (C_6H_5)P{=}CH_2$

e. $CH_3-\overset{O}{\underset{||}{C}}-CH_3 + H_2O \longrightarrow CH_3-\underset{\underset{OH}{|}}{\overset{\overset{OH}{|}}{C}}-CH_3$

45. The alcohol 4-bromo-1-butanol will react with heptanoic acid using an H⁺ catalyst to form
    a. Heptyl-4-hydroxybutanoate
    b. 4-Bromobutyl heptanoate
    c. Heptyl-4-bromobutanoate
    d. 1-Butoxyheptanal
    e. 4-Hydroxybutyl heptanoate

46. Which of the following compounds is **NOT** aromatic?

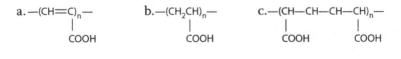

    a.   b.   c.   d.   e.

47. The structure of acrylic acid is $CH_2{=}CHCOOH$. What is the polymer structure of poly-acrylic acid?

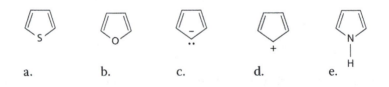

a. $-(CH{=}C)_n-$ with $COOH$ b. $-(CH_2CH)_n-$ with $COOH$ c. $-(CH-CH-CH-CH)_n-$ with $COOH$ $COOH$

d. $-(CH{=}CH-COO)_n-$

48. What is the product of the reaction shown?

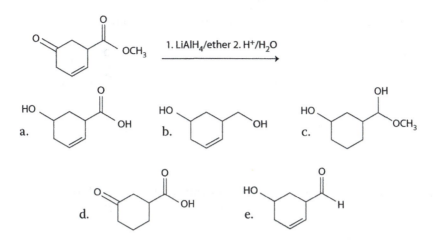

49. What is the structure of the biological dipeptide formed from tyrosine and serine? Tyrosine is 2-amino-3-(4-hydroxyphenyl)-propanoic acid; serine is 2-amino-3-hydroxypropanoic acid.

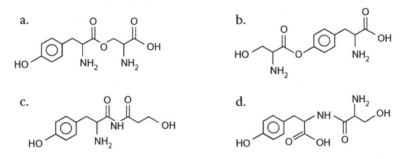

50. Which of the following reaction sequences will accomplish the following synthesis?

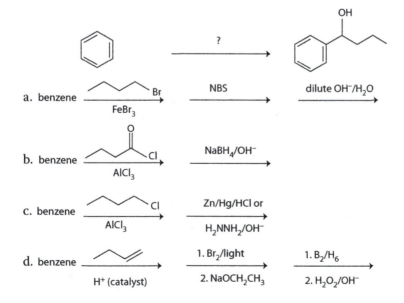

51. Which of the following molecules has the most acidic proton?

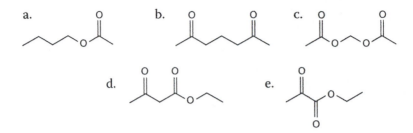

52. Which of the following reaction sequences will **NOT** accomplish the synthesis of the specific isomer **para**-nitrobenzoic acid starting from benzene?

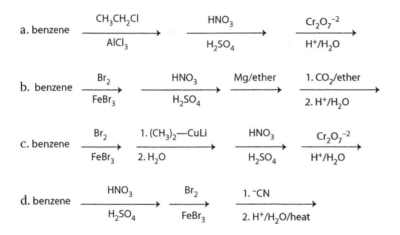

53. Which of the following reaction sequences will accomplish the synthesis shown?

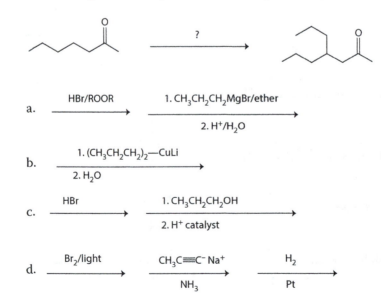

54. Which numbered carbon in the molecule shown will show the fastest rate for decarboxylation upon heating; that is, which carbon will most likely be removed as CO_2 upon heating? (Be careful reading the carbonyl structures.)
    a. Carbon #1
    b. Carbon #2
    c. Carbon #3
    d. Carbon #1 and carbon #2 would show equal rates
    e. Carbons #1, #2, and #3 would all show the same rate

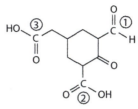

55. Which of the following reactants and synthetic sequences will produce the following molecule ⟶:

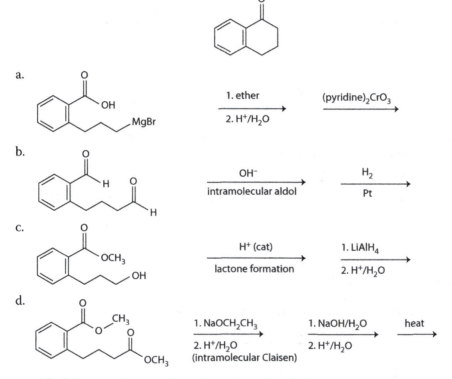

    e.   All of these syntheses will produce the molecule shown.

56. A certain compound contains only carbon, hydrogen, and oxygen. The IR of this compound shows a strong, broad peak at 3200–3400 cm^{-1} and a strong peak at about 1700 cm^{-1}. Which one of the following compounds mostly closely fits the IR spectrum described?

    a. $CH_3CH_2CH_2$—C—OCH_3
               ‖
               O

    b. CH_3CH=$CHCH_2$—O—CH_3

    c. $CH_3CH_2CH_2$—C—OH
               ‖
               O

    d. HO—$CH_3CH_2CH_2CH_2$—OH

    e. CH_3CH=$CHCH_2$—C—CH_3
                     ‖
                     O

57. Of the following three suggested reactions, which could be used to synthesize the β-ketoester shown below? (C_6H_5 = phenyl)

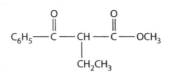

(1) C_6H_5—C(=O)—OCH$_3$ + CH$_3$COCH$_3$ $\quad$ 1. NaOCH$_3$ 2. H$^+$/H$_2$O
3. OH$^-$ (catalyst)/CH$_3$CH$_2$Br

(2) C_6H_5—C(=O)—OCH$_3$ + CH$_3$CH$_2$CH$_2$COCH$_3$ $\quad$ 1. NaOCH$_3$
2. H$^+$/H$_2$O

(3) C_6H_5—C(=O)—CH$_2$CH$_2$CH$_3$ + CH$_3$OCOCH$_3$ $\quad$ 1. NaOCH$_3$
2. H$^+$/H$_2$O

a. **1, 2,** and **3**   b. **1** and **2** only   c. **1** and **3** only   d. **2** and **3** only

58. Which set of reagents would accomplish the synthesis shown?

CH$_3$CH$_2$O—C(=O)—CH$_2$—C(=O)—OCH$_2$CH$_3$ $\xrightarrow{?}$ CH$_3$CH$_2$O—C(=O)—CH—C(=O)—OCH$_2$CH$_3$, CH$_2$CH$_3$

a. 1. NaNH$_2$ 2. CH$_3$CH$_2$OH
b. 1. NaOCH$_2$CH$_3$ 2. CH$_3$CH$_2$Br
c. 1. NaNH$_2$ 2. CH$_2$=CH$_2$
d. CH$_3$CH$_2$OH/H$^+$ (catalyst)

59. Which reaction sequence will convert benzene to produce propylbenzene?

a.   1. CH$_3$CH$_2$COCl/AlCl$_3$ 2. Zn/Hg/HCl

b.   CH$_3$CH$_2$CH$_2$Cl/AlCl$_3$

c.   1. Br$_2$/FeBr$_3$ 2. CH$_3$CH$_2$CH$_2$MgBr

d.   1. Cl$_2$/AlCl$_3$ 2. CH$_3$CH=CH$_2$/AlCl$_3$

60. What is the final product of the following reaction sequence?

C_6H_5—CH$_2$—Cl $\xrightarrow{\text{1. }(C_6H_5)_3P}$ $\xrightarrow{\text{2. }(CH_3)_3COK \text{ (strong base)}}$ $\xrightarrow{\text{3. }(C_6H_5)_2C=O}$?

a. $(C_6H_5)_3$COH
b. C_6H_5–CH$_2$–P(C_6H_5)$_3$
c. $(C_6H_5)_2$C=CHC$_6$H$_5$
d. $(C_6H_5)_3$COCH$_2$C$_6$H$_5$

ORGANIC CHEMISTRY COMPREHENSIVE EXAM: ANSWERS

1.	a	31.	e
2.	b	32.	b
3.	b	33.	b
4.	b	34.	d
5.	a	35.	a
6.	c	36.	b
7.	c	37.	d
8.	b	38.	a
9.	d	39.	a
10.	b	40.	a
11.	d	41.	d
12.	d	42.	e
13.	c	43.	c
14.	b	44.	a
15.	b	45.	b
16.	c	46.	d
17.	b	47.	b
18.	b	48.	b
19.	d	49.	d
20.	d	50.	b
21.	c	51.	d
22.	a	52.	d
23.	b	53.	b
24.	b	54.	b
25.	a	55.	d
26.	c	56.	c
27.	d	57.	a
28.	d	58.	b
29.	b	59.	a
30.	a	60.	c

Index

3